A Colour Handbook

Tomato Diseases

Identification, Biology and Control

Second Edition

Dominique Blancard

In collaboration with
Henri Laterrot, Georges Marchoux and Thierry Candresse

Consultant Editor (English edition) John Fletcher

Translated from the French by Denise McGee

Q éditions
Quæ

MANSON
PUBLISHING

Copyright © 2012 Manson Publishing Ltd
ISBN: 978-1-84076-156-6

First published in French as *Les maladies de la tomate*
Copyright © 2009 Éditions Quae, c/o INRA, RD 10, F-78026 Versailles Cedex, France
ISBN: 978-2-7592-0328-4

Ouvrage publié avec le concours du Ministère français chargé du la Culture –
Centre nationale du livre.
This edition is published with the help of the French Ministry of Culture

A CIP catalogue record for this book is available from the British Library.

For full details of all Manson Publishing Ltd titles please write to:
Manson Publishing Ltd, 73 Corringham Road, London NW11 7DL, UK
Tel: +44(0)20 8905 5150
Fax: +44(0)20 8201 9233
Website: www.mansonpublishing.com

International rights and coeditions: Gail Markham
Project manager: Paul Bennett
Copyeditor: Ruth Maxwell
Layout: Initial Typesetting Services
Printed by: Grafos SA, Barcelona, Spain

CONTENTS

■ Principal characteristics of pathogenic agents and methods of control 413

FOREWORD

Dr. John (Jay) Scott Warner,
Professor at the University of Florida

Much has changed with regard to the tomato plant since Dominique Blancard published his first book on diseases of the Solanaceae in 1987. Firstly, the scientific name *Solanum lycopersicum* L. has been proposed to replace *Lycopersicon esculentum* Mill. used for many decades. Indeed, the historical evidence shows that *Solanum lycopersicum* had been proposed by Linnaeus in 1753, a year before Miller's proposal to associate the tomato plant to the genus *Lycopersicon*. Phylogenetic studies support the idea that the tomato and its wild cousins *Lycopersicon* must be placed in the genus *Solanum*. Both names are still used in the literature, but *S. lycopersicum* is increasingly common.

The consumption of tomatoes has increased considerably since the time, several hundred years ago, when this species was regarded as toxic because of it belonging to the same family as the deadly nightshade. Today, the production of tomato is the fourth largest vegetable crop in the world. Its consumption is rising: it is over 12 kg per capita per year, with a maximum of over 100 kg in Greece and Libya. Part of the recent enthusiasm for tomatoes is due to reports in the medical literature of the 1990s, asserting that lycopene, the red pigment in tomatoes, is a potent antioxidant that reduces the risk of several cancers, particularly those of the gastrointestinal tract. So today, the benefits of lycopene for health are promoted on certain tomato processed products. The yield per hectare and the cultivated areas are also steadily increasing.

However, prices paid to producers have often not kept pace with increases in production costs, and producers face many challenges, including problems related to manpower and the availability of water or farmland due to urban expansion. In addition, a myriad of diseases and pests are spreading at an increasing rate in the globalized world, due to the expansion of international trade and tourist travel. It is a challenge for scientists to find solutions to often complex problems, posed by these pests and diseases.

On the positive side, the acceleration of progress in new technologies for crop improvement must be emphasized. At the forefront of this movement is the sequencing of DNA, which makes it possible to decode the genome at a rate unexpected 10 years ago. A global effort is being made to sequence the tomato genome and once this information is freely available, it should be a great resource for the improvement of this plant. The cultivated tomato has a narrow genetic base, and breeders have been frustrated by their inability to find molecular markers linked to traits of economic interest in this genome. One way to overcome this difficulty is to make crosses with wild relatives, in which polymorphisms are abundant. However, the limitation of this approach is that the traits of interest in the wild species are linked to other traits that have a negative impact on commercial varieties. This is called a 'genetic load' (linkage drag), and it often takes many years for breeders to separate the desired traits from those that are harmful. However, with the sequencing of the tomato genome, extensive opportunities will appear to find suitable molecular markers within groups of cultivated tomatoes. Of course, many features of interest, including disease resistance, are found in wild species, and researchers are introducing them in cultivated tomatoes. Knowledge of the tomato genome will not only eliminate the genetic problems more effectively by increasing the number of molecular markers near the gene or genes of interest, but will also accelerate the backcrossing processes because the recurrent parent genome can be selected. Even without having full sequencing of the genome, progress in the field of molecular markers has been astounding. In 1987, few markers were available. Isozyme markers have been used, but they were limited and sometimes not closely enough linked to be useful. Once the RFLP markers (restriction fragment length polymorphism) began to be used, they provided a much more dense coverage of the genome and have led to significant progress in the study of genetic linkage.

However, given the large number of plants to be screened, these markers are expensive to use in selection programmes. Then in the early 1990s, RAPD markers (random amplified polymorphic DNA) were developed by using random primers amplifying DNA sequences using PCR (polymerase chain reaction). Though useful, RAPDs markers failed to complete coverage of the genome, and many of these markers were difficult to use in the laboratory. Since then, a wide range of marking technologies has emerged, associated with acronyms such as AFLP, CAPS, COS, SNP, or INDEL, and their use has mapped many genes of interest and strengthened efforts for genetic improvements. If the development of breeder-friendly molecular markers has allowed breeders to have tools that make their work more efficient, they are by no means a panacea to meet all needs.

Development of molecular markers is costly and often takes several years. In the future, knowledge of the genome sequence should at least help reduce the time needed to develop useful markers. Nevertheless, the identification of genes for disease resistance will always require high quality data in terms of phenotypic characterization. This means that breeders and pathologists must have efficient methods for assessing the behaviour of plants to disease, whether in artificial conditions (glasshouse or heated units) or in natural conditions in the field. With such techniques, breeders can make significant progress, whether or not they have molecular markers.

Another relatively new biotechnological approach for developing plants resistant to disease or for improving some other criteria is genetic transformation. This method allows the insertion of genes from the donor species into the DNA of cultivated species. The plants obtained are referred to as 'GMO' (genetically modified organisms), and some of them (e.g. corn, soybeans, and cotton) are grown on a very large scale. GM varieties are not widely used in horticulture or for other crops considered minor, because of the associated regulatory costs and consumer fears. Genetic transformation has so far been limited to the implementation of single genes; quantitatively controlled characters important in selecting new varieties have not yet been taken into account. On the other hand, if one looks at the resitances available in the multiresistant tomato varieties obtained by conventional breeding, it is clear that almost all of these are controlled by a single gene. In fact, most are dominant genes, which have enabled the creation of F1 hybrid varieties that are so commonly used in commercial cropping as well as in amateur gardens.

Since 1987, the health situation of the tomato has evolved a lot. It was always fascinating but sometimes disturbing to see how certain diseases emerged in areas of production where they were previously absent. Sometimes it is easy to understand why a disease emerges, such as when an insect vector carrying a virus is introduced. In other cases, the disease is spreading dramatically with no known reason. There is no doubt that diseases are a threat to tomato crops wherever they are, and that is why breeders have worked so hard to improve its resistance. Some of the most important diseases are discussed in what follows.

In humid tropical regions, one of the most common diseases is bacterial wilt (*Ralstonia solanacearum* – the Latin name has changed since the last edition of this book by D. Blancard). Some resistant varieties are available and some of them are quite effective in controlling the disease. However, this bacterium affects a broad host range (over 200 plant species) and, in recent years, a potato strain infecting tomato has been found, a strain for which until now, no source of effective resistance has been found. Less pervasive, bacterial spot – caused by several species of *Xanthomonas* – is nevertheless a major disease. The nomenclature of the bacteria in question has changed, and four new types have been identified since 1990. In temperate regions, speck (caused by *Pseudomonas tomato*) was the main bacterial disease, and a second race of this bacterium has overcome the resistance used in some production areas. Bacterial canker (due to *Clavibacter michiganensis*) can be devastating when it is transmitted mechanically during cultivation operations.

The sweet potato whitefly (*Bemisia tabaci*) transmits a growing number of viruses in tropical regions of production, called 'Begomovirus' or 'Geminivirus'. The best known of them is the Tomato yellow leaf curl virus (TYLCV), but many of the Geminivirus in the New World also cause considerable damage to many crops in Central America or South America. In this area of production, it is often necessary to have a variety resistant to bacterial wilt and Geminivirus to produce tomatoes successfully. Tomato spotted wilt virus (TSWV) is an 'old' virus that still causes considerable damage in tropical and temperate regions. A 'new' viral problem, caused by Pepino mosaic virus (PepMV), occurs in glasshouses. It has implications in terms of quarantine for the virus, which survives well in soil and on tools, as it is mechanically and seed-transmitted during cultivation operations.

In fungi, new strains or races of established pathogens challenge tomato breeders. This is the case with *Phytophthora infestans*, responsible for blight, a disease of cold and wet climatic conditions. Strains of this pathogen have cohabited for several years and this has allowed those belonging to different compatibility groups to reproduce sexually. Numerous sporangia that are air-borne result in disease epidemics on single-gene resistance carrying varieties. Alternaria (*Alternaria tomatophila*) is one of the most common foliar disease for producers, but also for home gardeners. A race of *Verticillium dahliae*, responsible for Verticillium wilt, for which there is no resistance, causes crop losses in many growing areas. It is the same for race 3 of *Fusarium oxysporum* f. sp. *lycopersici*, the cause of Fusarium wilt, which continues to spread, requiring the use of resistant varieties that did not exist in 1987. With the phasing out of methyl bromide, *Sclerotium rolfsii* could become a serious threat to crops in many regions of production. There are also races of nematodes not controlled by available resistances. Resistances to nematode attack is highly inefficient when the soil temperature is high.

Insects also cause various types of damage on tomato. Feeding punctures caused by the sweet potato whitefly results in irregular ripening of the fruit. Bugs cause spots on the fruit and lead to bad taste; thrips cause unappetizing superficial golden lesions of the fruit; moths and caterpillars consume the fruits, making them unmarketable. Various mites may also be responsible for serious leaf damage, leading to substantial yield losses.

Abiotic diseases are sometimes confused with parasitic diseases, and may also cause significant losses. Some are due to erratic irrigation or rainfall, to inappropriate nutrition, to sudden fluctuations in temperature, and/or interactions between these parameters. Tomato varieties often differ in their susceptibility to these disorders. The selection of varieties tolerant to abiotic diseases is often difficult because symptoms do not always develop under the selection conditions. A breeder can unknowingly select a line sensitive to an abiotic disorder and only realise later that this ruins the variety.

It is obvious that many factors, whether parasitic or not, can damage a tomato crop. It is important to be able to identify the problem so that the various appropriate measures can be implemented to address them. This is where a book such as this one is so valuable. Although it is not always possible to determine the cause of an anomaly from photos, we can at least reduce the number of possibilities and therefore effectively aid the diagnosis and, ultimately find the solution.

Given the evolution of tomato diseases in the world and the availability of new resistant varieties or other means of control, this book will provide readers with updated knowledge, not available in the previous book. As a tomato breeder, I only hope that this new edition will not become obsolete too quickly, because we already have enough work! Enjoy the wealth of information contained in this book, it will help you diagnose and solve your problems on tomato. Best wishes for your next tomato crops and, for your own health and wellbeing, do not forget to eat plenty of tomatoes and other fruits and vegetables!

ACKNOWLEDGEMENTS

This book is the result of a passion for the tomato and its diseases, which began very early, as soon as I joined INRA. I was strongly influenced by my colleague and friend Henry Laterrot. It also embodies many observations and investigations in the field, often in collaboration with many fellow researchers, technicians, and growers. They are all warmly thanked.

All my gratitude also goes to Michel Clerjeau, who greatly influenced my interest in the study of diseases of vegetable crops and tomato.

I express my gratitude to Dr. John Warner Scott, world expert on tomato, for having written the Foreword for this book.

I would also like to thank everyone who contributed in various capacities to improve the quality of this work by:
– Their comments on the script (H. Laterrot, G. Marchoux, and T. Candresse) which enabled the completion of the book (S. Chamont);
– Valuable collaboration during various aetiological studies (A. Marais, X. Foissac, K. Gebre Selassie);
– Providing digital photographs or electron microscopy of fungi (A. Corbiere, J. Montarry for Photos **825** and **826**), viruses (B. Delécolle for Photos **883–890**), phytoplasma (M. Garnier for Photo **879**), insect vectors (J.-L Danet for Photo **880**), and diseases in the field (T. Zitter for Photo **309**, F. Bertrand for Photo **423**, and M. Davis for Photos **582–584**);
– A critical and constructive review (P. Castagnone, L. Delbac, L. Delière, X. Foissac, P. Gognalons M. Jaquemont, A.-I. Lacordaire, C. Manceau, M. Martinez, M. Piron, P. Pracros, Y. De Schepper);
– Their friendliness, availability, and efficiency in the realization of the book (service Editions Quae, especially D. Bollot, C. Colon, G. Perraud, and J. Veltz).

The Koppert company has largely contributed to the pages on pests. I express my gratitude for the the quality of the documentation made available to me. These are from the book by M. Malay and W. Ravensberg, *Connaître et Reconnaître : la Biologie des Ravageurs des Serres et de leurs Ennemies Naturels*; they correspond to Photos **84–86**, **89–91**, **98–100**, **293–295**, **368–371**, **391–403**, and Figures **17, 18, 24, 27, 32a, 34–37**, and **49**.

We would also like to thank all the advertisers and partners who have contributed to the financing of such a book: Koppert France, BASF Agro, Clause, De Ruiter Seeds France, Gautier Seeds, Rijk Zwaan, Sakata Europe, Vilmorin, and especially the Plant Health and Environment department of INRA.

INTRODUCTION

To identify a disease in the field, the practitioner must have academic training and extensive *knowledge* of various disciplines together with experience. One must remember that in Greek, *diagnôsis* means 'knowledge'. The extent of this knowledge is justified by the large number and diversity of the causes of diseases. Thus a good practitioner needs to be informed of all *symptoms* and *signs* caused by disease. But his expertise is not only related to the biology of pathogens (sources of contamination, modes of transmission, favourable weather) but also to understand the exact nature of the *pathological context*. The latter may be insufficient or inadequate, particularly in the context of a nonparasitic disease. Thanks to many observations and appropriate questions, he/she will also be able to define the environmental and cultural conditions of the disease development. In fact, the diagnostician is a real detective, using a procedure constructed from his knowledge and especially his experience of diagnosis. It consists of a *progressive sequences of observations* of the diseased plants and their different organs, of questions mainly intended to clarify the cultural context, and identifying possible *hypothetical causes* and *prioritizing* them.

This book attempts to formalize this most difficult art that represents the diagnosis of plant diseases. We learned and perfected the particular approach outlined in our previous works on diseases of tomato, cucurbits, tobacco, and salads. This included a knowledge and the intellectual processes being used by an experienced generalist phytopathologist and is now used when confronted with the identification of tomato diseases.

The objectives of this book are therefore many:
– Allow readers to identify interactively parasitic and nonparasitic diseases of tomato common in the world from both descriptions and photos, thus avoiding the many possible causes of confusion in diagnosis.
– *Make available a summary of recent knowledge* on almost all of common and serious pests and pathogens on tomato.
– Help them *choose*, knowingly, the *methods of protection* most relevant to control these pests;
– Contribute to their development in the diagnosis of tomato diseases to improve their level of expertise.

The section Diagnosing parasitic and nonparasitic diseases is designed as a true *diagnostic tool*, illustrated by more than 900 colour photographs and numerous illustrations to aid the observation of the plants. This tool can easily be used as the symptoms are grouped under simple headings. It is also educational and should ultimately enable the reader gradually to acquire the approach and the many factors needed to establish a reliable diagnosis.

We draw the attention of prospective readers to the benefits of using this method, taking into account the methodology defined in the following pages. Indeed, some readers with knowledge may tend to 'bypass' the proposed process. In this case, they may be confused by the organisation of the work which is not typically designed according to the nature of the diseases, but rather takes into account the symptoms that characterize them.

Once identification is complete, the reader will find a *detailed description* for the majority of pests and pathogens in the third section, Principal characteristics of pathogenic agents and methods of control. This section details the distribution and impact on tomato growing in the world, the main symptoms produced, and their morphological and biological characteristics. With this knowledge, readers will be able to choose the best suited control method and to define a protection strategy to be implemented immediately or during the next crop.

The first and third sections are designed to raise awareness of the tomato plant and its botanical family and its cultivation, but especially the potential offered by the Solanaceae in term of resistance to diseases.

HOW TO USE THIS BOOK

Faced with so many photos, the reader is often tempted to use the book by flipping through quickly and to make diagnoses by comparing symptoms on diseased plants to those which ressemble them in the book. The diagnosis procedure proposed here, although sometimes a bit tedious, is the result of a fairly extensive consideration of the difficulties of identifing a disease. Carefully following the procedure is the best guarantee of a reliable identification and will result in the gradual development of self-confidence in the diagnosis of parasitic and nonparasitic diseases of tomato.

Preparing the diagnosis

The identification of tomato diseases is not easy. Indeed, as this plant is affected by multiple diseases confusion abounds. To increase the odds of a correct answer, the following procedure should be adopted:

a) **Make the observations on diseased plants with more and less developed symptoms representing different stages of disease progression.**

It is essential to recover the whole plant, including the root system. It should be dug up carefully and the roots washed, as it is extremely important to clean them properly. This process is often overlooked by technicians and producers, but is important in the identification of many pests and pathogens, notably those soil-borne.

b) **Collect as much information as possible** (diagnosis often requires a proper investigation):
– **About the disease and its symptoms** (distribution in the plot and the plants, possibly on the parts of the plants – see pp. 36–41, rate of change in the crop and on individual plants, climatic conditions before its appearance or aiding its development);
– **On the plant** (including varietal characteristics and notably resistance to pests and diseases, origin and quality of seeds and plants);
– **On the plot** (location, soil characteristics, presence of waterlogged areas, previous cropping, addition of soil or manure);
– **On the agricultural practice** (fertilizers, irrigation management – method of irrigation, frequency and quantities applied, climate management in protected crops, application of pesticides on the crop or nearby – dose volume of water/ha, material used).

 Please note that additional questions regarding this information will often be needed during diagnosis to confirm or disaprove certain hypotheses.

Making the diagnosis

a) **Carefully determine the location and nature of the visible symptoms on diseased plants** and, subsequently, choose one or more of the available options. Symptoms may be observed on all organs of the tomato plant and their presence on these vary with disease. Some will affect only the fruit, others will occur on the fruit, but also occur on the leaves and stalk.

The structure of Section 2 of the book allows the reader to consult a wide range of symptoms seen in each of the organs of the tomato plant. Numerous links between different parts of this section of the book will help to identify both organ-specific disease as well as a disease that manifests itself in several plant organs. The organization is fairly simple and embodies the chronology of plant observation when a generalist phytopathologist is confronted with diseased plant.

First, the foliage is observed, which can be affected directly or show the effects of lesions already in place on another organ. The many observable symptoms are divided into four subsections, covering

The leaflets and leaves

Abnormal growth of plants and/or
form of leaflets and leaves

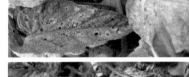

Discoloration of the leaflets and leaves

Spots and damage on leaves and leaflets

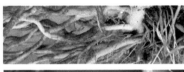

Wilting, desiccation, necrosis of leaflets
and leaves (preceded, accompanied or
not by a yellowing)

Roots and/or stem base

The stem

Fruits

abnormal growth and shape of the leaves, colour of leaves, spots on the leaves, and wilts, dessication, and leaf necrosis.

In many situations, the other organs of the tomato should be reviewed to complete a hypothesis or find others. With the same aim of simplicity and clarity, the symptoms observed on these organs are subdivided in three chapters: 'Symptoms on roots and stem base,' 'Internal and external symptoms of the stem' and 'Fruit symptoms'. The observation of soil organisms (roots, stem base) will necessitate washing the roots to observe them. After having taken care to cut the stem longitudinally or transversely at different levels it can be checked carefully along its entire length, both outside and inside. An important chapter has been devoted to fruits that are affected by many diseases, notably specific ones, occurring either in the field or after harvest during storage.

b) **Refer to the topics that best match the 'main' symptoms that is see on the plants.**

At the beginning of each chapter or subchapter are mentioned:
– The symptoms studied;
– The possible causes.

Several symptoms are suggested, they correspond to several possible diagnoses.

c) Now choose a more 'specific' symptom and go directly to the pages that describe it.

In many cases it will be wise to look at all the symptoms in the subheading or heading to avoid confusion.

At this moment in the diagnosis process, it is important to be very precise in the definition of the observed symptoms. In addition, each symptom will be associated with one or more 'possible causes'. They will need to be distinguished (a symptom corresponds to several possible hypotheses).

d) Determine the cause of the symptom.

To distinguish between the different hypotheses, it is suggested:
– To compare the symptoms observed on plants with those presented in the numerous photos;
– To use the **'additional diagnosis criteria'**;
– Not to hesitate in examining the symptoms described in the neighbouring pages, or even in other chapters or subchapters as advised.

At this stage of diagnosis, the nature of the disease that affects the crop should have been determined quite precisely, or at least narrowed down to a few possible hypotheses. If there is any doubt, a specialist laboratory should be consulted.

Drawings, photos, and comments are liberally included throughout this section of the book devoted to the diagnosis and should facilitate analysis. They are accompanied by symbols that define the purpose of the information given.

Diagnosis: defines the level of difficulty of diagnosis.

Helps the diagnosis: the precise nature and/or distribution of a symptom provides additional diagnostic criteria.

Shows and/or explains a symptom and its evolution.

Suggests the use of a lens, a binocular microscope, or a high power microscope.

Know the disease and select appropriate methods of control

Once the diagnosis is made, the reader should refer to the fact sheets constituting the third section of the book, summarizing the knowledge on tomato pests and pathogens. These review the following points:
– **Frequency and extent of damage** (summary of the worldwide distribution of the disease, its frequency and its impact on crops);
– **Symptoms** (short list of symptoms, list of the reference numbers of the pictures showing them in the second section of the book);
– **Biology, epidemiology** (description of the main stages of the life cycle of the pathogenic agent: survival, penetration, dissemination, favourable conditions for its development);
– **Methods of control** (listing the different methods of protection – prophylactic, chemical, genetic, biological – to apply during cultivation and during the next cultivation period).
The reader should read the note to the reader on p. 417.

In the case of nonparasitic diseases, the measures to be taken to limit their development will be derived from the possible causes. In many cases, poor climate control and/or unsuitable agricultural conditions will be responsible. To remedy this, it will be necessary to correct these mistakes and/or provide a better conditions for the plants. In this case the possible options will be described in detail in the second section of the book.

1 The tomato plant and its culture

Lycopersicon esculentum and related species[1]

• Origin, history of its expansion in the world

The tomato, unknown in the Old World until the sixteenth century and still very little used in the nineteenth century, became the star of the vegetables in the twentieth century, both in commercial cultivation and in home gardens. It is appreciated for its freshness and is the base or topping in all sorts of dishes, either raw or cooked. Its use in sauces is traditional, especially in Italy. The processing industry offers many different preparations: concentrate, juice, peeled tomatoes, crushed tomatoes, and so on. Because of its relatively high level of consumption, tomato accounts for a significant part in the intake of vitamins and minerals in the diet (*Table 1*).

Table 1 Chemical composition of tomato fruit (%)

Water			95	
Total dry matter	Soluble dry matter	Sugar (glucose, fructose)	55	
		Acids (citric, malic)	12	
		Minerals	7	79
		Carotenoid pigments*, volatile compounds, vitamins**	5	
	Insoluble dry matter (cellulose, pectic substancees		21	5

* Yellow orange pigments (betacarotene = provitamin A) or red pigments (lycopene).
** Vitamins C (18–25 mg/100 g of fresh fruit), B, K, E.

The arrival of the tomato in the Old World

Originally from South America, the tomato was domesticated in Mexico. Its introduction in Spain and Italy, and from there, into other European countries, was in the first half of the sixteenth century (Figure 1). Originally, it was cultivated by the Aztecs; its name comes from '*tomatl*' from the Nahuatl language spoken in the region of Mexico, and corresponds to *Physalis philadelphica*; the tomato itself, *Lycopersicon esculentum*, was called '*jitomatl*'.

The first mention of the tomato in the Old World is by the Italian botanist, Pietro Matthioli Andreas, in 1544. He described it as a species carrying "flattened and ribbed fruits, which go from green to golden yellow and which some consume fried in oil with salt and pepper, like eggplant and mushrooms".

A decade later, he mentions that there are yellow tomatoes and red tomatoes. The Italian name 'pomodoro' seems to confirm that the first tomatoes, at least those that arrived in Italy, produced yellow fruit. In the texts of the sixteenth and seventeenth centuries, tomato receives various names, including 'mala aurea', the Latin equivalent of 'pomodoro.' The name 'pomme d'amour' in French, with its equivalents 'love apple' in English, and 'Liebesapfel' in German, alludes to the aphrodisiac effect then attributed to this fruit.

In the Old World, the first graphical representations of the tomato are those of Rembertus Dodonaeus (Antwerp, 1553), Georg Oelinger (Nuremberg, 1553), again R. Dodonaeus (Antwerp, 1574, see Figure 2), and Castore Durante (Rome, 1585, see Figure 3). In this

---

[1]In 2006 Peralta *et al.* proposed a revision of the nomenclature of species in the genus *Lycopersicon* (Peralta *et al.* 2006. Nomenclature for cultivated wild and tomatoes. TGC Report, **56**:6–12), and some of the proposed changes are shown in Box 1. This revised classification is not unanimously accepted, so we have preferred to keep the old names of the different species of *Lycopersicon* that are discussed in this book.

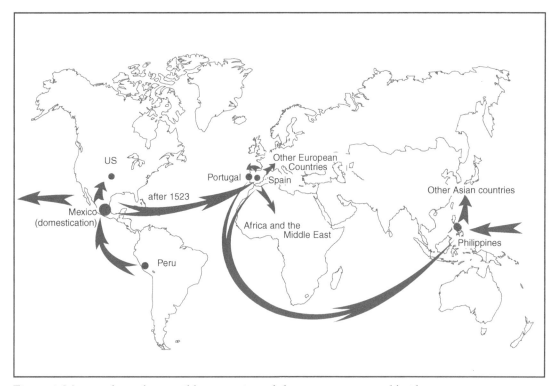

Figure 1 Map to show the possible expansion of the tomato crop worldwide.

part of the world, the tomato was viewed with suspicion. It was believed to be toxic, like other species of the family Solanaceae to which the tomato belongs, such as belladonna, the nightshade, or mandrake, a plant with magical powers. For a time the tomato was cultivated as an ornamental curiosity and is still sometimes used to decorate balconies, after which its fruit is then eaten. It appears that its use in food developed first in the form of sauces to complement cooking. Tomato consumption in the form of fresh fruit started in the Mediterranean and spread northward in the late eighteenth century.

Botanists have repeatedly altered the names of genus and species assigned to the tomato. In some earlier books, *Solanum esculentum*, *S. lycopersicum*, and *Lycopersicon lycopersicum* are used. The name finally selected until very recently (see Box 2), *Lycopersicon esculentum* Mill., was assigned by Philip Miller in 1754. The name of the genus *Lycopersicon* is Greco-Latin, and means 'wolf-peach'. The species name '*esculentum*' comes from Latin and means 'edible'. This does not mean edible foliage, or young green fruit as they contain toxic alkaloids (tomatine, solanine). These disappear during fruit development.

Figure 2 Drawing of a tomato plant published in the book *Purgantium aliarumque eo facietium historiae. libri III*, Antwerp 1574, by Rembertus Dodonaeus, sometimes cited under the names 'Dodoens', 'Dodonée', or 'Dodon'; physician and botanist to two Germanic emperors, born in Friesland (1517), and died in Mechelen (1585).

Figure 3 Drawing of tomato plants published in the book *Herbario nuovo che con figure rappresentano the bright piante che in tutta Europa nascono and nell'Indie, etc.*, Rome 1585, Castor Durante; Italian physician and botanist, physician to Pope Sixtus V, born in Gualdo, died in Viterbo (1590).

• Economic importance

The tomato is, after the potato, the most consumed vegetable in the world, either fresh or after processing. It is grown in all latitudes in a variety of conditions (climate, production, Photos 1–5), demonstrating useful variability and demonstrating the effectiveness of the work done by plant breeders.

World production of tomatoes has increased steadily during the twentieth century and has grown considerably over the past three decades. It rose from 48 million tonnes in 1978 to 74 million in 1992, 89 million in 1998, and reached 124 million in 2006. Among the 16 countries that produced 1 million tonnes or more, six are well above 5 million tonnes (*Table 2*). An estimated 30% of tomatoes produced are processed, but this percentage varies from one country to another.

Consumption of tomatoes per individual, whether fresh or processed, is increasing worldwide. The Mediterranean countries are large consumers, at all times of the year. In France, consumption of fresh tomatoes is around 13 kg/person/year, while the consumption in the form of processed products is reaching the equivalent of 22 kg of fresh tomato. These figures, from the Centre Technique Interprofessionnel des fruits et légumes (CTIFL), are based on domestic production, incorporating imports and exports. It is likely that demand in tomatoes will continue to grow, due to the human population increase, the shelf life that allows long-distance transport, the diversification of variety types, and the dietary changes that guide consumers towards products such as this vegetable.

Table 2 Production of tomatoes in 2006 (in million tonnes, FAO, 2007)

World production: 123.7 million tonnes

China	32.5	Mexico	2.9
USA	11.3	Russia	2.4
Turkey	9.9	Greece	1.7
India	8.6	Uzbekistan	1.6
Egypt	7.6	Ukraine	1.5
Italy	6.4	Morocco	1.2
Iran	4.8	Chile	1.2
Spain	3.7	France	0.74
Brazil	3.3	France + Overseas territories	0.76

• *Lycopersicon esculentum*: the species and its biology

The cultivated tomato, *Lycopersicon esculentum*, belongs to the Solanaceae family. It is a diploid species with 2n = 24 chromosomes, among which there are many single-gene mutants, some being very important for selection.

Figure 4 The main aerial parts of a tomato plant.

Two growth patterns

The variety 'Marglobe', taken as a reference by breeders, has a morphology representative of cultivated varieties, with only one stem and indeterminate growth pattern. After germination and emergence of two cotyledons, the stem grows and six to nine alternate leaves are produced before the onset of an inflorescence producing five to 12 flowers. Subsequently, three new leaves will form before a new inflorescence. Growth continues in series (sympodially) of three leaves and one inflorescence. This is called sympodial growth.

Side shoots grow from the axils of each leaf. These are also of indeterminate growth. In single-stem culture, they are manually removed by pruning.

A spontaneous mutation in Florida in 1914, resulting in a determinate growth pattern, was quickly exploited. It is a recessive mutation called 'self pruning' whose genetic symbol is 'sp'. Plants with determinate growth have a stack of shoots varying from two to five depending on the plant material in which the allele 'sp' is found. The number of leaves per shoot is three, then reduced to two, then one. The main stem and side branches are terminated by an inflorescence. The plant is limited in size and it rapidly stops growing. The determinate varieties are mainly used for mechanically harvested crops used in food processing.

Morphology

The leaves are imparipinnate with indented leaflets There are some varieties with very seldom indented leaves and nonserrated edges which are called 'potato leaf varieties'. This monogenic recessive characteristic is controlled by the allele 'c'.

The flowers of the cultivated varieties are grouped into simple or branched inflorescences (Figure 4). Their number is variable, ranging from five to 12. The flower is made of five to eight sepals, five to eight petals, five to eight stamens, and an ovary containing two to ten carpels (Figure 5). The stamens are fused into a cone surrounding the pistil, i.e. the ovary, style, and stigma. Each stamen releases the pollen it contains through a longitudinal slot located inside the staminal cone. This is called a dehiscent introrse of stamens. The pollen is received by the stigma that is inside, at the end of the cone of stamens. Pollen grains germinate on the stigma, and their pollen tubes penetrate the style to reach the ovary and the widely varying number of ovules depending on variety. The fertilized eggs form the seeds.

Tomato fruits, fleshy and tender, are actually berries, and vary in size, colour, and texture depending on the variety. The shape also varies (Figure 6), and their weight can vary from a few tens of grams to more than 1 kg. The colour, more or less dark green when immature, evolves during maturity to reach various shades according to the cultivars: cream, yellow, orange, pink, red, or brown. Some rare varieties are striped.

The tomato cycle from seed to seed, varies from 3.5 months to 4 months, depending on variety and environmental conditions. It takes 6–8 weeks from sowing seed to blossoming, and 7–8 weeks from blossom to ripened fruit with seed formation.

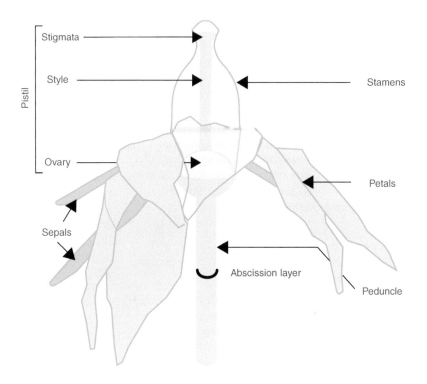

Figure 5 Section of a tomato flower.

1 Key distinguishing features of tomato varieties

The type of plant growth

• *Indeterminate growth*
  The plant produces seven to ten leaves and an inflorescence, then three leaves and a second inflorescence and continues this pattern indefinitely.

• *Determinate growth*
  The plant stops its development after two to five inflorescences; lateral shoots stop growing after one to three inflorescences.

The type of shoulder in mature fruit
Dark green shoulder when immature, becoming more or less like the whole fruit thereafter. Shoulder does not differ from the rest of the fruit that is therefore uniform in colour.

The shape and size of fruit
The fruit may have a quite different morphology and size depending on the variety: more or less large, flattened, slightly flattened, rounded, rectangular, cylindrical, elliptical, heart-shaped, obovate, oval, pear-shaped, ribbed, or smooth (Figure **6**).

Fruit colour
The fruits also reveal a distinct colour according to varieties: creams, yellows, orange, pink, red, brown, and showing darker stripes.

Resistance to pathogens and nonparasitic diseases
These are generally monogenic dominant characters mentioned in the seed catalogues, and are detailed in *Table 54*, p. 652.

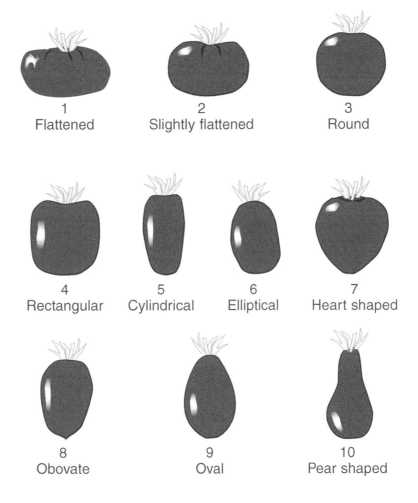

Figure 6 Morphology of tomato fruit (longitudinal section) selected by the International Union for the Protection of New Varieties of Plants (UPOV, 2001).

• Related species

The genus *Lycopersicon*, to which the cultivated tomato and its wild type *Lycopersicon esculentum* var. 'Cerasiforme' belong, includes nine species (see Box 2), all from western South America (Andean region of Peru and Ecuador), with the exception of *Lycopersicon cheesmanii* which is found in the Galapagos Archipelago (see Box 3).

Morphology

The nine species of *Lycopersicon* have bisexual flowers with the same constitution as that described above for *Lycopersicon esculentum*. They differ by the colour of the mature fruit, the number of leaves between the flower clusters, and their mode of reproduction.

Sepals, petals, and stamens of wild species always number five and the number of carpels is two.

Self-pollination is the rule among the cultivated tomato; however, certain environmental conditions, resulting in the lengthening of the style with the stigma coming out of the cone of stamens, allows cross-fertilization between neighbouring plants. Such cross-fertilizations are performed by pollen collecting insects, notably solitary bees (*Exomalopsis biliotti*) which are very common in the subtropics. The autogamy is less consistent in the wild species of *Lycopersicon*. Two species, *L. peruvianum* and *L. chilense* are cross-pollinated, due to an almost self-incompatibility.

The species *Lycopersicon pennellii* has long been classified in the *Solanum* genus as it differs from other *Lycopersicon* species, having free stamens (not fused into a cone), and also by a poricide dehiscence, i.e. by pores at the top of each of the five stamens. This species can be hybridized with the cultivated tomato; such hybrids have been classified as *Lycopersicon*.

All species of *Lycopersicon* are diploid with 12 basic chromosomes. Crosses between the cultivated tomato and other *Lycopersicons* are relatively easy, except with *L. peruvianum* and *L. chilense*. All hybridizations are performed by taking the cultivated plant as the female parent.

L. esculentum has a high apparent diversity (size, shape, colour of fruit, plants bearing) but a low genetic variability, so breeders are looking to the wild species to increase the variability of the crop. The tomato is the vegetable for which wild related species have been the most widely used in breeding programmes, notably as a source of pest and pathogen resistant genes. Their exploitation by breeders continues to grow not only as a source of disease resistance, but also for adaptation to various stresses (cold, salinity, drought), for the plant structure (two leaves sympodially), and to improve fruit quality (soluble solids, carotenoid content, including lycopene, and vitamin C).

2 Classification of tomato species

Nine species of the genus *Lycopersicon*

1. Interior of red ripe fruit; seeds of 1.5 mm or more

   1.1. Fruit diameter greater than 1.5 cm; usually serrated leaf

      1.1.1. Fruit of 3 cm or more; 2 or more carpels

         • *Lycopersicon esculentum* Miller (*Solanum lycopersicum* L.*)

      1.1.2. Fruit 1.5–2.5 cm; 2 carpels

         • L. *esculentum* var. '*Cerasiforme*' (Dun.) Gray.

   1.2. Fruit diameter less than 1.5 cm, usually about 1 cm; leaf edges usually wavy or plain

      • L. *pimpinellifolium* (L.) Miller (*S. pimpinellifolium* L.*) (Figure **7**)

2. Yellow or orange interior of ripe fruit; seeds 1 mm or less

   • L. *cheesmanii* Riley (*S. cheesmaniae* [L. Riley] Fosberg*) (Figure **8**)

3. Green or whitish interior of ripe fruit; seeds of variable size

   3.1. Sympodial with 3 leaves

      • L. *hirsutum* Dunal (*S. habrochaites* S Knapp & DM Spooner*) (Figure **9**)

   3.2. Sympodial with 2 leaves

      3.2.1. Inflorescences with reduced or absent bracts (subgenetic complex 'minutum')

         a) Small flower (corolla 1.5 cm or less in diameter); seeds of 1 mm or less

            • L. *parviflorum* CM Rick, Kesicki, Fobes & M. Holle (*S. neorickii* DM Spooner, GJ Anderson & RK Jansen*)

         b) Large flowers (corolla 2 cm in diameter); seeds of 1.5 mm or more

            • L. *chmielewskii* CM Rick, Kesicki, Fobes & M. Holle (*S. chmielewskii* [CM Rick, Kesicki, Fobes & M. Holle] DM Spooner, GJ Anderson & RK Jansen*)

      3.2.2. Inflorescences with large bracts

         a) Anthers fused into a tube, lateral longitudinal dehiscence (subgeneric complex *peruvianum*)

            Erect plants, peduncle more than 15 cm, tight flowers, straight staminal tube

            • L. *chilense* Dunal (*Solanum chilense* [Dunal] Reiche*)

            Spread out plants; peduncle less than 15 cm, flowers more widely spaced; generally curved staminal tube

            • L. *peruvianum* (L.) Miller (*S. peruvianum* L., but also S. *huaylasense* Peralta & Knapp or *S. corneliomuelleri* Macbr. [1 geographic race 'Misti nr. Arequipa'] or *S. arcanum* Peralta [4 geographic races 'humifusum', 'lomas', 'Maranon', 'Chotano-Yamaluc']*) (Figure **10**)

         b) Free anthers; poricide dehiscence

            • L. *pennellii* (Correll) D'Arcy (*S. pennellii* Correll*) (Figure **11**)

* Denotes name as per the new classification. The old nomenclature is used in this work.

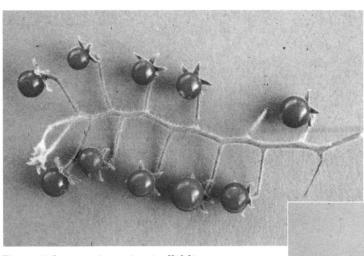

Figure 7 *Lycopersicon pimpinellifolium.*

Figure 8 *Lycopersicon cheesmanii.*

Figure 9 *Lycopersicon hirsutum.*

Figure 10 *Lycopersicon peruvianum.*

Figures 7–10 Fruits of some species of the genus *Lycopersicon* (*Solanum*) sometimes with a commercial tomato fruit for reference.

3 Areas of origin of the wild species of *Lycopersicon*

Lycopersicon esculentum var. '*cerasiforme*' occurs in the wild as *L. esculentum* in a great variety of ecotypes in very different environments, such as in Ecuador and in Peru. This species migrated and settled in other tropical areas.

Lycopersicon pimpinellifolium, prevalent in many forms in Peru coastal valleys, may have been cultivated in this area by pre-Columbian civilizations.

Lycopersicon cheesmanii comes in many forms in the Galapagos archipelago.

Lycopersicon hirsutum is present in the Andes, Ecuador, and Peru at a great range of altitude from 500 to 3300 m, in rather humid areas.

Lycopersicon parviflorum and *L. chmielewskii* form a sub-generic complex and are located in central Peru, at mid-altitude in the Andes. *Lycopersicon parviflorum* is located a little farther north than *L. chmielewskii*.

Lycopersicon chilense and *L. peruvianum* form the subgeneric complex *L. peruvianum* and have very many ecotypes in dry or temporarily dry areas along the coast of Peru and the Andean valleys of Peru and northern Chile to the Pacific. *Lycopersicon chilense* extends a little farther south than *L. peruvianum*.

Lycopersicon pennellii is found halfway up the west side of the Andes, in the central part of Peru. It occupies drier areas than any other *Lycopersicon* species.

Figure 11 *Lycopersicon pennellii*.

Cultural diversity of the tomato plant and phytosanitary consequences: from extensive to hyperintensive crop production

The culture of tomato has dramatically evolved in recent decades. Cultivation techniques have diversified and are now implemented in very different environmental conditions (Photos 1 5) and with constantly changing varietal types.

Different objectives are pursued by professionals: increasing yield, extending the production schedule, mechanizing cultivation operations, and improving fruit quality for both the fresh market and for tomatoes for industrial processing.

This broad diversification, the crop intensification, and world trade have helped to change, (sometimes turn upside down) the phytosanitary situation in the field, especially when new diseases have been introduced. For example, *Fusarium oxysporum* f. sp. *radicis-lycopersici* and *Oidium neolycopersici* were introduced into France during the 1980s, as were several viruses more recently: Pepino mosaic virus (PepMV), Tomato chlorosis virus (ToCV), and Tomato yellow leaf curl (TYLCV).

It should be recognized that the risks of one or more pathogens or pests developing in crops are not the same in different types of production, as shown in *Table 3*, p. 33.

• Open field staked cultures

The area of staked tomatoes in open fields in France have significantly declined in recent years. In this method of production, plants are grown on a single stem and are stopped at various heights after four to six trusses of fruit. The same is true for plants grown over two or three stems, usually trellised on horizontal wires holding the stem vertically or in a slightly incline, with a maximum height of 1.5 m.

• Open field cultivation on a grid system

Other crops that could be described as 'supported' are grown to benefit from the determinate growth of the plant. Wide mesh netting is used and forms a large mini-tunnel 0.4 m high and 0.6 m wide. The stems of the uncut plants grow through the netting and rest on it; the fruits are kept above ground and are not in contact with it, thereby limiting many rots.

• Open field cultures for industrial processing

In some areas, tomatoes are grown on the ground without any support, sometimes on ridges, with or under mulch, and are intended for the fresh market. Large flat areas of tomato crops are produced for industrial processing, with two to four manual harvests or, increasingly, a single mechanical harvest.

The irrigation of open field tomatoes is achieved by various methods. Irrigation by gravity requires lots of water and levelled land. Overhead irrigation can be used in nonlevel fields and very large areas, particularly in crops for industry. Trickle irrigation with low-flow drippers is still little used in open field cultures.

• Protected unheated crops

Generally, tomatoes grown under plastic tunnels, without heating, are planted in soil. They are grown in single or double rows, with a stem per plant held vertically by a string. Some production units grow the plants on mesh without pruning.

• Protected heated crops

Protection is usually provided by heated glasshouses, and culture is mainly without soil.

Substrates of various types are used:
– Containers or bags containing pozzolan, peat, or coconut fibre;
– Rolls of mineral wool or glass wool;
– Troughs with circulating nutrient solution.

The irrigation of the trays, bags, and rolls is accomplished by drip. The plants are trained by twisting them around strings hanging from horizontal wiring at a height of about 2.5 m. Plants are usually grown on one stem, but in some circumstances, they are grown with two or three stems, especially in the case of plants grafted onto rootstocks bringing extra vigour (see Section 4). Depending on the duration of production, the crop is grown vertically and the stem growth stopped by removal of the terminal growing point after 12 trusses, or stems are gradually inclined so that 20 trusses can be harvested.

Some farms use a variety of techniques of planting to ensure continuous production.

The conditions for fruiting are quite often borderline in protected crops, so pollination is frequently assisted by mechanical vibration of the flowering trusses or by the use of bumble-bees (*Bombus terrestris*) that pollinate by visiting the flowers to collect nectar.

1 A family garden in Africa.

2 A crop on mounds in a rice field in China.

3 Field production in the United States.

4 Traditional Chinese glasshouse; bamboo is widely used, particularly for staking plants.

5 A large glasshouse in the Netherlands; a forest of strings can support plants for several months.

Some examples of tomato cultivation in the world: from extensive to very intensive

Table 3 Estimated frequency of main tomato pathogens in different production systems in Europe[1]

	Culture Type		
	Field crop (staked, grill, industry)	Protected - unheated, in soil	Protected - heated, without soil
Air-borne fungi			
Alternaria tomatophila	++	+/–	–
Botrytis cinerea	+/– to +	+ to ++	++
Didymella lycopersici	+/– to +	+/– to +	+/–
Mycovellosiella fulva (Fulvia fulva)	+/–	+ (r)	+ to ++ (r)
Leveillula taurica	+/– to +	+/– to +	+/– to +
Oidium neolycopersici	– to +/–	+/– to +	+ to ++
Phytophthora infestans	++	+/– to +	+/– to +
Stemphylium vesicarium	–	+/– (r)	–
Soil-borne plant pathogens			
Colletotrichum coccodes	+	+ to ++	+
Fusarium oxysporum f. sp. *radicis-lycopersici*	+/–	+/– (r)	+ (r)
Fusarium oxysporum f. sp. *lycopersici*	– (r)	– (r)	– (r)
Macrophomina phaseolina	+/–	–	–
Phytophthora spp.	+/– to +	+/– to +	+/– to +
Pyrenochaeta lycopersici	++	+ to ++	+/–
Pythium spp.	+/–	+/–	+ to ++
Rhizoctonia solani	+	+	+/–
Sclerotinia sclerotiorum	+/– to +	+/– to +	+/– to –
Sclerotium rolfsii	– to +/–	– to +/–	–
Spongospora subterranea	–	– to +/–	– to +/–
Thielaviopsis basicola	– to +/–	– to +/–	– to +/–
Verticillium dahliae	– to +/– (r)	– to +/– (r)	– to +/– (r)
Nematode with galls	+/– to ++ (r)	+/– to ++ (r)	–
Nematode with cysts	–	+/–	–

Continued overleaf

[1] The proposed trends are estimates which are difficult to define as the disease occurrence in the field varies according to the production areas, the tomato varieties used, and their resistance to plant pathogens:

–: absent; +/–: rare to uncommon; +: common; ++: very common; (r) resistant varieties widely used and often controlling the disease.

	Culture Type		
	Field crop (staked, grill, industry)	Protected - unheated, in soil	Protected - heated, without soil
Air- and soil-borne bacteria			
Agrobacterium tumefaciens (tumours and root proliferation)	–	– to +/–	– to +
Clavibacter michiganensis subsp. *michiganensis*	+/– to +	+/– to +	+/– to +
Pectobacterium carotovorum subsp. *carotovorum*	– to +/–	+/– to +	– to +/–
Pseudomonas corrugata	– to +/–	+/– to +	+/–
Xanthomonas spp.	+	+/–	–
Pseudomonas syringae pv. *tomato*	+	+/– to +	–
Viruses			
Tomato mosaic virus (ToMV)	– (r)	– (r)	– (r)
Pepino mosaic virus (PepMV)	–	+/–	+/– to +
Cucumber mosaic virus (CMV)	+/– to +	+/–	+/–
Afalfa mosaic virus (AMV)	+/–	–	–
Potato virus Y (PVY)	+/– to +	+/– to +	+/– to +
Tomato spotted wilt virus (TSWV)	+/–	+/–	+/–
Tomato chlorosis virus (ToCV)	–	+/–	+/–
Tomato yellow leaf curl virus (TYLC)	+/–	+/–	+/–

2 Diagnosis of parasitic and nonparasitic diseases

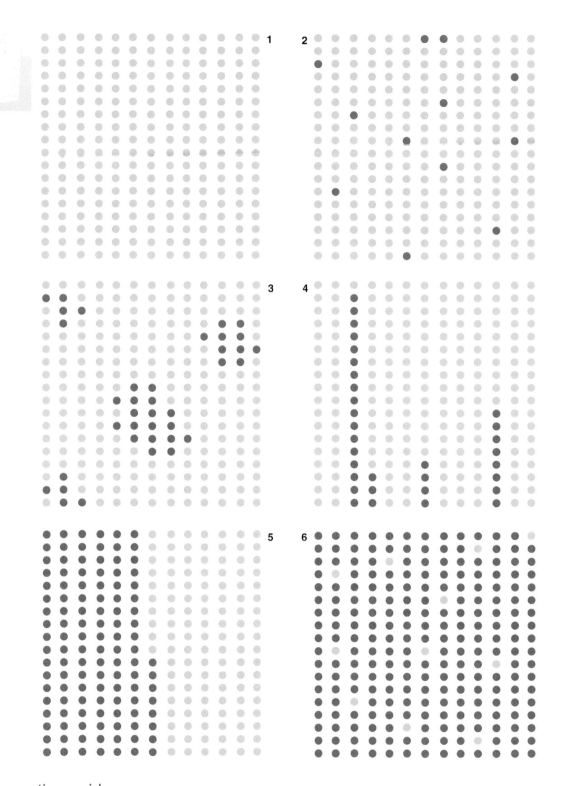

Observation guide

Figure 12 Distribution of diseased tomato plants within the crop.

1 Healthy crop.
2 Diseased plants scattered randomly.
3 One or more dispersed groups.
4 One or more rows of varying lengths of diseased plants.
5 A large group.
6 Widespread disease affecting most of the crop.

Observation guide

Figure 13 Location of foliar symptom(s) on the tomato plant(s) examined.

1 Healthy plant.
2 The terminal bud and the apex.
3 The young leaves (top of plant).
4 Patchy and random.
5 Leaves on one side of plant only (unilateral).
6 Old leaves (base of the plant).
7 All leaves (general).

Distribution of small areas of disease caused by
an air-borne pathogen with rapid development

Distribution of large areas of disease caused by
an air-borne pathogen with rapid development

6 In the open field, two areas of dried plants can
be observed, with a limited spread.
Phytophthora infestans (blight)

Distribution of area of disease caused by a
slowly developing soil-borne pathogen

7 Spray irrigation encouraged the development of
blight; the diseased area is now larger.
Phytophthora infestans (blight)

8 Many plants grouped in the same crop area have
stunted growth, chlorosis, and a gradual drying of leaves.
Rhizoctonia crocorum

**Some examples of distribution of diseased
plants in tomato crops**

Distribution in large sectors

10 The tomato crop appears divided into two parts: on the left a disease-free area, and on the right an area with plants severely affected by leaf necrosis. In fact, one of the two cultivated varieties was very sensitive to a phytosanitary treatment.
Phytotoxicity

In-line distribution of disease transmitted through pruning wounds

9 Many adjacent plants in a line have started to wither and gradually dry up. Now, many of them have died.
Clavibacter michiganensis subsp. *michiganensis* (bacterial canker)

Widespread distribution

11 In this amateur garden, all the lower leaves of the tomato plants are rolled.
Rolled leaflets (physiological leafroll)

Table 4 Examples of distribution of the major parasitic diseases in tomato crops in Europe

Pathogens	Distribution of diseased plants				
	Random	In group(s) + or − large	In line(s)	In sections	Widespread
Air-borne bacteria (*Pseudomonas syringae* pv. *tomato*, *Xanthomonas* spp.)	+/−	+		+ (if some resistant and sensitive varieties are grown together in the same plot)	*
Vascular and stem bacteria					
Clavibacter michiganensis subsp. *michiganensis*	+/−		*	.	*
Ralstonia solanacearum		+			
Pectobacterium spp.		+			
Pseudomonas corrugata	+/−	+			
Air-borne fungi					*
Phytophthora infestans	+/−	+			
Alternaria tomatophila	+/−	+			
Botrytis cinerea	+/−	+			
Mycovellosiella fulva		+			
Leveillula taurica		+			
Oidium neolycopersici		+			
Sclerotinia sclerotiorum		+			
Root and stem base soil-borne fungi					
Pythium spp.		+			* (in soil-less conditions)
Rhizoctonia solani		+			
Pyrenochaeta lycopersici		+			*
Fusarium oxysporum f. sp. *radicis-lycopersici*	+/−	+			
Sclerotium rolfsii	+/−	+			
Soil-borne vascular fungi					
Fusarium oxysporum f. sp. *lycopsersici*		+			
Verticillium dahliae, *Verticillium albo-atrum*		+			
Viruses transmitted by seeds and contact (PeMV, TMV)	+/−	+	+		*
Viruses transmitted by aphids (AMV, CMV, PVY)	+/−	+	+/−		
Viruses transmitted by whiteflies (ToCV, TICV, TYLCV)	+/−	+			*
Viruses transmitted by thrips (TSWV)	+/−	+			
Viruses transmitted by nematodes (TRSV, TBRV, ToRSV)		+			
Nematodes attacking the roots (*Meloidogyne* spp., *Pratylenchus* spp.)		+			*

+/− : sometimes observed distribution, especially at the beginning of attack.

+ : commonly observed distribution.

++ : distribution seen when favourable disease conditions persist or in plots where tomato has been grown for many years.

* : distribution seen when particularly favourable disease conditions persist.

Table 5 Distribution of principal nonparasitic dieases in tomato crops

Nonparasitic diseases	Distribution of diseased plants				
	Randon	In groups(s) + or − large	In line(s)	In sections	Widespread
Silvering	+	+			
Other genetic disorders (mutants, chimeras)	+				
Root asphyxia		+			⋆ (soil-less culture)
Lightning damage		+			
Nutritional disorders	+/−		+/−		+
Leaf roll					+
Oedema				+/−	⋆ (under shelter)
Blotchy ripening	+				⋆
Blossom-end rot	+/−	+			⋆
Various phytotoxicities	+		+	+	+
Air pollutants		+/−			+

+/− : distribution sometimes observed, especially at the early stages of the phenomenon.
+ : distribution commonly observed.
* : distribution observed when particularly disease favourable conditions persist.

12 Several leaflets on this leaf are rolled towards the upper side of the leaf; this is a good example of a leaf deformation due to a nonparasitic disease: physiological roll (leaf roll).
Physiological disease

14 The underside of the leaflet is covered with two types of spots: angular small brown to black spots and more extensive olive brown patches delimited in places by the veins.
Pseudomonas syringae* pv. *tomato* and *Mycovellosiella fulva

13 Inter-vein leaf yellowing is clearly visible on the leaflets; this symptom similar to a mosaic should be considered an abnormal discoloration of the leaf.
Pepino mosaic virus (PEMV)

15 The leaflets of this leaf are slightly chlorotic and wilted; the tip leaflets are drying out.
Clavibacter michiganensis* subsp. *michiganensis* and *Pseudomonas corrugata

Examples of leaflet symptoms

Symptoms on leaflets and leaves

To simplify diagnosis, the symptoms observed on the leaflets and leaves have been divided into four subsections:
– Plant growth irregularities and/or deformed leaflets and leaves (p. 47);
– Leaflet and leaf discoloration (p. 91);
– Spots on leaflets and leaves (p. 149);
– Wilting, necrosis, and dried leaflets and leaves (with or without yellowing) (p. 217).

Figure 14 Examples of possible distributions of one or more symptoms on tomato leaf.

1 Leaflets at the base of the plant.
2 Leaflets located on one side of the leaf (unilateral).
3 Some random leaflets.
4 Leaflets located at the top of the plant.
5 All the leaflets (general).

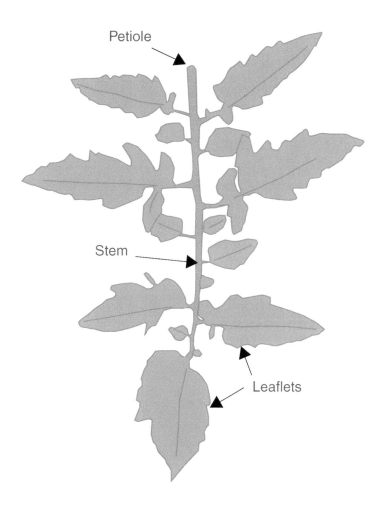

Figure 15 Appearance of healthy leaf and leaflets.

16 The apex of this plant shows very slow growth. The internodes are very short and the leaflets are tiny and very chlorotic.
Tomato yellow leaf curl virus (TYLCV)

18 The apex leaflets are smaller, more denticulate, blistered, and somewhat twisted.
Complex viruses

20 The newly formed leaves on this plant are very small and filiform.
Cucumber mosaic virus (CMV)

17 Several leaflets are rolled towards the top of the leaf.
Physiological roll

19 Regular perforations are visible on several leaflets.
Butterfly larvae

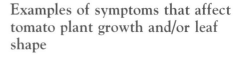

Examples of symptoms that affect tomato plant growth and/or leaf shape

Plant growth irregularities and/or deformed leaflets and leaves

Symptoms studied

- Plants with abnormal growth (dwarfed, stunted, and proliferated vegetation)
- Leaves, leaflets partially or totally distorted (blistered, curled, curved, rolled)
- Leaves, leaflets with abnormal proportions and shapes (smaller, filiform, serrated)
- Perforated, cut, shredded, mined leaflets

Possible causes

- Soil-borne fungi in the case of stunted plants (see Chapter on Irregularities and changes to the roots and/or stem base)
- Small leaves syndrome agent
- Several phytoplasmas
- Many viruses
- Beet curly top virus (BCTV)
- Cucumber mosaic virus (CMV)
- Tobacco mosaic virus (TMV)
- Tobacco rattle virus (TRV)
- Tomato aspermy virus (TAV)
- Tomato bushy stunt virus (TBSV)
- Tomato mosaic virus (ToMV)
- Tomato pseudo curly top virus (TPCTV)
- Many Begomoviruses

- Several viroids
- Various genetic abnormalities
- 'Self-determination'
- Nutritional disorders
- Physiological rolling
- Hail
- Oedema
- Various phytotoxicities
- Blind plants
- Leafminer larvae
- Butterflies caterpillars
- Aphids
- Bugs
- *Cuscuta* (Dodder)

Difficult diagnosis

Diseases causing deformations of the leaflets often have very similar symptoms, making their identification difficult. We therefore suggest you look at all the symptoms and photos in this sub-section. Moreover, these diseases also cause discoloration of the leaves. It is worth looking at this sub-section too on p. 91.

Symptoms induced by several viruses have been included in this sub-section. As the risk of confusion in diagnosis between different tomato viruses is high, we therefore strongly advise you to ask a specialist laboratory for a serological or molecular test to identify with certainty the responsible virus(es).

In many cases, faced with stunted plants and leaf deformation, we can only make hypotheses about the disease responsible, with the exception of some rare diseases with pronounced characteristics.

Figure 16 Appearance of some leaflet deformations.

1 Partially distorted leaflets sometimes serrated at the edge of the leaf.
2 Highly filiform leaflets.
3 Curved or slightly curled leaflet.
4 Perforated leaflet.

■ Plants with abnormal growth (dwarfed, stunted, and proliferated vegetation)

Possible causes

- Soil fungi in the case of dwarfed plants (see Chapter on Irregularities and changes in roots and/or stem base)
- Small leaves syndrome agent (see p. 72)
- Several phytoplasmas (see pp. 63 and 69): Description 28
- Many viruses
- Beet curly top virus (BCTV)
- Cucumber mosaic virus (CMV)
- Tobacco mosaic virus (TMV)
- Tobacco rattle virus (TRV)
- Tomato aspermy virus (TAV)
- Tomato bushy stunt virus (TBSV)
- Tomato mosaic virus (ToMV)
- Tomato pseudo curly top virus (TPCTV)
- Many Begomoviruses
- Several viroids: Description 48
- 'Self-determination'
- Nutritional disorders
- Physiological roll
- Hail
- Oedema (see p. 65)
- Various phytotoxicities (see p. 77)
- Blind plants
- Aphids (see pp. 63 and 210)
- *Cuscuta*

Additional information for diagnosis

In general, parasitic micro-organisms, pests, and several nonparasitic diseases have different effects on the different parts of the tomato plant, as seen in various chapters of this book. They particularly affect growth, including that of the aerial parts of the plants. The latter often have a modified appearance and structure in contrast with apparently healthy surrounding plants and this will draw attention to them.

Slowed to stunted growth

The growth of tomato plants can be altered to varying degrees depending on the nature of the affecting disease and how early the onset. Indeed, the development of plants is sometimes disrupted very early. This is the case, for example, in plots heavily contaminated by a soil fungus or by nematodes; the plants are very quickly affected by the inoculum in the soil which attacks their roots and/or results in vascular invasion which disrupt their development (see pp. 233 and 323). Thus, reductions in growth are recorded during early attacks by many fungi such as *Pyrenochaeta lycopersici*, *Thielaviopsis basicola*, *Rhizoctonia solani* to a lesser degree, and by nematodes such as *Meloidogyne* spp., *Pratylenchus* spp.

21 The plant on the right, affected by Cucumber mosaic virus (CMV), shows reduced growth in contrast with the adjacent, apparently healthy, plant.

22 This plant, affected by a phytoplasma shows many stunted axillary branches giving it a characteristic appearance.
Candidatus Phytoplasma solani

23 The youngest leaves of this isolated plant display particularly deformed leaflets; the terminal bud is completely suppressed. Note that this is the only plant affected.
Genetic abnormality

Examples of plants with
abnormal growth

Amongst other micro-organisms, viruses (see below and *Table* 6, overleaf), regardless of their modes of transmission and the nature of their symptoms, interfere with plant development (Photo **21**). The reduction in plant growth is all the more severe when plants become infected early in their development. It may be particularly pronounced if the virus is transmitted by the seeds and therefore multiplies in seedlings (TMV); it will also occur from contamination at the nursery stage or just after planting, hence the interest of protecting the plants during their propagation.

Examples of viruses that may limit the growth of tomato and cause growth abnormalities (they are classified according to the three major types of leaflet distortions observed) include:

• Viruses mainly responsible for stunting

– Beet curly top virus (BCTV)

– Tobacco rattle virus (TRV)

– Tomato aspermy virus (TAV)

– Tomato bushy stunt virus (TBSV)

– Tomato golden mosaic virus (TGMV)

– Tomato pseudo curly top virus (TPCTV)

• Viruses mainly responsible for filiform leaflets

– Cucumber mosaic virus (CMV)

– Tobacco mosaic virus (TMV)

– Tomato aspermy virus (TAV)

– Tomato mosaic virus (ToMV)

• Viruses mainly responsible for curved, rolled leaflets

– Beet curly top virus (BCTV)

– Cowpea mild mottle virus (CPMMV)

– Eggplant mottled dwarf virus (EMDV)

– Eggplant yellow mosaic virus (EYMV)

– Indian tomato leaf curl virus (IToLCV)

– Pepino mosaic virus (PepMV)

– Peru tomato virus (PTV)

– Potato yellow mosaic virus (PYMV)

– Sinaloa tomato leaf curl virus (STLCV)

– Taino tomato mottle virus (TToMoV)

– Tobacco leaf curl virus (TLCV)

– Tomato bushy stunt virus (TBSV)

– Tomato chlorosis virus (ToCV)

– Tomato golden mosaic virus (TGMV)

– Tomato leaf crumple virus (TLCrV)

– Tomato leaf curl virus (ToLCV)

– Tomato pseudo curly top virus (TPCTV)

– Tomato yellow dwarf virus (ToYDV)

– Tomato yellow leaf curl virus (TYLCV)

– Tomato yellow mottle virus (ToYMoV)

Less than ten viroids have been associated with natural disease outbreaks in tomato, including: Tomato planta macho viroid (TPMVd), Tomato apical stunt viroid (TASVd), Potato spindle tuber viroid (PSTVd), Tomato chlorotic dwarf viroid (TCDVd), Citrus exocortis viroid (CEVd synonymous Indian Tomato Bushy top viroid, I-TBTVd), Columnea latent viroid (CLVd). They all cause similar symptoms on the tomato (*Table* 7): dwarf plants, apical proliferation, narrower chlorotic, deformed apical leaflets with necrotic and/or brittle tissue (Photos **24** and **25**). It should be noted that some of them, especially the seed-transmissible ones, are now occurring in some production areas. At least seven phytoplasmas are reported on tomatoes occurring in many countries, often causing similar symptoms. Among them, *Candidatus* Phytoplasma asteris, *Candidatus* Phytoplasma aurantifolia, *Candidatus* Phytoplasma trifolii, *Candidatus* Phytoplasma solani, and *Candidatus* Phytoplasma australiense, have been associated with big bud symptoms. In the case of early attacks, the mature plants develop a rather unique bushy appearance due to the proliferation of poorly developed axillary shoots, producing small leaflets (Photo **22**). For other symptoms, see pp. 70 and 124.

Several nonparasitic diseases can cause the same effects on the tomato. Thus, some genetic disorders are the cause of 'wild' plants producing small, deformed and stunted, sometimes filiform leaflets (Photo **23**). Eventually, these plants, often isolated and with an unusual appearance, will contrast with the surrounding healthy plants (see p. 64).

Table 6 Main symptoms and modes of transmission of the viruses responsible for dwarf and stunted plants, with numerous erect axillary shoots

Viruses	Symptoms	Vectors
Beet curly top virus (BCTV) Description 45	The leaflets, with chlorotic and eventually blue tinged and prominent veins, are wrinkled and rolled upward while the stalks are curved downward. The leaf tissues are thicker and may have a pronounced chlorosis. The leaves are sometimes rough and eventually turn yellow. The proliferation of axillary buds gives the plants a stunted and bushy appearance.	Several species of leafhoppers, in a persistent and circulative manner: *Circulifer tonelluc* (USA), *C. opacipennis* (Turkey, Iran), *Agallia albidulla* (Brazil), *A. ensigera* (Argentina), *Empoasca decipiens*. BCTV is not seed-borne in tomato.
Tobacco rattle virus (TRV) Description 46	Dwarf plants with leaflets showing a well marked mosaic.	Several species of nematodes belonging to the genera *Trichodorus* spp. and *Paratrichodorus* spp. These nematodes remain infectious for several months or even years. Low seed transmission in some weeds.
Tomato aspermy virus (TAV) Description 38	Apex stopped, proliferation of numerous axillary buds giving the plant a bushy appearance. The mottled leaflets and leaves are heavily deformed and sometimes filiform. Fruit set is greatly reduced, and the rare fruits are small and deformed. Seed production is low to zero, unlike with CMV where the symptoms are similar.	Transmitted by 22 species of aphids (*Myzus persicae*) in a nonpersistent manner. One strain is seed transmitted in *Stellaria media*.
Tomato bushy stunt virus (TBSV) Description 32	Spots in chlorotic concentric rings appearing on the leaflets. These can also be chlorotic with anthocyanin and sometimes can be deformed. Necrotic streaks may develop on the stems. The apex may become necrotic; many axillary shoots form, contributing to the bushy appearance of diseased plants. These have a stunted growth and are often dwarf and bushy, producing small fruits.	Transmitted by contact, with water, and by pollen.
Tomato golden mosaic virus (TGMV) Description 42	Major leaf deformation, bright yellow mosaic of leaves. The dwarf plants have many axillary shoots giving them a bushy appearance.	Transmitted by whiteflies (*Bemisia tabaci*). The increased damage caused by this virus has coincided with the introduction of the whitefly B biotype.
Tomato pseudo curly top virus (TPCTV) Description 45	Chlorosis along the leaf edges, accompanied by a vein lightening. The leaflets are curved or rolled in a spoon shape. There is also a proliferation of axillary buds.	Transmitted by an insect *Micrutalis festinus* (Hemiptera) in a semi-persistent manner.

Table 7 Major symptoms and modes of transmission of viroids affecting tomatoes worldwide (Description 48)

Viroids	Symptoms	Transmission modes
Citrus exocortis viroid (CEVd)	In South Africa, plants are stunted and display leaf epinasty. In India, in addition to the above symptoms, leaf distortion and vein necroses were observed.	This viroid has rarely been found on tomato and its modes of transmission have not been specifically studied. Like other viroids, it is probably transmitted mechanically from plant to plant during cultural operations.
Columnea latent viroid (CLVd)	Plants are stunted with chlorotic and deformed leaves (epinasty), possibly with vein necroses.	There is little information on the mechanism of transmission of CLVd in tomato, but it appears that this viroid is transmitted by seed in this species. As with other Pospiviroid however, mechanical transmission from plant to plant during cultivation is very likely.
Potato spindle tuber viroid (PSTVd)	The plants are stunted and their leaflets show a pronounced epinasty, accompanied by vein necroses.	PSTVd is transmitted by vegetative propagation, leaf contact between plants, and by agricultural implements. Pollen and seeds are likely to harbour and transmit it.
Tomato apical stunt viroid (TASVd)	A severe stunting of plants accompanied by a shortening of the internodes is observed. The leaflets can display a variety of symptoms: severe epinasty, and leaf deformation, yellowing, vein necroses, brittle tissues. The fruits are small and discolored, pale red in colour.	This viroid is mechanically transmitted during cultivation. It can be transmitted by grafting, by seed, and by the bee *Bombus terrestris*.
Tomato chlorotic dwarf viroid (TCDVd)	Plants are stunted, with small chlorotic leaves and with petiole and vein necroses (see Photos **24** and **25**)	We currently have no information about the possible transmission of this viroid from tomato seeds.
Tomato planta macho viroid (TPMVd)	The symptoms are similar to those induced by PSTVd, but are more severe, including acute plant stunting, leaf epinasty, sometimes with pronounced vein necroses. Necrotic lesions may also be observed in stem. The plants produce many flowers and fruits, but the development of the latter is blocked early, their size does not exceed that of a large marble, hence the name 'male plant' given to the disease. Losses can be very severe.	This viroid seems to be mainly transmitted by contact between plants and/or tools and during cultivation.

24 The leaflets on this leaf are small, distorted, and chlorotic and have necrotic lesions.
Tomato chlorotic dwarf viroid (TCDVd)

25 This leaflet clearly shows inter-veinal chlorosis gradually leading to a necrosis of the leaf.
Tomato chlorotic dwarf viroid (TCDVd)

The physiological rolling of the leaflets (leaf roll) also changes the appearance of the plants when commonly present in a crop. The factors underlying its onset are presented on p. 65.

In some contexts of production, the apex and axillary buds may remain stunted or suddenly grow only moderately after initial normal growth. In this case, the size of old leaves often contrast with those of the most recently formed. In addition to a viral infection occurring during cultivation, accidental exposure to an herbicide can cause such damage. For more information about phytotoxicities, the section on partially or totally distorted leaves should be consulted, p. 77.

Note that plants can sometimes become determinate unexpectedly (self-determination or self-topping), the apex aborting at a final truss which is often before the fifth. The origin of this phenomenon is not well known, but possible explanations include environmental conditions to be determined or the particular sensitivity of certain varieties. In some years this nonparasitic disease often occurs simultaneously in many glasshouses, especially on crops that are vegetative and have undergone a very sudden light increase

Therefore, it is best to avoid producing plants which are too vegetative. Moreover, when the problem occurs, it will always be possible to use an axillary bud to allow the plants to continue to grow and provide other trusses. A similar phenomenon is described as 'blind plants', which also seems to occur during the early stages of plant growth. The latter have their growth halted following the loss of the terminal meristem. Again, a particular varietal susceptibility (genetic) or certain weather conditions (temperatures and/or high light intensities, type of artificial lighting) may give rise to the phenomenon. Their growth can be stopped at the seedling stage, just after the appearance of the cotyledons or of the first leaf, and restarts after a few days. On some plants, the phenomenon can persist and causes 'dual headed'

seedlings. Plants can be blind due to transplanting stress, and in particular the destruction of part of their roots. The same can occur after planting, during the formation of second, third, and fourth trusses. Some varieties are particularly sensitive, such as 'Rondelo', 'Revido', and 'Recento', especially under special conditions:
– When using auxiliary lighting for an extended period;
– In response to sudden changes in environmental conditions occurring after the plants are moved from the nursery (where the conditions can be very favourable) to the producers glasshouse (where climatic conditions can from time to time be borderline for tomatoes, especially for the roots).

To avoid this physiological condition, plants kept in the nursery prior to delivery to glasshouse growers should not be exposed to high temperatures during the last phase of production. As for self stopping which is probably the same condition (under another name), an axillary bud may be used without adversely affecting plant production. Note that in mild forms of this disease one or two leaves, at the most, appear between each truss.

As in many other plant species, nutritional deficiencies in tomato plants, though now rare in the field, disturb plant growth to a variable degree and cause leaf deformation, especially when the deficiency is severe. In this section only those that induce symptoms localized on the apex and young leaflets, sometimes prematurely stopping the development of plants, are considered (*Table* 8). This is the case of calcium, boron and, to a lesser extent, zinc deficiencies. The lack of potassium and phosphorous also inhibits the development of the tomato plant. To obtain further details on these nutritional disorders we strongly advise the reader to refer to the Nutritional disorders section (p. 125).

Table 8 Main symptoms of deficiencies disrupting plant growth and changing the size and shape of tomato leaflets and leaves

Deficiency	Symptoms
Potassium (K)	Marked chlorotic spots, beginning at the periphery of the leaf and progressing to inter-vein tissues. Subsequently they turn brown and necrotic. The leaf turns brown at the edges and curves downward. Plant growth is more or less reduced. Fruits may be softer, hollow, poorly coloured, and display a symptom of grey wall (see p. 368).
Phosphorous (P)	The leaflets are dark green and blue-tinged on the underside of the leaf, especially on the veins. Petioles and stems, sometimes very thin, show a similar discoloration. Plants are generally poorly developed and the fruits poorly coloured and hollow.
Calcium (Ca)	The edges of the leaf of young leaflets are light green and necrotic lesions gradually develop. With very marked deficiency, the terminal bud(s) brown, gradually, become necrotic and die. Plant development is thus more or less stopped. The fruits are soft and spongy with waterlogged tissue which becomes necrotic at the stylar end. (Blossom-end rot, see p. 387).
Boron (Bo)	Young leaflets near the apex are slightly chlorotic, necrotic, and fragile with a tendency to distort, roll. Plant growth is reduced and the internodes of the upper parts of the stem are shorter. The terminal bud(s), also necrotic, eventually die. Internal browning and blotches are visible in and on fruit.
Zinc (Zn)	The leaflets are notably small, thicker and tend to curve downward. Inter-veinal tissues are chlorotic and necrotic.

Cuscuta (Dodder)

These annual parasitic plants (classified in the genus *Cuscuta*) have been reported around the world on the most diverse hosts, cultivated or not. Many tens of species belonging to the Cuscutaceae family have been identified; some authors however, classify them within the Convolvulaceae. The main species reported on tomato is *Cuscuta pentagona* Engelm, the 'field dodder'.

Dodder seeds can be stored for several years in the soil. After germinating on the surface or emerging from the soil, plantlets without roots emerge which will gradually cover the tomato plant through twining stems, and parasitize the stems and leaves through suckers or haustoria. These provide access to the vascular system of the plant and so remove the minerals, metabolites, and water essential for their development, without recourse to photosynthetic activity. Infected tomato plants have a decreased growth rate, especially if they are parasitized early (Photo 26). In the latter case, they can even die.

Dodder produce small flowers variable in colour depending on the species (white, yellow, pink) grouped in clusters (Photo 27). Numerous seeds are produced (up to many thousands) which remain dormant in soil for many years, sometimes more than 20 years. The seeds, like many soil-borne pathogens, are easily distributed by water, soil particles, the equipment used for cultivation, and infested plants.

Controlling these parasitic plants is particularly difficult because of the presence of many dormant seeds in the soil and their wide host range. Among the vegetables that are hosts are asparagus, beets, carrots, eggplant, onion, melon, pepper, and potato. Many weeds can be parasitized (*Amaranthus* sp., *Chenopodium album*, *Convolvulus arvensis*). It is therefore necessary to implement additional measures such as the application of long crop rotations with nonhost plants (cereals, beans, cotton, maize, sorghum) and the use of pre-emergence or contact herbicides. Furthermore, seeds should not be introduced into a plot through irrigation water which is being used for seeds or seedlings. The first cases observed should be destroyed as soon as possible by burning the infested plants, preferably before flowering. It is desirable to plant tomatoes outside the periods of emergence of dodder seedlings. Cultivation

equipment, tools, and workers' shoes must be cleaned after each operation in an infested field. It should be noted that solarization does not completely eliminate the dodder seeds in the soil, composting seems more effective. Finally, several microorganisms may infect dodder, including *Alternaria alternata* and *Geotrichum candidum* on *Cuscuta pentagona*. Some varieties of industrial tomatoes ('Heinz 9492', 'Heinz 9553',' and 'Heinz 9992') have proved tolerant to attacks of this kind. On these varieties, the early stages of the dodder parasitism occur but subsequently many haustoria seem unable to penetrate into the stems. Thus, the development of dodder is greatly reduced as is the rate of parasitism.

With these few examples, it will become obvious that the diseases described in this section, whatever their origin, often cause (temporarily or

27 The flowers, whose colour varies depending on the species, are grouped into clusters and produce many seeds which can survive for long periods in the soil. **Cuscuta sp.**

26 Many dodder stems have invaded the foliage of this tomato plant, slowing its growth.

permanently) tomato plants to have stem apical conditions which distinguishes them from the surrounding healthy plants. These symptoms are sometimes very specific or, in contrast, common to several diseases, and in all cases they reflect a malfunction of the plants. It is therefore necessary to be particularly careful when looking for their cause.

If it has not been possible to identify the disease that disrupts the development of plants, many other parasitic and nonparasitic diseases covered in other sections of the book can sometimes cause a cessation of growth and stunting of tomato plants. They also induce other symptoms, much more characteristic, which should be investigated and analyzed.

Leaves, leaflets partially or totally distorted (blistered, curled, curved, rolled)

Possible causes

- Agent of the small leaves syndrome (see p. 64)
- Several phytoplasmas: Description 28
- Many viruses (mainly Begomovirus)
- Several viroids (see p. 53): Description 48
- Various genetic abnormalities
- Nutritional disorders
- Physiological rolling
- Intumescence
- Various phytotoxicities (see p. 77)
- Aphids (see p. 210)

Additional information for diagnosis

It is rather difficult to list and distinguish the symptoms included in the category of abnormal shapes of the leaflets of the tomato. Indeed, in many cases a leaflet or a leaf, can be both crimped, rolled, or wrinkled, which gives them a peculiar appearance. It is therefore very difficult to identify the origin(s) from the deformations alone. In this sub-section we have therefore grouped some of the major diseases which are commonly inducing leaf malformation. It will very often be necessary to refer to other sub-sections or sections of the book to complement and confirm your observations

28 Blistered and more denticulate leaflets.
Tomato mosaic virus (ToMV)

29 Shrivelled and crumpled leaflets.
Genetic abnormality

30 Slightly upturned leaflets at the top of the leaf.
Some leaflets are smaller and filiform.
Cucumber mosaic virus (CMV)

31 Rolled leaflets.
Leaf physiological roll

**Examples of partially or totally
deformed leaves or leaflets**

Various viruses

Viruses, in addition to causing multiple discolorations (mosaics, yellowing, and so on, see p. 97), induce changes in the shape of the blade which can be quite spectacular (see *Table 9*, p. 62, and list below). The nature and intensity of these changes vary depending on the virus and the earliness of their attacks, as well as on the strain present.

For example, in the case of outbreaks of tobacco and tomato mosaic virus (TMV, ToMV, Photo **32**), cucumber mosaic virus (CMV, Photo **34**), mottled and stunted aubergine virus (EMDV, see Photos **33** and **36**) and tomato spotted wilt virus (TSWV, see Photo **35**), new leaflets formed on plants are often blistered, swollen, and are sometimes twisted. Many viruses induce comparable symptoms. For example, Pepino mosaic virus (PepMV) makes the apex of the plants thinner and gives the leaflets the appearance of being variably blistered, curved up or down, and appearing like dull nettle leaves.

Several viruses, some of which are emerging in several countries, cause a variably rolled leaflet. When it starts, it is referred as 'curled leaflets' up the blade or 'leaf roll' ;when it expands; the blade can be fully wound on itself. This symptoms of leaf roll can occur alone or be accompanied by a yellowing, an anthocyanization of the petiole, sometimes with necrotic damage (see p. 163). Among the possible viruses responsible, it is worth noting numerous Begomovirus including the yellow leaf curl virus (TYLCV) and related viral species (Photos **37** and **38**). This particular virus is present in several Mediterranean countries and was introduced in France in recent years, either through contaminated plants, or by its vector (*Bemisia tabaci*) introduced on various ornamental crops. Other Begomovirus responsible for such symptoms, include tomato yellow mosaic virus (ToYMV) (Photos **39** and **40**) that exists in the Caribbean, especially in Guadeloupe and Martinique.

Examples of viruses that can induce the formation of curved, rolled leaflets include:

- Viruses primarily responsible for rolling
  - Tomato yellow leaf curl virus (TYLCV)
  - Tobacco leaf curl virus (TLCV)
  - Tomato yellow mosaic virus (ToYMV)
  - Eggplant yellow mosaic virus (EYMV)
  - Indian tomato leaf curl virus (IToLCV)
  - Potato yellow mosaic virus (PYMV)
  - Serrano golden mosaic virus (SGMV)
  - Sinaloa tomato leaf curl virus (STLCV)
  - Taino tomato mottle virus (TToMoV)
  - Tomato leaf crumple virus (TLCrV)
  - Tomato leaf curl virus (ToLCV)
  - Tomato yellow dwarf virus (ToYDV)
  - Tomato yellow mottle virus (ToYMoV)
  - Tomato yellow vein streak virus (ToYVSV)
  - Potato leafroll virus (PLRV)
- Viruses primarily responsible for stunted plants (see p. 51)
  - Tomato bushy stunt virus (TBSV)
  - Beet curly top virus (BCTV)
  - Tomato golden mosaic virus (TGMV)
  - Tomato pseudo curly top virus (TPCTV)
  - Tobacco rattle virus (TRV)
  - Tomato aspermy virus (TAV)
- Viruses primarily responsible for mosaics and/or leaf chlorosis (see pp. 95 and 113)
  - Tomato chlorosis virus (ToCV): see chlorosis
  - Tomato infectious chlorosis virus (TICV): see chlorosis
  - Cowpea mild mottle virus (CPMMV): see mosaic
  - Eggplant mottled dwarf virus (EMDV): see mosaic
  - Pepino mosaic virus (PepMV): see mosaic
  - Peru tomato mosaic virus (PTV): see mosaic

This list, as well as those viruses mentioned in the previous section, raises awareness of the potential of certain viruses to induce leaf deformation. To help to resolve the cause, the section Leaflet and leaf discoloration should be consulted, p. 91 and *Table 9* (see p. 62).

32 The leaflets are blistered and slightly shrivelled.
Tomato mosaic virus (ToMV)

34 In this example, the leaflets are narrower than normal, and are blistered and deformed.
Cucumber mosaic virus (CMV)

36 Some prominent blisters and a comma-shaped appearance characterize these infected leaflets.
Eggplant mottled dwarf virus (EMDV)

33 Many leaves and leaflets of this plant are both rolled and twisted.
Eggplant mottled dwarf virus (EMDV)

35 The apex leaves, with bronzed leaflets tend to roll downwards.
Tomato spotted wilt virus (TSWV)

Blistered, shrivelled leaflets, due to a viral infection

37

38

39

40

37 This tomato plant, which was affected very early in its development, shows a bushy appearance because of shorter internodes and smaller comma-shaped leaflets.
Tomato yellow leaf curl virus (TYLCV)

39 This plant displays a bushy appearance.
Tomato yellow mosaic virus (ToYMV)

38 The leaves of these smaller and slightly chlorotic leaflets curves gradually upward giving them a shape reminiscent of a spoon.
Tomato yellow leaf curl virus (TYLCV)

40 In addition, the leaflets are smaller and rolled upwards.
Tomato yellow mosaic virus (ToYMV)

Examples of spoon-shaped leaflets frequently induced by tomato Begomoviruses

Table 9 Major viruses associated with leaf rolling and/or leaf chlorosis

Virus	Symptoms	Vectors
Tomato yellow leaf curl virus (TYLCV) **Description 41**	The leaflets are of reduced size, spoon-shaped, sometimes rolled, and suffer from a progressive yellowing which eventually becomes general. Some affected tissues may become harder and/or with anthocyanin production. Whether affected early or late, growth is more or less stopped. Many axillary branches, with short internodes develop giving the plants a bushy appearance. The flowers fall in large numbers and the few fruits are small.	Transmitted by whiteflies: *Bemisia tabaci* (B biotype including *B. argentifolii*), in a persistent and circulative manner.
Tobacco leaf curl virus (TLCV) **Description 42**	Symptoms fairly comparable to those produced by TYLCV: rolling of leaflets, chlorosis of leaves. The latter takes on a wrinkled and deformed appearance. In older plants, the leaves are often brittle. The growth of those infected early is greatly reduced.	Transmitted by *Bemisia tabaci* in a persistent or circulative manner. Contrary to TYLCV, 3 other species of whiteflies can transmit: *Aleurotrachelus socialis, Bemisia tuberculata,* and *Trialeurodes natalensis*.
Tomato yellow mosaic virus (ToYMV) Syn. Potato yellow mosaic virus	As with many other Begomovirus, the leaflets are small, chlorotic and more or less curved, with excessive anthocyanin on the edges. Prematurely infected plants have reduced growth, and are more or less bushy in appearance. Fruiting is greatly reduced, and fruits are much smaller.	Transmitted by B biotype of *Bemisia tabaci, B. argentifolii,* in a persistent manner.
Eggplant yellow mosaic virus (EYMV) **Description 42**	Leaflets curve and roll gradually and have irregular chlorotic spots. Plants very stunted. Symptoms similar to those caused by TLCV and TYLCV, and Begomovirus.	*Bemisia tabaci*
Indian tomato leaf curl virus (IToLCV) **Description 42**	Small, chlorotic, curved, rolled leaflets. Stunted plants when infected early.	*Bemisia tabaci* (biotype B in particular) in a persistent manner.
Serrano golden mosaic virus (SGMV) **Description 42**	Inter-vein chlorosis of young leaflets, the apex of the plants may become necrotic. Deformed fruits.	*Bemisia tabaci,* in a semi-persistent and persistent manner. Not transmitted by seeds.
Sinaloa tomato leaf curl virus (STLCV) **Description 42**	Chlorosis and yellowing of young leaves and leaflets which roll and sometimes take on a red tinge due to excessive anthocyanin. Internodes very short, show reduced growth.	*Bemisia tabaci,* in a persistent manner.
Taino tomato mottle virus (TtoMoV)	The symptoms are identical to those of TYLCV.	Probably *Bemisia tabaci.*
Tomato leaf crumple virus (TLCrV) – 'chino del tomate virus' (CDTV) would be linked to the same viral species **Description 42**	Marked leaf rolling and deformities; the leaf appears wrinkled and shows chlorotic mottling.	*Bemisia tabaci.* Not transmitted by seeds, nor by contact.
Tomato leaf curl virus (ToLCV) **Description 42**	The leaflets are small, chlorotic and rolled. Symptoms identical to those of TYLCV.	*Bemisia tabaci,* in a persistent manner.
Tomato mottle virus (ToMoV) **Description 42**	A chlorotic mottle is present on young leaves; the older ones are more yellow and rolled. The plants infected early are stunted.	*Bemisia tabaci* biotype B, in a persistent manner
Tomato yellow dwarf virus (ToYDV) **Description 42**	Yellowing of the leaflets that roll gradually. The plants infected early are stunted.	*Bemisia tabaci,* in a persistent manner.
Tomato yellow mottle virus (ToYMov) **Description 42**	Yellow mosaic mottling on the leaflets; the leaf curls up and is deformed. The plants can become very stunted.	*Bemisia tabaci,* in a persistent manner.

Tomato yellow vein streak virus (ToYVSV) Description 42	Yellow mosaic on leaflets. The leaf can be wavy. There are some chlorotic streaks along the veins.	Transmitted by *Bemisia tabaci* (*B. argentifolii*) biotype B in a persistent manner.
Potato leafroll virus (PLRV) Description 38	Yellowing of the leaf starting on the edges; the leaflets roll up later and become rigid and tough. Early affected plants stay stunted.	Transmitted by more than 10 aphid species in a persistent circulative manner especially *Aphis nasturtii*, *Aulacorthum solani*, and *Myzus persicae*.

Various phytoplasmas

Several phytoplasmas on tomato have been reported in the world under the name 'stolbur' (*stolbur*, *big bud*, *aster yellows*). Although often associated with leaf yellowing and anthocyanin production (see p. 143), these micro-organisms cause a number of symptoms which appear mainly on young organs and alter, in particular, the plant structure and leaf shape:
– proliferation of rigid and straight axillary shoots with a reduced growth and short internodes where the plant appears bushy;
– small, deformed leaflets, which tend to roll, to curl, and give the plants a characteristic appearance (Photos **41** and **42**);
– flowers are particularly affected; often sterile and reveal all kinds of defects: smaller size, enlarged calyx, green colour of the petals which may have a disproportionately large, leafy appearance, with anthocyanin production;

– the few fruits formed have reduced growth and are poorly coloured.

It is to be noted that the symptoms caused by *Candidatus* Phytoplasma solani, which is transmitted by several leafhopper species, occur in some European countries in July and persist throughout the summer. They are observed mainly on field crops; however, it is not uncommon to find some in protected crops, particularly on plants located near the doorways. The severity of damage varies from year to year: some years only a few scattered plants are affected while some other years, many plants are affected (sometimes over 50% of the crop), seriously affecting the harvest.

These phytoplasmas are common in many countries, causing similar symptoms irrespective of the tomato genotype grown.

41 The apex of this plant show a rigid thick stem and very short internodes. The leaflets are small and spoon shaped.
Candidatus **Phytoplasma solani**

42 In addition to being small, the leaflets are chlorotic and can curve down.
Candidatus **Phytoplasma solani**

Nutritional disorders

Several nutritional deficiencies (*nutritional disorders*) may disrupt the development of the leaflets of the tomato. They usually manifest themselves by a characteristic abnormal coloration (see the sub-section on Leaf yellowing p. 125). Some of them also induce changes in the shape of leaves that can take various forms: edge slightly curved towards the lower face (potassium, zinc), narrower (phosphorous), deformed and rolled (boron).

 Faced with these symptoms, a deficiency should not be diagnosed without first consulting a specialist and conducting physical and chemical analysis of soil and/or vegetation.

Genetic abnormalities

Several genetic mutations or chromosomal aberrations (in the constitution of chromosomes or their number) (*Genetic abnormalities*) are responsible for phenotypic traits often undesirable for the grower. They most often cause colour and/or form changes of leaflets and leaves. Photo **43** shows such symptoms: the leaflets are variably 'wrinkled', rolled over their axis, and tissues are thicker. Note also that the plants are often sterile (hence the name 'sterile mutants') and frequently take on a bushy appearance. Several types of sterile plants may occur in culture. The most common type comes in the form of small-leaved plants, often distorted and curled, with fewer long hairs on stems, leaves, and flower stalks than normal plants (see also p. 54). Another type shows itself in the form of more vigorous plants, with darker foliage and broader leaflets; stems and tissues are often thicker (sometimes called '*bull plants*').

These 'chimeras' should be regarded only as curiosities, without consequences for culture. Indeed, in most cases, a very small proportion of plants is affected. They are categorized as physiological diseases, should not be confused with viral diseases or phytotoxicities, and they are not transmissible to surrounding plants.

A genetic abnormality called 'blind plants' affects the plants in the nursery. These occur after the formation of the cotyledons and two leaves. The apical meristem becomes nonfunctional which disrupts their development. Seeds which are too old, too heated, or have been exposed to volatile toxic compounds are thought to be the cause of these blind plants. Further information on the subject is available on p. 54.

Other genetic defects affecting the production of chlorophyll in tomatoes are described on p. 144. Several genetic disorders affect the tomato fruit and are widely detailed in the section on fruit (see p. 345).

43 This single plant shows foliage that contrasts greatly with that of surrounding normal plants. It appears most notably stunted. Its dark green leaflets are wider, thicker, deformed, and slightly rolled down and are sometimes shrivelled.
Genetic abnormality

Physiological rolling of the leaflets

The symptoms of leaf curling (leaf roll) may occur on young plants grown under cover and put in open field during wet periods. It is generally observed later, on plants with at least three trusses of fruits in summer. The lower leaves of plants grown in open fields are first and foremost affected, and to a lesser degree those grown under protection. The blades of affected leaflets gradually curve upward (Photos **44** and **45**); opposed edges of the blade can meet and the leaflets eventually are fully rolled up on themselves. Finally, the tissues are often thicker, crisp, and shiny. In particularly severe cases, a high proportion of leaves are affected, giving the plant a characteristic appearance.

The direct impact of these symptoms on the development and production of plants is often low. In severe cases, the more highly exposed fruit are more vulnerable to sunlight and burns can be observed (*sunscald*; see p. 385).

This physiological disorder is very common. It is influenced by climatic conditions, cultural practices, and varieties grown. It occurs mainly when plants are loaded with fruit and/or when under particular climatic or agricultural conditions which affect the foliage–root balance and/or the water supply: a period of prolonged drought, soil temporarily wet after heavy rains or short of oxygen (compacted), pruning too severe, excessive intake of nitrogenous fertilizers, and destruction of roots due to cultural practices. It is very frequently found in plants grown in the gardens of amateurs, but it is also found in commercial practice. Differences in sensitivity between varieties exist, but none of them is totally resistant to this nonparasitic disease.

A number of measures are recommended to avoid this problem:

– Planting in well-drained soil;

– Ensure a balanced fertilization, particularly avoiding excess nitrogen;

– Do not cultivate in dry weather;

– Irrigate regularly to keep soil moisture constant;

– Mulch the soil.

Note that the rolling of tomato leaflets has been associated with high levels of CO_2 under protection.

Oedema

The presence of numerous small water-soaked blisters (*oedema*), located along the veins of the underside of the blade, results in some cases in the shrivelling, or curling of the leaflets, especially on the lower part of plants (Photos **46** and **47**).

This nonparasitic disease occurs mainly in this severe form in crops grown under protection during sudden changes in night climate (high air humidity, low temperatures). The onset of this physiological phenomenon is very rapid, taking place within hours.

It can be seen in a milder form with some scattered 'swelling' across the petiole, with chlorotic spots visible on the leaflets (for details, see p. 197).

44 All the plants of this protected tomato crop show variably rolled leaflets, especially the lower ones.
Leaflets physiological roll

46 The leaflets of the lower leaves of this young plant are variably rolled.
Oedema

45 The edge of the leaf is curved upwards and rolls up gradually.
Leaflets physiological roll

47 Many water filled groups of cells, located mainly along the veins, are responsible for the blisters. This particular distribution is responsible for the rolling of the leaves.
Oedema

Aphids

Nearly 4700 species of aphids have been identified in plants around the world, about 900 in Europe. Among them, a smaller number affects Solanaceous vegetables including tomatoes:

– *Aphis craccivora* Koch
– *A. fabae* Scopoli
– *A. frangulae* Kaltenbach
– *A. gossypii* Glover
– *A. nasturtii* Kaltenbach
– *Aulacorthum circumflexum* (Buckton)
– *A. solani* (Kaltenbach)

– *Macrosiphum euphorbiae* (Thomas)
– *Myzus ascanicus* Doncaster
– *M. certus* (Walker)
– *M. ornatus* Laing
– *M. persicae* (Sulzer)
– *Rhopalosipmonimus latysiphon* (Davidson)
– *Smythurodes betae* Westwood

Following aphid proliferation, the leaflets sometimes curl (see p. 210). Note that many aphids are efficient vectors of several important tomato viruses (see *Table 23*, p. 207).

48 The leaflets of the apex of this plant are particularly small. They are also chlorotic.
Tomato yellow leaf curl virus (TYLCV)

49 The leaflets from many upper leaves of this plant vary from narrow to filiform.
Cucumber mosaic virus (CMV)

50 In addition to being filiform, many leaflets are also twisted.
Phytotoxicity due to a hormone

51 These variably wrinkled leaflets have edges more serrated than usual.
Genetic abnormality

Examples of leaflets with abnormal proportions and changed shapes

Leaves and leaflets with abnormal proportions and shapes (smaller, filiform, serrated)

Possible causes

- Agent of the small leaves syndrome
- Several phytoplasmas: Description 28
- Many viruses
  - Tobacco mosaic virus (TMV)
  - Tomato mosaic virus (ToMV)
  - Cucumber mosaic virus (CMV)
  - Tomato aspermy virus (TAV)
  - Many *Begomovirus*
  - Pepino mosaic virus (PepMV) (see p. 115)
- Several viroids: Description 48
- Various genetic abnormalities
- Nutritional disorders
- Various phytotoxicities

Additional information for diagnosis

A rather limited number of parasitic and non-parasitic diseases induce the formation of small, filiform leaflets. Thus it is appropriate to suspect and be fairly confident that the cause is an attack by a phytoplasma or of certain viruses.

Various phytoplasmas

Stolbur and aster yellows are diseases caused by phytoplasmas localized in the phloem vessels. These micro-organisms, now classified as bacteria, are transmitted and spread by leafhoppers.

In some European countries, the first symptoms often appear in late July and during August and primarily affect young and actively growing parts. The leaflets which are already formed take a yellow-green tint; the newly formed ones are yellow and smaller (Photo **53**) or sometimes produce anthocyanin (Photo **54**). They tend to roll, to curl up towards the top of the blade (Photo **52**), which gives the plants a characteristic appearance. There are often more numerous young shoots with erect flowers, showing reduced growth and sometimes shorter internodes, giving the plants a bushy look (Photo **55**).

When present, the flowers are particularly affected. They are most often sterile and reveal all kinds of anomalies (Photos **56–58**): smaller size, green colour (virescence), and leaf-like forms of the petals (phylody), disproportionate size of the sepals, which can sometimes stick together and cause an hypertrophy of the calyx (*big bud*).

The few fruits formed have a thick stem, grow slowly, are hard, and poorly coloured (Photo **59**).

The stolbur, caused by *Candidatus* Phytoplasma solani, while common in the open field, is unusual under protection where affected plants are usually located near the doorways. Other species of phytoplasma also attack tomato throughout the world, causing comparable symptoms.

Diagnostic confusion may occur, especially with the Begomovirus (see p. 119 and following). To identify stolbur with certainty, we suggest you focus your attention on the flowers: they reveal abnormalities which are characteristic of this disease. Also, remember that diseased plants in the open field crops are distributed randomly and when under protected are grouped, generally near the doorways

52 The leaflets are reduced in size, exhibit a varaible inter-veinal chlorosis, and tend to roll towards the top of the leaf. These symptoms may cause confusion with those caused by some Begomovirus.

53 Sometimes the leaflets can be tiny and very chlorotic; this symptom could be caused by a particular phytoplasma strain causing the disease known as 'little leaf'.

54 Under certain conditions, the affected organ tissue take a purple (blue-tinged) shade (anthocyanin) as seen on the leaflets of several upright branches of this plant.

55 The plant on the left, affected by stolbur, contrasts with the surrounding plants. Many axillary branches, with reduced size and/or rolled leaflets give this plant a characteristic appearance, often bushy.

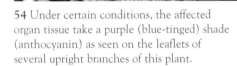

Examples of leaflets with abnormal proportions and shape caused by *Candidatus* Phytoplasma solani

56 The sterile flowers, with no petals, contrast greatly with normal flowers. Enlarged petioles give them an upright appearance. Note the absence of petals and reproductive organs.

57 The sepals of many flowers, partially fused, are overly broad and leafy; the petals have meanwhile remained green and smaller (virescence).

58 The leafy sepals of these flowers, like the rest of the vegetation contain anthocyanins; they have remained joined, which gives every flower an 'urn' appearance (big bud).

59 This affected fruit is of a smaller size and is very discolored. Its thick stalk contrasts with its small size.

Some examples of floral aberrations induced by *Candidatus* Phytoplasma solani

Agent of small leaves syndrome

This still poorly known problem, has been observed sporadically since 1986 in tomato crops in several US states (under the name 'tomato little leaf'), particularly in Florida, but also in Georgia, North Carolina, Ohio, Texas, and Maryland. It sometimes induces crop losses of up to 80%. It has also been reported in Poland, Italy, and probably in Indonesia.

First, the tomato growers generally observe a rather marked yellowing of inter-veinal tissues of the young leaflets. The veins are still very green, contrasting with the rest of the leaf. Subsequently, symptoms increase in severity and leaf deformation appears. The growth of new leaflets is particularly affected, hence the name 'little leaf disease'. The apex of affected plants is damaged and axillary bud growth is greatly reduced. The fruits are rather flattened and exhibit radial splits.

The origin of this disease is not yet definitely determined. It could be connected to a nonparasitic disease affecting tobacco named 'polyphylla' or 'frenching' causing very similar symptoms. Indeed, the tomato, like other plants such as petunias, eggplant, and squash, has been listed as sensitive to frenching for the last 60 years, without any complete description of symptoms or explanation for this nonparasitic disease. It could be induced by the effects of a nonpathogenic bacterium *Bacillus cereus* Frankland & Frankland, commonly found in soil. This could secrete a toxin and other products (like amino acids) that disseminate near the roots and are absorbed by the tomato. The nitrogen and protein metabolism of plants would be disturbed. An increase in the content of free isoleucine and other amino acids is frequently found in affected plants.

It should be noted that symptoms similar to those of the small leaves syndrome were also observed following toxicities due to thallium, manganese, and lead, and on plants subjected to the effects of metabolites produced by *Aspergillus wentii* Whermer and *Macrophomina phaseolina* (see p. 265).

In tobacco, the disease is mainly observed in wet and poorly ventilated soils, especially those that are warm, with a neutral or alkaline pH and sometimes inadequately fertilized, especially in nitrogen. It seems impossible to eradicate this physiological disease on this plant. During the next planting, one can always suggest drainage of the plot, and the avoidance of excessive addition of chemicals which increase the pH of the soil, ensuring a balanced fertilization and optimal irrigation to avoid situations of asphyxiation. It seems that some of these factors are valid in the context of the production of tomatoes.

Small leaf disease does not seem to occur in France, as opposed to frenching on tobacco that is frequently observed in different areas of production of this crop.

60 On this plant, mottled and blistered leaves alternate with filiform leaflets; these symptoms suggest infection by a virus.

62 Some leaflets may have a fern shape (fernleaf).

61 The leaves of virus infected leaflets only partially develops in width, which eventually gives them a variable filiform appearance.

63 In certain extreme situations, the leaf is reduced to the main vein, called a shoestring leaflet (shoestring).

64 Many young leaves of this plant are variably filiform, which gives it a characteristic appearance contrasting with the surrounding plants.

Cucumber mosaic virus (CMV)

Various viruses

It is not uncommon to see plants whose young leaflets do not develop as usual, in the open field but especially under protection. Their blades are normal in length but not in width, which gives them a different appearance.

Slightly narrower leaflets, or some restricted to the main vein, can be seen, with all variations in between. These symptoms are not the only ones on the plants: for instance mosaics, and other leaf deformation can suggest viral infection. A systematic search is advocated. For example, Photos **60–64** show Cucumber mosaic virus (CMV), the virus probably the most frequently associated with filiform leaflets, especially since many tomato varieties are resistant to TMV and ToMV.

It is important to know that such symptoms are not always evident in plants and are particularly influenced by climatic conditions. It is therefore not surprising to see filiform leaflets alternating with normal-shaped leaves, according to the climatic conditions.

A number of viruses can be implicated and some of their characteristics are listed in *Table 10*. These viruses cause other symptoms, see pp. 99–101.

Table 10 Examples of viruses inducing filiform leaflets and leaves on tomato

Virus	Symptoms	Vectors
Tobacco mosaic virus (TMV) Description 30	Slight mottling, mosaics on leaflets. Leaflets may be filiform in periods of low light. Virus is often incriminated in place of ToMV which is better adapted to tomato.	Transmitted by contact, hardly ever by seeds.
Tomato mosaic virus (ToMV) Description 31	Whitening of the leaf, mottling, leaflets showing green, yellow, or white patches of mosaic. The leaflets may also have raised tissue, be variably wrinkled, sometimes even filiform when plants lack light. Chlorotic rings are visible on green or mature fruit. Internal brown necrotic lesions are also found (*internal browning*).	Transmission is by contact and by seeds.
Cucumber mosaic virus (CMV) Description 35	Three types of symptoms are observed depending on the strain and climatic conditions: – Mottling, mosaic on young leaflets; – Deformation of the leaflets, extremely filiform, which have the appearance of a shoelace (shoestring) or that of a fern leaf (fernleaf); – Spots, necrotic lesions appear on leaflets and stem, extending to the apex of the plant. In addition, necrotic brown, raised, round lesions cover large areas of fruit.	Transmitted by aphids in a nonpersistent manner. Over 90 species are capable of acquiring and transmitting the virus, including: *Myzus persicae, Aphis gossypii, A. craccivora, A. fabae, Acyrthosiphon pisum.*
Tomato aspermy virus (TAV) Description 38	Apical growth is stopped with proliferation of numerous axillary buds giving the plants a bushy aspect. The mottled leaflets and leaves are heavily deformed and sometimes filiform. Fruit set is very reduced and the rare fruits formed are small and deformed. Seed production is low to zero (contrast with CMV).	Transmitted by 22 species of aphids (*Myzus persicae*) in a nonpersistent manner. Transmitted by seed in *Stellaria media.*

65 This leaflet has a partially deformed leaf with serrated edges and veins with a less pronounced branching angle.
Phytotoxicity (excess of a fruit setting hormone)

67 This leaf and its smaller and thick leaflets are markedly rolled and crumpled.
Phytotoxicity

66 Several leaflets of this young leaf are partially blistered, variably twisted and crumpled.
Phytotoxicity

68 The leaflets of the newly formed leaves are smaller and are more slender, variably curved and/or rolled, and the tissues are thicker.
Phytotoxicity

69 The leaves are not only filiform and thus shoestring shaped, but are also variably twisted.
Phytotoxicity

Examples of leaf deformation related to the effects of pesticides

Various phytotoxicities

Many pesticides used in agriculture may be the cause of phytotoxicity on tomato (chemical injuries). Damage, of varying severity, is sometimes associated with overdose or combinations of insecticides or fungicide. Those caused by herbicides are the most frequent and most damaging. Inhibitors of cell division, of fatty acids, and amino acids synthesis as well as plant hormones are the most involved in root and leaf malformation in tomato.

Symptoms associated with chemical injury may be of two main types: abnormal coloration (see p. 131) and leaf deformation, leading to a change in the plant's shape and structure associated with a reduction or a total cessation of growth. Photos 65–71 give a clear indication of the nature and intensity of symptoms induced by certain toxic plant compounds.

One can observe in particular:

– slow development of the younger leaves (Photo 73), up to a total cessation of growth; in these conditions, plants are stunted, especially if the phytotoxicity occurred early;

– partially or totally distorted leaflets, slightly more jagged or irregularly cut (Photo 65), mottled, blistered, and variably rolled (Photos 66 and 74);

– the distortion and/or variably rolled leaflets, which can be shorter, filiform, and exhibit strongly thickened tissue (Photos 68, 69, and 72);

– curved spoon- and/or comma-shaped leaflets;

– rolling of the entire blade;

– corkscrew and crumpled appearance of all leaves (Photo 67);

– the proliferation of rigid and thick shoots associated with fasciation occurring on the stem, peduncles, and sepals (Photos 70 and 71).

70 A number of axillary shoots grow; their leaves are smaller and chlorotic, with a thick stem and anthocyanin production.
Phytotoxicity

71 The flowers and fruits can be affected. Here, the sepals are out of proportion compared with the deformed young fruit; they are also fasciated and very thick.
Phytotoxicity

77

73 This plant's new leaflets are of reduced size, more denticulate and narrower; they also tend to curve upward.
Genetic abnormality

72 The first leaves of this tomato plant have a normal shape, in contrast with subsequent ones whose narrower and deformed leaflets give the affected leaves a very characteristic appearance.
Genetic abnormality

74 On this leaf, the leaflets are narrower, swollen, and roll toward the underside of the leaf; in extreme cases this gives them a crook shape.
Genetic abnormality

Examples of genetic abnormality

What to do after chemical injury?

In the presence of phytotoxicity, the following questions should be raised:
– Was the preceding crop treated with persistent herbicides?
– Were herbicide treatments made close to your crop?
– Has the spraying equipment been washed?
– Has the spraying equipment been maintained (cleaning, grading)?
– Was the correct product used at the right dose?
– Were the label recommendations on the package followed?
– Have incompatible products been mixed together or too many products used?
– Have applications been administered in good conditions, of wind and temperature in particular?

Irrigation water may have been polluted by a particular herbicide

Although there is no quick fix in this situation, the following measures can be adopted:
– Define the origin of the phytotoxicity;
– Prevent it happening again;
– Do not remove the plants immediately, let them grow normally and observe their development; the latter will crucially depend on the nature, dose, and persistence of the product in question and on the stage of plant growth, the type, and variety cultivated.
No other specific measures can be recommended.

In addition to previously reported deformations, phytotoxicity causes other symptoms (mainly variable yellowing). The sections Leaflet and leaf discoloration (p. 91), Spots on leaflets and leaves (see p. 149), and Wilting, necrosis and dried leaflets and leaves (see p. 217) should be consulted.

Genetic abnormalities

Several mutations or genetic aberrations (genetic abnormalities) cause tomato leaf abnormalities, which can be confused with those caused by certain viruses and some phytotoxicities. Photos 72–74 show the diversity and nature of symptoms associated with these genetic diseases.

For more information about these see p. 64.

75 The leaflets of this leaf have been eaten by a pest. There are numerous holes in the leaves, as well as areas which have only been superficially 'grazed'.

76 Holes of varying shape in the leaf and its periphery, riddle this leaflet which is also affected by inter-veinal chlorosis due to the effects of ToCV.
Butterfly caterpillars

77 Several fairly straight mines especially localized along the veins are visible on this leaflet.
Liriomyza sp.

Examples of perforated, cut, shredded, mined leaves or leaflets

Perforated, cut, shredded, mined leaflets

Possible causes

- Hail
- Various phytoxicities
- Leafminer larvae
- Caterpillars of butterflies

Additional information for diagnosis

Hail Following heavy rains accompanied by hail, it is not uncommon to observe considerable damage (hail injuries) in field production of tomato crops, especially in flat ground cultures. Indeed, this mode of production makes them particularly vulnerable to hail. By hitting the leaflets, hail causes holes and tears (Photo **78**), which can significantly alter the photosynthetic potential of the leaflets. Some leaves may be completely shredded. Note that these injuries can also promote the development of various pathogens. The hailstones are also the source of brownish necrotic lesions at the point of impact on the stem (Photo **79**), but mainly on fruits (see p. 383).

Various phytotoxicities

When applying pesticides on tomatoes, sometimes the spraying pressure may be temporarily too high. The young and tender leaflets may be pierced or even shredded by the high pressure (chemical injuries).

78 Several leaflets are perforated or jagged. There is also some hail damage along the stem.
Hail

79 There are localized lesions on the stem and cracks of a brownish colour. The colour is paler when superficial scarring occurs.
Hail

80 Rounded and well cut perforations are the signs of attack by moth larvae of the Noctuidae family.

81 The presence of perforations in the fruit are caused by the same pest.

82 It is not uncommon to observe the caterpillar(s) responsible for damage to the foliage.

83 Observing the butterflies is rarer. In this case, the white spot shaped like the Greek letter gamma allows identification of *Autographa gamma*.

Defoliating caterpillars

Moth caterpillars

The aerial parts of tomato are likely to be eaten by the caterpillar stage of several polyphagous Lepidoptera. These insects belong to various families, especially Noctuidae, and include *Autographa gamma* (L.), *Chrysodeixis chalcites* (Esper), *Helicoverpa armigera* (Hübner), *Lacanobia oleracea* (Linnaeus), and *Spodoptera exigua* (Hübner). These moths (loopers), and many others, are found in many production areas of the world where they sometimes cause considerable damage in both field and protected crops.

• Nature of damage

The larvae, especially the older ones, are responsible for feeding damage on the leaves. This results in the presence of numerous regular perforations located on the leaf or on its periphery (Photo **80**). Some of these larvae also attack fruit which are nibbled away or even holed mainly close to the stalk (Photo **81**). Galleries and many droppings are visible inside. In addition to inducing early ripening of fruit, these perforations enable the penetration of many decaying agents (see p. 389 and Photos **87** and **88**).

• Biology

Lepidoptera pass through four stages: egg, larva or caterpillar, pupa or chrysalis, and butterfly (Figure **17**).

Forms of conservation and/or alternative hosts: most of these butterflies hibernate through pupae or larvae, although the other stages of development can contribute to the winter survival of these insects. They have many hosts, both cultivated and wild.

Developmental stages (Figure **17**): the eggs (1) are variably coloured (transparent, white, brown, black, purple) and are less than one millimetre in diameter. They are deposited singly or in groups on the surfaces of leaves or other plant parts (Photo **84**). Subsequently, they hatch and

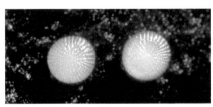

84 *Chrysodeixis chalcites* eggs.

85 Green caterpillar with a yellow line on each side of its body. ***Chrysodeixis chalcites***

86 Butterfly of *Chrysodeixis chalcites*.

83

give rise to caterpillars (2) 25–50 mm in length depending on the species, equipped with powerful jaws that allow them to eat plants, including leaves (Photos **82, 85, 89–91**).

These caterpillars vary in colour (green, sometimes becoming reddish brown with age) and go through several changes before pupation. The pupae (3) measure 2–2.8 cm and are red-brown; the sheaths of the legs and wings and abdominal segmentation are clearly visible within. The adults (4) are butterflies with two pairs of wings with a wingspan ranging from 25 to 45 mm. The fore and hind wings show a variable colour (reddish brown, brown, grey) and pattern (Photos **83** and **86**) depending on the species.

The duration of their life cycle varies with temperature, from 10 days to several weeks. Larvae are mobile and move easily from one leaflet to the other, as do the adults even more freely. The decrease in soil disinfection practices appears to contribute to the resurgence of this pest on vegetable crops among others.

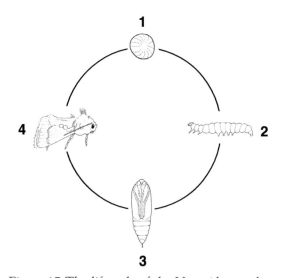

Figure 17 The lifecycle of the Noctuidae moth.

87 This widely punctured and eaten fruit is beginning to rot. Brownish excretions of the caterpillar responsible for this damage are visible in some places.
Caterpillar damage

88 A cross-section of a perforated fruit shows the damage caused by this green caterpillar.
Caterpillar damage

89 Caterpillar of *Autographa gamma* moth.

90 Caterpillar of *Lacanobia oleracea* moth.

91 Caterpillar of *Spodoptera exigua* moth.

Caterpillar damage

92 Many of the leaflets on this leaf show several sinuous mines.
Liriomyza sp.

94 The larvae developing in the leaf tunnel, forming sinuous galleries in the case of *Liriomyza bryoniae*, galleries along the veins and straight for *Liriomyza strigata*, and to a lesser degree for *Liriomyza huidobrensis*.

93 Tiny yellow feeding spots are mainly concentrated on the edges of this leaflet. They could serve as entry-points for certain pathogens.
Liriomyza sp.

95 The last stage larvae emerge from the leaf and are sometimes visible as barrel-shaped pupae.
Liriomyza sp.

Leafminers

Leafminers

Several polyphagous leafminers are likely to attack vegetable crops, and sometimes tomato crops that usually tolerate the presence of these pests. These insects are classified in the order Diptera and the family Agromyzidae. The main species of leafminer found in Europe include the tomato leafminer (*Liriomyza bryoniae* Kaltenbach, often erroneously referred as '*Liriomyza strigata* [Meigen]'), the American leafminer (*Liriomyza trifolii* Burgess), the South American leafminer (*Liriomyza huidobrensis* Blanchard and *Liriomyza strigata* Meigen), and the horticultural leafminer (*Chromatomyia horticola* Goureau). Note that the multiplication of these insects is often the target of biological control in glasshouses. In addition, many of them are classified as quarantine pests.

• Nature of damage

Many chlorotic punctures (Photo **93**) are first observed on the leaf; they may be very numerous and are made by females with their ovipositor. Mines (Photos **92** and **94**) appear later on the leaflets. The most affected leaves, sometimes carrying as many as 20 larvae per leaf may turn yellow, wilt, and dry out (Photos **95** and **96**). The photosynthetic activity, the plant growth, and the yields can be greatly reduced as a result of an infestation. The population control of these pests is often problematic due to their possible resistance to several insecticides which also eliminate beneficial biocontrol agents (parasitoid Hymenoptera). During very heavy infestations, particularly in the tropics, fruits may show some burns (see p. 385) because the many mined and deformed leaves no longer protect them from sunlight.

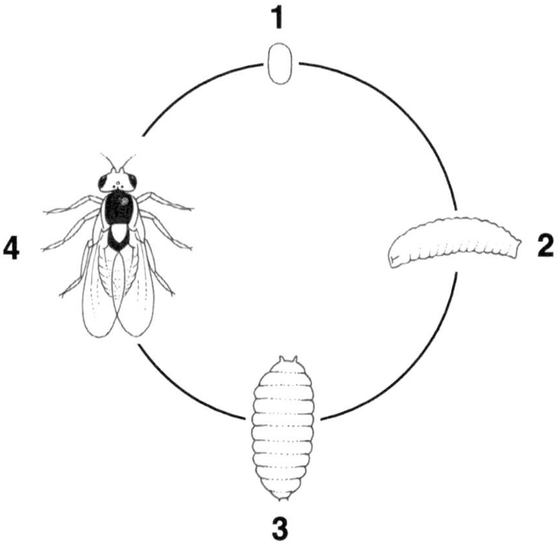

The damages from these flies Agromyzidae (leaf mining) should not be confused with that produced by *Tuta absoluta* Meyrick (Lepidoptera Gelechiidae), whose mines on leaves lead to white areas broader than those of leafminers. These areas, which eventually become necrotic, look rather similar to those caused by a severe attack of mildew. This Lepidoptera may also affect fruit and produce necrosis or holes on the stalk or other tomato parts. *Tuta absoluta* occurred in Algeria, Spain, Italy, Morocco, and Tunisia and has expanded into Corsica and France. Note that this pest can also parasitize other Solanaceae such as potato, pepino, eggplant, pepper, and various weeds (datura, jimson weed, yellow and black nightshade).

• Biology

Leafminers have six stages of development: egg, three larval stages, pupa, and adult.

Forms of conservation and/or alternative hosts: few adults are observed during the winter due to the entry of the pupae into diapause. As these leafminers are polyphagous they can multiply and live on many alternative cultivated hosts (cucumber, lettuce, melons, peppers, celery, bean, potato, chrysanthemum, gerbera) as well as on weeds present in or outside of the crop.

Developmental stages (Figure **18**): the eggs (1), cream coloured and oval shaped (those of *Liriomyza bryoniae* measure 0.12 × 0.27 mm) are deposited in tissues during egg-laying using the females ovipositor. A female can produce several hundred eggs which subsequently hatch into transparent larvae (2) 0.5 mm long (Photos **94** and **98**). These burrow galleries into the leaflets which are quite visible due to the presence of black excrement. The third stage white larvae, 2.5 mm long, pierce the leaf blade, leave the leaflets, drop into the folds of the plastic sheeting or onto the ground, and bury themselves in shallow depths. Thereafter, they change into barrel-shaped pupae (3) whose colour varies with age from yellow to dark brown (Photo **99**); blackened pupae are often infected. *Chromatomyia horticola* pupation takes place in the leaf: the puparium, covered by the leaf epidermis is then visible as a small protuberance at the end of the larval gallery. The adults (4) are small flies 2–3 mm long, yellow and black (*Liriomyza* spp. See Photos **97** and **100**) or dark grey (*Chromatomyia horticola*). The adult females, located on the upper side of the blade, puncture the leaf epidermis with their ovipositors, suck the plant juice (nutritional

Figure 18 The lifecycle of the leafminer.

96 When mines are numerous in the leaf, the latter may become necrotic and dry up.
Liriomyza sp.

97 Tiny leaf miners are often visible on the leaves.
Liriomyza sp.

98 Yellowish leafminer larva.
Liriomyza bryoniae

99 Tan to dark brown leafminer pupae.
Liriomyza bryoniae

100 *Liriomyza bryoniae* adult beetles are yellow and black. Females have a black mark on the abdomen. *L. trifolii* flies have a greyish-black colour, the head is yellow with red eyes. Yellow spots are visible on the thorax. Adults of *L. huidobrensis* are darker. Females have a black spot on the abdomen.

punctures) and lay their eggs (ovipositor punctures). Note that males without ovipositors also benefit from the feeding punctures.

The duration of the life cycle varies with temperature; for example, for *Liriomyza bryoniae* it is 41 days at 15°C and 17 days at 25°C. For other species the duration remains substantially the same; however, *L. trifolii* is more sensitive to low temperatures and the lifecycle of females changes from 14 days at 15°C to 7 days at 25°C.

Dispersion in the crop: the adults readily fly within the glasshouse, and even from one glasshouse to another, so spreading out in the crop. Newly infected plants (carrying eggs or very young mines) may also contribute to the spread of these insects.

Favourable conditions for development: insect population level is influenced by high light intensities, some rather vigorous host plants, and significant high humidities (80–90%) in particular.

Leafminers are usually heavily parasitized, mainly by chacidien Hymenoptera. Insecticide treatments, sometimes unjustified, are very harmful to the useful insect fauna and are often responsible for the outbreaks observed.

Many other pests consume the tissues of different tomato organs throughout the world. The main pests and the damage they cause to the tomato plant are shown in Figure 19.

Blapstinus spp. – darkling beetle
Eats the stem of seedlings at ground level.

Melothonta melothonta – cockchafer
Agriotes spp. – yellow worms
(wireworms, click beetle larvae)
Eat roots and stems of seedlings.

Scutigerella immaculata – garden symphylan
Eats seedlings before and after emergence.

Agrotis spp. – wireworm
Eats the stems of young plants at ground level, rarely the fruits and leaves at ground level.

Autographa californica – alfalfa looper
Eats the foliage and sometimes fruit.

Manduca sexta – tobacco hornworm
Manduca quinquemanculata – tomato hornworm
Eats the leaves, young stems, and green fruit.

Delia platura – seedcorn maggot
Eats seeds and seedlings before emergence.

Agrotis spp. – cutworm
Eats the stems of young seedlings at ground level, rarely fruits and leaves at ground level.

Epithrix spp. – flea beetle
Eats the stem and leaflets of seedlings, and the leaflets of adult plants.

Keiferia lycopersicella – tomato pinworm
Perforates green and ripe fruit. Sometimes produces mines on the leaves.

Helicoverpa zea – tomato fruitworm
Heliothis virescens – tobacco budworm
Spodoptera praefica (western yellow striped armyworm)
Eats the fruit, causing deep, moist cavities.

Figure 19 Other pests that can consume, perforate, or mine tomato seedlings, vegetation, and fruits.

101 Mottled leaflets variably blistered.
Tomato mosaic virus (ToMV)

102 Very strong inter-vein yellowing of several leaflets.
Magnesium chlorosis

103 Leaflets showing limited silvered patches.
Silvering

104 Leaflets with a bronze tint in contrast with the healthy surrounding leaves.
Aculops lycopersici (**tomato russet mite**)

105 Blue-tinged leaflets, with excessive 'anthocyanin'.
Candidatus Phytoplasma solani

Examples of discoloration of leaflets and leaves of tomato

Leaflet and leaf discoloration

Symptoms studied

- Mottled leaflets (mosaics and similar symptoms)
- Yellowish to white leaflets or leaflets with chlorosis
- Leaflets, leaves with other discolorations (silver, blue-tinge, bronze, dull)

Possible causes

- Various phytoplasmas
- Viruses responsible for 'mosaics'
- Alfalfa mosaic virus (AMV)
- Broad bean wilt virus (BBWV)
- Colombian datura virus (CDV)
- Cowpea mild mottle virus (CPMMV)
- Cucumber mosaic virus (CMV)
- Datura yellow vein virus (DYVV)
- Eggplant mosaic virus (EMV)
- Eggplant mottled dwarf virus (EMDV)
- Eggplant severe mottle virus (ESMoV)
- Moroccan pepper virus (MPV)
- Pepino mosaic virus (PepMV)
- Pepper veinal mottle virus (PVMV)
- Peru tomato virus (PTV)
- Potato virus Y (PVY)
- Ribgrass mosaic virus (RMV)
- Tobacco etch virus (TEV)
- Tobacco mosaic virus (TMV)
- Tobacco rattle virus (TRV)
- Tomato chlorotic spot virus (TCSV)
- Tomato mild mottle virus (TMMV)
- Tomato mosaic virus (ToMV)
- Tomato mottle virus (Tomov)
- Tomato ringspot virus (ToRSV)
- Tomato yellow vein streak virus (ToYVSV)
- Viruses responsible for line patterns
- Pelargonium zonate spot virus (PZSV)
- Pepper ringspot virus (PepRSV)
- Tobacco necrosis virus (TNV)
- Tobacco ringspot virus (TRSV)
- Tobacco streak virus (TSV)
- Tomato black ring virus (TBRV)
- Tomato bushy stunt virus (TBSV)
- Tomato infectious chlorosis virus (TICV)
- Tomato spotted wilt virus (TSWV)
- Viruses responsible for yellow mosaic
- Colombian datura virus (CDV)
- Cucumber mosaic virus (CMV)
- Pepino mosaic virus (PepMV)
- Potato yellow dwarf virus (PYDV)
- Tomato yellow mosaic virus (ToYMV)
- Serrano golden mosaic virus (SGMV)
- Sinaloa tomato leaf curl virus (STLCV)
- Taino tomato mottle virus (TToMoV)
- Tobacco etch virus (TEV)
- Tobacco mosaic virus (TMV)
- Tomato chlorosis virus (ToCV)
- Tomato golden mosaic virus (TGMV)
- Tomato infectious chlorosis virus (TICV)
- Tomato mosaic virus (ToMV)
- Viruses responsible for veins yellowing and rolling
- Curtovirus beet curly top (BCTV)
- Cowpea mild mottle 'virus' (CPMMV)
- Eggplant mottled dwarf virus (EMDV)
- Eggplant yellow mosaic virus (EYMV)
- Indian tomato leaf curl virus (IToLCV)
- Pepino mosaic virus (PepMV)
- Peru tomato virus (PTV)
- Tomato yellow mosaic virus (ToYMV)
- Sinaloa tomato leaf curl virus (STLCV)
- Taino tomato mottle virus (TToMoV)
- Tobacco leaf curl virus (TLCV)
- Tomato bushy stunt virus (TBSV)
- Tomato chlorosis virus (ToCV)
- Tomato golden mosaic virus (TGMV)
- Tomato leaf crumple virus (TLCrV)
- Tomato leaf curl virus (ToLCV)
- Tomato mottle virus (Tomov)
- Tomato pseudo curly top Topocuvirus (TPCTV)
- Tomato yellow dwarf virus (ToYDV)
- Tomato yellow leaf curl virus (TYLCV)
- Tomato yellow mottle virus (ToYMoV)
- Viruses responsible for excess anthocyanins
- Potato yellow dwarf virus (PYDV)
- Tomato yellow mosaic virus (ToYMV)
- Tomato spotted wilt virus (TSWV)
- Some viroids
- Various genetic abnormalities
- Silvering
- Nutritional disorders
- Excess salt
- Various phytotoxicities
- Temperatures too low
- Mites (see p. 199)
- *Aculops lycopersici* (russet mite)
- Thrips (see p. 171)

Observation guide

Figure 20 Main areas of tomato leaflet discoloration.

1 Healthy leaflet with uniform colour.

2 Occurring between veins.

3 Starting from the veins.

4 Occuring on one side of the leaf (unilateral).

5 On the top of the leaflet or leaf.

6 At the base of the leaflet or leaf.

7 At the edge of the leaf.

Difficult diagnosis

Many nonparasitic and parasitic diseases can be the cause of tomato leaflet discoloration, in particular a multitude of viruses, most nutrient deficiencies or toxicities, and phytotoxicities. The abnormalities most frequently observed on the leaves are mosaics or related symptoms, but mostly pronounced yellowing of the leaf to various degrees. Such discoloration can characteristic for a disease, but may be quite similar for many different diseases, making identification difficult. We suggest you carefully review all the pictures in this sub-section, and take time to consider the various hypothesises made in the Additional information for diagnosis section.

Note that these conditions also cause irregularities in leaf shape at the same time, so it is worth looking at the sub-section on such abnormalities to complement any diagnosis (see p. 47).

Symptoms of several viruses have been incorporated in this section. As diagnostic confusion between various tomato viruses is important, it is advisable, if in doubt, to ask a specialist laboratory to complete a serological or molecular test to identify with certainty the virus(es) responsible.

Discoloration

106 Mosaic deformation caused by Tomato mosaic virus (ToMV).

107 Mosaic with yellow patches due to Pepino mosaic virus (PepMV).

108 Chlorotic mosaic induced by Eggplant mottled dwarf Nucleorhabdovirus (EMDV).

109 Pseudo mosaic linked to a genetic defect: silvering.

Various mosaics on tomato leaves and leaflets (true mosaics or similar symptoms)

Mottled leaflets and leaves (mosaic and related symptoms)

Possible causes

- Various viruses
- Viruses responsible for 'mosaic'
  - Alfalfa mosaic virus (AMV)
  - Broad bean wilt virus (BBWV)
  - Colombian datura virus (CDV)
  - Cowpea mild mottle virus (CPMMV)
  - Cucumber mosaic virus (CMV)
  - Datura yellow vein virus (DYVV)
  - Eggplant mosaic virus (EMV)
  - Eggplant mottled dwarf virus (EMDV)
  - Eggplant severe mottle virus (ESMoV)
  - Moroccan pepper virus (MPV)
  - Pepino mosaic virus (PepMV)
  - Pepper veinal mottle virus (PVMV)
  - Peru tomato virus (PTV)
  - Potato virus Y (PVY)
  - Ribgrass mosaic virus (RMV)
  - Tobacco etch virus (TEV)
  - Tobacco mosaic virus (TMV)
  - Tobacco rattle virus (TRV)
  - Tomato chlorotic spot virus (TCSV)
  - Tomato mild mottle virus (TMMV)
  - Tomato mosaic virus (ToMV)
  - Tomato mottle virus (Tomov)
  - Tomato ringspot virus (ToRSV)
  - Tomato yellow vein streak virus (ToYVSV)

- Viruses responsible for yellow mosaic
  - Colombian datura virus (CDV)
  - Cucumber mosaic virus (CMV)
  - Pepino mosaic virus (PepMV)
  - Potato yellow dwarf virus (PYDV)
  - Tomato yellow mosaic virus (ToYMV)
  - Serrano golden mosaic virus (SGMV)
  - Sinaloa tomato leaf curl virus (STLCV)
  - Taino tomato mottle virus (TToMoV)
  - Tobacco etch virus (TEV)
  - Tobacco mosaic virus (TMV)
  - Tomato chlorosis virus (ToCV)
  - Tomato golden mosaic virus (TGMV)
  - Tomato infectious chlorosis virus (TICV)
  - Tomato mosaic virus (ToMV)
- Viruses responsible for line patterns
  - Pelargonium zonate spot virus (PZSV)
  - Pepper ringspot virus (PepRSV)
  - Tobacco necrosis virus (TNV)
  - Tobacco ringspot virus (TRSV)
  - Tobacco streak virus (TSV)
  - Tomato black ring virus (TBRV)
  - Tomato bushy stunt virus (TBSV)
  - Tomato infectious chlorosis virus (TICV)
  - Tomato spotted wilt virus (TSWV)
- Silvering (see p. 143)

Discoloration

95

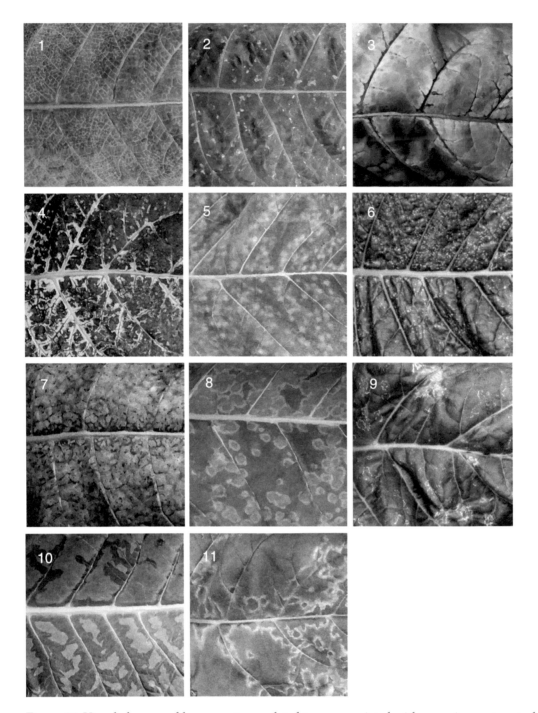

Figure 21 Visual glossary of key symptoms related to, or associated with, mosaics on tomato leaflets and leaves.

1 Vein clearing: clearing and yellowing of a high proportion of secondary veins, some of them may eventually turn brown and necrotic.

2 Yellow dots: tiny irregular chlorotic spots.

3 Vein necrosis: yellowing and necrosis of the main and secondary veins.

4 Yellowing spreading widely around the main veins.

5 Yellow spots: round yellow spots.

6 Etch: tiny necrotic lesions of various shapes (dots, lines, commas, short arcs) covering the leaf to a various degree.

7 Vein mottling: narrow and irregular bands of tissue adjacent to the veins remaining green.

8 Ring spots: spots with chlorotic rings.

9 Concentric ring: small and irregular necrotic rings, often concentric.

10 Vein banding: wide bands of homogeneous dark green tissue located on both sides of veins.

11 Line patterns: chlorotic, irregular and sinuous patterns partially covering the leaf.

Additional information for diagnosis

How to interpret viral symptoms

Symptoms that may vary in position and intensity are often grouped under the name 'mosaic'. Photos 106–111 and Figure 21 are representative of the range of leaf discoloration often associated with mosaic. When faced with these symptoms, the presence of a virus might be suspected, but the precise diagnosis of the virus(es) responsible will necessitate caution. Indeed, many viruses can cause a wide range of similar and confusing symptoms.

In addition, the sub-section Leaves and leaflet deformation should be consulted, as in addition to colour changes of the leaf, the viruses also cause some deformation of various importance.

⚠ It is not easy to assess symptoms like mosaic on tomato leaves, especially during sunny days. In this situation, the leaves should be studied when illuminated from behind, taking care to shade them.

110 This leaflet has a slight marbling; the discreet mosaic that affects it is difficult to see.
Potato virus Y (PVY)

111 Several yellow patches alternate with areas of normal colour. A mosaic is very visible on both leaflets.
Tomato mosaic virus (ToMV)

Discoloration

112 The young leaf is slightly mottled and blistered.

114 In addition to mosaics, several leaflets are smaller and have a more denticulate leaf.

116 The majority of the fruits of this truss have extensive chlorotic spots of varying degree.

113 Alternating bright yellow and green patches caused by an Aucuba strain of ToMV on tomato leaflets.

115 In cold conditions and low light the new leaflets can be filiform, and have a fern-like appearance (fernleaf).

Tomato mosaic virus (ToMV)

As mentioned above, many viruses can cause similar mosaic symptoms on tomato leaves. Several of them are described in *Tables 11* (p. 107) and *12* (p. 111). The most common, and often the most damaging, are described in more detail.

Tobacco mosaic virus (TMV) (Description 30)

This virus is reported a worldwide problem, particularly in the Solanaceae (tobacco, pepper, tomato). It occurs mainly in countries where the tomato varieties grown are not resistant to TMV, in either field grown or protected crops. This is no longer the case: it is now seen only occasionally, often on a few traditional varieties still sensitive. It should be noted that in recent years infection is increasing on tobacco. Note also that for a long time infections due in fact to Tomato mosaic virus (ToMV), a similar virus much more competitive in this species, were associated with TMV.

Depending on the strain, TMV may cause mild mottle, a mosaic, or even a particularly marked green to yellow mosaic. In winter conditions, leaf deformation may occur, with leaves and leaflets being filiform or fern leaf in shape. In some cases, the sepals and petals are wavy and the number of pollen sacs is reduced.

The nature and severity of symptoms may vary according to age and plant variety. Young and developing tissues are affected more severely. Plant handling encourages its transmission through contact.

Tomato mosaic virus (ToMV) (Description 31)

Long considered a strain of TMV, ToMV is a distinct viral species, also transmitted by contact. Present on every continent, this virus is found more frequently than TMV on tomato and pepper, both in field crops and under protection. The importance of ToMV has greatly diminished with the widespread use of resistant varieties of tomato. Recent experiments of introducing non-resistant crop varieties have shown however, that the very stable ToMV virus is still widely present in the field.

Symptoms of ToMV on tomato are quite varied:

– vein lightening followed by marked mottling or mosaic (Photo **112**). In the case of an Aucuba strain of ToMV, the alternating green, yellow, and/or white patches are particularly spectacular (Photo **113**);

– embossing and wrinkling of the leaflets and leaves (Photo **114**) that can be curved, reduced in size, and deformed. These can take a filiform or fern-like appearance especially in winter and under protection when the plants lack light (fernleaf, Photo **115**);

– flowers drop, mottling, or fruit discoloration (Photo **116**), with occasional presence of rings. Fruits can also show an internal necrosis of vascular tissues (internal browning), sometimes without any other symptom present on the plants (Photo **726**). The fruit may have a reduced size and have a bumpy surface, with yellow and/or necrotic rings;

– reduction in plant growth and yields, especially when the attacks have taken place early.

The intensity of these symptoms can vary depending on the nature of the strains, cultivar, stage of infection, temperature, intensity of light, and nitrogen and boron soil content. For example, high temperatures may reduce the intensity of symptoms on leaves. Conversely, resistant varieties with genes '*Tm-2*' or '*Tm-2^2*' under conditions of high temperature can produce necrotic reactions when infected with common strains of ToMV and TMV.

Mixed infections are very common, especially with CMV and PVY, in which case the symptoms can be much more serious. UK growers have reported a co-infection ToMV–PVX leading to the appearance of a very damaging tomato syndrome, called 'double streak'.

Remember that the ToMV is spread via the workers during cultural operations. This results in infected plants being frequently distributed in a line in the row(s) worked.

Discoloration

117 Several young leaflets show green mottling of differing intensities.

119 Some leaflets can also be affected to a variable extent by a yellow mosaic.

121 Several young leaflets show green mottling of differing intensities. Note that some are filiform.

118 Some leaflets are particularly deformed and blistered.

120 Brown to black necrotic lesions appear on several leaflets of this young leaf.

Cucumber mosaic virus (CMV)

Cucumber mosaic virus (CMV) (Description 35)

CMV transmitted by aphids, is present worldwide, especially in temperate areas and in Mediterranean climates. It is extremely polyphagous, affecting hundreds of different plant species. Its symptoms are observed more generally in the open field in summer, but it is not uncommon to see it under protection on plants near the vents or on plants in nurseries, even in winter.

This cucumovirus can cause three main types of symptoms:

– mottling or green to yellow mosaic on young leaves (Photos **117–119**);

– leaf deformation and reduction of the leaf that can sometimes be similar to that caused by ToMV, including fern-like leaves (fernleaf). The leaflets become filiform and are sometimes reduced to their main veins, then taking on the appearance of shoelaces (shoestring, see also section on leaves and leaflets deformation, see p. 75). Sometimes many consecutive levels of deformed leaves can be observed, separated by leaves of normal appearance. In some cases, all the apical leaves are affected (Photo **121**). Plant growth can be slowed or stopped. Fertility and size of forming fruit can be reduced;

– necrotic lesions sometimes converge on the leaf (Photo **120**, see also section on Wilting, necrosis, and drying of leaflets and leaves, on p. 217). These can then reach the petioles then develop on the stem in the form of brown longitudinal streaks that can completely encircle the stem and extend to the base and apex of the plant. With widespread necrosis, the plant dies quickly.

Fruits, as well as leaves, display quite varied symptoms: swelling, necrotic ring patterns, olive to brown in colour (Photos **122** and **123**).

As can be seen, the symptoms due to CMV are very diverse, due to several factors, including the existence of many strains and the presence of a necrogenic satellite RNA in some of them.

Note that it is not uncommon to find mixed infections between CMV and one or more other viruses: AMV, PVY, TSWV.

Discoloration

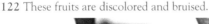

123 Raised circle and ring lesions partially cover the fruit.

122 These fruits are discolored and bruised.

124 These young leaflets have a discrete generalized mottling.

125 The leaflets can also be chlorotic to a greater or lesser degree and locally bronzed.

126 Brown necrotic spots are developing on this slightly curled leaflet.

127 Leaflets with many scattered necrotic lesions; locally inducing yellowing of the leaf.

Potato virus Y (PVY)

Potato virus Y (PVY) (Description 36)

PVY is a virus spread widely in the world but, unlike CMV, it is much less polyphagous with its host range rather limited to the Solanaceae. It is with CMV one of the major viruses of tomatoes, both in open field and protected crops in various European countries. Like CMV it is transmitted by several species of aphids.

The nature and intensity of symptoms caused by this virus on tomatoes are influenced by the timing of the attack, the cultivar, and environmental factors, but mainly the nature of the strain infecting the plant.

The 'common' strains cause discrete mottling, green mosaics on young leaflets (Photos **124** and **125**). Vein banding symptoms and pale green spots are visible on the leaf.

'Necrogenic strains' as their name suggests, induce brown leaf spots, sometimes slightly reddish, becoming fairly rapidly necrotic (Photos **126** and **127**). Longitudinal necrotic streaks may appear later on the petioles and stem. Eventually, this tendency to necrosis can affect the whole plant (see also pp. 162 and 223).

This virus causes no symptoms on tomato fruit.

As noted above, it is not uncommon to encounter co-infections, in field crops with CMV, and recently in glasshouses with *Tomato chlorosis virus* (ToCV). In this case, the symptoms may be more severe.

Pepino mosaic virus (PepMV) (Description 29)

The virus, introduced into Europe in the late 1990s, is, as PVY, almost entirely confined to the Solanaceae, with a much smaller host range. It occurs in glasshouse crops in a number of European countries. Its spread is largely by contact and by seeds. The spread of this virus in glasshouses can be dramatic. For example, in the Netherlands, approximately 70% of plants have been recorded as infected 6 weeks after the initial infection.

The nature and intensity of symptoms are quite variable and largely influenced by crop variety, the growth stage of plants, climatic conditions, and the time of year. During the summer (high light), the symptoms disappear and may even go unnoticed. They are much more marked in late winter and autumn, when the light intensity and temperature are more limiting.

Here we will only describe the mosaic-like symptoms appearing on leaves of affected plants. Vein clearing (Photos **131** and **132**), green to yellow mosaic in spots or patches (Photos **128–130, 133,** and **134**) are visible on the leaflets and are spreading gradually to the leaf or remain on the edge of it.

The leaves and leaflets can also be deformed, curved, and reveal swelling, with dark green enations on the leaf. The apical growth of the plants can be slowed down. Thus, the new leaflets are smaller, deformed, and more denticulate at the edge of the leaf, giving the leaflets and the top of the plant a look reminiscent of a nettle (nettle heading).

Other symptoms such as inter-vein chlorosis, yellow spots, or necrotic lesions on leaflets are described on pp. 116 and 165.

The flowers may turn brown and abort; partial browning of the calyx is sometimes found. The fruits are discolored, mottled (Photo **134a**), and sepals show chlorotic to necrotic patches, sometimes corky and slightly swollen (Photos **135** and **136**).

Mixed infections with other viruses common in glasshouses are possible. For example, co-infection between PepMV and ToCV has been seen in the same glasshouse and on the same plants. In this situation, and depending on time of year, identification of these two viruses by simple visual observation is very difficult, especially that of PepMV.

Discoloration

128–130 The three photographs show the diversity and evolution of mosaic-like symptoms caused by the Pepino mosaic virus.

Pepino mosaic virus (PepMV)

131 Vein clearing starts to spread, sometimes regularly.

132 This vein clearing has spread to a significant portion of the leaf tissues located between the veins.

133 A complete mosaic is sometimes visible on the leaf.

134 In some cases, a mosaic with bright yellow to white spots can be found. This is somewhat reminiscent of Aucuba mosaic described for ToMV.

Pepino mosaic virus (PepMV)

134a The fruit infected with Pepino mosaic virus (PepMV) is marbled and therefore has a slightly heterogeneous colour.

136 Chlorotic patches becoming gradually necrotic are visible at the base of each sepal, around the stalk.

135 On this truss, several green fruits have a defect in their sepals.

Pepino mosaic virus (PepMV)

Table 11 Common viruses responsible for mosaic or similar symptoms on leaflets

Virus	Symptoms	Vectors
Tobacco mosaic virus (TMV)	Slight marbling and variable mosaics marked on leaflets. These are sometimes filiform in periods of low light. The petals and sepals are sometimes deformed, there is also a decrease in the number of pollen sacs. This virus is often mistaken for ToMV which is better adapted to tomato.	Transmitted by contact, almost never by seeds.
Tomato mosaic virus (ToMV)	Vein lightening, mottling, mosaic on leaflets with green, yellow, or white patches. These are also crinkled, wrinkled, even filiform, when plants lack of light. Chlorotic rings are visible on green and mature fruit. Internal brown necrotic lesions are also found (see p. 368).	Transmission is by contact and by seeds.
Cucumber mosaic virus (CMV)	Three types of symptoms are observed according to the strains and climatic conditions: – mottling, mosaic on young leaflets; – deformation of the leaflets which have the appearance of a fern leaf or that of a shoelace when very filiform (shoestring); – spots, necrotic lesions starting on the leaflets, and extending to the stem and apex of the plant. Severe necrotic brown, round and raised lesions cover large areas of fruit.	Transmitted by aphids in a nonpersistent manner. More than 90 species are capable of hosting and transmitting including: *Myzus persicae, Aphis gossypii, A. craccivora, A. fabae, Acyrthosiphon pisum.*
Potato virus Y (PVY)	Symptoms vary depending on the strains: – Discrete mottling, green mosaic, vein banding, diffuse chlorotic spots (common strains); – Reddish brown necrotic spots, longitudinal necrotic lesions can be observed on the petioles and stem (necrotic strains).	Transmitted by about 40 different aphids in a nonpersistent mode (including *M. persicae, Aphis gossypii, A. fabae, Acyrthosiphon pisum, Macrosiphum euphorbiae*).
Pepino mosaic virus (PepMV)	Green to yellow mosaic sometimes on the edge of the leaf. The leaflets can be curved upwards or downwards and display blistering and enations. The leaflets at the apex are shrivelled and the apex is sometimes deformed. The sepals are superficially and locally altered. The fruits can be mottled.	Transmitted by simple contact, during the cultural operations, by pollinating bees, and by seeds.
Eggplant mottled dwarf virus (EMDV)	Lightening and yellowing of the veins in young leaflets. These are also smaller, deformed, and curled. The plants' growth might be reduced or even stopped. They end up being sterile, and growing fruits become badly discolored and ribbed or swollen.	This virus is probably transmitted by leafhoppers in the genus *Agallia.*
Broad bean wilt virus (BBWV)	Mosaics, arcs, and linear or concentric patterns on leaflets. These have a reduced growth and are slightly deformed. Necrotic lesions may appear on the leaf and spread to the petioles and stem.	Many aphids transmit the virus in a nonpersistent manner (including *Acyrthosiphon pisum, Aphis craccivora, A. fabae, A. nasturtii, Macrosiphum euphorbiae, M. solanifolii, Myzus persicae*). This virus is transmitted by seed in broad bean.
Peru tomato virus (PTV) Description 38	Whitening of the veins, mosaic on leaflets which are also deformed. Some strains cause necrotic lesions on leaves and stems. Mottling on fruit.	Transmitted by the aphid *M. persicae* in a nonpersistent manner.
Potato virus X (PVX) Description 32	Mottling, sometimes necrotic mosaic when accompanied by necrotic spots.	In tomato transmitted primarily through contact.
Tobacco etch virus (TEV) Description 37	Mottling, yellow mosaic of leaflets which may be deformed and wrinkled. In the case of early attacks on young plants, their growth may be reduced and the fruits are mottled and small.	Transmitted by several aphids in a nonpersistant manner (including *M. persicae, Aphis gossypii, A. craccivora, A. fabae, Macrosiphum euphorbiae*).

Discoloration

137 The affected plant is stunted. Its blocked apex contrasts with those of surrounding tomato plants.

139 Symptoms are sometimes visible on the intermediate to lower leaves: here, several leaflets are slightly smaller, deformed to a variable degree, and badly discolored.

138 Leaflets located at the apex are smaller than others, chlorotic, and deformed or blistered.

140 There is sometimes a vein clearing and yellowing, the discoloration gradually spreading to the leaf.

Eggplant mottled dwarf virus (EMDV)

Eggplant mottled dwarf virus (EMDV) (Description 45)

Little known and relatively limited in distribution to several Mediterranean countries, this Nucleorhabdovirus is common on tomato for many years since it has been detected for the first time on this plant in 1988, in the Pyrenees-Orientales region. Its presence on eggplant and pepper is more recent and dates from 1995. This virus affects mostly Solanaceae. Its vector, almost certainly an insect, is not known. The EMDV is not seed-borne in tomato crops.

Viral damage occurs in glasshouses in both winter and in summer, with rather different effects (a few plants to almost 50% of tomatoes affected) with a variety of symptoms:

During rather cold and grey periods, affected plants grow very slowly or are even stopped, giving them a stunted appearance (Photo 137).

The leaflets of the apex are chlorotic, deformed, and blistered (Photo 138). Plants can become unproductive due to the sterility of flowers;

In summer, there is irregular inter-vein yellowing (vein clearing) that can be widespread on the leaflets, forming a network in contrast with the rest of the leaves that retain their original green colour (Photos 139 and 140). Note that the leaves have a reduced size, are sometimes deformed and rolled (see section on leaf and leaflet deformation on p. 60). Some parts of the stems are slightly mottled (Photo 141).

The fruit, often smaller, are poorly coloured and have an irregular surface, dented or wrinkled (Photo 142).

Note that this virus may also be present sporadically in field crops and in all types of production.

141 Some parts of the stem may be mottled.

142 Poorly coloured and bruised fruits in a few rare trusses.

Discoloration

Pelargonium zonate spot virus (PZSV) (Description 46)

A Mediterranean virus of very limited distribution (mainly reported in Italy, Spain, and France), the PZSV affects a rather limited range of natural hosts including tomato in addition to pelargonium, artichoke, and several species of chrysanthemum. In France, it has been reported in glasshouses in some regions.

The PZSV, still an unclassified virus, causes rather characteristic symptoms. As its name suggests, it induces on the leaflets chlorotic sinuous and/or circular and concentric lines. These symptoms eventually become necrotic. The same type of symptoms can be found on parts of the stem (Photo **143**).

Some fruits are covered with spots, with chlorotic rings of variable width (Photo **144**), which eventually become necrotic. These fruit symptoms can be confused with those caused by TSWV, and to a lesser extent by CMV.

Note that the plants affected early may remain dwarf.

143 Sinuous and/or concentric chlorotic lines are clearly visible on the leaflets and stems.

144 Fruits show translucent to yellow concentric spots and rings.

Broad bean wilt virus (BBWV) (Description 38)

Widely distributed in the world and rather polyphagous, this Fabavirus is very common in crops of beans, lentils, and chickpeas in the Mediterranean region. It is reported on pepper and tomato in Europe, particularly Italy, Spain, and France. In France, BBWV is rather rare, and it also affects lettuce. More than 20 species of aphids provide transmission in the nonpersistent manner.

Its symptoms include mosaics on the leaflets (Photo **145**).

Chlorotic sinuous lines, sometimes concentric, also appear. The leaves have reduced growth and may be deformed. Under certain conditions, brown necrotic lesions develop on the leaf and gradually affect the petioles and stem. The intensity of symptoms may vary according to variety.

145 Broad bean wilt virus (BBWV) causes leaf deformation accompanied by a mosaic and sinuous and concentric chlorotic patterns.

Table 12 Rare or sporadic viruses causing mosaic or similar symptoms on leaflets

Virus	Symptoms	Vectors
Colombian datura virus (CDV) Description 38	Young leaflets with mosaics, poorly coloured fruit. The plants have reduced growth.	Aphids (*Myzus persicae*), in a nonpersistent manner.
Cowpea mild mottle virus (CPMMV) Description 42	Light symptoms: mottling in Nigeria, slight chlorosis of leaflets in Israel.	Whiteflies *(Bemisia tabaci)* in a semi-persistent way. Unlike other Carlavirus, it is not transmitted by aphids.
Datura yellow vein virus (DYVV) Description 47	Vein clearing, mottling, chlorosis of young leaflets. Leaf deformation.	Unknown mode of transmission.
Eggplant mosaic virus (EMV) Description 45	Leaf deformation accompanied by vein clearing of young leaflets, various mosaics from green to yellow, sometimes white. Necrosis and drying of the leaf. Stunting of plants affected at an early stage. Poorly coloured fruit, with mottling or spots.	Several strains transmitted by a beetle: *Epithrix* sp. The methods of transmission are still poorly understood.
Eggplant severe mottle virus (ESMoV) Description 38	Severe mosaic on leaflets.	Aphids (*Aphis craccivora, M. persicae*), in a nonpersistent manner.
Groundnut ringspot virus (GRSV) Description 44	Mottling and mosaic on young leaflets.	Thrips (*Frankliniella occidentalis, F. schultzei*).
Moroccan pepper virus (MPV) Description 47	Mottling, chlorotic spots on leaflets which are also deformed. Dwarfing of plants.	Unknown mode of transmission, transmission from contaminated soil is possible.
Pelargonium zonate spot virus (PZSV, non-classified virus) Description 47	Leaflets with patterns, sinuous lines, chlorotic rings becoming wider. Eventually, these lesions become necrotic. First chlorotic rings, concentric on stem and fruit, becoming necrotic thereafter. Growth of affected plants may be reduced, resulting in dwarfism.	Mode of transmission is unclear. Some pollen consuming insects are suspected, and transmission by water or nutrient solution is possible. The virus may be transmitted by pollen particularly in tomato.
Potato yellow dwarf virus (PYDV) Description 45	Chlorotic spots gradually extending, vein clearing on young leaflets.	Leafhoppers (*Agallia constricta, A. quadripunctata, Aceratogallia sanguinolenta*) in a multiplying circulative way.
Ribgrass mosaic virus (RMV) Description 32	Mottling and green mosaic on leaflets. Fruit can reveal internal tissue necrosis (internal browning, see p. 368).	Transmission by contact. This virus may have been detected in the water of certain rivers.
Tobacco vein banding mosaic virus (TVBMV) Description 38	Mosaic sometimes with dark green edging along the veins.	Several species of aphids (*M. persicae* is the most effective), in a nonpersistent manner.
Tobacco vein mottling virus (TVMV) Description 38	May infect some tomato varieties in the US without causing symptoms.	Several species of aphids (*M. persicae, M. nicotianae, A. craccivora, Macrosiphum euphorbiae*), in a nonpersistent manner.
Tobacco yellow dwarf virus (TYDV) Description 45	May infect some tomato varieties in Australia without causing symptoms.	Leafhoppers (*Orosius argentatus*), in a persistent manner.
Tomato mild mottle virus (TMMV) Description 38	Mottling, mild mosaic on young leaflets. Reduced yield.	Aphids (*M. persicae*), in a nonpersistent manner.

146 Several chlorotic spots, sometimes delineated by the veins, are scattered on these leaflets.
Leveillula taurica

148 On this leaflet, a chlorotic halo is easily visible that surrounds the few brown spots present.
Pseudomonas syringae **pv.** *tomato*

150 A yellowing of the veins spreads gradually to the leaflet.
Phytotoxicity

147 Some leaflets, only on one side, have a yellow colour. This unilateral yellowing is often due to vascular disease.

149 Basal leaflet shows yellowing which gradually spreads. Only the veins remain green.
Nutritional disorder (magnesium deficiency)

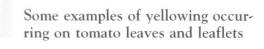

**Some examples of yellowing occur-
ring on tomato leaves and leaflets**

Partially or completely yellow (chlorotic), sometimes white, leaflets and leaves

Possible causes

Parasitic fungi and bacteria of the roots, stem base, stem, and vessels of tomato (see related sections, especially the sections on Vascular disease, pp. 323–339).

- Various phytoplasmas (Description 28)
- Various viruses (see also section Mottled leaflets and leaves p. 95)
- Viruses responsible for yellow mosaic and/or yellowing (see p. 111)
  - Colombian datura virus (CDV)
  - Cucumber mosaic virus (CMV)
  - Pepino mosaic virus (PepMV)
  - Potato yellow dwarf virus (PYDV)
  - Tomato yellow mosaic virus (ToYMV)
  - Serrano golden mosaic virus (SGMV)
  - Sinaloa tomato leaf curl virus (STLCV)
  - Taino tomato mottle virus (TToMoV)
  - Tobacco etch virus (TEV)
  - Tobacco mosaic virus (TMV)
  - Tomato chlorosis virus (ToCV)
  - Tomato golden mosaic virus (TGMV)
  - Tomato infectious chlorosis virus (TICV)
  - Tomato mosaic virus (ToMV)
- Viruses responsible for leaf yellowing and for some rolling (see also p. 62)
  - Beet curly top virus (BCTV)
  - Cowpea mild mottle virus (CPMMV)
  - Eggplant mottled dwarf 'virus' (EMDV)
  - Eggplant yellow mosaic virus (EYMV)
  - Indian tomato leaf curl virus (IToLCV)
  - Pepino mosaic virus (PepMV)
  - Peru tomato virus (PTV)
  - Tomato yellow mosaic virus (ToYMV)
  - Sinaloa tomato leaf curl virus (STLCV)
  - Taino tomato mottle virus (TToMoV)
  - Tobacco leaf curl virus (TLCV)
  - Tomato bushy stunt virus (TBSV)
  - Tomato chlorosis virus (ToCV)
  - Tomato golden mosaic virus (TGMV)
  - Tomato leaf crumple virus (TLCrV)
  - Tomato leaf curl virus (ToLCV)
  - Tomato mottle virus (Tomov)

- Tomato pseudo curly top virus (TPCTV)
- Tomato yellow dwarf virus (ToYDV)
- Tomato yellow leaf curl virus (TYLCV)
- Tomato yellow mottle virus (ToYMoV)
- Some viroids (Description 48)
- Nutritional disorders
- Excess salt
- Various phytotoxicities

The partial or generalized yellowing of one or more leaflets and leaves, called 'chlorosis', is a symptom not specific to a given disease, as it is frequently observed on tomato. It may be very variable, and confined to a small area in the form of a spot (Photo **146**; see p. 189) or with a dark spot surrounded by a yellow halo of varying intensity (Photo **148**; see e.g. p. 172);

– It sometimes affects one side of a leaflet or a leaf; this yellowing is described as 'unilateral' and can characterize vascular disease (Photos **147** and **149**; see p. 329 and following);

– It can develop from the veins (Photo **150**) or when between veins is called inter-vein chlorosis (Photo **147**).

Yellowing can also start with the young leaves of the apex or with the old leaves from the base of plants. It sometimes spreads to the whole plant, and is of different intensities from light green to bright yellow, sometimes progressing to white or the excessive production of anthocyanin on leaflets.

This symptom is indicative of a malfunction in the plant, frequently as a result of:

– One or more parasitic attacks occurring either directly and locally on the leave (e.g. air-borne disease), or on other plant parts, especially on the roots or in and on the stem;

– Virus attacks (particularly yellowing) or non-parasitic diseases, such as nutrient deficiencies or phytotoxicities.

 It is often difficult to determine the cause of yellowing: a very careful approach to diagnosis is advised.

151 Several leaves of this plant have chlorotic leaflets to varying degree.

153 In addition to the yellowing of the leaf, small reddish brown lesions are sometimes present.

152 A few inter-vein chlorotic spots are visible by viewing the leaflet with back light; this leaflet is showing the first symptoms of this virus.

154 Gradually, the yellowing is increasing, while many veins remain green.

155 In some situations, there is a very bright localized or generalized yellowing of the tissues.

Tomato chlorosis virus (ToCV)

Additional information for diagnosis

Tomato chlorosis virus (ToCV) (Description 39)

Transmitted by several species of whiteflies in a semi-persistent manner, this new virus has recently started to affect tomato crops in a limited number of countries: USA, Spain, Portugal, Italy, Greece, and France. In France ToCV was described in 2002 for the first time. It is now well established in various Mediterranean countries.

Affected plants are distributed randomly in the crop, and intermediate and lower leaves become chlorotic locally.

Irregular chlorotic spots appear between the veins of the leaflets and spread progressively (Photo 152); small reddish brown changes are sometimes visible on the yellowed leaf (Photo 153). Yellowing eventually appears on the upper leaves, raising concerns of a nutritional problem (Photo 151). After a few weeks, some plants develop a fairly sustained and generalized inter-vein chlorosis, but the leaflet veins remain dark green (Photos 154 and 155). The plants eventually grow old prematurely; the old leaves thicken and become brittle, curl, and sometimes dry.

The symptoms caused by ToCV are almost identical to those induced by Tomato infectious chlorosis virus (TICV). Their identification is difficult, especially as mixed infections are possible as is the case in Spain. Note that the TICV was very occasionally observed in southern France in 2003.

Pepino mosaic virus (PepMV) (Description 29)

This virus, recently introduced into northern Europe, produces a wide variety of symptoms on tomato. In addition to various mosaics described on p. 104, this virus can induce chlorosis of the leaf (Photos 156 and 157) and limited size, with angular bright yellow spots (Photos 158 and 159 and p. 196). The latter seems the most common and best known symptom in France. It occurs mostly in summer, the least favourable time for this viral disease. Other symptoms such as nettle-shaped leaves and necrotic lesions, are also described on p. 165. The fruits are also affected, exhibiting marbling (Photo 160).

Eggplant mottled dwarf virus (EMDV) (Description 45)

As described in the section Mottled leaflets and leaves p. 95, in winter this virus can cause a yellowing of the apex of the leaflets which are deformed and blistered (Photo 161). Affected plants exhibit very slow or even stopped growth, developing a stunted appearance (Photo 162). The sterility of the flowers makes them unproductive.

Discoloration

115

156 Several leaflets are slightly chlorotic.

158 Summer conditions intensify yellowing and make it more localized.

157 On close inspection of the leaflets, they progressively turning pale yellow between the veins.

159 In this case, yellow spots of slightly irregular shape cover very limited sectors of the leaf.

160 Fruit can be mottled and so poorly coloured.

Pepino mosaic virus (PepMV)

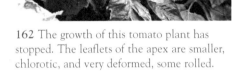

161 A number of apex leaflets are chlorotic, or intense yellow; note that some are also blistered.

162 The growth of this tomato plant has stopped. The leaflets of the apex are smaller, chlorotic, and very deformed, some rolled.

Eggplant mottled dwarf virus (EMDV)

Many Begomovirus attack tomato. Symptoms and vectors of a number of them are shown in *Tables 12* (p. 111) and *13*. The main leaf symptoms caused by Begomovirus are shown in the Photos *163–166*. Only the symptoms of two will be described here as they are very representative of this group of viruses.

Table 13 Main viruses causing leaf chlorosis, and sometimes, leaf roll

Virus	Symptoms	Vectors
Pepino mosaic virus (PepMV)	Green to yellow mosaic sometimes located on the edge of the leaf. The leaflets can curved upwards or downwards, and display blistering and enations. Those of the apex are shrivelled and the latter is sometimes deformed. The sepals are superficially and locally altered and fruits can be mottled.	Mechanically transmitted by casual contact, during the cultural operations, by pollinating bees, and by seed.
Tomato chlorosis virus (ToCV)	Irregular chlorotic mottling, inter-vein yellowing in patches at first, later affecting the entire leaf of lower leaves. Then, the most severely affected, become hard and brittle. Plants grow less vigorously and fruit maturity is delayed.	Transmitted by whiteflies (*Trialeurodes vaporariorum*, *Bemisia tabaci* biotypes A et B = *B. argentifolii*) in a semi-persistent manner.
Tomato infectious chlorosis virus (TICV)	Chlorosis, with some inter-vein yellowing of leaflets. Older leaves may also show a reddening and necrosis of the leaf. Leaf deformations are also visible, notably a leaf roll. Early affected plants are less vigorous. Fruit production can be greatly reduced.	Transmitted by *T. vaporariorum* (but not by *B. tabaci*) in a semi-persistent manner.
Tomato yellow leaf curl virus (TYLCV) and related viral species and/or associated with *yellow leaf curl* syndrome	Leaflets are small, curved like a spoon, sometimes rolled. They gradually turn yellow and the yellowing eventually spreads to the entire portion of the affected foliage. Some affected tissues may become harder and/or with anthocyanin. The plant's growth is more or less stopped at the time of infection. Many axillary branches, with short internodes, develop and give the plants a bushy appearance. The flowers fall in large proportions and the rare fruits are small.	Transmitted by *B. tabaci* (notably biotype B, *B. argentifolii*), in a persistent and circulative way.
Tobacco leaf curl virus (TLCV) (also note existence of the following species: Tobacco leaf curl Japan virus, Tobacco leaf curl Kochi virus, Tobacco leaf curl Yunnam virus, Tobacco leaf curl Zimbabwe virus) Description 42	Symptoms are fairly comparable to those produced by TYLCV – rolling leaflets and chlorosis. The leaf takes on a wrinkled and deformed appearance. In older plants, the leaves are often brittle. The growth of those infected early is greatly reduced.	Transmitted by whiteflies *B. tabaci*, *B. tuberculata*, *Aleurotrachelus socialis*, *Trialeurodes natalensis*, in a persistent or circulating manner. The TLCV is not congenitally transmitted to the progeny of its vectors. It is neither transmitted by seeds nor by contact, nor experimentally by mechanical inoculation.
Tomato yellow mosaic virus (ToYMV) Syn. Potato yellow mosaic virus (PYMV) Description 42	Like for many other viruses, the leaflets are small, chlorotic, and curved, with anthocyanin on the edges. Plants infected early have a stunted growth, and can have a bushy appearance. The few fruits formed are much smaller.	Transmitted by whiteflies: biotype B of *B. tabaci* (*B. argentifolii*) in a persistent manner.

163 Smaller leaflets, with the edge of the leaf rolling upward.

164 Small, spoon-shaped, chlorotic leaflets.

166 Yellowing leaflets gradually showing anthocyanin colouration.

165 Curled and slightly blue-tinged leaflets (with anthocyanin).

Nature of the main symptoms associated with Begomoviruses

167 This plant, rather bushy, has many axillary branches with very short internodes and much smaller leaves.

169 This viral disease sometimes gives plants an appearance reminiscent of that caused by stolbur. Moreover, some flowers have also dropped.

168 The leaflets are essentially rolled upward and toward the upper surface of the leaf.

170 On this plant, the leaflets are very chlorotic and their upward pointing leaves give them a spoon-shaped appearance.

171 Sometimes the leaflets turn blue-tinged especially along the veins and at the leaf edges.

Tomato yellow leaf curl virus (TYLC)

Tomato yellow leaf curl virus (TYLCV) (Description 41)

This serious Begomovirus, described for the first time in Israel on tomato decades ago, is now found in many countries in the form of different strains or even species[1]. It is found in Spain and Italy, together with its vector, *Bemisia tabaci*.

Plants infected by TYLCV have decreased growth rates, ultimately they are stunted and dwarf (Photos **167** and **169**). The leaflets are often smaller and tend to curl upwards and thereafter towards the interior of the leaf (Photo **168**). They are also slightly chlorotic, or intense yellow in the advanced stages of symptom development. In some cases, they look like a spoon, hence the original name of this virus (Photo **170**).

Subsequently the leaflets often develop a firmer texture and a blue tinge on the veins and surrounding tissues visible on the underside of the leaf (Photo **171**).

The flowers sometimes drop prematurely and the fruits still on the plants are smaller.

Symptoms caused by this virus are not very specific. It is therefore appropriate to be wary of possible confusion with many other viral diseases caused by Begomovirus (see *Table 13*, p. 118) but also with stolbur, physiological leaf roll, some phytotoxicities, and phosphorus deficiency in particular. The symptoms of these diseases are listed under this heading.

Tomato yellow mosaic virus (ToYMV) (Description 42)

This Begomovirus present in the Caribbeans, notably Guadeloupe and Martinique, was also called Potato yellow mosaic virus (PYMV), which could be confusing. It is also known in Brazil and Venezuela and is transmitted by the B biotype of the whitefly *Bemisia tabaci*.

The symptoms it produces are very comparable to those of TYLCV: the affected plants can have a bushy appearance as a result of reduced growth and shorter internodes (Photo **172**). The leaflets are smaller and rolled up (see section on abnormal plant growth and/or leaflets and leaf deformation, p. 61). They have a yellow mottling linked to the progressive development of irregular and inter-vein chlorosis of the leaf (Photos **173** and **174**). Excessive anthocyanin is also found later (see Photo **165**, and *Table 13* on the symptoms caused by Begomovirus).

Affected plants are usually not very productive. Indeed, flower drop often occurs. Fruits still on the plant have a reduced quality, are poorly coloured and deformed (Photo **175**).

---

[1] The emergence of the yellow leaf curl syndrome in many countries has led to the description of several strains of TYLCV: TYLCV-Ch (China), TYLCV-Is (Israel, including also strains from Egypt and Lebanon), TYLCV-Ng (Nigeria), TYLCV-Sar (Sardinia), TYLCV-SSA (Saudi Arabia South), TYLCV-Tz (Tanzania), TYLCV-Th (Thailand), and TYLCV-Ye (Yemen). Subsequently, these strains and new species have been reclassified and internationally recognized through the study of the percentage of identical genomes: Tomato yellow leaf curl virus (TYLCV, including the original virus strains as well as Israel, Almeria), Tomato yellow leaf curl Sardinia virus (TYLCSV), Tomato yellow leaf curl Malaga virus (TYLCMalV, which would be a recombinant between TYLCV and TYLCSV), Tomato yellow leaf curl China virus (TYLCCNV), Tomato yellow leaf curl Kanchanaburi virus (TYLCKaV), Tomato yellow leaf curl Thailand virus (TYLCTHV).
The following 'potential' viruses are not yet fully validated internationally: Tomato yellow leaf curl Nigeria virus (TYLCNV), Tomato yellow leaf curl Kuwait virus (TYLCKWV), Tomato yellow leaf curl virus Saudi Arabia (TYLCSAV), Tomato yellow leaf curl Tanzania virus (TYLCTZV), Tomato yellow leaf curl Yemen virus (TYLCYV). Note that we find a comparable situation for ToLCV. Other viral species associated with yellow leaf curl syndrome can be found in the Description 42.

172 The plant has a rather bushy appearance linked to a marked reduction in the growth of young leaves.

173 The leaflets are smaller and display a chlorotic mottle.

174 In addition to chlorosis, the leaflets also tend to curl upward.

175 The fruits are discolored, deformed, and bruised.

Tomato yellow mosaic virus (ToYMV)

Table 14 Other viruses causing leaf chlorosis, and sometimes, leaf roll

Virus	Symptômes	Vecteurs
Indian tomato leaf curl virus (IToLCV) Description 42	Small, curved, rolled leaflets with chlorosis. Dwarf plants when infected early (symptoms similar to those caused by Begomovirus TLCV and TYLCV).	Transmitted by whiteflies *Bemisia tabaci* (notably biotype B), in a persistent manner.
Serrano golden mosaic virus (SGMV) (or Pepper golden mosaic virus) Description 42	Inter-vein chlorosis of young leaflets; the apex of the plants may become necrotic. Deformation of fruits.	Transmitted by *B. tabaci*, in semi-persistent and persistent manner.
Sinaloa tomato leaf curl virus (STLCV) Description 42	Chlorosis and yellowing of young leaves and leaflets that curl and sometimes take on an anthocyanin tinge. The plants, with very short internodes, show reduced growth.	Transmitted by *B. tabaci*, in a persistent manner.
Taino tomato mottle virus (TtoMoV) Description 42	Symptoms similar to those caused by TYLCV.	Probably transmitted by *B. tabaci*.
Tomato leaf crumple virus (TLCrV) Chino del tomate virus (CDTV) may be linked to the same viral species. Description 42	Marked leaf deformation and roll; the leaves appear wrinkled and have a chlorotic mottle.	Transmitted by *B. tabaci*. Note that this virus is not transmitted by seeds, nor by contact.
Tomato leaf curl virus (ToLCV)[a] Description 42	The symptoms are identical to those of TYLCV: small, chlorotic, and curled leaflets.	Transmitted by *B. tabaci*, in a persistent manner.
Tomato mottle virus (ToMoV) Description 42	Mottling with chlorosis on young leaves, the older leaves are more yellow and curled. The plants infected early will remain dwarf.	Transmitted by biotype B of *B. tabaci*, in a persistent manner.
Tomato yellow dwarf virus (ToYDV) Description 42	Yellowing of the leaflets that curl gradually. The plants infected early will remain dwarf plants.	Transmitted by *B. tabaci*, in a persistent manner.
Tomato yellow mottle virus (ToYMov) Description 42	Mottling with yellow mosaic on leaflets, the leaves curl and are deformed. The plants remain dwarf.	Transmitted by *B. tabaci*, in a persistent manner.
Tomato yellow vein streak virus (ToYVSV) Description 42	Yellow mosaic on leaflets; the leaves can be wavy. There are chlorotic streaks along the veins.	Transmitted by biotype B of *B. tabaci* (*B.argentifolii*), in a persistent manner.
Potato leafroll Polerovirus (PLRV) Description 38	Yellowing of the leaves begins at the periphery, the leaflets curl and eventually become stiff and tough. The plants infected early will remain dwarf.	Transmitted by more than 10 aphid species such as *Aphis nasturtii*, *Aulacorthum solani*, *Myzus persicae* in a persistent circulative manner.

[a] Note also the existence of at least 15 viruses similar to ToLCV: Tomato leaf curl Bangalore virus (ToLCBV), Tomato leaf curl Bangladesh virus (ToLCBDV), Tomato leaf curl China virus (ToLCCNV, Tomato leaf curl Sudan virus (ToLCSDV), Tomato leaf curl Indonesia virus (ToLCIDV).

176 The leaflets are small and chlorotic and the flowers are sterile.

178 In addition to yellowing on these leaflets, the edges curve downward.

177 The considerably slowed growth of the apical leaves of this tomato plant gives it a very characteristic appearance.

179 In this case, the fruits are very small and poorly coloured, with slight anthocyanin on the observable aberrant flowers.

180 In this plot particularly affected by stolbur, plants are bushy with variable anthocyanin.

Candidatus Phytoplasma solani – stolbur

Various phytoplasma

Reported in many countries, several phytoplasmas causing stolbur (stolbur, big bud, aster yellows) are likely to disrupt the development of tomato young organs, altering the appearance of the plants and leaf shape (see section on abnormalities of plant growth and/or leaflets and leaf deformations p. 69 and Photos **52–59**). The leaflets and leaves of diseased plants are smaller and are chlorotic (Photos **176–178**), sometimes with excess anthocyanin (Photo **180**).

The flowers are particularly affected (see p. 71), often sterile, and reveal several different abnormalities: reduced size, enlarged calyx, petals of green colour that can have a disproportionate size, with leaf-like parts, and anthocyanin.

The few fruits formed have a reduced growth and are poorly coloured (Photo **179**).

Discoloration

Nutritional disorders

Like many crops, tomato needs various mineral elements for growth throughout its development cycle. When these are in excess or when deficient, nutritional disorders occur. Tomato grows in an optimum range of temperatures between 16 and 32°C.

In addition to not always being easy to grow, its nutrition can be influenced by various factors:

– the richness and balance of soil nutrients (soil tests are needed);

– the type of tomato grown and the variety chosen;

– planting density;

– the mode of production (cultivation in soil or soil-less, substrate type);

– the nature of the irrigation system and its delivery;

– any other factor disturbing plant growth.

It is thus not uncommon to be confronted in crops with plants with inadequate or excessive nutrients. These conditions are part of nonparasitic diseases and grouped under the term 'nutritional deficiencies'. These show quite frequently on tomatoes by yellowing of a variable nature and distribution (symptoms shown by Photos **181–188**). *Tables 15* (overleaf) and *15a* (p. 129) summarize the few deficiencies sometimes encountered in tomato, describe symptoms that may be encountered, and clarify the physiological functions that might be disrupted.

Remember that currently, when talking about deficiencies, there is often a tendency to equate true deficiencies with induced deficiencies.

True deficiencies (too little element in the soil) are increasingly rare. Their visual diagnosis is very difficult because, with few exceptions, the symptoms they produce are discolorations, leaf yellowing of varying intensity, which are very difficult for a nonspecialist to assess.

In most cases we are dealing with induced deficiencies (elements present but not available), which does not make their diagnosis any easier. In addition to discovering the nature of the deficiency, the cause(s) must be determined. These can be diverse, such as poorly managed irrigation (too much or not enough water), too low or too high a soil temperature or pH, or root systems in poor condition.

When faced with such symptoms, growers must avoid the temptation of assuming a deficiency without consulting a specialist and carrying out the necessary tests (physical and chemical analyses) on soil and vegetation.

⚠ Deficiencies occur particularly in crops which have been fertilized indiscriminately without any previous soil analysis.

Table 15 Symptoms of deficiencies first appearing on young leaflets which can later affect the entire foliage of the tomato

Deficiency	Symptoms	Main changes in physiological functions and *favourable conditions for the expression of the deficiency*
Calcium (Ca)	The periphery of the leaf of young leaflets is light green and necrotic lesions gradually develop. In cases of very marked deficiency, the terminal bud(s) become brown, necrotic, and die. Plant development is thus curtailed or even blocked. The fruits show moist alterations becoming necrotic at the stylar end (apical necrosis, blossom-end rot, see p. 387).	Ca is a onstituent of the cell wall in the form of pectate and calcium oxalate, which is involved in elongation and cell division, influences the cell pH, the structural stability, and permeability of cell membranes. *Acidic soil, rich in K^+, NH_4^+, Mg^{++}, with excessive or too low humidity or temperature.*
Boron (Bo)	Young leaflets near the apex are slightly chlorotic, necrotic, and fragile and tend to warp and to roll. Plant growth is slowed, the internodes of the upper parts of the stem are shorter. The terminal bud(s), also necrotic, eventually die. Internal browning and mottling are visible in and on fruit.	Bo is involved in sugar transport across cell membranes and in the synthesis of cell wall constituents. It affects transpiration, development, cell elongation, protein synthesis, and interacts with auxins. *Acidic, alkaline, sandy soils, with little water, and low organic matter.*
Manganese (Mn)	Young leaflets undergo inter-vein chlorosis; the discoloured tissues are scattered with small necrotic lesions but their veins remain green. In the case of severe deficiency, older leaves eventually turn yellow too.	Mn is associated with the progress of oxygen during photosynthesis, with electron transport systems, and is a constituent of multiple enzyme systems. *Alkaline soils.*
Sulphur (S)	The young leaflets are light green and slightly smaller. This slight chlorosis can become widespread to the entire plant. Stems and petioles can have anthocyanin. This deficiency is relatively rare in tomato.	S is a component of two amino acids essential for protein synthesis, involved in the formation of vitamins, hormones, and in oxidation–reduction reactions, and constituent of several coenzymes and certain lipids. *Sandy soils, poor in organic matter, leached.*
Iron (Fe)	Inter-vein chlorosis of young leaflets beginning at their base and gradually reaching their ends. The veins stay fairly green and contrast with inter-vein tissues. Eventually, the tissues can take a white tinge and chlorosis reaches the entire plant (Photos **181–184**).	Essential in the synthesis of chlorophyll, Fe is involved in nitrogen fixation, photosynthesis, the transfer of electrons, and in several enzyme systems including controlling respiration. *Alkaline soils, compacted, with a high content of heavy metals, damaged roots.*
Zinc (Zn)	The leaflets are noticeably small, with chlorosis and necrosis between the veins which are thicker and tend to curve downward. The strongly affected plants have a limited growth.	Zn is an essential metal component of several enzymes involved in electron transfer and in the synthesis and degradation of proteins. It is involved in the regulation of plant growth through auxins. *Acidic and alkaline soils, leached, poor in organic matter.*

181 An inter-vein chlorosis is gradually developing on this leaf.

182 At first the yellowing is slight, diffuse, making the veins stand out.

183 The iron deficiency, which can for example result from temporary asphyxia, is now very marked, the inter-vein yellowing is very intense.

184 Eventually, the leaf is almost completely discolored, or even white; only the veins retain their original green colour.

Example of a deficiency, in this case iron, starting with young leaves (apex of plant)

185 Many older leaves show a highly visible intervein chlorosis.

187 Thereafter, chlorotic areas extend between the veins.

186 A few inter-veinal chlorotic spots, mainly on the periphery of the leaf, are the first signs of magnesium deficiency.

188 In the end, the yellowing spreads throughout the leaf, and some areas eventually become necrotic. The veins remain green.

Example of a deficiency, in this case magnesium, starting with the old leaves (base of plant)

Table 15a Symptoms of mineral deficiencies first appearing on lower leaves which can later affect the entire foliage of the plant

Deficiency	Symptoms	Main changes in physiological functions and *favourable conditions for the expression of the deficiency*
Potassium (K)	Chlorotic spots, beginning at the periphery of the leaves and gradually extending to inter-vein tissues. They turn brown and necrotic. On the edges, the leaves turn brown and curve downwards. Plant growth is stunted to a variable degree. Fruits may be softer, hollow, and irregular in size, poorly coloured and show mottling symptoms (greywall) (see p. 368).	K is needed for osmotic regulation of the plant (ionic balance of the cells, fluid retention in the tissues, transport of water), pH regulation of cells (to achieve desirable plant growth through thicker cell walls). *Light, sandy, acidic, poorly fertilized, heavily leached soils, excess Ca, Mg, or N.*
Magnesium (Mg)	Inter-vein chlorosis starting at the periphery of young leaflets. In late stages, the yellowed tissues eventually become necrotic while the veins tend to retain their green colour (Photos **185** and **188**). The fruits may have a green stem base (see p. 369).	Mg is a constituent of chlorophyll and a component of many enzymes. *Acid, sandy soils, rich in K^+, NH_4^+ and Ca^{++}, plants heavily laden with fruit, asphyxia or lack of water, poor root renewal.*
Nitrogen (N)	The leaves are light green, the oldest exhibit a more pronounced yellowing and can become necrotic and fall. Plant growth is limited. Stems and petioles are rather rigid.	N is involved in protein synthesis (a component of amino acids, proteins, nucleic acids), a constituent of chlorophyll, and an important nutrient at fruit production. *Soils poor in organic matter, leached, poorly fertilized, plants grown in too small a volume of substrate.*
Phosphorus (P)	The dark green leaflets eventually show a purplish colouration on the undersides of the leaves, especially on the veins. Sometimes the petioles and very thin stems have a similar hue. The plants are generally small with erect leaves and curved leaflets. The fruits are hollow and poorly coloured. Older leaves may drop.	P is a constituent of enzymes, proteins, phospholipids and nucleic acids, vital for the life and reproduction of the plant. *Compacted, acidic or alkaline, poorly fertilized soils, plants growing at too low temperatures, and in low light.*
Molybdenum (Mo)	Yellowish brown necrotic lesions developing at the periphery of the older leaves whose edges curl up. The growth of vegetation and root system can be reduced. This deficiency is unusual on tomato.	Mo is a metal component of two enzyme systems: nitrate reductase and nitrogenase. *Acidic, alkaline soils, soils draining too freely.*
Copper (Cu)	Inter-vein chlorosis starting at the periphery of the leaves. Subsequently, tissues may become necrotic and dry and the edge of the leaves curl up. The plants wilt at times, and their growth is reduced. This deficiency is uncommon in tomato.	Cu is an element involved in many enzyme systems, in the formation of the cell walls, electron transport, and oxidation reactions. *Organic, acid, sandy, alkaline, leached soils.*

Discoloration

Excess salt

Although the tomato is considered relatively tolerant to salt, damage associated with excess can be seen in certain production areas, notably those that are arid and semi-arid. Affected plants are often smaller and less vigorous. The leaflets are chlorotic (Photo **189**). The lower leaves are particularly affected and are of a stronger yellow hue and defoliation can result. Differences in sensitivity between cultivars can be observed.

189 This young tomato plant has a stunted growth, and its leaflets are rather chlorotic, especially those of lower leaves.

Genetic abnormalities

Many genetic abnormalities are known in tomato. They are mainly responsible for leaf deformation in the Solanaceae (see p. 79), but also change the colour of the leaves to yellowish (Photos **190** and **191**), blue-tinged (see p. 146), or silver (see p. 145).

More information on these genetic aberrations can be found on p. 64.

191 On closer inspection, these leaflets show a fairly strong yellow inter-vein colouring.

190 This plant with chlorosis and prematurely blocked growth, is in sharp contrast with the surrounding normal plants.

Various phytotoxicities

Among the pesticides used in agriculture, herbicides (characteristics are presented in *Tables 16* (p. 133) and *17* (p. 136) and to a lesser extent insecticides and fungicides may cause significant damage to tomato. The phytotoxicities (pesticide injuries) they induce are classified as nonparasitic plant diseases, and are unfortunately not uncommon in the field. It is therefore wise never to exclude this possible diagnosis.

Examples of phytotoxic discoloration caused by pesticides and particularly herbicides on tomato leaves and leaflets are presented in Figure 22 and Photos **192–208**). These discolorations include greenish or pale leaflets, yellowing, whitening, excessive anthocyanin, dull and/or bronzed leaves. Yellowing is by far the most common symptom and may have a different intensity and distribution on the leaves and leaflets:

– yellowing in spots, in distinct patches, sometimes progressing to tissue necrosis;

– yellowing of veins and adjacent tissues;

– yellowing of young leaves of the apex;

– diffuse yellowing of leaves between the veins;

– diffuse yellowing of the entire leaf;

– rapid drying and yellowing between the veins;

– homogeneous yellowing of the leaf, sometimes progressing toward tissue whitening;

– whitening of the leaf.

It should be remembered that the use of a herbicide on or near crops is never totally harmless. The risks of causing a phytotoxicity can never be totally eliminated.

Other pesticides such as insecticides and fungicides, alone or in combination, and substances such as fertilizers, may also cause phytotoxicities on tomato. They can cause similar yellowing as well as other symptoms detailed on pp. 76, 168, and 226.

The origin of a phototoxicity is difficult to determine. Indeed, the grower often refutes the possibility of having made a mistake or suffered some harm as a result of misuse of materials. The study of the distribution in time (date of symptom onset and evolution) and space (distribution of diseased plants in the plot and evolution) of symptoms induced by this phytotoxicity can, in most cases, confirm the cause.

Distribution of symptoms over time

The delay between the application of the phytotoxicity-inducing product and the onset of symptoms is variable:

– very short (the relation of cause and effect), immediately after application of a pesticide to the crop or nearby (sprays);

– longer in the case, for example, of poor cultural practice, such as a previous annual or perennial crop cleared with a residual or poorly washed-out herbicide following a dry winter, a perennial crop treated for several years (leading to accumulation of herbicide in the soil), or following application of straw or manure from straw made from treated crops.

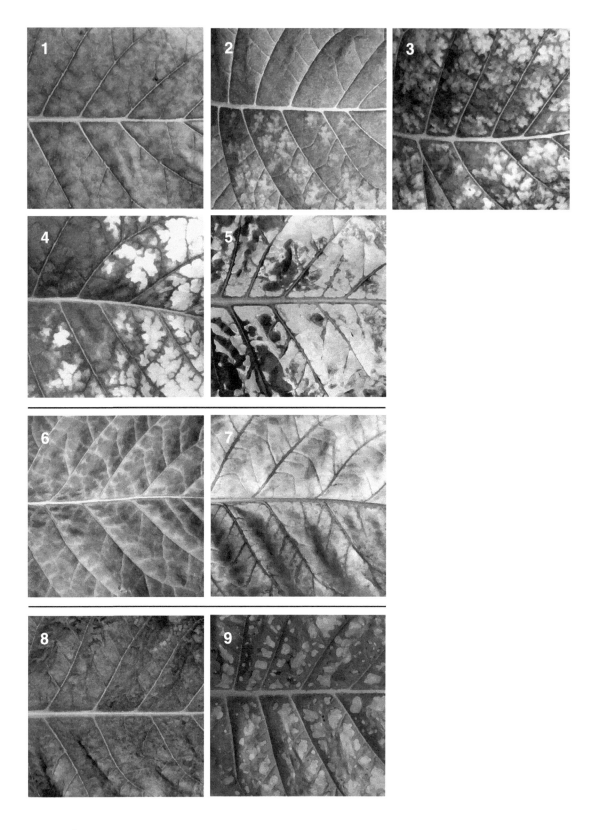

Figure 22 Simplified visual glossary of the development of some symptoms of phytotoxicity on leaves.

1–5 Inter-vein yellowing.

6 and 7 Vein yellowing.

8 and 9 Inter-vein drying.

Table 16 Characteristics of the main types of herbicides used in agriculture (see note to the reader p. 417)

Families of herbicides (examples)	Main plant symptoms	Main cell changes and altered functions
Photosynthesis inhibitors: amides, benzothiadiazones, bis-carbamates or phenylcarbamates, hydroxybenzonitriles (bromoxynil, ioxynil), phenyl-pyridazines (pyridate) pyridazinones, triazines (simazine, atrazine)[a] triazinones (metribuzin), uracil (bromacil), substituted ureas (diuron, linuron), bipyridiles (diquat, paraquat). Foliar herbicides for the most part, by contact, with low to high translocation according to the herbicide used.	Presence on the leaflets of large and bright inter-vein patches becoming necrotic and gradually drying, wilts.	Alteration of cell membranes (plasma, tonoplast). The assimilation of CO_2 is stopped. Inactivation of the electron transfer system. Increased membrane permeability. Decrease in chlorophyll content and carotenoids.
Carotenoids synthesis inhibitors: furanones, pyridinecarboxamide, pyrrolidones, isoxazoles, triketones, triazole (aminotriazole). Root herbicides and/or with low to high plant translocation according to the herbicide used.	Slow growth, stunted plants. Yellowing, whitening of the leaflets, leaves, and stems.	Selective action on chloroplasts that are then devoid of their internal lamellar system. Inhibition of the synthesis of carotenes. Then disappearance of chlorophyll, which is no longer protected. Reduction of nitrogen metabolism. Inhibition of the activity of certain enzymes (phytoene desaturase, 4-hydroxy-phenyl pyruvate dioxygenase).
Cell division inhibitors, antimitotics: benzamides (propyzamide), toluidines (pendimethalin, butralin, trifluralin), carbamates (carbetamide, chlorpropham), benzamides. Soil incorporated herbicides, penetrating through the roots, rarely or little by the leaves, with low to high migration, acting through stem, metabolic poisons reminiscent of colchicine.	Stunting and swelling of meristems. Root tips shaped as clubs. Inhibition of lateral root formation; if formed, they are short, thick, and solid.	Inhibition of mitosis, blocked in metaphase. Absence of microtubules. Inhibition of dihydropteroate synthase.

[a] Persistent herbicides which cause concern because of their involvement in water pollution.

Discoloration

192 Marked inter-vein patches characterize the effects of the photosynthesis inhibiting family of herbicides e.g. **Diuron**

194 On this young tomato plant, all the leaflets of the apex are markedly chlorotic, the tissues rapidly becoming necrotic thereafter.
Ioxynil (photosynthesis inhibitor)

193 Drying on the edges of the leaf is clearly visible on these leaflets. **Metribuzin (inhibitor of photosynthesis)**

195 Some herbicides can cause a yellowing of the leaves. **Ioxinyl (photosynthesis inhibitor)**

Examples of leaf yellowing and other symptoms related to the effects of herbicides

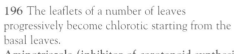

196 The leaflets of a number of leaves progressively become chlorotic starting from the basal leaves.
Aminotriazole (inhibitor of carotenoid synthesis)

197 The growth of this plant is reduced, the internodes are shorter, and the leaves smaller. Grossly affected leaflets, initially chlorotic, eventually become white.
Aminotriazole (inhibitor of carotenoid synthesis)

Herbicide damage

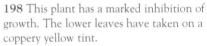

198 This plant has a marked inhibition of growth. The lower leaves have taken on a coppery yellow tint.
Pendimethalin (antimitotic)

199 The leaflets, reduced in size, may have a bronze-green tint, especially on the veins.
Pendimethalin (antimitotic)

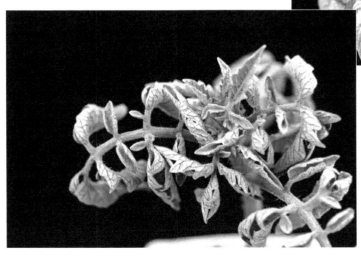

200 The leaflets, reduced in size, tend to curve upward and anthocyanin is present, especially on the veins visible on the underside of the leaves.
Propyzamide (mitotic)

Table 17 Characteristics of the main types of herbicides used in agriculture (see note to the reader p. 417)

Types of herbicides (examples)	Main plant symptoms	Main cell changes and altered functions
Inhibitors of lipid synthesis: aryloxyphenoxy-propionates, cyclohexan-ediones, benzofurames (ethofumesate), acetamides (napropamide), chloroaceta-mide (alachlor, propachlor), oxyaceta-mides, thiocarbamates. Soil incorporated herbicides, absorbed by the leaves and roots, and whose trans-location is small to large.	Inhibition of plant germination and emergence. Reduced growth of plants that are stunted. Leaf deformation and appearance of necrosis at the edges of the leaves.	Meristems are damaged. Decrease in epicuticular waxes. Inhibition of fatty acid synthesis. Increased permeability of the cuticle. Increases in root water uptake. Inhibition of gibberellic acid synthesis. Inhibition of protein synthesis. Inhibition of enzymes (elongase, and GGPP cyclization enzymes).
Herbicides disrupting the regulation of auxin IAA (indoleacetic acid): benzoic acid (dicamba), phenoxy acids alcanoides aryloxyacids (2,4-D, 2,4-MCPA), picolinic acid (picloram, triclopyr), quinolinecarboxylic acid. Systemic herbicide absorbed by the leaves and roots, persistent, translocated in the phloem to the meristem, acting through stem, with very important uptake.	Slight wilting of plants. Many deformations and fasciation of the entire plant. Thickening of the stem and petioles, formation of adventitious roots on the stem.	Stopping the activity of primary meristems and activating of secondary meristems. Hypertrophy of vascular tissues. Excretion of H^+ ions and entry of potassium (K) and water. Activation of wall hydrolases. Increase in DNA and protein (enzymes) synthesis. Increased production of ethylene.
Amino acids synthesis inhibitors amino phosphonates (glufosinate-ammonium, glyphosate), imidazolinone, sulfonylurea (chlorsulfuron, rimsulfuron), sulfonyl-amino- carbaryl triazolinones, triazolopyrimidines. Systemic foliar herbicides typically symplastic quickly degraded in the soil, with modes of action independent of photosynthesis, with leaf or root penetra-tion and low to important translocation.	Plants slow to develop. The leaflets of the apex are variably chlorotic and deformed (curved, crinkled). Portions of affected leaf tissue eventually become necrotic.	Disorganization of the chloroplasts of apical parts. Decreased chlorophyll content. Diminution of auxin content in treated tissues. Disruption of potassium, calcium, and magnesium absorption. Inhibition of several enzymes and therefore of the formation of several amino acids: glutamine synthase (glutamine), enolpyruvyl shikinate, phosphate synthase (phenylalanine, tyrosine, tryptophan), acetolactate synthase.

201 Growth at the apex of this plant is totally suppressed. This symptom is not unlike those caused by certain strains of phytoplasmas.
Butralin (antimitotic)

203 Many young leaflets are deformed, chlorotic, and thicker.
Pendimethalin (antimitotic)

202 Several leaflets of the lower leaves of this stunted plant are curled and with anthocyanin.
Chlorpropham (antimitotic)

204 On this stem, numerous roots are present, giving it a swollen appearance.
Propymazide (antimitotic)

Herbicide damage

205 A yellowing starting at the veins is very visible on these leaflets.
Metribuzin (photosynthesis inhibitor)

207 The leaves show a dull yellow colour after treatment applied in unsuitable conditions.
Dichlorvos (insecticide)

206 Many leaflets show yellowing followed by whitening at their base.
Herbicide

208 Some leaf edge yellowing has appeared on numerous leaflets after several applications through the irrigation system to control *Fusarium oxysporum* f. sp. *radicis-lycopersici*.
Benomyl (fungicide)

Pesticide damage

Distribution of symptoms in space

The distribution of plants treated with a phytotoxic product may vary depending on the composition of the phytotoxic compound, its mode of application, and its location on the plant (Figure **23**).

If the phytotoxic compound is applied to the foliage (foliar herbicide, overdosed or poorly applied fungicide or insecticide) the distribution of diseased plants may be:

– General and uniform (4);
– Beginning of line (2);
– On one side of plants.

If the compound is present in the soil as residue (root herbicide), the distribution of affected plants may be:

– Widespread and more or less homogeneous (4);
– Randomly distributed throughout the plot (1).

Differences in sensitivity exist between varieties and types of tomato. If several varieties are grown at the same time, the occurrence (3) of diseased plants and healthy plants may vary between varieties.

Furthermore, we also recommend you observe any weeds still present in the culture

or any other plants grown in close proximity that may have suffered the same phytotoxicity and thus express the same symptoms. If so, this partly confirms the hypothesis of a nonparasitic disease, and probably of phytotoxicity, especially if other information supports this possibility.

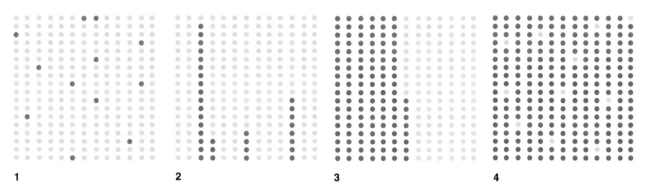

1   2   3   4

Figure 23 Possible distributions of plants showing phytotoxicity related symptoms in a tomato crop.

 If phytotoxicity is suspected, the following questions should be asked:

– Was the previous culture weeded with a residual herbicide?
– Were herbicide treatments made close to the crop?
– Has the processing equipment been properly rinsed?
– Is the spraying equipment well maintained (cleaning, grading)?
– Has the right product been used at the right dose?
– Has the product not been used too frequently (cumulative effect)?
– Have the recommendations on the label been met?
– Has the chemical been incorrectly stored (high temperatures, various stress)?
– Have incompatible products, or too many products, been mixed?
– Have applications been held in poor conditions (high winds, temperatures too low or too high)?

Remember that irrigation water may be contaminated by a herbicide.

What to do after phytotoxicity?

Although there is no quick fix in this situation, the following measures can be adopted:
– Define the origin of phytotoxicity;
– Prevent its recurrence;
– Do not remove the plants immediately. Grow normally and observe their development, as they may not be killed. In fact, everything depends on the nature, dose, and persistence of the product(s) concerned, and the stage of growth, the cultivation method, and variety of tomato plants. Hopes of recovery are always a possibility.

No other specific measures can be recommended.

The phytotoxicities cause other symptoms, in addition to the yellowing, described in Sections on abnormal growth of tomatoes and/or changes to leaflets and leaves (see p. 76), spots and damage on leaves and leaflets (see p. 168) and wilting, necrosis and dried leaflets and leaves (see p. 226).

209 In addition to tiny necrotic lesions, the leaflet shows a slightly bronzed colour.
Tomato spotted wilt virus (TSWV)

210 The stem and the leaf undersides of these small plants are blue-tinged.
Temperatures too low

211 A number of leaflets, with dull metallic to silver colours, stand out well against surrounding normal leaves.
Silvering

Examples of leaflets with discoloration other than yellowing

Leaflets, leaves with other discolorations (silver, bronze, blue-tinge, dull)

Possible causes

- Various phytoplasmas (see pp. 63, 69, and 125): Description 28
- Various viruses
- Tomato spotted wilt virus (TSWV): Description 43
- Alfalfa mosaic virus (AMV): Description 33
- Tomato yellow leaf curl virus (TYLCV): Description 41
- Tomato yellow mosaic virus (ToYMV)
- Tomato bushy stunt virus (TBSV)
- Potato yellow dwarf virus (PYDV)
- Several other Begomovirus
- Silvering
- Temperatures too low
- Various genetic abnormalities (sterile mutants, chimeras)
- Nutritional disorders
- Various phytotoxicities
- Mites (see p. 199)
- *Aculops lycopersici* (russet mite)
- Thrips (see p. 171)

Additional information for diagnosis

Various viruses

Tomato spotted wilt virus (TSWV, Description 43) is widely distributed worldwide in temperate and subtropical areas where it is on the increase since the early 1980s. It has a wide range of potential hosts and is transmitted by at least nine species of thrips.

Symptoms on the foliage of tomato can take various forms:

– leaf deformation with an apical curvature of the apex, a blockage of vegetation;

– a contrasted mosaic;

– spots and chlorotic lesions becoming necrotic, rings becoming gradually necrotic (see p. 164);

– chlorosis and bronzing of the leaves (Photos 212 and 214) or veins (Photo 215), accompanied by rings (Photo 213), small dark lesions becoming necrotic (see p. 164) also visible on petioles and stem;

– excessive anthocyanin on the leaves.

The fruits are also affected. They can be bronzed in large areas and with concentric chlorotic rings (Photo 217). Dry necrotic lesions, corky cracks, are sometimes visible (Photo 218). Early contamination leads to a reduction in the number and size of fruit; with late contamination, the fruits grow normally but are poorly coloured and can be deformed.

212 These young chlorotic leaflets reveal a slightly bronzed hue and numerous tiny and brownish lesions.

214 Many dark inter-vein changes which can cover the leaf, thereby increasing its bronze hue.

213 Some spots, chlorotic rings taking a dark to brown hue, are also visible.

215 The leaflets can take on a blue tinge, especially on the veins.

216 Brown elongated necrotic lesions (streak) are also visible on the petiole.

Tomato spotted wilt virus (TSWV)

Several other viruses described in the preceding pages may cause excessive production of anthocyanin in leaf tissues, particularly in the terminal phase of development of certain symptoms such as yellowing. Among these viruses are:

– Beet curly top virus (BCTV, Description 45) with smaller, crinkled, curled leaflets and blue-tinged veins;

– Tomato bushy stunt virus (TBSV, Description 32), characterized by the bushy appearance of infected plants, with chlorotic older leaves and excessive anthocyanin (see p. 51);

– many Begomovirus such as Tomato yellow leaf curl virus (TYLCV, Description 41), Tomato yellow mosaic virus (ToYMV, Description 42) (see pp. 118 and 123).

Various phytoplasmas

More detailed descriptions of the symptoms of multiple phytoplasmas in tomato are mentioned on pp. 63, 69, and 125 (stolbur, big bud, aster yellows). These micro-organisms can induce an excessive production of anthocyanin in the leaflets and/or the veins, in the stems, and even in the inflorescences and flowers (Photos **219** and **220**).

Silvering

This nonparasitic disease, also known as 'chimera', 'head silvering' when it affects the leaves of the apex of tomato, is relatively well known by glasshouse growers (Photo **221**).

This condition primarily affects protected crops, mainly in northern latitude areas of production. Note that these symptoms occur at any stage of development of the tomato and are characterized by:

– spots, patches of varied size, going from a grey-green colour to silver (Photos **222** and **223**);

– leaf deformation to various degree (leaflets of reduced size and/or denticulate, blistered, see Photo **223**);

– presence on the stem of streaks or stripes, variable in number and width (Photo **224**);

– partial or total sterility of the flowers located in areas affected by the silvering (although they have a normal appearance);

– sometimes deformed fruits with silvery green streaks becoming pale yellow at maturity when partially affected (Photo **225**).

This disease is only very damaging when the meristem of the apex is affected, resulting in symptoms on a whole plant and causing head silvering.

217 Chlorotic rings sometimes concentric can cover the fruit.
Tomato spotted wilt virus (TSWV)

218 In addition to rings, dry lesions, with corky cracks, are sometimes present on the fruit.
Tomato spotted wilt virus (TSWV)

The silvering is caused by abnormal development of palisade tissue of the leaf, occurring at a very early stage of development. This is reflected later by the formation of large intercellular spaces in the tissues of the leaf, giving the latter a typical silver colour.

Although all factors affecting the expression of this disease are not clearly defined, some of them give a better understanding and control:

– early sowing of winter crops are more sensitive than those made later;

– temperature seems the most important environmental factor to induce and influence positively or negatively this phenomenon. Indeed, the silvering occurs mainly as a result of temperature drops in the glasshouses, especially at the level of the apex of the meristem. This may be the case, for example, when a heat shield is removed too quickly, causing the cold air retained by the screen to reach the crop. The same occurs as the result of excessive ventilation of a closed tunnel, inducing over-evaporation of the plants and cooling of their apexes. Note that low glasshouses are particularly vulnerable, especially if they are poorly illuminated. Under these conditions, the apexes can experience temperatures of around 10°C. Conversely, high temperatures induce less silvering;

– many cultivated varieties are now resistant to silvering through the introduction of the gene 'Wi' in tomato. This situation is unfortunately not widespread in all cultivars, and it is not uncommon to observe this problem on older varieties on which the damage can sometimes be substantial (up to over 10% plants expressing symptoms). Note that this resistance is also missing in some new varieties in the types pink, orange, and cherry. In addition, the acquisition of resistance to powdery mildew could sometimes be at the expense of resistance to the silvering.

For example, silvering is particularly evident in plastic tunnels in southern Europe. During the summer, temperatures quickly reach 30°C in the morning. The sudden opening of these tunnels at both ends intensifies plant transpiration and causes a temperature lowering, positively influencing the expression of this physiological disorder.

To avoid this genetic disorder, resistant varieties should mainly be used. In the case of susceptible varieties, it is desirable to control the climate of the glasshouses to avoid too sudden drops in temperature. If plants show a silver head, it may be wise to remove the head and restart with an unaffected axillary shoot.

A limited number of other nonparasitic diseases are likely to induce excessive anthocyanin affecting the whole tomato plant. For example, Photos **226** and **228** illustrate the symptoms caused by a genetic condition other than the silvering, and the symptoms due to low temperatures on tomato (Photos **227**, **229**, **230**). Some deficiencies (phosphorous, sulphur, see p. 125) and some phototoxicities (see p. 131) also cause excessive anthocyanin on the leaves.

219 The leaflets, reduced in size, are blue tinged when affected by *Candidatus* Phytoplasma solani, responsible for stolbur.

220 There is the same colour on the flowers, which are also abnormal.
***Candidatus* Phytoplasma solani**

221 This plant affected by apex silvering, has many small leaves and leaflets, light green, dull to silver, contrasting with the normal lower leaves.

223 This leaflet shows raised irregular patches of dark green colour, contrasting with the rest of the silvered leaf. This symptom should not be confused with a mosaic due to a virus.

222 The silver in this case is unilateral, it affects only the leaflets on one side of this leaf.

224 Pale green to bright yellow longitudinal stripes of various widths, are sometimes visible on the stem.

225 Greenish to yellow longitudinal streaks can be seen on some fruit, which are deformed.

Silvering

226 Some sectors of several leaflets are chlorotic and blue-tinged.
Genetic abnormality

228 In some cases, only the leaflets on one side of the leaf have a blue tinge, a situation already observed with silvering.
Genetic abnormality

227 The stem and the underside of the leaflets of these young tomato plants, growing in a poorly heated nursery, have a blue tinge.
Climatic injury (temperatures too low)

229 On older leaflets, this blue tinge is mostly observed in the veins and along the leaf edges.
Climatic injury (temperatures too low)

230 The effects of cold on the growing parts of a plant can be catastrophic. In this case, the growth of the apex leaves was stopped. The leaflets are smaller, curled and deformed, with excessive anthocyanin.
Climatic injury (temperatures too low)

Aculops lycopersici

The mite *Aculops lycopersici* (Tryon) (tomato russet mite) and member of the Eriophyidae family, attacks several plants of the Solanacea family: tomatoes, potatoes, eggplant, pepper, tobacco, datura, petunia. It can be found in almost all production regions of the world, both in open fields and under cover.

• Nature of damage

The underside of the leaflets of the lower part of the plants takes an oily to metallic hue. Subsequently, the leaves display a bronze colour hence the origin of the disease name (Photos 231 and 232). Similar symptoms can be observed on quite significant segments of the stem (Photo 233) and on the petioles; the flowers may abort. When the attack is not controlled, plants scorch and dry out and may eventually die.

The fruits are also affected. They are discolored, and are often smaller at maturity with corky or cracked patches (Photos 234 and 235).

This damage is caused by the feeding punctures of this mite which breeds rapidly on tomato.

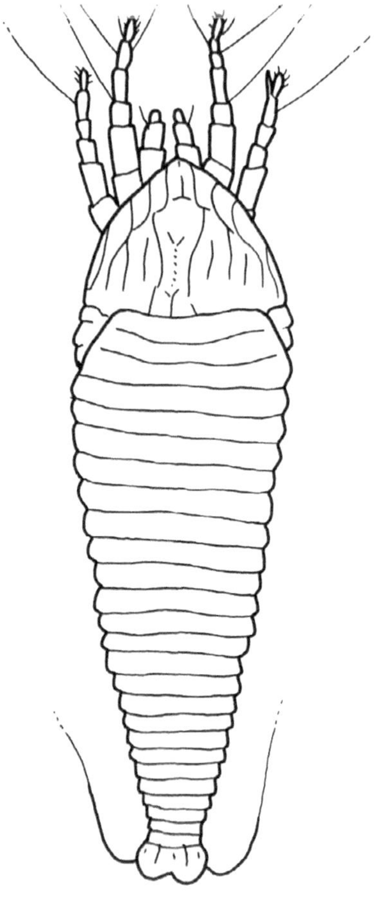

• Biology

Several developmental stages can be observed: egg, larvae, and adult.

Forms of survival and/or alternative hosts: this mite can overwinter on various surrounding weeds.

Developmental stages: the eggs, round and white, are deposited on the leaflets and the stem. The first instar larva hatches, white to yellowish, with the same appearance as the adult. These appear after the second instar. Conical in shape, they measure between 0.15 and 0.20 mm long and have a white to grey-brown or orange colour (Figure 24).

The duration of the life cycle of *Aculops lycopersici* generations fluctuates particularly according to weather conditions, between less than 1 week to more than 2 weeks.

Dispersion in the crop: this mite, which cannot fly, is released into the crop by wind, animals, and insects, but also by the workers and their tools during the farming operations.

Favourable conditions for development: the optimal development conditions of this mite are a temperature of approximately 27°C and a humidity of 30%, i.e. hot, dry weather. It also seems to cope fairly well with unfavourable weather.

Discoloration

Figure 24 Adult *Aculops lycopersici*, (tomato russet mite).

231 Many slightly chlorotic leaflets take on a metallic and bronzed shade.

234 Irregular, corky and superficial spots or patches, are visible on the fruit.

232 The underside of the leaflets appears slightly bronzed. The fruits are severely affected.

233 The stem shows a light bronze tint on one sector.

235 At maturity, the fruits are smaller, sometimes entirely corky and cracked.

Aculops lycopersici
(tomato russet mite)

SPOTS ON LEAFLETS AND LEAVES

Symptoms included in this section

- Small brown, beige spots – often necrotic in advanced stages
- Brown, beige spots – variably extending, often necrotic in advanced stages
- Yellow spots (chlorosis) of variable size
- Spots with powdery, downy patches, presence of mould

Possible causes

- *Alternata alternata* f. sp. *lycopersici*
- *Alternaria tomatophila*
- *Botrytis cinerea*
- Hyperparasitic fungi
- *Corynespora cassiicola*
- *Cristulariella moricola*
- *Helminthosporium lycopersici*
- *Leveillula taurica*
- *Mycovellosiella fulva*
- *Oidium lycopersici*
- *Oidium neolycopersici*
- *Penicillium* sp.
- *Phoma destructiva*
- *Phytophthora infestans*
- *Pseudocercospora fuligena*
- *Septoria lycopersici*
- *Stemphylium* spp.
- *Clavibacter michiganensis* subsp. *michiganensis*
- *Pseudomonas syringae* pv. *tomato*
- *Pseudomonas syringae* pv. *syringae*
- *Xanthomonas* spp.
- Various viruses (alone or in complex)
- Alfalfa mosaic virus (AMV)

- Beet western yellows virus (BWYV)
- Cucumber mosaic virus (CMV)
- Impatiens necrotic spot virus (INSV)
- Parietaria mottle virus (PMoV)
- Pepino mosaic virus (PepMV)
- Pepper veinal mottle virus (PVMV)
- Potato virus Y (PVY)
- Potato yellow dwarf virus (PYDV)
- Tobacco necrosis virus (TNV)
- Tobacco streak virus (TSV)
- Tobaco ring spot virus (TRSV)
- Tomato black ring virus (TBRV)
- Tomato chlorotic spot virus (TCSV)
- Tomato ringspot virus (ToRSV)
- Tomato spotted wilt virus (TSWV)
- Tomato torrado picorna-like viruses (ToTV)
- Mites
- Sooty mould
- Thrips
- Indefinite nonparasitic cause
- Oedema
- Various phytotoxicities
- Incompatibility reaction to cladosporium

Diagnosis sometimes difficult

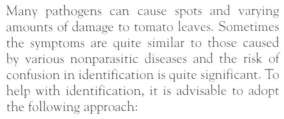

Many pathogens can cause spots and varying amounts of damage to tomato leaves. Sometimes the symptoms are quite similar to those caused by various nonparasitic diseases and the risk of confusion in identification is quite significant. To help with identification, it is advisable to adopt the following approach:

– observe several spots on leaflets, leaves, and plants in an attempt to assess the exact nature and development with time (size, colour, presence or absence of a halo, location on the plant, and distribution over the leaves; check the observation guide on p. 150);

– always carefully examine the upper and under sides of leaflets in order to note the possible presence of fungus fructifications or other signs that aid the diagnosis;

– then consult all sections of the book on spots in order to identify those most closely matching the observations. Note that these sections have been defined in part according to the spot size. Be aware that the spot size is likely to evolve over time, influenced by ambient humidity, especially when caused by phytopathogenic fungi or bacteria. Thus, during particularly wet conditions, their growth may be abnormally fast, while following dry periods, their development will be more limited than usual, which may mislead and refer the observer to the wrong section.

Figure 25 Possible appearance, development, and distribution of spots and changes on tomato leaves and leaflets.

1 Small brown dots or spots, without halo.

2 Small brown spots with yellow halo.

3 Powdery, fluffy spots.

4 Yellow spots or patches sometimes with well defined borders.

5 On the left, rounded brown spots; on the right, angular yellow spots delimited by veins.

6 Spots with brown and necrotic patches, sometimes with concentric patterns or chlorotic halo; some merge.

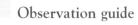

Observation guide

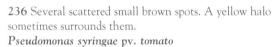

Spots

236 Several scattered small brown spots. A yellow halo sometimes surrounds them.
Pseudomonas syringae pv. *tomato*

238 Many yellow spots with diffuse edge.
Mycovellosiella fulva

240 This leaflet is partially covered by a dusty white powder.
Oidium neolycopersici

237 Two types of rather large spots are visible on this leaflet:
– a dark brown to black spot with concentric lines, surrounded by a yellow halo (**Alternaria tomatophila**);
– some more extensive brown discoloration demarcated by a darker border and sometimes localized on the leaf edges (*Phytophthora infestans*).

239 A beige to light brown necrotic area extends to the tip of this leaflet and a yellow edge marks the progression of necrosis.
Botrytis cinerea

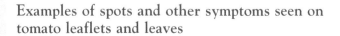

Examples of spots and other symptoms seen on tomato leaflets and leaves

151

241 Small brown spots, round or irregularly shaped, with a fairly pronounced yellow halo.
Pseudomonas syringae pv. *tomato*

243 Numerous necrotic lesions, small-sized, rather linear or ring-shaped.
Tomato spotted wilt virus (TSWV)

242 Grey to brown dull spots, of various shapes.
Stemphylium sp.

244 Tiny brown lesions, with chlorosis on the edges, mainly located near the veins.
Phytotoxicity

Examples of small necrotic spots

■ Small brown, beige spots often necrotic in advanced stages

Possible causes

- *Alternata alternata* f. sp. *lycopersici*: Description 9
- *Corynespora cassiicola*: Description 9
- *Helminthosporium lycopersici*
- *Mycovellosiella fulva*: Description 4
- *Phoma destructiva*
- *Septoria lycopersici*: Description 9
- *Stemphylium* spp.: Descrption 8
- *Penicillium* sp.
- *Clavibacter michiganensis* subsp. *michiganensis*: Description 24
- *Pseudomonas syringae* pv. *tomato*: Description 21
- *Pseudomonas syringae* pv. *syringae*: Description 21
- *Xanthomonas* spp.: Description 22
- Various viruses (alone or in complex)
- Alfalfa mosaic virus (AMV): Description 33
- Beet western yellows virus (BWYV): Description 34
- Cucumber mosaic virus (CMV): Description 35
- Impatiens necrotic spot virus (INSV): Description 44
- Parietaria mottle virus (PMoV): Description 44
- Pepino mosaic virus (PepMV): Description 29
- Pepper veinal mottle virus (PVMV): Description 38
- Potato virus Y (PVY): Description 36
- Potato yellow dwarf virus (PYDV): Description 45
- Tobacco necrosis virus (TNV): Description 46
- Tobacco streak virus (TSV): Description 44
- Tobaco ring spot virus (TRSV): Description 46
- Tomato black ring virus (TBRV): Description 46
- Tomato chlorotic spot virus (TCSV): Description 44
- Tomato ringspot virus (ToRSV): Description 46
- Tomato spotted wilt virus (TSWV): Description 43
- Tomato torrado picorna-like viruses (ToTV, see p. 149)
- Oedema (see pp. 65 and 197)
- Various phytotoxicities
- Thrips

In a culture, the tomato plant may have various spots to a greater or lesser extent, especially on younger leaves as these are often softer and therefore more susceptible. These spots are often first oily, sometimes slightly chlorotic. Subsequently, they are a light or darker brown and become necrotic. Their exact identification is, in many cases, difficult.

This section and *Table 20* (p. 167) summarize the major parasitic or abiotic diseases that may be suspected. Risk of confusion is significant: it is advisable also to check the next section. Moreover, in some cases, microscopic observations will be needed to detect structures, fungal spores on and in damaged tissues: it will then be necessary to contact a specialized laboratory.

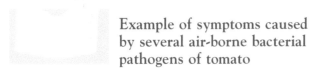

245 A well marked yellow halo surrounds brown spots, limited in size, visible in several leaflets. Note that a fruit displays black pinhead lesions reminiscent of fly specks.
Pseudomonas syringae pv. *tomato*

247 Brown necrotic lesions, especially localized at the edge of the leaves, affecting many leaflets.
Pseudomonas syringae pv. *syringae* (bacterial speck)

246 Many of the spots covering these leaflets are necrotic. Cankerous corky lesions, 3–4 mm in diameter, are visible on one of the fruits.
Xanthomonas sp. (bacterial spot)

248 Numerous small whitish cracks cover part of the upper side of this leaflet.
Clavibacter michiganensis subsp. *michiganensis* (bacterial canker)

249 This fruit has many localized white spots which are darker at the centre. These lesions have the appearance of a bird's eye.
Clavibacter michiganensis subsp. *michiganensis* (bacterial canker)

Example of symptoms caused by several air-borne bacterial pathogens of tomato

Additional information for diagnosis

Leaf bacterial diseases

At least seven phytopathogenic bacteria may cause small leaf spots on tomato including: *Pseudomonas syringae* pv. *tomato*, *P. syringae* pv. *syringae*, *Xanthomonas* spp., and *Clavibacter michiganensis* subsp. *michiganensis*.

Pseudomonas syringae pv. *tomato*, responsible for bacterial speck, causes small, irregular, brown spots 1–2 mm in diameter (Photos **245, 250–252**). These spots are gradually surrounded by a defined yellow halo (Photo **253**). If conditions are particularly wet, they expand, fuse together and affect whole sectors of the leaf. Brown elongated lesions may develop on petioles, stems, and inflorescences initially, causing, in the latter case, flowers to fall. The fruits present particularly characteristic tiny black spots, similar to flyspecks (Photo **254**).

Pseudomonas syringae pv. *syringae* (syringae leaf spot) causes small brownish to black leaf spots fairly comparable to those of bacterial speck (Photo **247**). A yellow halo sometimes surrounds these spots. This bacterium is also associated with leaf necrosis, particularly that localized at the periphery of the leaf.

Several species of *Xanthomonas* have relatively recently been associated with bacterial spots: *X. euvesicatoria*, *X. vesicatoria*, *X. perforans*, and *X. gardneri*. These bacteria cause small moist spots at first, soon becoming brown to black and not exceeding 2–3 mm in diameter (Photo 246 and 256). They also feature a yellow halo, which is slightly more discreet. Some elongated moist brown symptoms are also visible on petioles and stems (Photo **257**). The lesions on the fruit are very characteristic and are translucent at first, then cankerous becoming gradually corky (Photos **255, 258,** and **259**). Be aware that similar symptoms can be caused by hail damage (see p. 382).

Clavibacter michiganensis subsp. *michiganensis*, is a vascular bacterium responsible for bacterial canker which causes whitish cankerous spots on the leaflets (Photo **248**). Similar symptoms – spots like a bird's eye – are found on the fruit (Photo **249**). See p. 325 for more information on this disease which also invades the vascular system of tomato and causes much larger damage of a different nature.

Thus, these bacteria cause, with the exception of *Clavibacter michiganensis* subsp. *michiganensis*, almost identical foliar symptoms on all aerial organs of the tomato, except the fruit. When faced with brown spots on leaflets, the fruits should be observed very carefully. Other diagnostic criteria are available in *Table 20* (p. 167).

Note that two other bacteria, *Pseudomonas viridiflava* and *Pseudomonas cichorii* in Florida cause bacterial blight on tomato foliage during summer in very wet growing conditions.

Spots

250 Numerous brown to black spots are scattered over a large part of these leaflets.

252 The spots are brownish, rather irregularly rounded.

251 The spots are also visible on the underside of the leaf. A fairly marked yellow halo surrounds some of them.

253 The affected leaflets eventually turn yellow and wither.

254 Tiny rounded black pustules, generally less than 3 mm in diameter, similar to flyspecks, are visible on green fruit.

Pseudomonas syringae pv. *tomato*
(bacterial speck)

255 On this plant, the leaflets, the remaining petioles, and fruit show necrotic lesions of limited size.

257 More elongated, brown to black lesions are seen on stems and petioles.

256 On the leaflets, brown spots are scattered, some slightly angular. There may be chlorosis near these lesions.

258 Round brownish spots, 4–5 mm in diameter, develop on the most exposed parts of the fruit. An oily halo may surround them.

259 Ultimately, the lesions are corky and cracked (this symptom is reminiscent of scab on apple).

Xanthomonas spp.
(bacterial spot)

260 Dull brownish spots are randomly distributed on the leaflets.

262 As it ages, the leaf gradually turns necrotic and sometimes cracks. Note that some spots have converged at the tip of the leaflet.

261 Developing spots are small and brownish. Their centre gets lighter and takes a greyish colour, hence the name of this fungus.

263 In conditions of high humidity, the size of some spots may be abnormally large. This situation is particularly encountered in the tropics and may lead to confusion with other foliar diseases. Note that the leaves gradually turn yellow and may dry out.

Stemphylium spp.
(grey leaf spot)

Stemphylium spp.

Several species of *Stemphylium* are reported on tomato leaflets especially in various humid production areas of the world: *Stemphylium solani*, *S. lycopersici* (syn. *S. floridanum*), *S. vesicarium*, and *S. botryosum* f. sp. *lycopersici*.

These fungi cause the same symptoms on all the Solanaceae, namely spots developing on all leaflets, very rarely on petioles and stem, never on fruits. These spots are brownish at first, then accompanied by a slight chlorosis at the periphery (Photos **260** and **261**) and measure about 2 mm in diameter. Their centres often gradually become lighter, dry out, and can crack (Photo **262**). They also have a dull appearance. In wet conditions, they can spread considerably and/or coalesce, giving them a less characteristic appearance (Photo **263**).

The identification of *Stemphylium* grey leaf spot is quite difficult because its symptoms are highly reminiscent of those of the previously described bacterial diseases, but also of those of several fungal diseases that are discussed in the next section (*Alternaria*, *Septoria* spp.).

The only way to identify it with certainty is to look for the spores of *Stemphylium* spp. on the underside of the leaf spots using a binocular microscope and/or a light microscope. These are quite characteristic and substantially different (by their appearance and size) depending on the species involved (Photos **264–269**). If such instruments are not available, *Table 20* (p. 167) should be consulted and/or a specialized laboratory contacted that will make the observations.

Note that among the species of *Alternaria* spp. pathogenic on tomato, only two induce foliar spots: A. *tomatophila*, which affects all aerial parts of this plant (see p. 173), and the specialized form of A. *alternata* (*Alternaria alternata* f. sp. *lycopersici*). The latter mainly attacks the rare susceptible varieties which do not possess the incompletely dominant allele 'Asc' in the homozygous state. For these varieties, the leaf symptoms are secondary to a canker present in the lower part of the stem, from which spreads a toxin. This results in chlorosis of the leaf and the development of widespread inter-vein brown to black, necrotic, angular spots.

Lesions also form on the fruit; see p. 175 for guidance.

Also, a *Helminthosporium*, *Helminthosporium lycopersici* Roldan (and another species that does not appear to be synonymous: *Helminthosporium lycopersici* Maubl. & Roger), has very occasionally been reported in various places, including Africa, the Philippines, and India. It may be responsible for leaf spots occurring on tomato seedlings. Spots of a few millimetres in diameter, rather circular, grey to brown, form on the leaflets and are the cause of partial defoliation. It has also been associated with rot on fruit. In general, brown and short conidiophores, bearing brown, elongated conidia which have 7–14 sections, are visible on the lesions (Figure **26**).

Spots

264

265

266

267

268

269

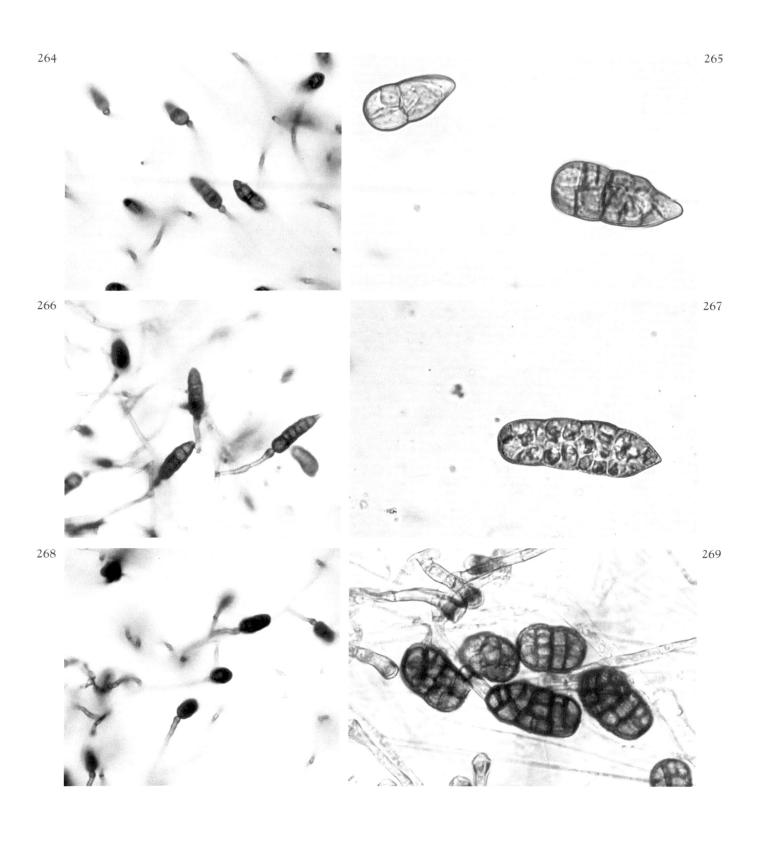

 See captions opposite

Identification of the spores of different *Stemphylium* species present on leaf spot (grey leaf spot)

264 There are a few brown and pointed conidia among brown conidiophores (130–200 × 4–7 μm).
Stemphylium solani

266 These conidiophores have cell walls and are olive coloured and with a swollen apex. They measure 75–300 × 3–5.5 μm. Conidia are also pointed but they are longer than those of *S. solani*.
Stemphylium lycopersici **(S. floridanum)**

268 Conidiophores are quite comparable to those of the previous species, they have single conidia, brown in colour, which are rounded at their ends.
Stemphylium vesicarium

265 Muriform conidia are very brown when mature, rounded at one end, and pointed at the other end. Their average dimensions are about 48 × 22 μm.
Stemphylium solani

267 The conidia are olive coloured and also muriform. They are pointed at one end and are longer (19.9–62.2 × 4.6–23 μm) with more numerous and pronounced cells.
Stemphylium lycopersici **(S. floridanum)**

269 Heavily melanized, multicellular conidia are rather rectangular and measure 14–41 × 9–26.6 μm.
Stemphylium vesicarium[1]

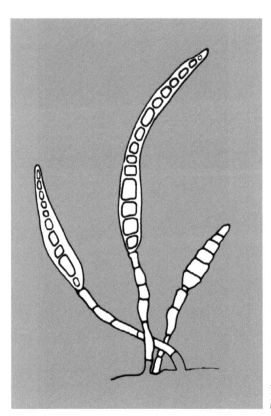

Figure 26 *Helminthosporium lycopersici* fructifications.

---

[1] Identification of *Stemphylium solani* and *S. lycopersici* is not difficult; this is not the case with *S. botryosum* f. sp. *lycopersici*. Indeed, many observations made on strains from different areas of production, including the Mediterranean, show that it may be confused with *S. vesicarium*, a species with slightly less elongated and more melanized conidia.

270 This plant, whose growth is stopped, displays numerous small necrotic lesions on the leaflets of young leaves.

272 Brown lesions cover the whole of this leaf; some, rather elongated lesions are visible on petioles.

271 These brown and diffuse necrotic lesions frequently have a fairly widespread distribution on the leaflets and leaves, which differentiate them from those of bacterial diseases.

273 Sometimes the lesion distribution is related to the main veins. In this example, numerous brownish necrotic lesions have spread from the larger veins.

274 The network of secondary veins is also affected, and this is readily seen on the underside of this leaflet.

Potato virus Y (PVY) (necrotic strain)

Various viruses

Among the symptoms caused by viruses are spots or lesions of various kinds. In this section we have chosen three viruses whose symptoms show the nature and type of small lesions (equivalent to spots) that viruses can cause on the tomato leaflets. A fairly comprehensive list of viruses capable of causing lesions on the leaf of tomato and some of their characteristics are listed in *Tables 18* and *19* (p. 166). The sections dealing more specifically with these micro-organisms should also be checked on pp. 59, 97, 141, and 223.

Table 18 Common viruses causing spots, rings, and chlorotic and/or necrotic lesions on leaflets and leaves of tomato

Virus	Symptoms	Vectors
Alfalfa mosaic virus (AMV) Description 33	Spots, necrotic lesions starting from the base of the leaflets and spreading to the whole leaf (see p. 224). The veins themselves are sometimes necrotic. These develop on the stem, and extend over several centimetres affecting the apex. A yellow mosaic of the aucuba type can sometimes be found on tomato leaflets. Brown necrotic lesions, are visible on the fruit. Fruits may also be lumpy.	Transmitted by several species of aphids in the nonpersistent manner: *Myzus persicae, Aphis fabae, A. gossypii, A. craccivora*.
Cucumber mosaic virus (CMV) Description 35	The nature of the symptoms varies with strains and conditions and include: – spots and necrotic lesions, which start on the leaflets and reach the stem and the apex of plants; – mosaic and leaf deformation (see p. 101). Severe necrotic brown lesions, round and raised, which cover large areas of fruit.	Transmitted by aphids in the nonpersistent manner. More than 90 species are capable of acquiring and transmitting it, including: *Myzus persicae, Aphis gossypii, A. craccivora, A. fabae, Acyrthosiphon pisum*.
Potato virus Y (PVY) Description 36	Symptoms vary depending on the strain: but include reddish brown necrotic spots, necrotic longitudinal streaks which can be seen on the petioles and stem (necrotic strains [Photos **270–274**]). Discreet mottle, green mosaic, vein banding, diffuse chlorotic spots (common strains) (see p. 103).	Transmitted by dozens of aphids in the nonpersistent manner, including: *Myzus persicae, Aphis gossypii, A. fabae, Acyrthosiphon pisum, Macrosiphum euphorbiae*.
Tomato spotted wilt virus (TSWV) Description 43	Spots, rings with chlorosis or brown colour, becoming gradually necrotic on the leaflets (Photos **275–277**). Reddish brown patches are also visible. The leaves can be mottled (with mosaics) and/or chlorotic and necrotic; they bronze gradually and ultimately become necrotic. Brown elongated lesions (streaks) may appear on petioles and stem. The tip of the stem sometimes curves. Large patches, rings, arcs partially cover the fruits (Photo **278**). These can be locally discolored, bronzed, and somewhat deformed.	Transmitted by thrips in the persistent manner. Nine species are vectors, including: *Frankliniella occidentalis, F. fusca, F. schultzei, Thrips palmi, T. tabaci, Scirtothrips dorsalis*.
Pepino mosaic virus (PepMV) Description 29	Necrotic spots on lower leaves and extended 'scorched' zones (Photos **279–282**). Brown, corky lesions on stems and petioles. Mosaic and leaf deformation, see p. 103.	Transmitted by casual contact, during the cultural operations, by pollinating bees, and by seed.

Spots

275 These leaflets with chlorosis show small dark brown necrotic lesions especially in the immediate vicinity of the veins.

277 In some situations, regular brown rings form on the leaf. Sometimes they are concentric and complete.

276 On this leaflet, the changes are fewer, larger, brown to brownish.

278 Some large white rings and corky cracks are visible on the fruit.

Tomato spotted wilt virus (TSWV)

281

282

279 The reduced sized leaflets are mottled with a scattering of small discrete brown lesions.

281 Longitudinal brown corky lesions are visible along the petiole.

280 The lesions are small, reddish-brown, and near the veins.

282 Some fairly extensive areas of the leaf are 'scorched', giving the impression of having been burned.

**Pepino mosaic
virus (PEMV)**

Table 19 Main viruses causing spots, rings, and chlorotic and/or necrotic lesions on leaflets and leaves of tomato

Virus	Symptoms	Vectors
Beet western yellows virus (BWYV) Description 34	Small chlorotic spots, inter-vein yellowing of the leaflets. Older leaves are deformed.	Transmitted by several aphid species in the persistent manner (mainly *Myzus persicae*, *Aphis gossypii*, *Macrosiphum euphorbiae*).
Impatiens necrotic spot virus (INSV)	Spots and brown inter-vein necrotic lesions.	Thrips: *Frankliniella occidentalis*, *F. fusca*.
Paretaria mottle virus (PMoV) Description 44	Spots, necrotic lesions, mainly at the base of the leaflets. Longitudinal necrotic lesions on stems, terminal bud death. Deformation of and chlorotic rings on fruits, later becoming brown and corky and causing the deformation. Strain TI12 isolated in Italy, causes plant dwarfing.	Probably transmitted by thrips and pollen.
Pepper veinal mottle virus (PVMV) Description 38	Chlorotic to necrotic spots, and mosaic on leaflets. These are also deformed; longitudinal necrotic lesions may develop on the stem.	Transmitted by several aphids in the nonpersistent manner (*Aphis gossypii*, *A. craccivora*, *Myzus persicae*).
Potato yellow dwarf virus (PYDV) Description 45	Chlorotic spots gradually spreading, inter-vein whitening on young leaflets.	Transmitted by leafhoppers in the circulative propagative manner (*Agallia constricta*, *A. quadripunctata*, *Aceratogallia sanguinolenta*).
Tobacco necrosis virus (TNV) Description 46	Necrotic rings on leaflets.	Transmitted by the fungus *Olpidium brassicae*.
Tobacco ring spot virus (TRSV) Description 46	Chlorotic spots and rings located near the veins of the leaflets. These gradually become necrotic. Mottling can be seen on the leaf.	Transmitted by several nematode species of the genus *Xiphinema*. This virus is reported to be transmissible by pollen in other plant species.
Tobacco streak virus (TSV) Description 44	Spots with chlorotic rings on leaflets, becoming progressively necrotic, especially near the veins. Necrotic streaks are visible on the stem, and extend to the branches. The flowers, also necrotic, may fall. As the leaflets, the fruits are sometimes covered with necrotic rings.	Transmitted by several species of thrips (*Thrips tabaci*, *Frankliniella* spp.). It is also transmitted in seed in beans, and by several weeds such as *Datura stramonium*, *Chenopodium quinoa*, *Melilotus albus*. Mechanical transmission via contaminated pollen.
Tomato black ring virus (TBRV) Description 46	Spots of varying sizes, chlorotic and necrotic rings. The leaflets can be deformed with chlorotic veins. Plant growth is sometimes reduced if infection occurs early in the growth of the plant.	Transmitted by nematodes: *Longidorus* spp. (*L. elongatus*, *L. attenuatus*). It is transmitted by seed in several species, as well as by pollen.
Tomato chlorotic spot virus (TCSV) Description 44	Chlorotic spots on leaflets. Yield reductions.	Transmitted by thrips (*Frankliniella occidentalis*, *F. fusca*).
Tomato ringspot virus (ToRSV) Description 46	Small necrotic lesions, mottling on leaflets. Systemic necrotic lesions.	Transmitted by nematodes (*Xiphinema americanum* and its subspecies). Seed-borne in other species.

Table 20 Some diagnostic criteria for identifying key biotic or abiotic agents responsible for small brown leaf spots on tomato leaflets in Europe

	Pseudomonas syringae pv. *tomato*	*Xanthomas* spp.	*Stemphylium* spp.	Various viruses	Various phytotoxicities
Presence	+	+	+/–	+	+
Frequency on tomatoes	Fairly common in open fields and under protection in cold conditions.	Fairly common in open fields.	Very rare, *S. vesicarium* observed very occasionally under protection.	Quite common: – PVY + + (open fields or under protection); – TSWV + (open fields or under protection); – PEMV + / – (under protection).	Fairly frequent in open fields but more common under protection.
Localization of symptoms on the plant	Randomly on all aerial organs.	Randomly on all aerial organs.	Primarily on the leaves.	On young leaves (systemic aspect of the disease), on the stem, fruit.	On the young leaves on one side of the plants.
Possible presence of other symptoms	–	–	–	+ (see pp. 97, 141, and 223)	+ (see p. 77 and p. 131)
Period of symptoms onset	Spring, summer	Summer	Summer	Spring, summer	All seasons
Speed of symptom progress in wet periods	++	++	+	–	–
Presence of specific structures on spots	–	–	+ Conidiophores and multicellular muriform conidia muriformes.	–	Deposits sometimes occur.
Possible varietal resistance	+	–	+	– (+ for TSWV in particular)	– Differences in sensitivity between genotypes are possible.

Spots

167

Various phytotoxicities

It is necessary to be particularly careful when small lesions are detected on tomato leaflets. Indeed, the leaf tissues, especially when they are young or from plants grown under protection, are sometimes soft and fragile. Also, some products (especially pesticides) can cause problems when used under certain conditions and be phytotoxic to tomato (chemical injuries). Various necrotic lesions can then appear on the leaflets, visible on both the upper and lower sides of the leaves (Photos **283** and **284**). It is possible to confuse these symptoms with some spots described in this section. If in doubt, it is wise to confirm the diagnosis by consulting a specialist laboratory. For more information about the phytotoxicities, see pp. 77 and 131.

Penicillium sp.

In situations where controlling whiteflies in particular is not satisfactory, it is not uncommon to see beige to brown spots on leaflets associated with a mould in addition to sooty mould (see p. 207 and Photos **285** and **288**). Their size is rather limited and their location very specific. In addition, larvae and pupae and many black parasitized whiteflies (Photo **286**) are visible on the underside of the leaf. Some of them are covered by sporulation of a fungus (Photo **287**). The latter is a *Penicillium* sp. using whitefly pupae as a nutrient base to establish on the leaflets before colonizing them.

These spots are harmless to plants as they will be more affected by the presence of large populations of whiteflies than by the moulds. If it is not too late, the control of these pests is the only satisfactory control measure.

283 A multitude of small spots, reddish brown to brown and irregularly shaped, cover the upper side of this leaflet. They are surrounded by yellow halo of differing intensity.

284 On the underside of the leaf, the lesions are duller and have a more metallic appearance.

Example of leaf spots induced by phytotoxicity (chemical injuries)

285

286

287

288

Spots

285 Rounded spots, with well-defined edges which are beige to brownish, are visible on this leaflet.

287 Examination with a dissecting microscope shows some tufts characteristic of *Penicillium*.

286 A greyish to blue mould is present on the underside of the spots.

288 Ultimately many leaflets display some brown spots, some with chlorosis, partially dried and/or covered with sooty mould.

Penicillium **sp.**

289

290

292
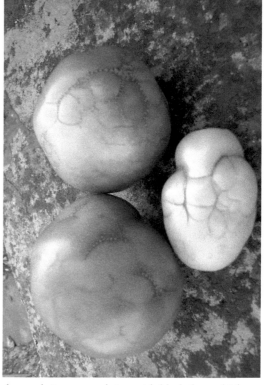

291

289 Many necrotic lesions of varying size cover this leaflet.

291 Several diffuse white halos, rather circular, are visible on the green fruit. Note the presence of a punctiform lesion in each of them corresponding to the oviposition site of *F. occidentalis*.

290 The lesions have a beige to metal tint, with black dots, which are *F. occidentalis* droppings. Note that a few tiny thrips are visible in places.

292 Three fruits affected early on by thrips. They reveal many linear patterns, superficially corky, sinuous to circular, which are the cause of the partial deformation of some tomatoes.

Frankliniella occidentalis — thrips

293 Two eggs are visible under the leaf.

294 A larva walks on the underside of this leaflet.

295 Adult *F. occidentalis* are 0.8–1 mm long. These can cause brown/beige spots which can be extended and are often necrotic at the later stages.

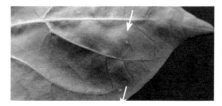

Thrips

Thrips, which belong to the order of Thysanoptera and the family Thripidae, are distributed worldwide, present in tropics to polar regions, depending on the species. Several of them attack vegetable crops, particularly tobacco thrips (*Thrips tabaci* Lindeman) and californian thrips (*Frankliniella occidentalis* Pergande). The first has been a problem for many years in Europe. They have been gradually supplanted in importance by the latter that has caused significant damage on many crops, especially in glasshouses, since the early 1980s (cucumber, pepper, melon, eggplant, chrysanthemum, rose). Both thrip species may attack tomatoes. Note that they are vectors of several serious viruses, particularly Tomato spotted wilt virus (see Descriptions 43 and 44). There are other thrips species that attack tomato including, *Thrips palmi* Karny, *Heliothrips haemorrhoidalis* (Bouché), *Frankliniella* spp.

• Nature of damage

Silver lesions, irregular in shape and size, appear on the leaf, and they gradually become necrotic and take on a beige shade (Photos **289** and **290**). These lesions are also dotted with tiny black spots which are the faeces of thrips (Photos **289** and **290**). The leaflets tend to be affected by chlorosis and take on a dull shade. Flowers might fall. Oviposition sites of *F. occidentalis* on young fruit appear as brown dot lesions, which are surrounded by a broad sub-epidermal whitish ring (Photo **291**). Sometimes deformed fruits show linear corky lesions (Photo **292**).

• Biology

The thrips development includes six stages: egg, two larval stages, two pronymphs stages, and adult stage (Figure 27). The lifecycle varies with temperature and host plant; for example, for *F. occidentalis* it varies from 34 days at 15°C to 13 days at 30°C.

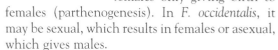

Figure 27 Lifecycle of the thrip.

The larvae and adults survive on alternative hosts or they may take refuge in places with mild temperatures. In this respect, they may use plant debris in glasshouses, and have been found to burrow up to 8 cm deep. They may also hibernate in field crops, including in various *Allium*.

Developmental stages: reniform eggs (1, see Photo **293**) are deposited on the aerial organs, especially the leaves of tomato. Once formed, the larvae (2 and 3, see Photo **294**), which are highly mobile, will feed on the underside of leaves. At the end of the second instar (3), they drop to the ground and pupate. The first pronymph stage (4) is characterized by the appearance of wing primordia. The second pronymph stage (5), clearer, also shows more substantial wing primordia and long antennae curved toward the rear of the body. Adults (6) have two pairs of well developed wings.

Thrips tabaci and *Frankliniella occidentalis* (Photo **295**) are morphologically quite similar. Only microscopic observation of the adults can distinguish them. Their reproduction may be different depending on the species. It is asexual in *T. tabaci*, with unmated females only giving birth to females (parthenogenesis). In *F. occidentalis*, it may be sexual, which results in females or asexual, which gives males.

Thrips feed by sucking the contents of epidermal cells. Damaged tissues quickly become necrotic.

Dispersion in crops: Thrips are spread easily in crops, either by air currents and/or actively by flying. Workers can contribute during cropping operations. These insects can be spread by seed or plants of other affected species.

Favourable developmental conditions: The development of thrips is especially influenced by the plant species, the temperature, and the relative humidity in the crop.

297 A few large spots, brownish with a pale edge, are clearly visible on several leaflets.
Phytophthora infestans

296 Brown spots of varying size and with a well defined edge affect a significant part of this leaflet. A well marked yellow halo surrounds them.
Alternaria tomatophila

298 A brown symptom, starting on the tip of the leaflet, affecting a significant part.
Botrytis cinerea

Examples of necrotic spots

Brown and beige discoloration, often with necrosis in advanced stages

Possible causes

- *Alternata alternata* f. sp. *lycopersici* (see p. 309): Description 9
- *Alternaria tomatophila*: Description 1
- *Botrytis cinerea*: Description 2
- *Corynespora cassiicola*: Description 9
- *Cristulariella moricola*
- *Phoma destructiva*
- *Phytophthora infestans*: Description 7
- *Septoria lycopersici*: Description 9
- *Pseudocercospora fuligena*: Description 9
- Indeterminate nonparasitic affliction
- Various phytotoxicities (see p. 168)

As in the previous section, the spots described in this section are oily or even water soaked at first, and tend to expand thereafter. They may also be slightly chlorotic and/or with a yellow halo. In late stages, they often become lighter or darker brown and become necrotic.

The cause of these spots is also difficult to determine, because they can be confused with those caused by various fungal pathogens. Microscopy will often be necessary to examine the morphology of fungal structures in and on damaged tissues, in which case a specialized laboratory should be contacted.

Also, the previous section should be checked.

Spots

Additional information for diagnosis

Alternaria tomatophila

This fungus causes early blight, mainly on the lower leaves of tomato and very frequently affects field crops.

Initially, the spots may have a rounded shape, and are of limited size. Their colour is dark brown to black (Photos **299** and **300**). Thereafter, they gradually spread and a well defined yellow halo appears (Photo **301**). They may be up to several millimetres in diameter and have discrete concentric rings of a darker brown colour (Photo **303**). When they appear near the main veins, the distal leaf yellows first, then gradually becomes necrotic (Photo **299**). During severe outbreaks associated with periods of high humidity, the numerous lesions greatly reduces the leaflets' photosynthetic ability.

Note that conidiophores and conidia of the fungus can be seen on the underside of the leaves, on the affected tissues, with the aid of a binocular or light microscope (Photo **302**).

Alternaria tomatophila also attacks other aerial parts of tomatoes such as flower stalks (Photo **304**), petioles (Photo **306**), and stems, which show dark brown to black lesions. These are rather elongated, show concentric rings, and are lighter in the centre. When the stem is 'girdled' the distal branch withers.

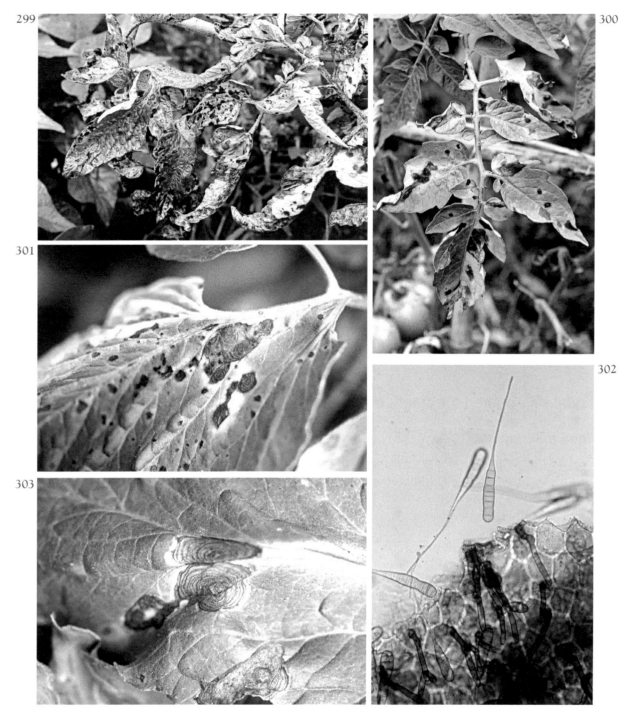

299 All these leaflets are covered with brown spots to a various degree.

301 The spots, black to brown, converge; some have a vivid yellow halo.

303 Darker concentric rings are clearly visible within the spot.

300 On some leaflets, spots localized near the main veins cause the yellowing and necrosis of large areas of the leaf.

302 The conidiophores and conidia of A. *tomatophila* are observable by light microscopy on the damaged tissues. The spores of this multicellular and elongated fungus are single with a filiform hyaline extension.

Alternaria tomatophila
(early blight)

304 A moist and black lesion that developed on the inflorescence is responsible for the fall of several flowers.

306 On stems, lesions are elongated and brighter in the centre; they also show concentric rings.

305 A brownish necrotic sepal colonized by A. *tomatophila*. The latter will now be able to colonize the green fruit.

307 Note on two of the fruits the circular lesions around the peduncular end of the fruit. They are concave and covered with a compact black mould.

308 A brown spot with concentric rings developed on one side of the fruit. Its location is unusual.

Alternaria tomatophila
(early blight)

Figure 28 Several circular spots of *S. lycopersici* spot are on this leaflet. The centres are pale while the periphery remains brown. Black dots, pycnidia, are visible

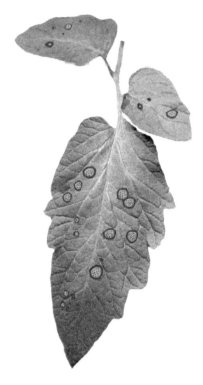

Septoria lycopersici
(Septoria leaf spot)

Figure 30 A few rare brown spots, with a well defined edge and surrounded by a yellow halo and concentric patterns are visible on this leaflet. These spots are similar to those caused by *Alternaria tomatophila*.

309 When numerous, the spots may converge and cause damage to extensive areas of the leaf. Note that a yellow halo, sometimes marked, surrounds the lesions.

Figure 29 *Septoria lycopersici* forms black globular pycnidia, within which multicellular hyaline and elongated spores are formed.

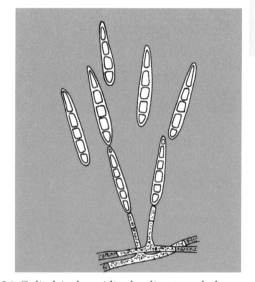

Figure 31 Cylindrical conidia, hyaline to pale brown, are easily visible on the damaged tissues. Note that they have many pseudo-walls.

Corynespora cassiicola – **target spot**

176

Green or ripe fruits are also affected. The fungus first colonizes the sepals and/or the peduncle scar (Photo **305**); colonized tissues gradually turn black. Subsequently, the fungus reaches and invades the fruit, causing extensive circular concave lesions, sometimes with a ridged surface and a rather hard texture (Photo **307**). In the end a dense black mould covers them, which is the sporulating structures of *A. tomatophila*. Rarely, lesions may appear on the side of fruit (Photo **308**; see also p. 398).

Note that *Phoma destructiva* Plowr. (*Diplodina destructiva* [Plowr.] Petr., [1921]) sometimes causes irregular dark small spots on leaflets which can be confused with those caused by *A. tomatophila*. Subsequently, they expand and become necrotic. Concentric patterns are also visible, as well as tiny brownish globular structures – the fructifications of the fungus (pycnidia). Severely affected leaves eventually turn yellow and wither. Large black lesions may develop on the stem and fruit. For more information about this pathogen, see p. 396.

Another fungus, generally in the subtropics on large scale commercial crops, causes large concentric leaf spots (zonate leaf spot). This is *Cristulariella moricola* (Hino) Redhead (syn. *C. pyramidalis* Waterman & Marshall). It is mainly reported on tomato in several states of the US. Unlike *Alternaria tomatophila*, concentric rings are symmetrical without chlorosis, and lesions are not located on the lower leaves. The fruit can rot and drop prematurely.

Septoria lycopersici

Present on virtually every continent, this fungus causes symptoms that are sometimes confused with those caused by *Alternaria* spp. Like the latter, Septoria leaf spot occurs primarily on older leaves (Figure **28**). Small circular spots, moist at first, appear on the leaves. They expand slowly (2–6 mm diameter) and go brown. A chlorotic halo can surround them (Photo **309**). Their centre becomes necrotic rather quickly and takes a tanned to white hue. Tiny black fructifications of the fungus (pycnidia) can be seen within the affected leaf tissue through a microscope (Figure **29**).

If conditions are particularly wet, the disease spreads to the young leaflets. Some spots con-verge, resulting in severe damage to whole sectors of the leaf and ultimately in leaf fall. Darker and less extensive similar lesions are reported on other aerial parts of tomates, including the stems and fruit stalks. Symptoms on fruit have not been described.

Contrary to early blight, spots caused by Septoria are more regular and numerous. Their centres are clearer and do not show concentric patterns. The primary lesions of this disease should not be confused with those caused by the two tomato aerial bacterial diseases: tomato speck and spot in tomato (see p. 155).

It occurs in Italy and Germany as well as in eastern Europe.

Corynespora cassiicola

Generally confined to tropical production areas, this fungus is the cause of small, moist lesions on the upperside of the leaf that may be similar to those induced by *Alternaria* spp. They have a radial growth and are sometimes limited locally by a vein. They are rather circular, can reach 2 cm in diameter, and are tinged with a visible yellow halo. Concentric patterns are also present in some spots and it is this symptom that gives the disease its common name of target spot (Figure **30**).

Brown longitudinal lesions also occur on stems and petioles. They are sometimes completely surrounded, resulting in desiccation of leaflets and leaves.

Indeterminate necrotic spots

310 Numerous rounded brown metallic spots and concentric patterns, have developed between the veins of the leaf. Note that they have a uniform size.

311 In some cases, the leaf lesions are more extensive with chlorosis.

312 Large irregular and necrotic spots have formed on the leaf in places where fertilizer particles were deposited.

313 Spray droplets of a contact herbicide locally damaged the leaf, causing several necrotic spots which are light beige, round, and well defined.

Various phytotoxicities (chemical injuries)

The fruit symptoms vary depending on their stage of development:

– On young fruit, the lesions are small, light brown, have a darker edge, and have a dry consistency;

– On mature fruit, large circular spots develop. They are slightly depressed, and eventually turn brown and split in the centre. The fungus spores prolifically on the lesions, giving them a dark grey to black colour (Figure **31**).

As with Septoria leaf spot, the primary symptoms of target spot should not be confused with those of speck and bacterial spot (see p. 155). Note that this fungus is not reported in Europe.

Indeterminate necrotic spots

For many years, field and under protected tomato growers have occasionally observed dark spots, which soon turn brown and become necrotic similar to those caused by *Alternaria tomatophila*. They also have concentric patterns and a chlorotic halo (Photos **310** and **311**). Numerous plants and leaflets are affected. However, contrary to early blight and other similar foliar diseases, stems and fruits are not affected. Surprisingly, the spots always appear simultaneously in several crops in the same area but sometimes several miles away and in other centres of agricultural activity. Note that laboratory investigations have failed to identify a pathogen

The disease pattern is quite characteristic of a nonparasitic problem occurring in a limited sequence in time but with much more extensive distribution. Various studies have been conducted to investigate causes (climatic problem, pollution related damage) but so far the identify of this physiological disorder has not been determined.

Note, however, that these spots and how they appear is somewhat reminiscent of a disease occurring in tobacco and named Drought or Drought spots. On tobacco, there is sometimes brownish inter-veins necrosis on the lower leaves of many plants. They often occur simultaneously in several crops in the same production area. Their origin, which is still poorly understood, could be a very temporary disruption of the plant's water supply, which would immediately be shown on the leaves by their partial desiccation.

This information may not solve the problem (there is none better at present), but is designed to save unnecessary fungicide treatments, in particular for the control of blight.

Spots

Various phytotoxicities

Some agrochemicals can cause a local scorch of the leaf tissues. This is particularly the case when the leaves are accidentally in contact with fertilizers (Photo **312**) or herbicides sprays (Photo **313**). Often, only the most exposed leaflets or those located on one side of the plants are affected. The symptoms are particularly marked when the product responsible is retained on the plant surface for longer periods

See also pp. 77 and 131 which deal with other damage related to phytotoxicities (chemical injuries).

Table 21 Some diagnostic criteria for identifying key biotic or abiotic agents responsible for necrotic spots on leaflets of tomato

	Alternaria tomatophila	*Botrytis cinerea*	*Didymella lycopersici*	*Phytophthora infestans*	Various phytotoxicities
Frequency in the crops	Very common, mainly in open fields.	Very common under protection, less common in open fields.	Quite rare in open fields, more common under protection.	Very common in open fields and under protection.	Rare.
Symptoms distribution on the plant	First on the lower leaves then on all aerial parts.	Randomly on the lower to middle eaves and on all aerial parts.	Randomly on the lower to middle leaves and on all aerial parts.	On quite recent leaves, on the stem, and fruits.	On all the leaves, with the youngest being more sensitive, on one side of the plants and/or on the most exposed leaflets.
When symptoms develop	Spring, summer.	All year.	Spring, summer.	Late spring and summer.	All year.
Speed of symptoms in wet periods	++	+++	+	++++	−
Presence of specific structures on spots	Conidiophores and multicellular club-shaped conidia (Photo **302**).	Grey mould (Photos **319** and **320**).	Black globular structures: pycnidia (Photo **538**).	White down consisting of sporangiophores and of hyalin sporangia (Photo **335**).	Deposits sometimes occur.
Possible varietal resistance	−	−	−	+	− Differences in sensitivity. etween genotypes are possible.

Botrytis cinerea

This fungus, common on vegetables in glasshouses and open field when conditions are wet, is able to attack almost all aerial parts of the plant. It is ubiquitous and found in all production areas of the world. It occurs both in propagation (see p. 243) and during cultivation. It also poses some problems after harvest during fruit transport and storage (see p. 397).

On leaflets, it causes lesions on the side of the leaf, most frequently at the periphery and end. They are brown and moist in early stages, quickly become necrotic and become beige to brownish; some of them show clearly visible concentric patterns (Photo **314**). When they develop from the tip of the leaf in the shape of a flame, they spread very quickly, gradually invading and destroying the leaflets (Photo **315**). A yellow border is sometimes visible at the periphery. If weather conditions are favourable, *B. cinerea* colonizes the petiole and invades the other leaflets of the leaf (Photo **316**) and eventually the stem. In particularly humid glasshouses, stem infection through the petioles can be very common (Photo **317**), causing considerable damage.

This fungus colonizes the tomato in various ways. It can directly penetrate the cuticle of a leaflet or a fruit (Photo **321**), can enter through a pruning wound on a stem, or use various nutrient bases such as senescent floral parts (Photo **322**) or plants debris. It is an opportunist benefitting from its saprophytic potential. Moreover, once it has invaded a tissue, it grows very quickly as a grey mould of various density, consisting of conidiophores and conidia (Photos **318–320**); transmission by contact of diseased with a healthy parts frequently occurs.

Other symptoms occurring in other tomato parts can be observed in pp. 305 and 397.

Spots

Didymella lycopersici

As with *Botrytis cinerea*, this fungus can attack most parts of the tomato. It is only found in some production areas worldwide, such as several European countries and North Africa. In Europe, it occurs both in the open field and under protection. This is not a very common pathogen, but once established on a farm it persists and can sometimes be severe.

314 A necrotic, irregular, and well defined spot is invading the middle of this leaflet. It is beige to brown with discrete concentric blackish patterns.

316 Once established on a leaflet, *B. cinerea* reaches the petiole and can colonize other leaflets of the leaf that eventually will be completely affected. The fungus may then invade the stem, causing a canker.

315 A flame-shaped brownish lesion with a chlorotic halo is present on the affected leaflet.

317 On this plant, located in a glasshouse damp area, several layers of leaves are affected.

Botrytis cinerea –
grey mould

318 Water persisting longer at the tip of the leaf has allowed *B. cinerea* mycelium to penetrate the cuticle directly, and a necrotic lesion has developed. In addition to the dark brown concentric patterns, the lesion is partially covered with a diffuse gray mould.

319 A pruning wound enabled the penetration of *B. cinerea* into the stem. A dark brown canker with a well defined edge, is gradually encircling it. Again, a rather dense grey mould covers part of the lesion.

320 Flowers, particularly senescent petals, are especially susceptible to *B. cinerea*. The latter colonizes them quickly in wet season, causing a brown rot covered with the characteristic grey mould.

321 Several whitish rings a few millimetres in diameter are scattered over the fruit. These are actually halos resulting from *B. cinerea* aborted infections. A tiny brownish penetration puncture can be seen in the centre of these rings which are commonly called 'ghost spots'.

322 Several leaflets attacked by *B. cinerea* have enabled it to colonize one of these two fruits. A wet, soft rot affects parts of the fruit.

Botrytis cinerea – grey mould

323 Several spots are visible on this leaflet.

326 *D. lycopersici* is the cause of this wet canker, brown to dark brown, which has developed from a pruning wound.

324 The spots, brown at first, have taken on a beige tint while the veins are blackish.

325 If a lesion appears on the tip of the leaf, it gradually spreads and takes the shape of a flame, but should not be confused with the symptoms caused by *B. cinerea*.

327 A large slightly brown lesion spreads from the peduncular scar of the fruit. It is slightly concave and the skin is wrinkled. The presence of melanized hyphae and black pycnidia explains the dark colour visible in the centre of the lesion.

Didymella lycopersici – Didymella leaf spot, canker

Under protection, lesions on leaves (Didymella leaf spot) are reminiscent of those caused by *B. cinerea*. They appear on the leaf, sometimes at its periphery. The spots are initially brown to beige and moist, and spread (Photos **323** and **324**). Veins present in the affected areas tend to be brown to black (Photo **325**). Concentric patterns are sometimes seen.

Symptoms can appear on other parts of the tomato, especially on the stem at ground level (see p. 287) or higher up, at pruning wounds. In the latter case, dark moist brown lesions spread and gradually encircle the stem (Photo **326**), ultimately killing the distal end. The leaves quickly turn yellow, wilt, and wither. *Didymella lycopersici* often colonizes the fruit from the stem end. A brownish circular lesion extends fairly quickly. Tissues tend to crumble slightly and to darken, because of colonization by the melanized mycelium of the fungus, and the formation of many fruit bodies (Photo **327**). Whatever the organ attacked, tiny black and globular structures are visible on the damaged tissues. These are the pycnidia, the asexual spore producing structure of this fungus (see Photo **538**, p. 287).

More photos of symptoms occur on pp. 300 and 398.

Phytophthora infestans

This fungus is responsible for tomato late blight. It is feared by tomato and potato producers in many world production areas particularly because it affects all types of production, from the more intensive to the more extensive, including plants in the gardens of amateurs.

Phytophthora infestans can infect all aerial parts of the tomato. On the leaflets, it is responsible for poorly defined moist spots that grow fairly quickly, giving locally a light green to greenish brown colour to the leaf (Photos **329** and **330**). Significant sections of the leaf are affected, and tissues soon become necrotic (Photo **328**). These spots are frequently surrounded by a margin of water soaked tissue where sometimes a discrete white down forms. When conditions are particularly wet, the symptom progression on the foliage can be spectacular; many leaves turn brown and wither fairly quickly. Many of the plants eventually die.

Moist, brown to black lesions, sometimes extending over several centimetres, appear on the petioles and stems (Photo **333**) and circle them, contributing to shoots and flower death (Photo **334**).

The fruits infected at an early stage are often deformed and have very characteristic brown mottle, which spreads rather slowly and has a diffuse edge. If attacks occur later, the mottles are more uniform and often divided into concentric circles with looped edges (Photos **335** and **336**). A whitish down is occasionally visible on their surface. This is the asexual form of *Phytophthora*, and hyaline sporocystophores bearing colourless lemon-shaped sporangia can be seen (Photo **337**).

Note that the first symptoms of *P. nicotianae* on fruit (the causative agent of soil-borne Buck-eye rot) are sometimes confused with those due to *P. infestans*. Features of *P. nicotianae* infection are shown in Photos **732–736**. In contrast, *P. nicotianae* only attacks the fruit and the stem base of the tomato.

Spots

328 Several large necrotic spots are visible on various leaflets.

331 Many leaflets eventually wither and die, exposing the fruit to possible sunburn. These have irregular browning.

332 In addition to leaves, stem parts are surrounded by blackish lesions.

329 The spots with a rather pale edge, appear ill-defined on the upper side of the leaf; the damaged tissues are dark green in colour and rapidly become necrotic.

330 Under the leaflets, the spots are still moist and more diffuse.

Phytophthora infestans
(late blight)

333 Some stem parts are also affected Brown to black lesions around the stem may be several centimetres long.

334 A brown to black lesion is present on flower stalks, leading to progressive necrosis of several flowers.

337 In some damaged tissues, it is not uncommon to see hyaline sporangiophores emerging through the stomata for example, and bearing lemon-shaped sporangia.

335 Fruit affected at the green stage show lesions which are irregular, brown, and rough. A white mould covers them in part.

336 Some fruits invaded by *P. infestans* turn completely brown. Sometimes they eventually rot, mainly because of various micro-organisms, the secondary invaders.

Phytophthora infestans
(late blight)

338a A multitude of yellow dots partially cover this leaflet.
Tetranychus urticae

Examples of yellow spots on leaflets

338b Several yellowish spots, with diffuse edges, dot this leaflet.
Mycovellosiella fulva

339 Both leaflets reveal bright yellow spots, rather angular, whose centre becomes gradually necrotic.
Leveillula taurica

◼ Yellow spots with chlorosis

Possible causes

- *Leveillula taurica*: Description 5
- *Mycovellosiella fulva*: Description 4
- *Pseudocercospora fuligena*: Description 9
- Various viruses
- Pepino mosaic virus (PepMV, see p. 103)
- Pepper veinal mottle virus (PVMV)
- Potato yellow dwarf virus (PYDV)

- Tomato chlorotic spot virus (TCSV)
- Tomato bushy stunt virus (TBSV)
- Tomato spotted wilt virus (TSWV)
- Oedema
- Various phytotoxicities
- Incompatibility reaction to cladosporium
- Mites

Additional information for diagnosis

Mycovellosiella fulva

This globally widespread fungus is responsible for leaf mould, a name associated with an earlier Latin name: *Cladosporium fulvum*. It mainly affects glasshouse-grown tomatoes, and is sometimes found in open fields in warm and highly humid production areas.

Mycovellosiella fulva mainly attacks the leaflets which first show pale green spots on the upper side (Photo **340**). They reach several millimetres in diameter and have a blurred outline; they eventually turn yellow. On the underside of the leaf, a purplish brown to grey velvet covers the spot (Photo **342**) and is the conidiophores of the pathogen emerging through stomata with brownish conidia (Photos **343** and **344**). In the case of severe attacks, the fungus can sporulate on the upper side of the leaf (Photo **341**). In this case, the leaves turn completely yellow and dry out. The lower leaves are affected first then the disease gradually spreads to the younger leaves.

Spots

189

340 Several pale green to yellow spots, with diffuse edges, partially cover this leaflet. Some of them are bounded by the veins.

342 On the underside of the leaf, the spots are covered with a fairly typical velvety olive-brown coating.

341 *Mycovellosiella fulva* may colonize the upper side of the leaf during favourable climatic conditions. In this case, the leaf quickly becomes chlorotic and eventually necrotic.

343 Using a dissecting microscope, the numerous fructifications constituting the velvety coating can be easily observed.

344 Infrequently seen brown conidiophores and numerous irregular conidia (which are also brown and sometimes septate) are seen on the leaf, along with a few epidermal hairs.

Mycovellosiella fulva
(leaf mould)

Note that a fungus, *Hansfordia pulvinata* (Berk. and MA Curtis) Hughes can attack M. *fulva*; hyperparasitism results in the development of white mould which gradually covers the sporing structures of the pathogen (Photos **345–347**). *H. pulvinata* is sometimes seen in glasshouses. In the tropics it may occur in the field.

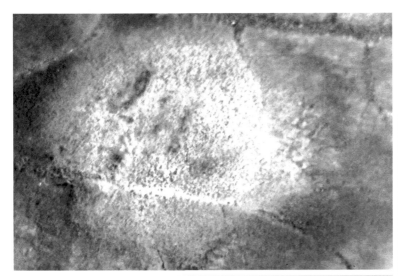

345 A white colony is progressively covering a leaf mould spot.
M. *fulva* parasitized by *H. pulvinata*.

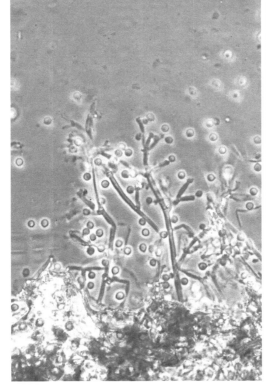

346 All the spots of this leaflet are now invaded by the hyperparasite fungus. On some, the velvety olive-brown coating is still visible.
M. *fulva* parasitized by *H. pulvinata*.

347 *Hansfordia pulvinata* produces branched conidiophores bearing denticles at the tips of branches. The conidia are globular and 4–7 µm in diameter.

Hansfordia pulvinata – **hyperparasite**

348 Several pale green to yellow spots, with diffuse edges, are scattered on many leaflets.

349 A more detailed observation of the upper side of the leaf reveals spots with diffuse edges and a yellow tint. They are sometimes defined by veins, which gives them an angular appearance.

350 On the underside, the yellowish spots are more diffuse.

351 At a more advanced stage, the spots are well defined and sometimes necrotic. Some leaflets are dried up.

352 Eventually the invaded tissues die. The spot centre then become gradually necrotic while concentric brown lesions sometimes appear.

Leveillula taurica (powdery mildew)

Leveillula taurica This fungus is responsible for tomato powdery mildew. Reported in a significant number of production areas, it is well adapted to Sahelian climate during the dry season as well as those of the Mediterranean basin. It is also reported in several countries in Asia and the Americas. It is now common on tomato in southern European countries, but also on many other vegetable crops such as peppers and artichoke. It occurs in field crops and under protection.

Leveillula taurica, penetrates the leaf and develops within it which is unlike the vast majority of powdery mildews where development is on the surface. Its symptoms are different from those of other known powdery mildews on tomato (see p. 203). Light green to yellow spots of varying intensity appear on the upper side of the leaf (Photos **348–350**), with diffuse edges more or less bounded by the veins, which give them an angular appearance (Photo **353**). A discreet white mould is seen on some spots on the underside of the leaf (Photo **353**): These are conidiophores and conidia emerging through stomata and this is characteristic of this fungus (Photos **354** and **355**). Eventually, the centre of the spots becomes necrotic and appears brownish. Concentric patterns are sometimes visible within the necrotic tissues. In favourable conditions many spots cover the leaflets and eventually they can completely dessicate.

This fungus does not attack other aerial parts of the tomato.

353 On the underside of this leaflet, many spots, well defined by the veins, are covered by the white fructifications of the imperfect form of *L. taurica*.

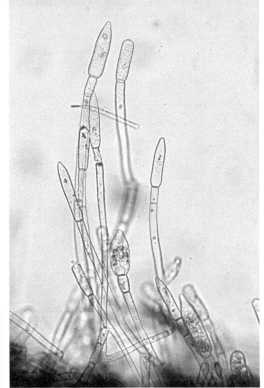

354 When the conditions are very favourable, *L. taurica* can grow on the upper side of the leaf, and several downy white spots then appear.

355 The white down is the numerous slender and septate conidiophores of the fungus. At their ends, the conidia are formed. Note that the terminal conidium is characteristic because it has a pointed apex, called 'spearhead'.

Leveillula taurica
(powdery mildew)

Figure 32 Chlorotic spots covered with a velvety matt black coating are present on the leaf.

356 This fungus occurs on the leaf tissues and conidia and conidiophores are produced (3.5-5 × 70 × 2.5–5 μm and 40–120 μm, respectively).

Pseudocercospora fuligena (Cercospora leaf spot, black leaf mould)

Pseudocercospora fuligena

Most likely to occur in tropical and humid areas of Asia, Africa, and Central America, this fungus is responsible for the disease known as 'black leaf mould' (Cercospora leaf mould). It is characterized by chlorotic spots especially visible on the upper side of the leaf.

Note that these can be confused with those induced by *Mycovellosiella fulva*. Initially pale yellow, they have a poorly defined outline and gradually spread. Sporing structures which change progressively from white to black, eventually cover the spots. Eventually, large blotches with a matt black velvety down cover the leaves and are visible on the under and upper sides of the leaflets (Figure 32). In very favourable conditions, many leaflets and leaves may become necrotic but remain on the plants.

Note that this fungus can colonize the petioles, the stem, and the fruit stalk.

This fungus is not prevalent in Europe, but attacks tomato in some countries like Somalia along with *L. taurica* and *M. fulva*. In these countries, it is necessary to be especially careful with diagnosis. Furthermore, *Cercospora diffusa* has been described on tomato in Cuba.

Spots

Table 22 Some diagnostic criteria for identifying the three main fungal diseases causing yellow spots on tomato leaflets

	Mycovellosiella fulva	*Leveillula taurica*	*Pseudocercospora fuligena*
Frequency	Common, mainly under protection.	Uncommon, under protection or in open fields.	Absent (in France)
Symptoms onset period	All year.	Mainly from early summer onwards.	–
Presence of specific structures on the spots	Rather dense down, olive-brown to purple. Brown conidiophores and budding conidia. Photos **343** and **344**.	Discreet white down. Slender conidiophores with a few hyaline conidia; the terminal one has a pointed apex. Photo **355**.	Dense blackish down. Brownish conidiophores bearing elongated multicellular conidia. Photo **356**.
Possible varietal resistance	+	+	–

Various viruses

Several viruses cause various symptoms including chlorotic spots. This is particularly so with Pepper veinal mottle virus (PVMV), Potato yellow dwarf virus (PYDV), Tomato chlorotic spot virus (TCSV), Tomato bushy stunt virus (TBSV), Tomato spotted wilt virus (TSWV), and Pepino mosaic virus (PepMV). *Table 22a* present a few characteristics of viruses with the greatest economic impact. For TSWV, see pp. 163 and 221.

Table 22a Examples of viruses that cause yellow spots on tomato leaflets

Virus	Symptoms	Vectors
Pepino mosaic virus (PepMV) **Description 29**	Clearly defined spots with chlorosis. These yellow spots are the most characteristic symptom of this virus (Photo **357**). Green to yellow mosaic is sometimes located on the edge of the leaf. The leaflets can be curved up or down, and display some blistering and enations. Those of the apex are shrivelled and are sometimes distorted. The sepals are superficially and locally affected. The fruits are mottled.	Mechanically transmitted by casual contact (during the cultural operations for example), by pollinating bumblebees, and by seeds.
Tomato bushy stunt virus (TBSV) **Description 32**	Chlorotic rings and arcs appearing on the leaflets which are also chlorotic. Necrotic streaks may develop on the stems. Affected plants stop growing and are often stunted and bushy, producing small fruit.	Transmitted by contact and by water. This virus has been detected in running and in stagnant water in several European countries. Transmitted by pollen in the tomato and other species.

357 Several bright yellow spots on this leaflet are very characteristic of a viral infection with Pepino mosaic virus (PepMV).

Oedema

This physiological disorder of tomato occurs mainly in protected crops in periods of high humidity (in late winter, early spring, or autumn) and/or as a result of abrupt climate change particularly at night (high humidity, low temperatures). It often affects many plants spread over the entire crop, or those plants in the wettest cold places. Low light, poor air circulation, and high densities of plants are likely to favour oedema.

Under these conditions, chlorotic spots appear along the veins of the lower leaves (Photo **358**). These spots form when the water balance between root uptake and transpiration from plants is disrupted, the latter absorbing more water than they can eliminate. Then water saturation of leaf tissues follows which is at the cause of oedema. Small groups of cells located on the underside of the leaf become saturated with water, sometimes leading to skin disruption. Oedema with a moist to oily appearance develops (Photo **359**).

Subsequently, the cells burst and brown corky tissues forms (Photo **360**). During particularly wet conditions, the presence of numerous swellings causes the deformation and rolling of numerous leaflets (see p. 65).

Although sometimes very dramatic, this nonparasitic disease is normally harmless to tomato crops especially if corrections are made to the glasshouse climate at the onset of the problem. Measures to prevent oedema include:

– Use a well-drained soil or substrate;

– Avoid too high planting densities;

– Reduce the moisture levels by heating and ventilating early in the morning;

– Control the irrigation, especially during cloudy and humid periods. It should be stopped in late afternoon, before nightfall in order to prevent the soil or substrate remaining too wet.

Spots

Genetic incompatibility

Growers sometimes see the appearance of yellow and diffuse leaf spots which quickly becoming necrotic, on the lower leaves of certain plants (Photo **361**). These spots are more pronounced on the underside of the leaf (Photo **362**). They merge and can cover the leaves entirely. Particularly affected leaflets tend to roll.

This genetic disorder (autogenous necrosis) is the result of an incompatibility between a dominant gene 'Cf', for resistance to leaf mould, and

a recessive necrotic gene, 'ne'. In conditions of high temperature and brightness, this symptom can result.

This problem is sometimes found on some cultivars and is well known by breeders. Although often without crop effects, fruit size may be reduced during severe attacks. There are no control measures for this disease but growers are well advised to cultivate a different variety for the next crop.

358 Some discrete spots with chlorosis are visible near some veins.

361 On the upper side of this leaflet, several confluent yellow, and necrotic spots, are visible.
Genetic incompatibility

359 Many groups of cells swollen with water have formed on the underside of this leaflet. Some of them look moist to oily.

360 The cells eventually burst and the tissues are changed to a beige to light brown colour, forming necrotic lesions. The leaf is also now distorted.

362 The symptoms of genetic incompatibility are easier to see on the underside of the leaflets. The leaf is chlorotic and irregular brown necrotic lesions are visible between the veins.
Genetic incompatibility

Oedema

Mites

Mites, like insects, are arthropods and belong to the Arachnida class (spiders). Adults have four pairs of legs, unlike insects which have only three. Many mite families include plant pests; the most well known are the Tetranychydae. Species in this group are commonly referred to as 'mites', or yellow, red, or green 'spiders'. Note that some species feed on tomato, including *Tetranychus urticae* (Koch) which is the most widely reported mite on this crop. Also called spider mite because of the webs it forms on plants, it is cosmopolitan and highly polyphagous and has been reported on nearly 2000 plant species. It can be responsible for significant damage particularly on protected crops of both ornamentals and vegetables. Cases of resistance to acaricides have been reported with this mite. Other species of phytophagous Tetranychidae and predatory Phytoseiidae mites can occasionally occur on tomato.

• Nature of damage

Tiny scattered chlorotic spots appear on the leaflets (Photos **363** and **364**) which gradually turn yellow and take on a dull colour (Photos **365** and **366**). During severe attacks, some leaves may wilt and wither. Delicate silken webs are visible in the vegetation (Photo **367**).

• Biology

The spider mite passes through five developmental stages: egg, larva, protonymph, deutonymph, and adult (Figure **32a**). The cycle time varies with temperature: about 7 days at 30°C and 36 days at 15°C.

Forms of survival and/or alternative hosts: it is the females in diapause that overwinter. Diapause occurs when the temperature is low or food supplies diminish.

Developmental stages: eggs (1) are laid mostly on the underside of the leaflets. They are round, tiny (0.14 mm) and translucent at first and then become opaque, and eventually turn yellow close to hatching. Larvae (2), which have three pairs of legs are very pale at first, then take on a greenish hue. They have two red eye spots and two dark spots in the middle of their bodies. The protonymph (3) is larger, with four pairs of legs and its colour varies from light green to dark green; they also have two more contrasting spots on their bodies. Deutonymphs (4) are larger still but the same colour. It is in the adult stage (5) that males can be distinguished from females. The male is more active, smaller, and narrower than the female and a variable colour (light yellow to orange, dark yellow to brown). They are oval and measure 0.5–0.6 mm in length. On tomatoes, *T. urticae* females are often reddish, although the colour varies (orange, yellow, light green to dark green or red.

Spots

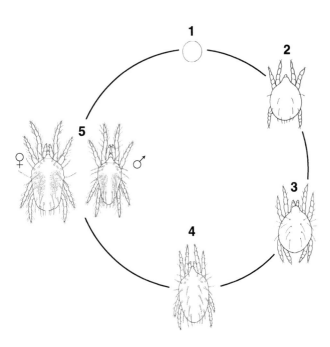

Figure 32a Life cycle of the mite.

363 Tiny lesions with chlorosis are visible in several places on this leaflet.

364 The chlorotic lesions are now much more numerous and cover the entire leaf.

365 The leaf, covered in lesions, becomes yellow, takes on a dull hue, and eventually dries up.

366 The yellowing and dullness of the leaf is even more pronounced.

367 On this delicate web, numerous mites are present.

Tetranychus urticae –
spider mite

The larvae, nymphs and adults, are often found on the underside of the leaf, and feed by biting and sucking the contents of plant cells. Feeding punctures are sometimes very numerous, and the cause the symptoms.

Mites are dispersed in crops by plant to plant contact; they may also fall on the floor and spread to other plants or walk along the crop wires. They are also transported by workers, equipment, and tools, or scattered by plant movement.

Favourable development conditions: strongly growing shoots of the plant are more favourable to the development of mites. Similarly, hot, dry periods are also favourable to their development. Factors which may disadvantage the crop (low temperatures, applications of insecticides) may favour the development of *T. urticae*.

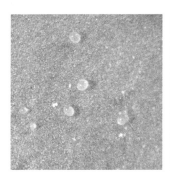

368 Tiny mite eggs.

369 Different stages of *T. urticae*.

370 Red *T. urticae*.

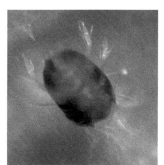

371 Green *T. urticae*.

372 A few powdery spots, white and circular, are visible at the tip of this leaflet.
Oidium neolycopersici

373 Numerous downy and olive-colored spots cover the underside of this leaflet.
Mycovellosiella fulva

374 Several moulds have developed on this leaflet; it is also chlorotic and slightly sticky.
Sooty mould

375 Several areas of the leaf, defined by the veins, are covered by a dense white felting.
Leveillula taurica

Examples of spots with powdery, velvety patches on tomato leaflets

Spots with powdery, velvety patches; presence of mould

Possible causes

- *Botrytis cinerea* (see p. 181): Description 2
- *Mycovellosiella fulva* (see p.189): Description 4
- *Leveillula taurica* (see p. 193): Description 5
- *Oidium lycopersici*
- *Oidium neolycopersici*: Description 6
- *Penicillium* sp. (see p. 168)
- *Phytophthora infestans* (see p.185): Description 7
- *Pseudocercospora fuligena* (see p.195): Description 9
- *Botryosporium* sp.
- Sooty mould

Additional information for diagnosis

A mould, sporulating on tomato leaflets generally indicates the presence of a pathogenic fungus, developing from the plant tissues. These signs are frequently preceded by symptoms such as chlorotic spots (*Leveillula taurica*, *Mycovellosiella fulva*, *Pseudocercospora fuligena*), brownish concentric lesions (*Botrytis cinerea*), and water-soaked patches (*Phytophthora infestans*). The pathogens develop on the tissues in a second phase, after colonization of the tissue.

In a more limited number of cases, both leaf symptoms and fungal pathogen sporulation appear at the same time. This is especially true of several external powdery mildews and of fungi in the sooty mould complex which are associated with the development of several pests. They are the subject of this section

Oidium neolycopersici

For the last two decades, a new tomato powdery mildew has emerged in many countries (several countries in Europe, Africa, North and South America, and Asia), sometimes causing considerable damage. It affects both crops under protection and in the field. It was introduced into Europe in the mid 1980s in young tomato plants imported from the Netherlands. Subsequently, it spread to all production areas, attacking tomato crops throughout the year.

Unlike *Leveillula taurica* (the internally colonixing powdery mildew responsible for chlorotic spots), this external powdery mildew immediately produces white powdery spots covering the upper side rather than the underside of the

tomato leaflets (Photos **376** and **378**). This white mildew is the mycelial network colonizing the leaf surface, topped by numerous coniodiophores producing isolated hyaline conidia (Photo **379**, Figure **33**) or sometimes pseudo-strings of 4–6 spores when relative humidity is high. Such spots may also be observed on the stem (Photo **377**). Fruits do not seem to be affected.

Affected tissues eventually become chlorotic or locally brown and become necrotic (Photos **380** and **381**). During severe attacks, the leaf is completely covered by the fungal mycelium and some leaflets turn yellow and become entirely necrotic (**Photo 382**).

Spots

203

376 Many isolated white powdery colonies are scattered over the upper side of the leaflets of several leaves.

379 *Oidium neolycopersici* produces mainly single ellipsoidal to ovoid hyaline conidia, measuring 22–44 × 10–20 μm.

377 Colonies are also visible in places on the stem.

378 The colonies have converged and the white felting now covers the entire leaflet.

Oidium neolycopersici
(powdery mildew)

The identification of this powdery mildew has led to much confusion and several names have been incorrectly attributed to it. Recent research has led to a better understanding of its systematic classification and the name 'Oidium neolycopersici L. Kiss, sp. nov' has been adopted. In several countries where this pathogen is present, several other powdery mildews responsible have already been described on tomato, and classified (*Erysiphe orontii. E. polygoni, E. cichoracearum*). Herbarium specimens have shown that in some cases, these were misidentified and were *O. neolycopersici*. In this respect its presence in Asia dates back at least 50 years.

Another species of *Oidium* (*O. lycopersici* Cooke & Massee, Grevillea) was first reported in 1888 in Australia where it has been present for many years. It also produces powdery colonies on both the upper and lower sides of the tomato leaf, as well as on the petioles and the calyx of fruit. This fungus always forms conidia in chains of 3–5 spores (Figure 33).

In conclusion, powdery white spots on tomatoes are the likely sign of an attack of *Oidium neolycopersici* because it is the most common species around the world. The absence of conidia chain formation at the tips of conidiophores will confirm its identity. However, in wet conditions this fungus is capable of producing pseudo-chains and this may be a source of confusion in its identification.

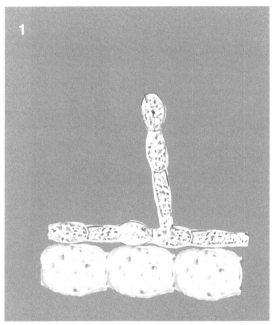

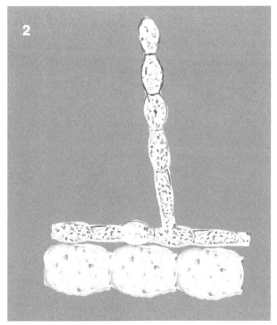

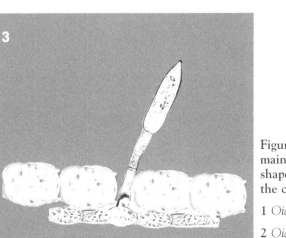

Figure 33 Identification of the three main tomato powdery mildews: note the shape and number of conidia present at the conidiophores' tips.

1 *Oidium neolycopersici*

2 *Oidium lycopersici*

3 *Leveillula taurica*

380 Although this biotrophic parasite does not kill its host quickly, colonized tissues becomes chlorotic.

381 Tissues are increasingly yellow and turn a diffuse brown shade.

382 *Oidium neolycopersici* inexorably colonizes this tomato plant, and several affected leaves are partially necrotic, others completely dry.

Oidium neolycopersici
(powdery mildew)

Sooty mould

Many insects, notably aphids, whiteflies and scale insects collect large amounts of sap to meet their protein needs. This forces them to reject excess sugar – in large quantities – in the form of honeydew which then contaminates many parts of the colonized plants surfaces.

This sweet honeydew is an ideal medium for fungi that live on leaf surfaces and it is their growth that is recognised as sooty mould. The colour of the mould varies with the fungus involved which include *Alternaria* spp. *Cladosporium* spp. *Capnodium* sp., *Penicillium* sp., etc (see Photos **283–286**). Sooty mould is harmful because it can interfere with plants photosynthesis. Only the control of pest populations can prevent it.

Table 23 Main viruses affecting tomatoes, transmitted by insects

Possible vectors	Main transmissible viruses
Aphids (*Myzus persicae, Macrosiphum euphorbiae, Aphis craccivora, A. gossypii, A. fabae, Acyrthosiphon solani, A. pisum*)	NP: Alfalfa mosaic virus (AMV); Cucumber mosaic virus (CMV); Potato virus Y (PVY); Tobacco etch virus (TEV); Broad bean wilt virus (BBWV); Colombian datura virus (CDV); Eggplant severe mottle virus (ESMoV); Pepper veinal mottle virus (PVMV); Peru tomato virus (PTV); Potato leafroll virus (PLRV); Potato yellowing virus (PYV); Tobacco vein banding mosaic virus (TVBMV); Tobacco vein mottling virus (TVMV); Tomato aspermy virus (TAV); Tomato mild mottle virus (TMMV). P: Beet western yellow Luteovirus (BWYV).
Whiteflies (*Trialeurodes vaporariorum, Bemisia tabaci, B. argentifolii*)	P: Tomato chlorosis virus (ToCV); Tomato infectious chlorosis virus (TICV); Tomato yellow leaf curl virus (TYLCV); Tomato yellow leaf curl Sardania virus (TYLCSV); Cowpea mild mottle virus (CPMMV); Eggplant yellow mosaic virus (EYMV); Serrano golden mosaic virus (SGMV); Tobacco leaf curl virus (TLCV); Tomato leaf curl virus (ToLCV); Indian tomato leaf curl virus (IToLCV); Sinaloa tomato leaf curl virus (STLCV); Tomato mottle virus (ToMoV); Taino tomato mottle virus (TToMoV); Tomato golden mosaic virus (TGMV); Tomato leaf crumple virus (TLCrV); Tomato yellow dwarf virus (ToYDV); Tomato yellow mosaic virus (TYMV); Tomato yellow mottle virus (ToYMoV); Tomato yellow vein streak virus (ToYVSV).
Thrips (*Thrips tabaci, Frankliniella occidentalis*)	P: Tomato spotted wilt virus (TSWV); Groundnut ringspot virus (GRSV); Impatiens necrotic spot virus (INSV); Paretaria mottle virus (PMoV); Tobacco streak virus (TSV); Tomato chlorotic spot virus (TCSV).
Leafhoppers and other insects (*Agallia albidulla, A. ensigera, A. constricta, Aceratogallia sanguinolenta, Circulifer tenellus, C. opacipennis, Micrutalis festinus, Orosius argentatus*)	P: Beet curly top virus (BCTV); Eggplant mosaic virus (EMV); Potato yellow dwarf virus (PYDV); Tobacco yellow dwarf virus (TYDV); Tomato pseudo curly top virus (TPCTV).

NP : viruses transmitted in a nonpersistent manner.
P : viruses transmitted in a semi-persistent or persistent manner.

383 On this leaflet, several fungi causing sooty mould are forming colonies on almost all of the surface. Many colonies are still light in colour, some have taken on an olive-brown hue.

385 The sooty mould is now well developed. Many dark green colonies cover the fruit.

384 These fruits are covered with honeydew on which the various fungi comprising the sooty mould complex will grow.

386 In an outbreak of whiteflies, the development of sooty mould is important as it gradually withers the plants; their growth is reduced, and many leaflets and leaves yellow and wither.

Sooty mould

388 In the presence of a large aphid population, the leaflets can turn yellow, shrivel, or even become partially necrotic.

387 Several green aphids are active under this leaflet.

389 A multitude of whiteflies cover these leaflets.

390 In addition to the sooty mould, many white scales at different stages are visible on the stem, petioles, and to a lesser degree, the fruit of this tomato plant.

In the presence of sooty mould, remember to look for the evidence of pests illustrated in Photos 386–390

Aphids

Several species of aphids can form colonies on young tomato leaflets at the propagation stage: *Macrosiphum euphorbiae* (Thomas) (Photo **393**), *Myzus persicae* (Sulzer) (Photo **391**), *Aphis gossypii* Glover (Photo **392**), and *Aulacorthum solani* (Kaltenbach). These polyphagous insects belong to the Hemiptera order, Sternorrhyncha suborder, and Aphidoidae superfamily. They grow fairly frequently on tomatoes in the form of colonies, but they do not have the same dynamic development as whiteflies. On tomatoes, they are a special problem because of their ability to transmit several viruses (see *Table 23*, p. 207, and Descriptions **33–38**). In addition, chemical control of these pest populations is often problematic due to their possible resistance to several insecticides.

• Nature of damage

Feeding punctures are the cause of chlorotic dots and can deform the young leaflets (Photo **388**). A reduction in the growth of young shoots or even plants occurs. In addition to colonies of aphids, there are very often white caste skins and the presence of honeydew on the surface of tomato aerial parts, on which sooty mould grows (sooty mould, see Photos **383–386**). The latter may have several consequences, including photosynthesis and leaf respiration reduction, as well as fruit uncleanness which are thereby unmarketable.

• Biology

These insects have a rather complicated life cycle, and the adults of some species change host in winter. There are eggs, populations foundresses, and adults – among them, winged or not viviparous females, and winged males. The lifecycle time varies depending on the species, the nature of the host plant and its condition, and climatic conditions.

Forms of survival and/or alternative hosts: the eggs are often laid in particular on many weeds at the beginning of winter, allowing these insects to overwinter. They can, of course, survive in heated glasshouses, in particular in the form of viviparous females.

Developmental stages: eggs laid on different herbaceous or woody hosts, hatch and give rise to the founders. Subsequently, during a long period, there are viviparous females within colonies.

Young larvae are formed, which feed on the sap immediately and moult four times before giving birth to adults. White cast skins on vegetation betray the presence of aphids in the crop. Adults are winged (1) or not (wingless, 2) (Figure **34**). Each individual may give birth to 40–100 offspring depending on the host and the climatic conditions in particular.

Larvae and adults, often found on the underside of the leaves, feed with their rostrum. The excess sugar in sap is released in the form of honeydew.

Dispersion in culture: a few contaminated plants scattered in the crop are the initial source. These are wingless aphids which may move to neighbouring plants. Once the winged adults emerge (in outbreaks), they spread in the crop or in fields nearby. Plants and workers can contribute to their dissemination.

Favourable conditions for development: these insects are favoured by mild temperatures and the glasshouse summer conditions.

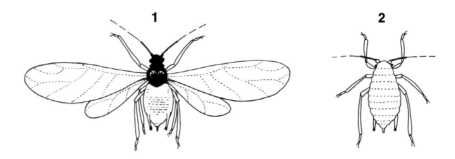

Figure 34 Winged (1) and wingless (2) aphids.

391 The green peach aphid, *Myzus persicae*, is a globally dispersed and important pest on tomatoes, but also on cucumber, potato, and tobacco. The wingless forms are smaller than the winged ones, and their colour is variable.

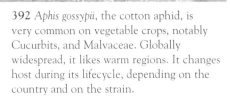

392 *Aphis gossypii*, the cotton aphid, is very common on vegetable crops, notably Cucurbits, and Malvaceae. Globally widespread, it likes warm regions. It changes host during its lifecycle, depending on the country and on the strain.

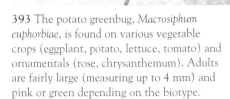

393 The potato greenbug, *Macrosiphum euphorbiae*, is found on various vegetable crops (eggplant, potato, lettuce, tomato) and ornamentals (rose, chrysanthemum). Adults are fairly large (measuring up to 4 mm) and pink or green depending on the biotype.

394 *Aulacorthum solani*, the foxglove and potato striated aphid, feeds well on tomatoes, but also on lettuce, peppers, eggplant, and beans. The adult is green, its head and thorax are dark.

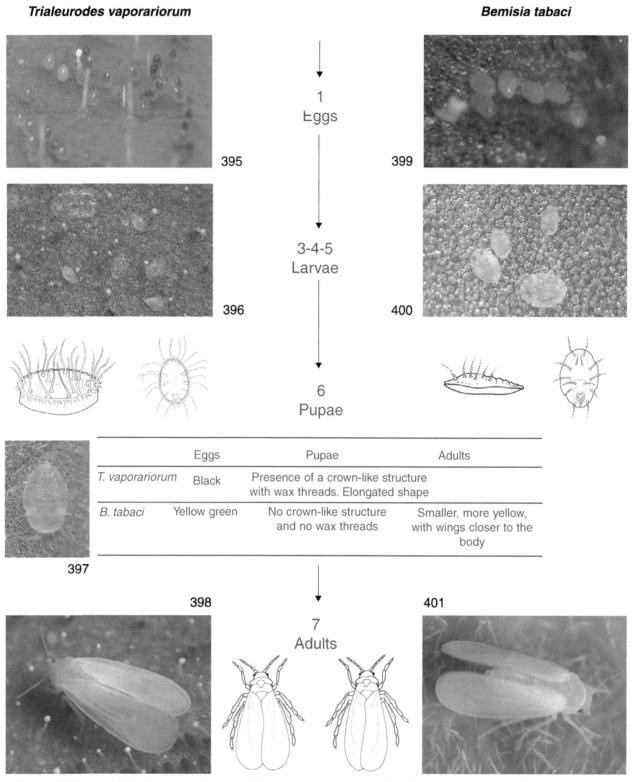

Trialeurodes vaporariorum

Bemisia tabaci

1
Eggs

395

399

3-4-5
Larvae

396

400

6
Pupae

	Eggs	Pupae	Adults
T. vaporariorum	Black	Presence of a crown-like structure with wax threads. Elongated shape	
B. tabaci	Yellow green	No crown-like structure and no wax threads	Smaller, more yellow, with wings closer to the body

397

398

401

7
Adults

Figure 35 Principal developmental stages of the two major tomato whiteflies, and their identification criteria.

Whiteflies

Two polyphagous whiteflies occur on tomato in France: *Trialeurodes vaporariorum* (Westwood) and *Bemisia tabaci* (Gennadius). The latter species is represented (in some populations) by a particularly formidable B biotype, also known under the name *B. argentifolii* (Bellows & Perring), which develops on over 900 hosts. These whiteflies belong to the Hemiptera order and Aleyrodidae family. *Trialeurodes vaporariorum*, also called the 'glasshouse whitefly', is a native of Central America. It has been a problem in Europe, particularly under protection, for many years. *Bemisia tabaci*, the sweet potato whitefly, described for the first time in Greece, is emerging in recent decades in many parts of the world. It has become a problem everywhere for at least three reasons:

– Its high population levels are responsible for significant damage;
– Its vector potential allows the transmission of numerous viruses (more than 111, see *Table 23*, p. 207, and Descriptions 39–42);
– Resistance of some of its populations to one or more insecticides is making the effectiveness of chemical control uncertain.

• Nature of damage

As with aphids, the many feeding punctures caused by whiteflies results in a slowdown in plant development.

Honeydew is also produced in large quantities and is subsequently colonized by sooty mould covering the surface of tomato aerial parts (Photos **383–386**), causing leaf yellowing. In addition to reducing leaf photosynthesis and respiration, sooty mould stains the fruit and causes discoloration, making it unfit for marketing. *B. tabaci* may also cause discoloration (irregular maturity), or spots on the fruit (Photo **374**).

• Biology

Whiteflies have three stages of development occurring on the underside of the tomato leaflets: egg, four larval stages, and adult (Figure **36**). The length of the complete lifecycle varies with temperature. It fluctuates for *T. vaporariorum* from under 20 days at 27°C to over 40 days at 14°C (over 50 days for *B. tabaci*).

Forms of survival and/or alternative hosts: these insects do not have a suitable stage for the winter phase. They only survive if their hosts do not die. Note that the eggs can survive temperatures below 0°C for several days. Whiteflies readily remain on many cultivated hosts, but also on various weeds; these should therefore be carefully removed.

Developmental stages: eggs (1 and 2) (Photos **395** and **399**) are mainly deposited on the underside of the top leaflets. They are white, oval, and have a diameter of 0.25 mm. In the following days, they become darker. Between 7 and 10 days later, the larvae hatch (3) (Photos **396** and **400**); these are oval, flat, measure 0.3 mm and have antennae and well developed legs. They lose the legs thereafter, so are immobile and eat with their rostrum. The second instars are flattened, transparent, and measure 0.37 mm. The third and fourth instars are quite similar, but with lengths of 0.51 and 0.73 mm (4 and 5). At the last larval stage, the insect secretes wax. When its red eyes appear it is called a puparium (6) (Photo **397**).

Subsequently, the whitefly develops and takes a white shade. Adults (7) (Photos **398** and **401**) have two pairs of wings of different sizes according to sex – 1.1 mm for females and 0.9 mm for males. The body and wings are covered with a characteristic waxy white powder. Slight morphological criteria (see opposite) can differentiate the two sexes. Larvae and adults, often found on the leaf underside, feed through their rostrum which acts as suction pump. The excess sugar in sap is released in the form of honeydew, especially by large larvae.

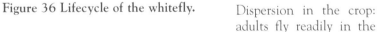

Figure 36 Lifecycle of the whitefly.

Dispersion in the crop: adults fly readily in the glasshouse and spread gradually from a few initially colonized plants. Spread increases as the plant size and temperature increase. The widespread distribution of plants of horticultural species has contributed significantly to the spread of *B. tabaci* in the world.

Favourable development conditions: these insects are favoured by mild temperatures and glasshouse summer conditions. Their lifespan is between 10 and 20 days on tomato, but fluctuates depending on temperatures. *B. tabaci*, has temperature requirements higher than those of *T. vaporariorum* and does not survive temperatures below 0°C.

402 On the lower part of the stem, the remains of the first truss is still present, and covered by dense white and mealy material. Note the presence of sooty mould on some leaflets. Looking more closely at the affected areas, one can see the larvae and females of the white to pinkish scale insects.

403 View of a pinkish female scale insect, on average 4 mm long.

Pseudococcus viburnii –
scale insects

Scale insects

Scale insects are sucking biting insects belonging to the Hemiptera order and the Coccoidea super-family. Tomatoes are especially affected by some scale insects belonging to the Pseudococcidae family.

Scale insects of this family have no horny shield (sort of shell), and the body is usually covered with a white mealy secretion. They also have prominent lateral waxy filaments on the periphery of the body, and a cottony secretion containing eggs can sometimes be seen at the end of the abdomen. In the context of the tomato the most important mealybug is *Pseudococcus viburni* (Signoret) (syn. *Pseudococcus affinis* [Maskell]) which was introduced in the Netherlands in the early 1990s and in France from 1997. This scale insect is now found in a significant number of glasshouses located in various European production areas.

• Nature of damage

Plant growth is reduced because of the many feeding areas of the larvae and females, especially on the stem. Honeydew and sooty mould cover the surface of tomato aerial parts (Photos **383** and **386**). Sooty mould, reducing photosynthesis and respiration, is the cause of leaf yellowing and deformation. It also contaminates or changes the fruit colour, making them unmarketable (Photo **402**).

• Biology

Pseudococcus viburni has several developmental stages: egg, three larval stages, and adult (Figure 37).

Forms of survival and/or alternative hosts: *P. viburni* survives fairly well in the cold and spends the winter in the nondiapause pupa state in the soil. It can be maintained on other alternative hosts, including potato, apple, lemon, and grape.

Developmental stages: eggs (1) are laid within a waxy white ovisac. After hatching, the larvae of the first stage (2) disperse on infested plants. After the second larval stage (3), this insect will form two successive false nymphs in which males undergo a metamorphosis. These have the appearance of small ephemeral winged midges. Females (4) (Photo **403**), measuring 4 mm long, do not undergo metamorphosis and therefore do not change form. The pinkish colour of their body is hidden by white mealy wax.

Dispersion in culture: this scale is primarily spread by plant material. Workers in cultural operations and animals contribute to its spread.

Favourable development conditions: *P. viburni* develops well in glasshouse conditions.

Spots

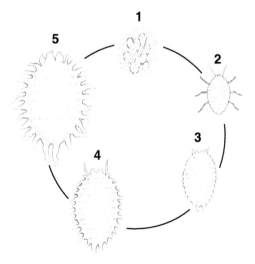

Figure 37 Lifecycle of the scale insect.

404 Leaves of the apex of the plant wilted suddenly.

406 The wilting of the leaflets is accompanied by a strong yellowing of the leaf.

405 The partial wilting of these leaflets is more insidious. Some necrotic and irregular inter-vein patches are the only symptoms reflecting a disrupted water supply of the plant.

407 After yellowing and withering, some sectors of the leaf have dried out, taking a particular form, in the shape of an upside down 'V'.

Examples of wilting, yellowing, drying, and leaf necrosis

WILTING, NECROSIS, AND DRIED LEAFLETS AND LEAVES (WITH OR WITHOUT YELLOWING)

Possible causes

- Pathogenic agents, other pests, nonparasitic diseases responsible for roots, crown, and stem damage (see the sections on these plant parts, pp. 231 and 297)
- Pathogenic agents, other pests responsible for spots and other leaf symptoms (see the section Spots on leaflets and leaves, p. 149)
- Various viruses
  - Tomato spotted wilt virus (TSWV): Description 43
  - Cucumber mosaic virus (CMV): Description 35
  - Alfalfa mosaic virus (AMV): Description 34
  - Potato virus Y (PVY): Description 36
  - Tomato torrado virus (ToTV)
  - Viral complexes
- Para-acetic acid (PAA)
- Allelopathy
- Root asphyxia
- CO_2
- Excess salinity
- Lightning
- Nutritional disorders (see p. 125)
- Various phytotoxicities
- Air pollutants
- *Aculops lycopersici* (see p. 147)
- Bugs (see p. 381)

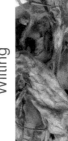

Analyzing and understanding wilting

- Typically, when one to several tomato leaves wilt and sometimes turn yellow or dry out, growers tend to focus their attention on these leaf symptoms and look for a cause(s). In many cases, the origin of these nonspecific foliage symptoms should be sought elsewhere. It is more commonly the roots, stem base, or the outside or inside of the stem that is the primary cause of the symptom accounting for the wilt.

Wilting

408 If there is yellowing, wilting, or drying out on the leaflets …

… Have the right approach! Also observe very carefully:

409 The stem over its entire length, inside and outside.

411 The roots.

410 The stem base and lower part of the stem.

If the observations are conclusive, consult the section on the plant part concerned.

Changes which occur in these plant parts can disrupt the water and nutrient absorption (roots) and transportation (crown, stem, and vessels). The resulting lack of water in plants will eventually cause the plant to wilt (temporarily or irreversibly), and dry out partially or completely.

The appearance of wilting and speed of development will depend essentially on three factors:

– the development stage of the plants. Seedlings develop irreversible wilt more readily than adult plants, and these symptoms occur more frequently nearing harvest time, when plants are laden with fruit;

– the nature and degree of the development at the onset of the symptom. For example, early in the attack, when the damage is not too great, wilting starts slowly and may briefly be reversible, especially at night when the leaflets evaporate less and become turgid again;

– climatic conditions. High temperatures and a windy period increases plant evapotranspiration and cause their early wilting. Conversely, under conditions of low evapotranspiration, for example during a cold, wet spring, plants may not wilt at first, despite a disease problem occurring on the plant parts mentioned earlier (stem base rot, root rot, vascular damage). When they do, it will often be too late to intervene.

In some cases, wilting only affects parts of the leaves or plants, characteristic of one or more particular diseases. This will become obvious later, especially with vascular diseases (see p. 323).

Foliage is a good indicator of the plant's water status. Wilting and drying of one or more leaves should generally be considered a signs of disruption of the water status, whose cause(s) should be sought by carefully examining the root system, crown, and stem along its entire length.

Several diseases cause wilting, drying, or similar symptoms on tomato, which do not always induce other specific symptoms on other parts of the plant and thus can be more difficult to identify.

Furthermore, foliar necrosis is included in this section. This symptom is not always easy to distinguish from drying out as a result of wilting as it often occurs during the final stages of disease development associated with lesions which affect the leaf. For this reason the possible causes include a multitude of pests and nonparasitic diseases that often have nothing to do with each other. They are provided for information only, and their symptoms can be found in other subsections of the Diagnosis section of this book.

Wilting

412 Many of the leaves of the apex are completely dry or have a quite noticeable bronze tint.

414 The symptoms are now more advanced; large intervein necrotic areas of a rusty brown colour are clearly visible.

413 Inter-veinal tissue gradually becomes chlorotic and then necrotic.

415 On this plant, the evolution of necrotic lesions on the leaflets can be seen. First inter-veinal and limited in size, they eventually spread to the entire leaf that takes on a bronze colour and ends up drying out.

Tomato spotted wilt virus (TSWV)

Additional information for diagnosis

Viruses and viral complexes

Several viruses or some of their strains, alone or together, are capable of causing wilting, drying out, and necrotic lesions of varying sizes on leaflets of tomato.

Some characteristics of the main viruses capable of inducing such symptoms are listed in *Table 24* (p. 223). Some of them also produce other more characteristic symptoms on leaves and other parts of affected plants. The sections on Leaflets and leaf discoloration (p. 91), on leaflet and leaf deformations (p. 47), and Spots on leaflets and leaves (p. 149) should also be consulted.

Among these viruses, of particular note are:

– Tomato spotted wilt virus (TSWV), responsible for the tomato spotted wilt, is very polyphagous and has been troublesome for several years on many vegetable and flower crops. The introduction of a more efficient vector (*Frankliniella occidentalis*) and the close proximity of sensitive flower and vegetable crops are the cause of this situation. Tomato crops are also affected and severe attacks are found throughout the year in some European countries. This particular virus is the cause of the scorched and bronzed appearance of the leaflets (Photos **412–415**) and in certain situations, their wilting and drying out;

– Cucumber mosaic virus (CMV), especially its necrotic strains, causing sometimes extensive necrotic lesions converging on the leaves and extending to the petioles and stems and affecting the fruit (Photos **416–419**). Although CMV is widely distributed in several countries in Europe, necrotic strains are more sporadic and occur mainly in field crops;

– Alfalfa mosaic virus (AMV) is less common than CMV. It causes necrotic lesions on the leaflets, which develop rapidly and lead to the leaves drying out (Photo **420**). The fruits show brown necrotic lesions (Photo **421**). It affects mainly field crops;

– necrotic strains of Potato virus Y (PVY) are the cause of necrotic spots on the leaf which, when numerous, may cause the drying of entire leaflets.

Note that some virus complexes affecting tomato sometimes cause much more severe symptoms. In such cases it is not unusual to see some very necrotic lesions (Photo **422**) leading to drying of all or part of plants.

Wilting

416 Several necrotic lesions accompanied by chlorosis are spreading on this leaflet.

417 On this plant, infected for some time, CMV caused necrosis and drying of many leaflets and several leaves. Several petioles and portions of the stem are also affected.

418 Some brown to black necrotic streaks are visible on the stem.

419 On some trusses, the fruits show irregular necrotic lesions of whitish colour.

Cucumber mosaic virus (CMV)

Table 24 Main characteristics of common viruses responsible for wilted or necrotic leaves on tomato

Virus	Symptoms	Vectors
Tomato spotted wilt virus (TSWV) **Description 43**	Spots, chlorotic to brown rings, becoming gradually necrotic on the leaflets; reddish brown patches are also visible (p. 164). The leaves can have mosaic and/or be chlorotic and bronze gradually, becoming necrotic over time. Brown elongated lesions (streak) may appear on petioles and stem. The end of the stem sometimes curves. Large patches, rings, arcs, partially cover the fruits (Photo **412**). These can be locally discolored, bronzed, and deformed.	Transmitted by thrips species, in the persistent manner (*Frankliniella occidentalis, F. fusca, F. schultzei, Thrips palmi, T. tabaci, Scirtothrips dorsalis*).
Cucumber mosaic virus (CMV) **Description 35**	Spots, necrotic lesions causing necrosis of one or more leaflets (Photo **416**). These necrotic lesions subsequently spread to the stem and the apex of the plant (Photos **417** and **418**). Severe necrotic lesions beige to brown, round and raised, covering large areas of fruit (Photo **419**, see also p. 408). Mosaic and leaf changes see pp. 60 and 101.	Transmitted by aphids in the nonpersistent manner (*Myzus persicae, Aphis gossypii, A. craccivora, A. fabae, Acyrthosiphon pisum*).
Alfalfa mosaic virus (AMV) **Description 33**	Necrotic lesions starting at the base of the leaflets, becoming generalized to the leaf (see Photo **420** and p. 319), sometimes affecting the veins. They extend on the stem and several centimetres may be affected as well as the apex. Brown necrotic lesions are visible on fruits that are deformed (Photo **421**). A yellow mosaic of aucuba type can also be found on the leaflets in some conditions.	Transmitted by aphids in the nonpersistent manner (*Myzus persicae, Aphis fabae, A. gossypii, A.craccivora*).
Pepino mosaic virus (PepMV) **Description 29**	Necrotic spots on lower leaves, extensive scorched areas. Brown, corky lesions on stems and petioles. Mosaic and leaf changes (pp. 115 and 165).	Transmitted by casual contact, during the cultivation operations, by pollinating bees, and by seed.
Potato virus Y (PVY) **Description 36**	Symptoms vary depending on the strain: – Reddish brown necrotic spots (Photo **422**). Longitudinal necrosis can be observed on the petioles and stem in the case of necrotic strains (see p. 162); – Discrete mottling, green mosaic, vein banding, diffuse chlorotic spots for the common strains (see p. 102).	Transmitted by about 40 aphids in the non-persistent manner (including *Myzus persicae, Aphis gossypii, A. fabae, A. pisum, Macrosiphum euphorbiae*).
Impatiens necrotic spot virus (INSV) **Description 44**	Brown spots and necrotic inter-vein lesions.	Thrips (*Frankliniella occidentalis, F. fusca*).
Parietaria mottle virus (PMoV) **Description 44**	Spots, necrotic lesions, mainly at the base of the leaflets. Longitudinal necrotic lesions on stems, terminal bud death. Alterations and rings with chlorosis on fruit, browning later, becoming corky and leading to fruit deformation.	Probably transmitted by thrips and pollen.
Tomato ringspot virus (ToRSV) **Description 46**	Small necrotic lesions, mottling on leaflets. Systemic necrotic lesions.	Nematodes (*Xiphinema americanum* and its sub-species). Seed-borne in other plants.

Wilting

420 Necrotic lesions begin to form at the base of many of these young leaflets. They will eventually extend to the entire leaf.
Alfalfa mosaic virus (AMV)

421 The fruits are covered with irregular concave blackish necrotic lesions to varying degree.
Alfalfa mosaic virus (AMV)

422 Brownish necrotic lesions are spreading to all the inter-veinal tissues on these leaflets.
Potato virus Y (PVY)

A new virus, not belonging to a known genus, *Tomato torrado virus* (ToTV) was observed in tomato in the region of Murcia in Spain in 2001, and in the Canary Islands in 2003. It is causing severe necrotic leaf lesions.

In affected crops, initially necrotic spots surrounded by a pale green to yellow halo appear at the base of the leaflets (Photo **423**). Subsequently, the necrotic lesions spread and become widespread, giving the affected plants a burned appearance. Longitudinal necrotic lesions develop on the stem and the fruit show necrotic damage in spots, browning, and arcs, accompanied by splits and deformation. Plant growth is reduced and yields greatly diminished.

Note that this virus, with icosahedral viral particles and measuring 28 nm in diameter, could be transmitted by whiteflies. Moreover, during the aetiology of this emerging viral disease, other viruses were isolated from diseased plants, including PepMV. These co-infections, as well as other unknown factors could affect the nature of symptoms.

This virus has been reported on tomato in 2003 and 2004 in Poland, in the Wielkopolska region, associated with large populations of whiteflies.

It will be necessary to be particularly vigilant in Europe if it is confirmed that this virus is transmitted by whiteflies; it could cause epidemics similar to those caused by TYLCV, for example.

Whatever the symptoms, if there are any doubts about the diagnosis, a specialized laboratory that will perform specific tests for viruses should be contacted.

Wilting

423 Necrotic lesions are visible on the part of the leaf at the base of several leaflets.
Tomato torrado virus (ToTV)

Various phytotoxicities

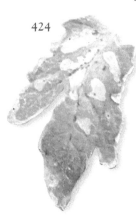

424

Some pesticides, notably herbicides, used in adverse conditions or accidentally applied to tomatoes may be responsible for phytotoxicities (chemical injuries), resulting in wilting, drying, and leaf necrosis. Thus, the leaflets of tomato plants may show:

– marked inter-vein patches on lower leaves, spreading, becoming necrotic and gradually drying out;

– necrosis and drying of the periphery of young and old leaves;

– wilting and drying of the leaves at the stem apex.

Photos **424–426** illustrate some of the symptoms induced by herbicides, particularly those that inhibit photosynthesis (see *Table 25*).

To help confirm this hypothesis, other parts of the book with further information on phytotoxicities should be consulted, especially pp. 77 and 131.

425

Table 25 Some characteristics of photosynthesis inhibiting herbicides used in agriculture

Contact herbicides, mostly foliar, with low to significant translocation depending on the herbicide.	Major physiological functions affected	Major cellular alterations and disrupted functions
Amides, benzothiadiazones, biscarbamates or phenylcarbamates, hydroxybenzonitriles (bromoxynil, ioxynil), phenyl-pyridazines (pyridate), pyridazinones, triazines (simazine, atrazine)[a], triazinones (metribuzine), uracil (bromacil), substituted ureas (diuron, linuron), bipyridiles (diquat, paraquat).	Presence on the leaflets of wide distinctive inter-vein patches, becoming gradually necrotic and withering, wilting (Photos **424** and **425**).	Alteration of cell membranes (plasma, tonoplast). Assimilation of CO_2 is stopped. Inactivation of electron transfer system. Increase in membrane permeability. Decrease in chlorophyll content and carotenoids.

[a] Herbicides feared for their persistence and their involvement in water pollution.

426 All the leaflets of this leaf show a gradual inter-vein drying.
Phytotoxicity

Other compounds, phytotoxic under certain conditions, may also induce temporary wilting on tomato, followed by a reduction in the size of all plant parts. This for example occurs with para-acetic acid (PAA), a disinfectant used in hydroponics with low impact on the environment. The induced transient wilt may be due to increased electrolyte loss by the roots, a reduction of oxygen consumption by the roots, and the destruction of root tips.

Please note that too high concentrations of CO_2 in glasshouses can be responsible for chlorosis and necrosis of older leaves or leaflets, particularly at the stem apex.

Lightning Damage may occur on tomatoes following storms. In such cases, growers see circular zones of varying widths of wilted plants which dry up gradually. Some of them, near the centre of these zones, can be completely destroyed. These areas are in fact the point of impact of lightning in the culture. Note that the stems show several changes, described on p. 343.

The identification of this nonparasitic damage should not cause any problems for at least two reasons:

– The association of cause and effect is immediate, the symptoms appear immediately after a stormy episode in the region;

– The affected plants are in clearly defined sectors.

Unfortunately, lightning induced damage is often irreversible and no action can improve the condition of the affected plants.

Root asphyxia

Water stresses (of different frequency and types) that affect both soil and soil-less crops, are collectively called 'root asphyxia' (drowning, water wilt). Note that they can occur very occasionally, e.g. after heavy rainfall or irrigation pipe leaks or as a result of defective solenoid valves. In such cases, the water, which can submerge a crop by accumulating in troughs and by saturating a substrate, results in transient stress. Often the apex leaves brighten, fade, and curve to varying degrees. If these asphyxiating conditions persist the leaves eventually turn yellow and become necrotic.

Water stresses may be unintentionally repeated over time at each irrigation. This applies, for example when a producer does not fully control the plants' irrigation and tends to use too much water at each watering. Watering tomato plants is a critical operation which when done carelessly can lead to crop damage. A succession of root asphyxia leads to symptoms similar to those described above. In this case, they may be preceded by more limited plant growth, and wilting predominantly on the lower leaves. Similar situations are also found in some heavier soils as well as in troughs.

Whatever the nature and frequency of water stresses, the effects on the plants are similar. The tomato, like many other plants, tolerates poorly its root system being totally submerged for an extended period of time. Such a situation leads to a lack of oxygen: and root decay follows which accounts for the wilting. This soil phenomenon may sometimes be amplified by the proliferation of anaerobic micro-organisms that contribute to the decay of the roots.

In general, significant damage is mainly observed in poorly drained fields, with heavy soil and during periods of hot weather that promote anaerobic micro-organisms, accelerate plant transpiration, and thus wilt.

In plots where asphyxia occurs fairly regularly, it will be necessary to drain the soil. Be wary also of some substrates which tend to be more liable to becoming anaerobic. In all cases, it will be necessary to correct the frequency and quantity of water used for each irrigation. Keep in mind that the tomato has important root regenerative capacity that should enable the grower to overcome some asphyxiating situations by earthing up plants, or by controlling irrigation.

Wilting

227

Air pollutants

Air is normally made up of 78% hydrogen and 21% oxygen, the majority of the remaining percent being composed of water vapour and CO_2. It may also include air pollutants (phytotoxic air pollutants) that are causing plant symptoms which are fairly comparable to those caused by diseases or pests, nutritional deficiencies, phytotoxicities, and sometimes even by climatic stresses. Thus, the symptoms observed on plants are of three types: necrotic damage, yellowing and other colour changes of the tissues, and reduced growth of some plant parts or the whole of the plant.

These pollutants are classified into two groups:

– Pollutants from one or more localized sources (a factory plant, a volcano) which affect the plants in close proximity (sulfur dioxide, nitrogen oxides, chlorine compounds, ammonia, ethylene, volatile herbicides such as 2,4-D, and dust);

– General air pollutants from a variety of dispersed sources such as car engines, and affecting the atmosphere over large areas (ozone, peroxyacetyl nitrate).

Note that when these gases can remain unchanged in the atmosphere, they are referred as 'direct pollutants'. They can also react with other normal or abnormal air components, and causing secondary pollutants. These are dependent on mechanisms such as photochemical reactions, the formation of free radicals, and oxidation giving rise to new compounds. These reactions, which are still poorly understood, are the source of oxidizing compounds (ozone) and photochemical compounds (smog).

Tomatoes are particularly sensitive to several inorganic pollutants (ozone, sulfur dioxide, chlorine) and organic pollutants (peroxyacetyl nitrate, ethylene). Their effects can be quite different depending on the production context and the pollutants involved (see also *Table 26*).

Ozone, which is an active form of oxygen, leads to the emergence of a multitude of small dark lesions between the main veins on the upper side of the leaf. These eventually whiten the upper side of the leaflets or result in the appearance of large necrotic patches. Young plants and newly mature leaves appear more susceptible to ozone. Very young leaflets and old leaves may be resistant. Ozone is the main air pollutant in the US.

Sulphur dioxide is the source of light, dry, and papery spots, located between the veins or on the leaf edges. During chronic pollution, inter-veinal leaf spots which are red-brown to black, are sometimes found on both the upper and lower sides of the leaves. The more developed leaves are the most sensitive.

Peroxyacetyl nitrate (PAN) causes discoloration in spots or stripes on the leaflets, and the underside of the leaf is silver to bronze coloured. This change in colouration of the leaf is in fact due to absorption of PAN at the stomata. This gas then diffuses into mesophyll cells and destroys them. Air pockets are formed between the lower epidermis and palisade cells and are the cause of the metallic appearance of the leaflets.

Note that several factors influence the level of damage of these pollutants, such as their critical toxicity, concentration, time of appearance, the factors influencing their distribution, the sensitivity of plant species (direct pollutants), the plants development stage, the crop variety, and production conditions. The most damage is often observed during periods of hot, calm, humid weather, with high atmospheric pressure. Under these conditions, air inversions happen between the hot air that takes the place of the cold air at ground level, thus allowing the accumulation of pollutants near the plants.

Table 26 Characteristics of two major air pollutants which can affect tomato

	Ozone	Peroxyacetyl nitrate (PAN)
Disrupted functions in plants	– Modification of membrane permeability. – Loss of turgidity of guard cells. – Reduced gas exchange. – Acceleration of starch hydrolysis.	Collapse of mesophyll cells whose place is filled with air.
Symptoms observed on plants	– Small recessed necrotic spots, ranging from white to dark brown or even black or red depending on species. – Early senescence of tissues that may yellow and become necrotic. – Turgor loss of and tissues whitening. – Premature leaf and fruit drop. – Reduced plant growth and yields.	The undersides of the leaves have a shiny or metallic appearance, and a silvery to tan colour (silver leaf).
Other features	The most damaging air pollutant. Nitrogen oxides and hydrocarbons are precursors.	Gasoline vapours, combined with O_3 or NO_2 in differing proportions produce PAN, a phytotoxic photochemical compound. Damage is seen in very sunny areas with high traffic related emissions of hydrocarbons, near large towns, and in the presence of fog. Damage is now observed in northern Europe.

To identify the cause of air pollution symptoms is not easy particularly when they occur for the first time. In any case one should identify the nature of the compound, its concentration, and its geographical distribution in the region. The hypothesis of a toxicity linked to air pollutant will need to be considered, particularly in production areas adjacent to large cities, smelters, refineries, airports, highways, incinerators, factories, and various production plants (bricks, pottery, cement).

The solutions to the problems of air pollutants are complex and extend far beyond the production of tomatoes. Factors involved include the awareness of public authorities, the introduction of measures to regulate emissions from motor vehicles and aircraft, stopping or limiting emissions at source, the selection of tolerant plants for critical areas, the use of less susceptible cultivars and so on.

Wilting

Allelopathy

The previous crop can be a stimulating or on the contrary inhibiting factor for the growth of other plants. These effects, referred to as allelopathy, are due to the production of chemicals released into the soil, exuded by the roots or from the decomposition of crop debris. When harmful, these substances may be responsible for significant effects on plants and their development.

For example, this type of phenomenon is reported in the literature, notably in the US where the tomato is grown near walnut trees. In this case, in summer, the plants closest to the trees may wilt and die suddenly. The furthest tomato plants do not die but may be stunted and/or wilt. Note

that the woody tissues of the stem sometimes brown, which leads to some confusion with the symptoms of Fusarium and Verticillium wilt (see pp. 335 and 339). This phenomenon is called 'walnut wilt'. It is due to the toxic effects of juglone, a substance emitted by the walnut tree roots (and several other trees of the Juglandaceae family) and is also toxic to many plants such as potato, asparagus, peas, and cabbage. Note that other vegetables may be more resistant (beets, beans, carrots). This substance is found in leaves, shells, bark, and roots of these trees and released into the soil where it is absorbed by the roots of susceptible plants. Juglone is quite stable in soil and its effects persist for several months or even years, even after the trees are removed.

Such weeds as *Cardaria draba* are also reported as responsible for the phenomenon of allelopathy on several vegetables in Jordan, leading to reduced germination and growth of tomato, cabbage, and onions.

The identification of such phenomena, rare and poorly understood, is not easy. It is therefore important not to omit them in diagnosistic hypotheses. Thus, it will be necessary to be aware of some previous crops, the presence of certain weeds, or the proximity of 'dangerous' plantations when establishing a tomato crop. The level of organic matter in soil must be kept high because the toxins adhere to it instead of being absorbed by the plants. Moreover, the activity of microbial populations capable of metabolizing toxins into nontoxic compounds, is also favoured by high levels of organic matter.

Excess salinity

The tomato is relatively tolerant of excess salinity (salt injury). Despite this, like many plants, they may display a number of symptoms in extreme situations: reduced growth, yellowing foliage (see p. 130), necrosis and drying on the edges of the leaves (Photo **427**). This last symptom is particularly evident in the presence of excess chlorine.

427 Several leaflets display a peripheral scorching, due to an excess of salt.
Phytotoxicity

Symptoms on roots and stem base

Symptoms studied

- Yellowing, browning, blackening of roots, sometimes accompanied by stem base symptoms
- Other root symptoms (superficial suberization, small black spots, cysts, galls)
- Various symptoms at the stem base and lower stem

SYMPTOMS OF ROOTS AND/OR STEM BASE

Possible causes

- *Acremomium strictum*
- *Alternaria alternata* f. sp. *lycopercisi*
- *Calyptella campanula*
- *Colletotrichum coccodes*
- *Didymella lycopersici*
- *Fusarium oxysporum* f. sp. *radicis-lycopersici*
- *Humicola fuscoatra*
- *Macrophomina phaseolina*
- *Olpidium brassicae*
- *Ozonium texanum* var. 'parasiticum'
- *Phytophthora* spp.
- *Plectosporium tabacinum*
- *Pyrenochaeta lycopersici*
- *Pythium* spp.
- *Rhizoctonia crocorum*
- *Rhizoctonia solani*
- *Rhizopycnis vagum*
- *Sclerotinia minor*
- *Sclerotinia sclerotiorum*

- *Sclerotium rolfsii* (*Athelia rolsii*)
- *Spongospora subterranea* f. sp. *Mediterranea*
- *Thanatephorus cucumeris*
- *Thielaviopsis basicola*
- *Agrobacterium radiobacter* biovar 1
- *Agrobacterium tumefaciens*
- *Globodera tabacum* (cyst nematodes)
- *Meloidogyne* spp. (galls nematodes)
- *Pratylenchus* spp. (semi-endomigratory nematodes)
- Other nematodes dependent on roots (*Rotylenchulus reniformis, Belonolaimus longicaudatus*)
- *Orobanche* spp.
- Root asphyxia
- Corky stem base and tap root
- Excess salinity
- Soil pH is too acidic
- Various phytotoxicities

Very difficult to diagnose

Deterioration of the roots, stem base, and lower stem have been grouped deliberately in the same section as it is not always easy to separate these three parts of plants when making a diagnosis. Indeed, the fungi attacking the roots are sometimes able to colonize the stem base and the bottom part of the stem. Similarly, stem base fungi sometimes progress downwards towards the roots.

The identification of diseases affecting these three plant parts is particularly difficult because they frequently cause similar symptoms. Confusion may arise, so one needs to be very observant and follow carefully the advice given throughout this section. If there are any doubts, a specialized laboratory should be contacted which will be equipped to observe or isolate the pathogen/pests attacking the tomato parts that are underground or close to the ground.

Roots and stem base

428 Healthy root system (from a soil-grown plant), with white roots, after being carefully lifted and washed thoroughly.

430 When the roots are severely affected, browning can occur in the vessels in the tap root (1) and also at the stem bases (2). In this photo, vascular browning is very pronounced on the diseased plant on the left. The adjacent healthy plant reveals nothing special. Its vessels (3) and pith (4) are intact.

429 Excessive watering and *Pyrenochaeta lycopersici* cumulative effects are causing yellowing, browning, and corkiness of part of the root system.

431 Two main parts are easily distinguished on the root, the cortex outside (1) and central cylinder (2) inside. Where severely degraded, the cortex detaches from the central cylinder and breaks in places. The root is characteristically shaped like a pearl necklace.

432 Some parasitic diseases cause browning of vessels of the root central cylinder (1), as shown in longitudinal section.

Observations of tomato root systems with a variety of symptoms

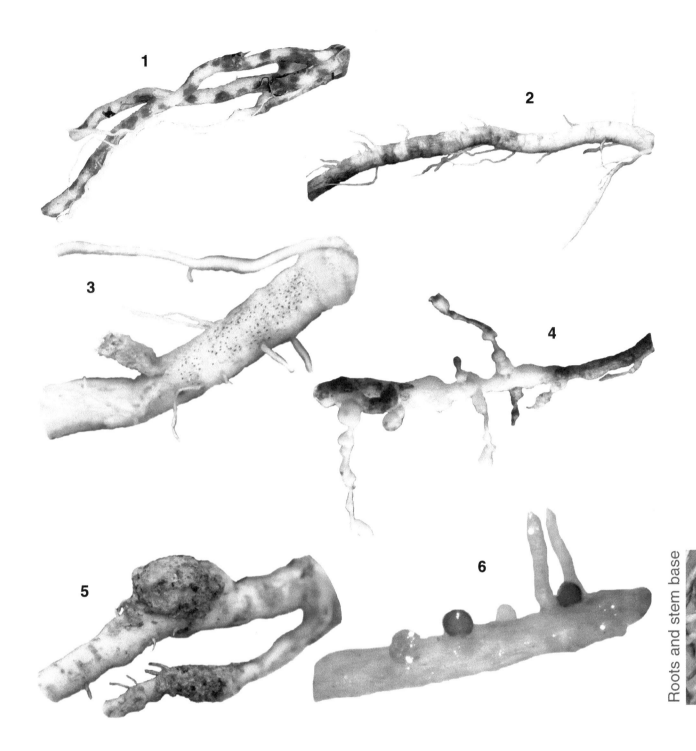

Figure 38 Main symptoms on roots.

1 Root yellowing and browning
2 Suberized root described as 'corky'.
3 Black spots on roots.
4 Roots with galls.
5 Tumour deforming a root.
6 Roots with cysts of differing maturity.

Figure 39 Examples of stem base lesions.

1 Slightly sunken flame-shaped canker.
2 Moist canker, dark brown to black, with diffuse edge.
3 Dry canker encircling the stem, beige, with a well defined outline.

Observe the roots carefully and in a good condition

To assess the health of a plant root system, it is essential to make observations of the roots in the best possible condition. This is often poorly undertaken by many producers and technicians, but is necessary in order to make a correct diagnosis.

Soil culture

First, the root system(s) must be recovered. To do this, they must be lifted very carefully otherwise the damaged roots, which are the weakest, will be lost and these are the most interesting for diagnosis. It is then necessary to wash the roots carefully and thoroughly with water to remove soil particles that often mask some symptoms.

Once the roots are prepared they can be examined (Photos **428–432**) carefully, preferably with a magnifying glass in order to look more closely at some structures that are characteristic of the presence of one or more pathogens/pests. The main changes and abnormalities found on roots are shown in Figure **38**.

If lesions are evident on the stem base and/or lower part of the stem, the root system should also be checked, and *vice versa*. Indeed, the stem base may also have a number of lesions which are shown in Figure **39**. The section on Internal and external symptoms of the stem, p. 297 should also be consulted.

Soil-less culture (hydroponics)

In hydroponic cultivation, observing the root system poses less of a problem although the roots may sometimes be more difficult to retrieve. Simply open a bag of substrate or clear a channel to view a portion of the roots. The health status of these will be representative of the other roots buried within the substrate (Photo **433**). It may also be advisable to consider extracting some of the 'water roots'. In cultures in NFT (nutrient film technique), the absence of substrate makes observation easier. The condition of the stem base base should also be examined (Photo **434**).

 The symptoms shown here are fairly typical of a root problem but they are insufficient to identify the exact cause(s). They only allow the root system to be implicated in the plant's dysfunction.

433 A 'window' made in a block shows that many roots visible on the side are brown and rotten.

434 The stem base of the tomato is frequently affected; it is the stem portion located at ground or substrate level. In this case, a moist black lesion, caused by *Didymella lycopersici*, surrounds the lower part of the stem.

Roots and stem base

237

435 Poor germination and the rapid collapse and disappearance of seedling are features of damping-off. In this case, the population of etiolated seedlings is very sparse following an attack by *Pythium ultimum*.

437 The part of stem at ground level shows a blackish lesion; soft and succulent tissues have literally melted.
Pythium ultimum

439 A window in this rock wool slab shows numerous yellowish roots mixed with healthy roots.
Pythium group F

436 The epi- and hypocotyl of the seedling show dark brown lesions.
Pythium ultimum

438 Some *Phytophthora* spp. are particularly aggressive on tomato roots of soil-less cultures. Necrotic lesions and brown rot indicate their presence.
Phytophthora cryptogea

***Pythium* spp., *Phytophthora* spp. (damping-off, root rot, and basal stem canker)**

YELLOWING, BROWNING, BLACKENING OF THE ROOTS, SOMETIMES ACCOMPANIED BY STEM BASE SYMPTOMS

Possible causes

- *Acremomium strictum*
- *Colletotrichum coccodes*: Description 10
- *Fusarium oxysporum* f. sp. *radicis-lycopersici*: Description 11
- *Humicola fuscoatra*
- *Olpidium brassicae*: Description 18
- *Ozonium texanum* var. *"parasiticum'*: Description 18
- *Phytophthora* spp.: Description 12
- *Plectosporium tabacinum*
- *Pyrenochaeta lycopersici*: Description 13

- *Pythium* spp.: Description 12
- *Rhizoctonia crocorum*: Description 18
- *Rhizoctonia solani*: Description 14
- *Sclerotium rolfsii* (*Athelia rolsii*): Description 16
- *Thielaviopsis basicola*: Description 17
- *Pratylenchus* spp. (semi-endomigrating nematodes): Description 51
- Other root nematodes
- Root asphyxia (see p. 227)
- Excess salinity
- Plough pan (see p. 277)

The majority of nonparasitic and parasitic diseases affecting tomato roots cause, at first, a diffuse yellowing, then browning (localized or generalized), and finally necrosis and disappearance of many roots and rootlets. In severe cases, the root system is completely damaged and the vessels located at the stem base can turn slightly yellow and brown (Photo **430**). Symptoms sometimes develop towards the stem base and the base of the stem, causing cankers (see section Various changes of stem base and stem portion close to the ground,

p. 281). These symptoms can be observed both on seedlings in the nursery and on adult plants during cultivation. In the first case, they will be responsible for damping-off.

In addition to the symptoms described above, a number of root pests induce other relatively characteristic changes or signs that make their identification easier. In many cases, laboratory tests will be needed to determine their exact nature.

Additional information for diagnosis

Pythium spp.

Pythium spp. are mainly known for attacking tomatoes early in their growth at the propagation stage or after planting in the field, by affecting the seeds or seedlings before or after germination (damping-off). They are sometimes rampant on adult plants under special conditions of soil cultivation such as hot and very humid conditions

in the tropics. Soil-less crop development on substrates and NFT has allowed these organisms to settle in these production systems and to cause root damage (Pythium root rot), sometimes in spectacular ways during the plant production cycle.

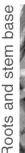

Roots and stem base

239

Symptoms are mainly seen on the plant parts in the soil or substrate (roots, stem base), but they can also develop on the tomato foliage or fruits.

In nurseries, *Pythium* spp. can inhibit seed germination or disrupt the development of young seedlings soon after colonizing the roots and/or stem base (Photo **436**). Affected tissues, often succulent, tender, and very sensitive, are moist and rapidly turn brown, rot, and die. Some roots, but mainly the stem at the soil or substrate surface, may be severely constricted (Photo **437**). Subsequently, it is quite common to see seedlings wilting, turning yellow, withering and/or collapsing and eventually disappearing[1] (Photo **435**).

Note that air-borne attacks on seedlings are possible when conditions are particularly wet and rainy. Thus, moist and brown lesions may appear on leaves, petioles, and stem and lead to the plant's death. Such symptoms have been reported in Florida and are associated with *P. myriotylum*.

Other pathogens may be responsible for symptoms on seedlings (see Figure **42**); some of their characteristics are listed in *Table 27* (p. 243). As they have similar effects on the roots and/or stem base, it is not easy to distinguish between them with the naked eye.

The effects of *Pythium* spp. may persist beyond the nursery to developing plants after planting and also on adult plants. The symptoms resulting from the direct effect of *Pythium* spp. must be considered separately from those related to their impact on the functioning of plants. The first symptoms mainly concern the roots and stem base. In many situations, yellowing (Photo **439**), pronounced and widespread root browning, and the disappearance of many rootlets are seen. Moist and dark lesions may develop on the stem base and spread a few centimetres (*P. aphanidermatum*, *P. myriotylum*).

These symptoms can have various consequences: the appearance of apical necrosis on fruit, a reduction in plant growth and sometimes yield, some rapid and reversible wilting, drying, and plant death. They are mostly found in soilless cultures but also in field crops in tropical production[2].

Note that damage synergy has been observed between *P. aphanidermatum* and the root-knot nematode *Meloidogyne incognita*.

Green and ripe fruits in contact with or near the ground can especially be affected by certain species of *Pythium*. These lesions are moist and change rapidly, causing fruit liquefaction and collapse. A white cottony mycelium may be visible on lesions[3].

It is only the observation of oospores (or sometimes of other structures such as sporangia or chlamydospores) in and on the affected tissues or isolation on artificial media, which will confirm the diagnosis without any doubt (Photos **454** and **835–842**).

Figure 40 Main structures characteristic of *Pythium aphanidermatum*.

Pathogens involved are:

[1] *Pythium aphanidermatum* (Figure **40**), *P. dissotocum*, *P. deliens*, *P. inflatum*, *P. irregulare*, *P. kunningense*, *P. periplocum*, *P. spinosum*, *P. ardicrescens*, *P. ultimum*, *P. ultimum* 'sporangiiferum'.

[2] *Pythium aphanidermatum*, *P. dissotocum*, *Pythium* groupe F, *P. irregulare*, *P. myriotylum*, *P. ultimum*.

[3] *Pythium aphanidermatum*, *P. ultimum*, *P. myriotylum*, *P. arrhenomanes*, *P. acanthicum*.

Phytophthora spp.

Several species of *Phytophthora* can affect tomato (damping-off, *Phytophthora* root rot) by causing symptoms on the roots and stem base (see p. 283), sometimes on the foliage (see p. 185) and on fruit in contact with the ground (see p.374).

Root contamination frequently take place early, often in the nursery or after planting, in open field and soil-less cultures. They are the cause of symptoms that occur before those caused by the other agents of root necrosis. Whatever the species involved, light brown lesions appear on the young roots, gradually extending to the whole root system and causing root rot (Photo **438**). Rot sometimes reaches the stem base that turns a brown to black colour. Plants may wilt and wither.

Root and/or stem base *Phytophthora* include:
– *P. nicotianae* (Figure **41**), widespread in the world and often known as '*Phytophthora parasitica*'. It attacks both the seedlings in the nursery and the roots, stem base, and fruit of mature plants (see pp. 283 and 374). On roots, it causes brown and moist lesions, the root system completely turns brown and finally rots (Photo **438**). The roots are sometimes circled by local lesions that are quick to cause the death of the distal part. This organism can reach the stem base and cause a diffuse brownish lesion. Symptoms from soil contamination are mostly found on plants in nurseries or in the weeks after field planting. It affects plants in all systems, gardens of amateurs, intensive farming in open fields or under protection, and tomatoes for industry. Rootstocks resistant to major soil-borne pathogens/pests are also affected (see p. 284). It is found in soil-less cultures, introduced with the plants, and its incidence is limited; it is in no way comparable with that of *P. cryptogea*;

– *P. cryptogea*, which is also severe in the weeks after planting, causes a brown stem base lesion which gradually encircles it. The root system also rots (Photo **438**). Subsequently, the leaves eventually turn yellow, wilt, and wither. This *Phytophthora* is described in several countries on tomato roots growing in nutrient solutions in soil-less culture. It is occasionally seen in soil-less cultures, causing substantial root losses and mortality of adult plants;

– *P. arecae* was reported in the Netherlands in 1960 on tomatoes grown in poorly drained soils and sometimes over fertilized. Symptoms caused by this fungus occurred in the weeks after planting and were mainly localized on plant roots. The latter had rot, sometimes extensive, leading to rapid wilting of plants;

– *P. capsici*, very common on vegetables, causes seedling damping-off, moist soft dark brown stem base cankers, and root rot causing their disintegration. Vascular tissues nearby may be a brownish tinge (seen in soil-less cultures in South Africa);

– *P. citricola* was considered responsible for stem base rot and wilting of seedlings in Italy in 1960;

– *P. mexicana* causes damping-off and increased mortality of seedlings. It mainly causes lesions on fruit;

– *P. erythroseptica* was involved in root and stem base rot occurring on tomatoes grown in Australia. Plants also exhibited reduced growth and wilting. It is also reported in soil-less cultures in Bulgaria.

Note that *P. cactorum* (Lebert & Cohn) J. Schröt. and *P. drechsleri* Tucker have very occasionally been reported on tomato.

Roots and stem base

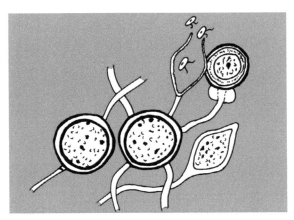

Figure 41 Main structures characteristic of *Phytophthora nicotianae*.

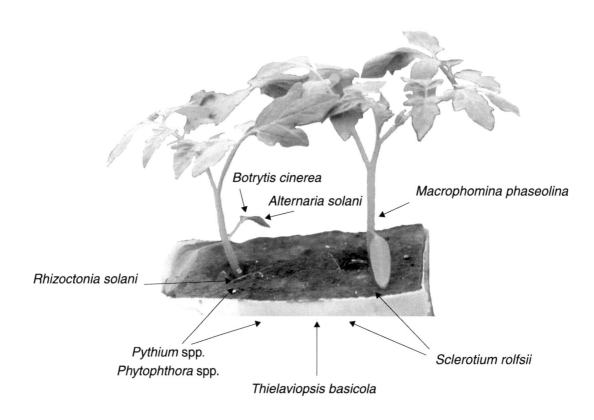

Figure 42 Main fungi attacking tomato at propagation stage and potential locations of the attacks on seedlings.

Table 27 Key features of fungi or similar organisms responsible for attacks on tomato seeds and seedlings in traditional or intensive nurseries

Pathogen	Symptoms	Attack frequency	Differentiating structures on, in, or close to damaged tissues
Roots and/or stem base			
Pythium **spp.** *Phytophthora* **spp.** (damping-off)	Soft wet brown rot on the stem base and roots.	Frequently.	Oospores, sporangia, and chlamydospores (Photos **454** and **455**).
Sclerotium rolfsii (damping-off, stem base rot)	Moist brown rot rapidly surrounding the stem at ground level and extending to the roots.	Mainly in traditional nurseries in tropical areas.	Abundant white mycelium covering the base of plants and soil. Smooth spherical sclerotia (1–3 mm), first white, then gradually brown (Photos **541** and **542**).
Thielaviopsis basicola (black root rot)	Roots partially or completely brown to black.	Rare.	Groups of-barrel-shaped chlamydospores (Photos **450** and **460**); Sometimes phialides producing endoconidia (see Description 17)
Stem base and/or stem			
Rhizoctonia solani (damping-off, stem base rot)	Reddish brown lesions on hypo- and epicotyl before emergence. After emergence, reddish brown symptoms starting at ground or substrate level, surrounding the seedling and causing its death.	Quite frequently.	Compartmentalized mycelium, hyaline to brown, with a constriction at branches, developing on tissue surfaces (Photos **441** and **457**).
Botrytis cinerea (grey mould)	Beige to brown discoloration starting from senescent cotyledons.	Frequently.	Grey mould with numerous branched conidiophores bearing powdery conidia (Photo **319**).
Alternaria tomatophila (early blight)	Brownish to black lesions, can be elongated, well defined, with concentric patterns.	Now infrequent, mainly in a few extensive nurseries.	Brown multicellular conidia with a long filiform appendage (Photo **302**).
Macrophomina phaseolina (charcoal rot)	Blackish canker below the cotyledons at the time of emergence.	Reported only in a few countries.	Small sclerotia with a diameter less than 1 mm (Photo **492**).

Roots and stem base

4 In the presence of root browning, these questions should be addressed!

By simply answering the following questions, an explanation of the
problem can sometimes be obtained (any positive answer to one question will confirm it).
– Was excessive irrigation used during propagation or when planting out?
– Are the diseased plants located in the wetter parts of the area?
– Were the tomatoes planted out at a time when the ground was still cold and
  damp?
– Is the soil well drained?
– Has too much fertilizer been added before planting or during cultivation?

440 A high proportion of the root system is
rotten. Injured tissues, rather wet, show a reddish
brown to dark brown colour. The cortex has
disappeared in some places.

441 The brown and branched mycelia of *R. solani* are
clearly visible on the root.

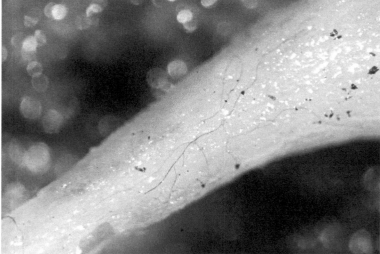

Rhizoctonia solani (damping-off,
Rhizoctonia root rot, basal stem canker)

Olpidium brassicae

This aquatic fungus is not thought pathogenic on tomato; however, it is so on seedlings of tobacco which is also a solanaceous plant. Yet it is not uncommon to observe its sporangia and chlamydospores, or 'resting spores' in the cells of the epidermis and cortex of tomato roots (Photo **456**), whether grown in soil or soil-less. The quantities found in the soil-less crop roots are sometimes very large and question the role of this parasitic root fungus. Is it a simple biomarker of roots during decomposition or does it influence senescence? Note that it is also the vector of two very damaging viral diseases on salads: big vein disease and orange blotch.

Environmental conditions including temperature and oxygenation of soil, the substrate or nutrient solution, and especially the root condition, certainly influence its behaviour. This fungus is well adapted to aquatic life and has, as *Pythium* spp. and other aquatic fungi, mobile zoospores allowing it to spread easily in water and in the nutrient solutions of soil-less culture.

Rhizoctonia solani

This soil fungus is very common and is widespread in the world, indeed is almost ubiquitous in vegetable plots. It can be considered as a biomarker of so called 'sick' soil. It is therefore not surprising that it affects tomato, on which it can cause various symptoms on nearly all organs.

As with *Pythium* spp., *Rhizoctonia solani* particularly attacks seedlings after planting causing damping-off. Both before and after emergence, brown lesions, reddish brown to black, are visible on the affected seedlings. Once the latter are out of the ground, the changes are mainly localized on the stem base, causing the death of young seedlings.

This fungus is also capable of attacking the plants at their developing stage and even adults plants. Again, it can affect the roots and stem base of the tomato. On the roots, the nature of the symptoms noted seems influenced by the cultural context and in particular the soil moisture. It is not uncommon to see the fungus associated with slightly superficially suberized tomato roots, rather 'dry', on which many rootlets are missing. In this case, it is not always easy to know if it is merely a biological marker of root systems which have suffered climatic or agricultural stress. It is also found on root systems with moist reddish-brown lesions and/or rotting of the same nature (Rhizoctonia root rot, see Photo **440**). In this situation, the degraded cortex can detach in places.

Note that *R. solani* can be observed alongside *Colletotrichum coccodes*, *Pyrenochaeta lycopersici*, the phytophagous nematodes on soil-grown tomato roots. It is sometimes encountered in soil-less cultures, in complex with *Pythium* spp., *Fusarium oxysporum* f. sp. *radicis-lycopersici*, and *Colletotrichum coccodes*.

It also affects the stem base (basal stem canker) and many aerial organs of tomato. Therefore, pp. 293 and 411 should also be consulted. The perfect form of this fungus, *Thanatephorus cucumeris*, also causes damage to this plant (see p. 311).

Whatever the nature and location of symptoms, the compartmentalized hyaline to brown mycelium of *R. solani*, progressing on or near the damaged tissues indicates its presence (Photos **441** and **457**).

Roots and stem base

442 Under this rockwool cube, a significant proportion of roots are brown and rotten.

444 Longitudinal cuts made in the tap root and other roots reveal a browning of the xylem.

446 A moist and brown canker, slightly concave, appears at ground or substrate level and gradually surrounds the stem. Flame-shaped, it can be covered by a fairly typical pink mucous-like sporulation.

443 Numerous lesions, rounded and localized, progressively cover the roots; many fine roots are rotten and have disappeared.

445 The vascular browning can extend in the stem over several centimetres above the stem base. The vessels take on a brown colour, sometimes very dark.

Fusarium oxysporum **f. sp.** *radicis-lycopersici* – FORL
(Fusarium crown and root rot)

Fusarium oxysporum f. sp. radicis-lycopersici (FORL)

This devastating fungus, now present in many countries, is capable of attacking the roots and stem base of tomato (Fusarium crown and root rot) in field and in protected crops. It was introduced into Europe during the 1980s from contaminated horticultural substrates and spread rapidly to all areas of production. Before the advent of resistant varieties it was a real threat to protected crops, especially soil-less ones.

This *Fusarium* is primarily a pathogen of the tomato root system (Photo **442**) on which it initially causes dark brown to brown localized lesions on the cortex (Photo **443**). Subsequently, they spread while others appear. Eventually, the roots turn almost entirely brown, rot, and decompose. In addition to the cortex, which is particularly affected, the xylem may turn brown: as seen in a longitudinal section of the taproot and a few large roots (Photo **444**).

This vascular browning then reaches the stem vessels (Photo **445**), which are not all affected. Often, a brown streak, a few millimetres wide, extends into the stem for a few centimetres. Subsequently, browning affects a larger portion of the stem vessels and can be found up to at least 30 cm above the ground or substrate.

When the disease is fairly advanced, a moist and brown lesion, slightly depressed and clearly outlined, appears on the stem portion located at ground or substrate level. This canker has a particular configuration when it surrounds the stem base: it develops on one side of the stem, giving it a flame shape. It can be covered by a pink to salmon mucous made of the many sporodochia formed by FORL on the injured tissues (Photo **446**).

In very wet conditions, the cortical tissues of the lower stem is completely rotten and detaches in places.

Of course, these various effects lead to the wilting (reversible initially) of the apical leaflets and leaves and/or the yellowing of the plant's lower leaves. These foliar symptoms often occur at the approach of harvest, on hot days, and at a time when the plants begin to be laden with fruit.

Other symptoms can be found on p. 285.

Roots and stem base

447 On the left hand module there are few roots and those remaining are in the process of browning and rotting. The right hand module is apparently not attacked by *T. basicola*.

449 These three plants were attacked by *T. basicola*. In addition to reduced growth, their root systems are affected to various degree, and are smaller.

451 A multitude of moist localized lesions, pale brown to blackish, cover the cortex of these roots. In places, they are more extensive and surround the roots over several centimetres. The cortex eventually rots and falls off the central cylinder.

448 Many rootlets are missing and browning is visible on the root system of the adult plant.

450 *Thielaviopsis basicola* is easily identified by careful observation of the roots. In many cases, brown barrel-shaped chlamydospores are easily visible.

Thielaviopsis basicola (black root rot)

Colletotrichum coccodes

Best known as an agent of fruit anthracnose, this fungus is often found on the roots of soil or soil-less cultivated tomatoes (black dot).

In field crops it attacks the root cortex, which becomes dull, brown, and gradually decomposes. In soil-less cultures, the roots tends to discolor and decompose. The central cylinder is sometimes invaded, and it takes on a local black colour. In the cortex it produces black microsclerotia which reveal its presence on the roots (Photo **459**). This will be examined in detail in the next section (p. 263).

Thielaviopsis basicola

This soil fungus is responsible for black root rot on several vegetable crops (beans, eggplant, salads). It causes damage and deformities on chicory roots, which makes it particularly damaging on this vegetable. It has only occasionally been described on tomato, mainly in the US.

Although it is present in many soils, its damage seems to have long gone unnoticed. In recent years it has been seen on plants for the production of processing tomatoes (Photo **447**) and on adult plants from extensive crops (Photos **448** and **449**). In all cases, young plants or adults had a poor growth, sometimes wilting during the hottest part of the day. Careful observation of the root system has always shown roots devoid of lateral roots, some necrotic and rotting, brown to black in colour (Photo **451**); however, it is never as severe as the diseases it causes on tobacco.

This fungus, little known on tomato and difficult to isolate *in vitro* on a nutrient medium, is never considered in the diagnostic hypothesis by agronomists. Its involvement in tomato pathology may be underestimated. Only microscopic observations of affected root fragments can show the chlamydospores typical of *Thielaviopsis basicola* on and in the cortical cells (Photos **450** and **460**).

Calyptella campanula (Nees ex-Pers.) WB Cooke ss WB Cooke

This fungus with carpophores, classified among the basidiomycetes, has occasionally been mentioned in the early 1980s, mainly in the UK in soil-grown tomato cultures. It caused root browning, slowly evolving into root rot accompanied by leaf wilting when the early fruit matured (Calyptella root rot). Vascular tissues of the lower part of the stem also showed a slight browning. The fungus was quickly identified by its characteristic fructifications – small yellow-orange carpophores – appearing in summer on the ground around the plant and near the stem. The damage was not negligible: in a culture, 16% of the plants died from attacks by C. *campanula*.

This decay, which manifested itself especially in damp soil, on heavily irrigated plants, near harvest time, no longer seems important on tomato in the UK, or in any other country where tomatoes are cultivated. Note that this basidiomycete appears able to colonize the roots of many plants such as eggplant and potato.

Roots and stem base

5 Humicola fuscoatra Traaen (Humicola fuscoatra var. 'fuscoatra').

This fungus has been reported and studied on tomato primarily in a limited number of countries – first in the Netherlands, where it was isolated in the early 1980s from affected roots of soil-less-grown plants. The roots of diseased tomatoes had a brown and corky cortex, which was gradually breaking down (hence the origin of the name of the problem, 'corky root'). A decade later, it was associated with the same symptoms, in similar cultural contexts and in several crop production areas in the Americas (Ontario, Quebec, British Columbia, Colorado). In all cases, many 'aleuriospores' were visible in the cells of affected roots (Photo **461**). Artificial inoculations carried out to confirm the pathogenicity of this fungus on tomato have always been negative. *H. fuscoatra* does not seem capable of inducing symptoms on tomato roots. Subsequently, several other hypotheses have been advanced to explain the symptoms of corky root, including a lack of oxygen to roots and salinity problems.

It is also a widespread saprophytic soil fungus, able to break down cellulose and it has been isolated from the rhizospheres of several soil-grown plants.

Humicola sp aleuriospores have frequently been seen in recent years in soil-less tomato crops on or in cells of the cortex of partially or completely decomposed roots (Photos **452** and **453**). Roots, although often brown and altered, did not always show any particular suberization. In all situations, the presence of several other potentially parasitic fungi on roots could explain the root symptoms observed. In addition, inoculations made with *Humicola* on tomato seedlings failed to induce any root symptoms.

It seems that this fungus is an opportunistic colonizer of root cortical cells which are towards the end of their life. It is probably more common than one might think on roots of soil-less tomatoes. Indeed, the mycelial development *in vitro* is rather slow, which does not help its isolation. Therefore, it goes easily unnoticed if root examinations are not thorough.

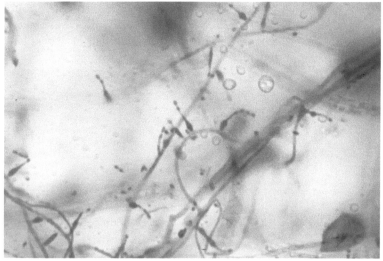

452 *Humicola* sp. produces discrete phialides *in vitro* which are perpendicular to the mycelium and produce chains of. conidia.

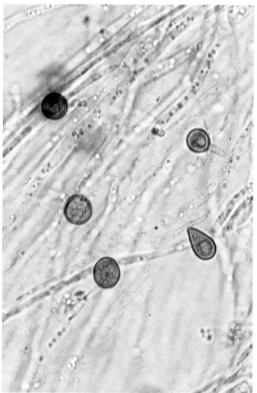

453 Aleuriospores are also formed *in vitro*. At first hyaline, they melanize gradually. They are no different from those observed on the roots (see Photo **461**).

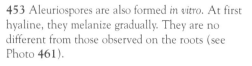

Humicola sp.

Table 28 Key features of fungi (or similar organisms) associated with root browning in soil or soil-less-grown mature tomato plants

Pathogen	Frequency on roots and cultural systems	Parasitic potential		Structures allowing identification on and in the roots	Other features
		Soil	Soil-less		
Pythium spp. (damping-off, root rot)	Very common, especially in soil-less culture.	+/–	+/– to ++	Sporangia and/or oospores (Photo **454**).	Several species are possible according to the cultivation, the country, and climatic conditions. They have different virulence on tomato in soil-less culture (Description 12).
Phytophthora spp. (damping-off, root and crown rot)	Less frequent, mainly in soil sometimes in soil-less culture.	++	+++	Sporangia (Photo **455**) and/or oospores.	*Phytophthora parasitica* is a common species in soil culture. It is able to attack the tomato rootstocks. It sometimes occurs in soil-less culture. Other less common species such as *P. cryptogea* can also occur (Description 12).
Rhizoctonia solani (damping-off, root rot)	Frequent, particularly in soil-grown culture	+	+/–	Hyaline to brown mycelium developing on the tissue surface (Photo **457**).	Frequently observed on the roots; its parasitic role is sometimes difficult to assess. It also attacks tomato rootstocks of KNVF type (Description 14).
Fusarium oxysporum f. sp. *radicis-lycopersici* (FORL)	Frequency varies according to the production area. Present in both soil and soil-less culture.	++	+++	Phialides, conidia and chlamydospores in and on cortical cells (Photo **458**) (unreliable).	FORL mainly affects soil-less crops. It is in sharp decline since the advent of resistant varieties.
Colletotrichum coccodes (black dots)	Quite frequent, in soil, rather than soil-less culture.	+	++	Black sclerotia, sometimes covered with acervuli displaying brown seteae and producing cylindrical conidia (Photo **459**).	Fungus with an often underestimated pathogenicity. It is also capable of attacking the tomato rootstocks of KNVF type.
Thielaviopsis basicola (black root rot)	Rare.	+	+/–	Chlamydospores and sometimes endoconidia (Photo **460**).	Found both in the nursery and in the field, this fungus does not appear to have significant parasitic potential on tomato (Description 17).
Calyptella campanula (Calyptella root rot)	Very rare.	+	?	Sporophores sometimes forming around plant base.	A pathologist's curiosity, this basidiomycete has mainly been reported on tomatoes grown in soil in the UK.
Humicola fuscoatra (corky root)	Unknown frequency.	+/–	+/–	Brown to black aleuriospores visible on and in the roots (Photo **461**).	Associated with brown and/or corky roots in soil-less-grown crops. Its pathogenicity has never been confirmed (see Box 5). Not to be confused with *Pyrenochaeta lycopersici*.
Plectosporium tabacinum	Very common in soil-less culture.	+/–	+/–	Bicellular conidia visible in cortical cells (Photo **461a**).	Fungus with unclear pathogenicity on roots in soil-less culture (see Box 6).
Olpidium brassicae	Present in soil culture in damp and asphyxiating conditions, very common in soil-less culture.	?	?	Sporangia and chlamydospores visible in root cells (Photo **456**).	Virus vector fungus on other species; present in very large amounts in the cortex of senescent roots. It seems to be a 'biomarker' of roots undergoing decomposition (Description 18).

+/–: Isolated on roots but no specific pathogen potential; +: induces relatively harmless root symptoms; ++: causes damaging lesions on roots; +++: particularly damaging.

Roots and stem base

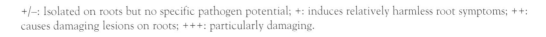

251

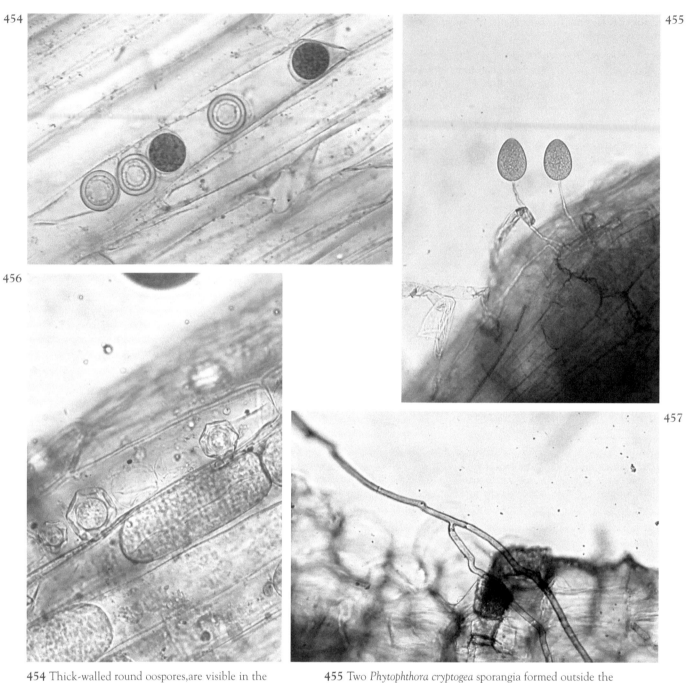

454 Thick-walled round oospores,are visible in the cortex cells; they often reveal the presence of *Pythium* spp. in the roots.

456 *Olpidium brassicae* sporangia (1) and chlamydospores (resting spores) (2) have developed in several cortical cells.

455 Two *Phytophthora cryptogea* sporangia formed outside the necrotic and decayed root, taken from a soil-less culture. Usually, the *Phytophthora* spp. sporangia form more frequently inside tissues.

457 *Rhizoctonia solani* can easily be spotted on colonized roots. A brown compartmentalized mycelium, sometimes with a constriction at the base of its branches, covers the roots in places.

***In situ* features of the main fungi affecting the roots of seedlings and/or mature plants of tomato (in soil and soil-less systems)**

458

459

461

460

461a

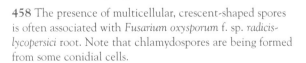

458 The presence of multicellular, crescent-shaped spores is often associated with *Fusarium oxysporum* f. sp. *radicis-lycopersici* root. Note that chlamydospores are being formed from some conidial cells.

460 The chlamydospores of *Thielaviopsis basicola* are brown and are shaped like barrels stacked on each other. They are easily visible on and in cortical cells of rotted roots.

459 *Colletotrichum coccodes* black microsclerotia develop on the roots. Acervuli are produced with cylindrical conidia and setae (bristles) very characteristic of the genus (see also Photo **491**).

461 The dark brown aleuriospores of *Humicola* sp. dot this root.

461a Many bicellular conidia more or less fill some cortical cells.
Plectosporium tabacinum

Roots and stem base

462 The leaflets inoculated with
P. tabacinum, are slightly rolled and display
some coalescing necrotic spots. Small dry
and elongated cankerous lesions are also
visible on the petioles.

463 The stem cankers are dry, sunken in the
centre. Surrounding tissues may turn brown.
P. tabacinum

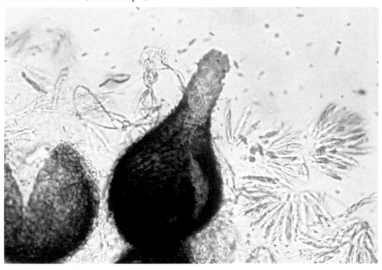

465 Several slender phialides, slightly swollen at
the base, form hyaline conidia in which droplets are
clearly visible.
P. tabacinum (anamorph)

464 A perithecium with a lighter collar is located close to asci
clusters containing hyaline and bicellular.ascospores.
M. cucumerina (teleomorph)

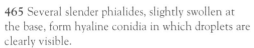

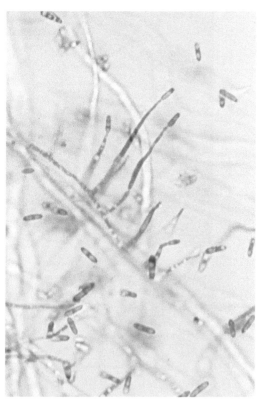

Plectosporium tabacinum,
Monographella cucumerina

6 *Monographella cucumerina* (Lindf.) Arx, *Plectosporium tabacinum* (Van Bema) Mr E. Palm, W. Gams & Nirenberg

This fungus has been isolated from many soils in the world and on the roots of the most diverse hosts. It is particularly pathogenic on Cucurbits (in field and soil-less), on basil, lupin, sunflower, peanut, sugar beet, and campanula.

It has been observed on and isolated from tomato roots grown in soil, but mostly from soil-less roots. Soil-less cultural context conditions are particularly favourable for its development. It is isolated on the roots with very high frequencies at certain times of the year, higher than that of *Fusarium oxysporum*. Its parasitic status in soil-less cultures is still unclear. Inoculations made in this context of culture failed to show any virulence on roots of mature tomato plants.

However, this fungus is known as a pathogen on this plant. In Australia, *Plectosporium tabacinum* has already been isolated from diseased seedlings with small leaf spots. These measured 1–4 mm in diameter, were dark before lightening in the centre. The periphery of the leaf was folded and rolled. These spots could coalesce and be the cause of the leaflets dying. The effects progressed along the veins and reached the petiole and whole leaf.

On the stem, small, longitudinal, concave, and discolored lesions were observed on the part buried in the ground. A slightly strangled brown area could appear on the part of the stem above the ground. Note that similar symptoms have also been described on tomato seedlings produced in glasshouses in Italy in 1974. It is the same in Canada, where cankerous alterations of the stem, beige, dry and longitudinal are attributed to *P. tabacinum* and *Acremomium strictum* W. Gams. In addition to the canker, browning and collapse of the pith can be found on the lower part of the stem.

P. tabacinum was also mentioned in Egypt in 1981 as responsible for root rot, yellowing, and wilting of foliage followed by tomato seedlings dying.

Inoculations performed on young tomato plants with strains isolated from the roots have confirmed the parasitic potential of this fungus on leaflets and on stem, and have created symptoms quite similar to those previously described (Photos **462** and **463**). Its virulence on roots could not be clearly demonstrated.

Note that *P. tabacinum* is also used as a bioherbicide on some weeds, especially during the early stages of seedling development (*Galium spurium*). It is also considered a nematophagous fungus as it has been isolated from egg masses of *Meloidogyne hapla* parasitizing tomato roots in Belgium.

The biology of this fungus is not well known, but it appears to survive easily in many soils and plant debris. It is spread through wind and water.

It has a teleomorph, *Monographella cucumerina* (Lindf.) Arx (*Plectosphaerella cucumerina* [Lindfors] W. Gams), rarely observed in nature. It is characterized by slender bottle-shaped perithecia with a dark base and a lighter neck, measuring 200–300 µm long by 90–110 µm wide. The asci are cylindrical (50–65 × 6.5–7.5 µm) with a thin wall. They contain eight ellipsoid ascospores, hyaline and bicellular (10.8–12 × 2.8–3 µm) (Photo **464**).

The anamorph, *Plectosporium tabacinum (Microdochium tabacinum* [JFH Beyma] Arx [1984], *Fusarium tabacinum* [JFH Beyma] W. Gams [1968]), is characterized by slightly base swollen phialides, solitary or whorled, sometimes forming on branched conidiophores (12–30 × 3–4.5 µm). Conidia are united and bicellular, hyaline, containing lipid droplets, ellipsoid, slightly curved and pointed (8.2–13.5 × 2.2–3 µm) (Photo **465**). The dimensions given may vary depending on the culture medium used.

Roots and stem base

466 This large group of slow growing plants, with chlorosis and gradual drying out, indicates the presence of *R. crocorum* in a tomato crop.

468 The cortex of larger roots gradually breaks down and eventually disappears, only the central cylinder persists.

470 On the root surface some black dots, sclerotia, are scattered throughout the cortex.

467 The roots of affected plants are either a brown colour when young, or a plum to purple colour when larger and therefore older.

469 A cankerous symptom with a purple colour is sometimes visible at the base of the stem; well delineated, it extends over several centimetres.

Rhizoctonia crocorum (*Helicobasidium brebissonii*)
(violet root rot)

256

Rhizoctonia crocorum (Helicobasidium brebissonii)

This fungus, damaging in many countries and highly polyphagous, is known as purple Rhizoctonia root (violet root rot), affecting mainly asparagus, beet, and carrot. It has a diverse host range, the tomato occasionally being part of it, infrequently found in glasshouse and field crops.

On the tomato it may be the cause of locally severe root destruction. Affected plants, spread into groups (Photo **466**), are slow growing, with chlorosis, and may eventually wilt and wither. Photos **467** and **469** show the many purplish brown lesions on root and stem base cortex. These are covered by a purple to wine coloured surface mycelium, sometimes isolated but more often condensed into tortuous cords (Photo **471**) or in the form of a sheath. This agglomerated mycelium darkens when evolving, and the soil strongly adheres to it. The inner part of the root cortex eventually rots completely, leaving only the external foundations and the central cylinder (Photo **468**). In addition to the mycelium, small clumps of mycelium are seen, called 'tuberoid bodies' if rather elastic, or 'miliary bodies' when aggregated (Photos **470** and **472**). The term 'sclerotia' is used to describe these structures that contribute to the perpetuation of *R. crocorum* in the soil. The mycelium of this *Rhizoctonia* is similar to that of *R. solani* but it is coloured pink to purple with age (Photo **473**).

471 In addition to miliary bodies wine-coloured mycelium strands are visible on the root.

472 Sclerotia consist of clusters of hyphae.

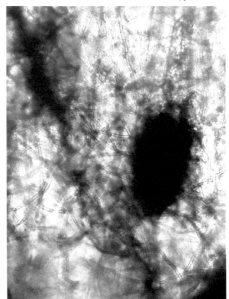

473 The mycelium of *R. crocorum* is characteristic of *Rhizoctonia*: large cells of the mycelium whose branches form an angle of 45°, with a constriction of the mycelium at the base of the branches.

Roots and stem base

Pratylenchus spp.

Numerous ectoparasitic nematodes are found on tomato. They are listed in *Table 30* (p. 269). They cause root necrosis (lesion nematodes) and restrict root growth, and consequently that of the plants.

Other phytophagous nematodes, migratory endoparasites (see Photo **474**), also attack tomatoes. Among them, *Pratylenchus* spp. appear more damaging. They enter the cortex of the root and gradually destroy many cells, inducing abundant reddish brown to black lesions (Photo **474**). Tissue colouring is due to release of phenolic substances resulting from the action of nematodes' hydrolytic enzymes. As for ectoparasite nematodes, root effects disturb the plants' growth: plants are often small and their leaves turn yellow and/or sometimes wilt.

Observation of dissected root fragments in water shows *Pratylenchus* adults or larvae, with a well defined stylus (Photo **475**) and some eggs, which can confirm the diagnosis. Note that several species of *Pratylenchus* have been observed on tomato: *P. penetrans*, *P. neglectus*, *P. scribners*, and *P. zeae*.

Generally, the impact of these nematodes on tomato is not well known, it does not seem important.

The damage caused by other nematodes, with cysts or galls, are reviewed on pp. 268–273.

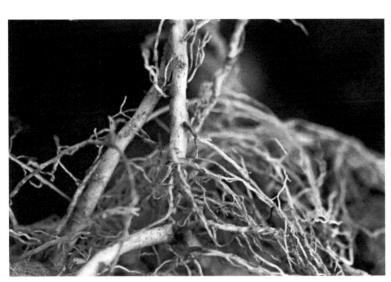

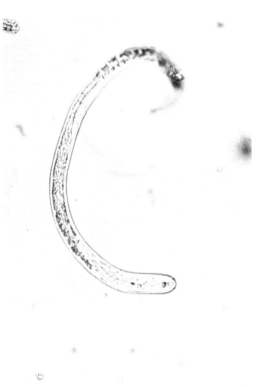

474 The roots attacked by parasitic endomigrating nematodes have many moist lesions, yellowish to reddish brown. Rotted rootlets disappear and the affected cortex can be fully decomposed.
Pratylenchus penetrans

475 *Pratylenchus penetrans* is a rather short nematode with a visible mouth stylus.
Pratylenchus penetrans

476 These roots are brown and corky in places. Sheaths and cracks are found locally.
Pyrenochaeta lycopersici

477 Tiny black dots cover part of the roots, which are also brown in places.
Colletotrichum coccodes

478 Rounded galls, white or brownish, are clearly visible on the root.
Meloidogyne **sp.**

480 Brown cysts cover the roots.
Globodera tabacum

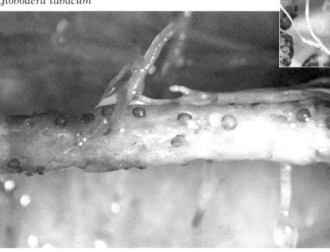

479 Tumour-like growths distort the roots along their entire length.
Agrobacterium tumefaciens

Examples of root symptoms

OTHER ROOT SYMPTOMS (SUPERFICIAL SUBERIZATION, BLACK SPOTS, CYSTS, GALLS)

Possible causes

- *Colletotrichum coccodes*: Description 10
- *Macrophomina phaseolina*: Description 18
- *Pyrenochaeta lycopersici*: Description 13
- *Rhizoctonia crocorum* (see p. 257)
- *Rhizoctonia solani*: Description 14
- *Rhizopycnis vagum*
- *Spongospora subterranea* f. sp. *mediterranea*: Description 18
- *Agrobacterium radiobacter* biovar 1: Description 23

- *Agrobacterium tumefaciens*: Description 23
- *Globodera tabacum* (cyst nematodes): Description 49
- *Meloidogyne* spp. (gall nematodes): Description 50
- Other root nematodes
- *Orobanche* spp.

Additional information for diagnosis

Pyrenochaeta lycopersici

This fungus which causes tomato corky root disease (brown and corky root; Pyrenochaeta corky root), also attacks other Solanaceae, and various Cucurbits and salads. It is generally present in market gardening soils and affects both field and protected crops. In the latter, it makes it necessary to disinfect the soil or to use resistant rootstocks. It was found in the past in some soil-less cultures on pozzolan or on peat in glasshouses where the underlying soil was uncovered. It has now virtually disappeared from these cultures.

On all hosts, *P. lycopersici* causes the decay and disappearance of many rootlets (Photo **481**). Brown symptoms are also visible on young roots (brown root rot). The larger roots show dry and corky lesions (Photo **483**). Of limited extension at first, they gradually spread, eventually circle the roots and form sheaths over several centimetres (Photo **482**). These have sometimes the appearance of the bark of a tree. They can be very corky with cracks (Photo **484**). Note that this fungus can produce typical pycnidia (Photos **485** and **486**).

Roots and stem base

261

481 This root system, although slightly affected, is still devoid of rootlets. Dark corky sections are quite visible on the larger roots.

483 A superficial cortex suberization in localized units, can be seen on the roots. Affected tissues, rather dry, are brown to light brown.

485 *P. lycopersici* rarely forms pycnidia on the roots. These display brown setae located at the ostiole which easily identify this fungus.

482 The severity level of *P. lycopersici* on the root system is much more acute: the corky sheaths are much more advanced and pronounced.

484 The suberization quickly becomes widespread. Some corky sheaths are cracked. Note that the central cylinder shows no browning, unlike in other diseases.

Pyrenochaeta lycopersici
(corky root)

Its impact on tomatoes may be significant in heavily contaminated soils. The limitation of the root system and the many lesions result in reduced plant vigour. Wilts that are reversible during the night can be become permanent during warm periods close to harvest time, perhaps and are often preceded or accompanied by leaflet chlorosis as well as premature leaf drop.

Pyrenochaeta lycopersici is reported in many countries in cool conditions but is also common in all types of tomato production including in the gardens of amateurs. It is not uncommon for it to be associated with other root problems such as those caused by *Colletotrichum coccodes*, *Rhizoctonia solani*, and gall nematodes. *Rhizoctonia solani* is sometimes associated with root systems slightly suberized on the surface, so beware of possible confusion (see p. 245)!

Note that another fungus, *Rhizopycnis vagum* DF Farr, has recently been isolated from the roots of field-grown tomatoes in several plots in central and southern Italy. The diseased roots showed symptoms very similar to those caused by *P. lycopersici*, especially cortex suberization. Previously, *R. vagum* was best known for causing considerable damage to the roots of Cucurbits, including melon and watermelon in the US, Mexico, Guatemala, Honduras, and more recently in Spain. Note that the microsclerotia can be observed on the secondary roots of affected plants.

Colletotrichum coccodes

This fungus is widespread and pathogenic on a wide range of hosts and is primarily responsible for tomato fruit anthracnose (see p. 391), and also attacks its root system (black dots). As for *P. lycopersici*, it is present in many soils where vegetable crop rotations are inadequate, and sensitive crops such as peppers, potatoes, and especially tomatoes are grown too frequently. This situation helps to increase the rate of inoculum gradually in the soil. It is most damaging on roots of plants grown under protection, both in soil and in soil-less cultures.

The root systems are rather dull and have few rootlets and lateral roots (Photo **487**), the latter atrophied or decayed. Some lesions, grey to brown at first, can be extensive, and are also visible in the cortex of large roots (Photo **488**). These gradually become darker, decompose and become partially covered in black microsclerotia, sometimes carrying black bristles (visible with a magnifying glass, see Photos **489** and **491**). Ultimately, the cortex, rotten and dotted with sclerotia, easily detaches from the central cylinder; the latter is colonized in the late course of the disease, taking a blackish hue and showing many microsclerotia. Note that in soil-less culture, the tissues also tend to discolor (Photo **490**).

Roots and stem base

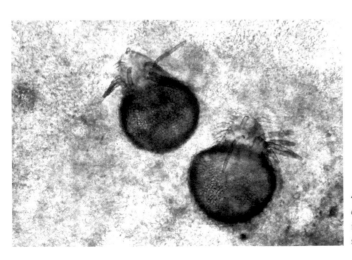

486 These two pycnidia of *P. lycopersici* are characterized by the presence of several setae implanted around the ostiole of these fungal structures.

487 The lack of rootlets, the roots dullness, their dirty grey to brown colour characterize developing symptoms of *C. coccodes*.

490 In soil-less cultures, roots colonized by *C. coccodes* can sometimes have a different appearance: the cortical browning may be less severe, the tissues are rather discolored. Microsclerotia and acervuli with black setae are still visible.

488 Closer examination shows brown necrotic patches covered with black dots.

489 In the most affected root zones the cortical tissues are brown, rather moist, and decompose slowly. The microsclerotia of *C. coccodes* are now clearly visible.

491 The microsclerotia are bristling with black hairs (setae) that are within the acervuli, and are characteristic of the genus *Colletotrichum*.

Colletotrichum coccodes (black dot)

This fungus is considered by some authors as a secondary invader of decomposed old roots. Note that it is frequently observed on roots with other pest or disease problems (*Pyrenochaeta lycopersici*, *Rhizoctonia solani*, *Fusarium oxysporum* f. sp. *lycopersici*, *Pythium* spp., *Meloidogyne* spp.) and these could aid its colonization. The significant attacks reported on some farms, both in soil and soil-less cultures however, raises questions on this viewpoint. Recent attacks observed on several rootstocks in several Mediterranean countries, have confirmed its parasitism.

Colletotrichum coccodes may be capable of causing brown circular lesions with a yellow halo on leaflets, and may also affect the stems.

Macrophomina phaseolina

This soil fungus, highly polyphagous and widespread, affecting more than 500 different hosts. It occurs mainly in hot conditions accompanied by dry periods which weaken the plants. It can attack young seedlings (see p. 243) and also adult plants. On the latter, it gradually invades the roots during the season, especially the cortex that eventually becomes necrotic, decomposes and is detached from the central cylinder (charcoal rot). It can also reach the stem base and then colonize the stalk. In this case, the pith is particularly affected. Diseased plants are less vigorous, their lower leaves may yellow, wilt, and wither. Scores of black and irregular structures – microsclerotia – form on damaged tissues. These can be seen particularly in the cortex (Picture **492**) or in the central cylinder when the latter has disappeared (Photo **493**). They should not be confused with those of *C. coccodes*.

Although present in various countries and able to attack crops such as sunflower, this fungus has never been seriously reported on tomato. In contrast, it does affect peppers. Moreover, it is not uncommon to isolate M. *phaseolina* on the roots of several vegetable crops (melon, various Solanaceae) but whether or not it is pathogenic on these plants is not known.

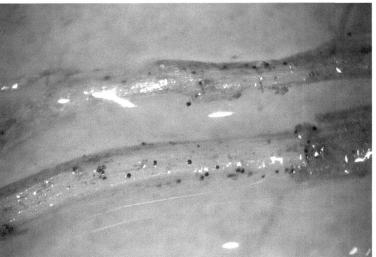

493 The microsclerotia are also visible on the central cylinder of roots lacking cortex.

492 A multitude of irregular black masses can be seen on the root cortex, with brown and decomposed tissues.

 *Macrophomina phaseolina* (charcoal rot)

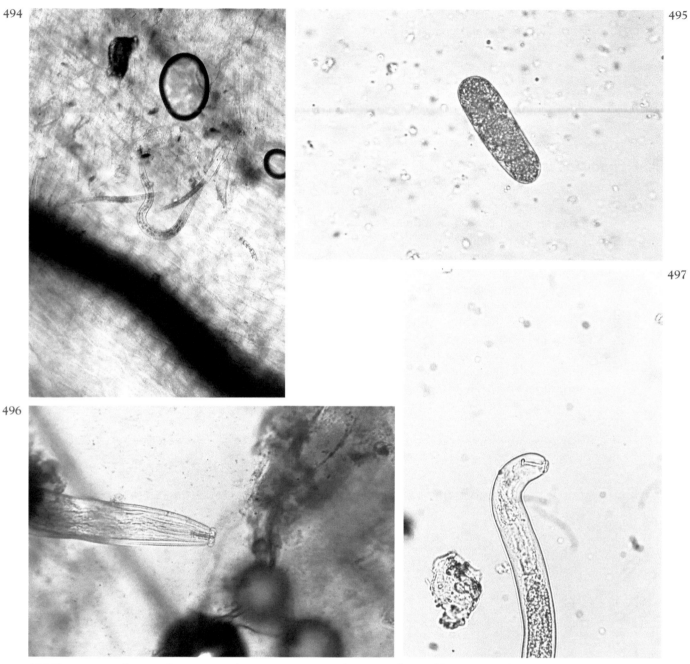

494 The presence of nematodes in or near the roots is confirmed by observation of tiny worms using a dissecting microscope or a higher powered microscope.

495 Eggs are often visible in or near the damaged tissues and characterize the presence of nematodes in the roots.

496 Saprophytic nematodes have no stylus.

497 Phytophagous nematodes are equipped with a hollow mouth stylus allowing them to penetrate the cells to absorb the content.

How to differentiate easily between root saprophytic nematodes and phytophagous nematodes

Table 29 Characteristics of the main pathogens affecting roots in tomato crops

Pathogen	Frequency on roots and cultural systems	Impact		Identyfying structures on and in roots	Other features
		Soil	Soil-less		
Pyrenochaeta lycopersici (corky root)	Very frequent especially in soil cultures.	+++	+	Often no structure, very rarely some pycnidia adorned with black seteae around the ostiole (Photo **486**).	Some varieties and rootstocks are resistant (Description 13).
Colletotrichum coccodes (black dots)	Quite frequent in soil, less frequent in soil-less.	++	+	Black sclerotia, sometimes topped by acervuli with brown seteae and producing cylindrical conidia (Photo **491**).	Fungus with often underestimated pathogenicity. It is also able to attack tomato rootstocks of KNVF type.
Macrophomina phaseolina (charcoal rot)	Rare, essentially in soil.	+	−	Black and irregular microsclerotia (Photo **492**).	The hot, dry summers seem to encourage its expression on the roots of several Solanaceae.
Globodera tabacum (cyst nematodes)	Uncommon, primarily in soil.	++	−	Brown cysts (Photos **498–501**).	Severe attacks are sometimes seen (Description 49).
Meloidogyne spp. (root-knot nematodes)	Frequent, essentially in soil cultivation.	+++	+	Galls (Photos **503–505**).	Some varieties and rootstocks are resistant (Description 50).
Agrobacterium tumefaciens (crown gall)	Rare, only in soil-less cultivation.	−	+	Tumours (Photos **508–511**).	Uncommon, and when present has little effect on the crop.
Spongospora subterranea f. sp. *mediterranea* (Spongospora gall)	Rare, in soil.	+	−	Tumours (Photo **512**).	Never seen in the last few years.

−: Not observed or isolated from roots; +/−: isolated from roots but no particular parasitic potential; +: harmless root changes; ++: damaging root changes; +++: particularly damaging root changes.

Phytophagous nematodes in the soil, also called 'eelworms', are transparent tiny worms. Unlike saprophytic nematodes, parasitic nematodes have a visible mouth stylus which they use and induce root lesions (Photos **494–497**).

Dozens of species belonging to different genera have been reported on tomato (*Table 30*, p. 269), in two main groups according to their mode of root colonization: ectoparasites and endoparasites (Figure **43**).

Roots and stem base

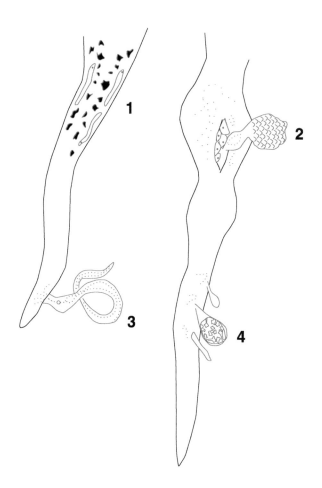

Figure 43 Location and behaviour of the main nematode groups.

1 Migratory endoparasitic nematodes.
2 Sedentary endoparasitic knot nematodes.
3 Ectoparasitic nematodes.
4 Cyst nematodes.

Table 30 Characteristics of the main nematodes parasitic on tomato roots

	Ectoparasitic	Endoparasitic migratory	Endoparasitic sedentary	
			Cyst	Gall
Symptoms	Root necrosis, reduced growth of roots and plants.	Many brown to reddish lesions on the roots, rot of the root system.	Root necrosis, presence of numerous brown cysts.	Swollen roots, largely composed of galls.
Main genera and species reported on tomato	Dolichodorus heterocephalus Belonolaimus gracilis B. longicaudatus Helicotylenchus digonicus H. dihystera H. indicus H. pseudorobustus H. tunisiensis H. varicaudatus Hemicycliophora arenaria H. similis Hoplolaimus indicus Longidorus elongatus L. maximus Paratrichodorus christiei P. minor Paratylenchus projectus Radopholus similis Rotylenchus buxophilus Rotylenchulus reniformis Tetylenchus joctus Trichodorus spp. Tylenchorhynchus acutus T. brassicae T. capitatus T. claytoni T. leviterminalis Tylenchus filiformis Xiphinema americana X. diversicaudatum	Aphelenchoides ritzema-bosi Ditylenchus dipsaci Pratylenchus neglectus P. penetrans P. scribneri P. zeae	Globodera rostochiensis G. tabacum sensu stricto Heterodera schachtii H. trifoli	Meloidogyne acronea M. arenaria M. chitwoodi M. ethipica M. floridensis M. hapla M. incognita M. javanica M. mayaguensis M. minor Nacobbus batatiformis N. serendipiticus
Incidence	Little known.	Low, occasionally serious.	Low, occasionally serious.	High.
Simplified parasitic process	Perforation of root cells with their stylus and removal of cytoplasmic contents. In principle, these nematodes do not penetrate the roots.	Penetration and destruction of cells in the root cortex.	Penetration of young roots by the young larvae which feed on the cells. Hypertrophy of females which remain connected to the roots by their head.	Root penetration and secretion of salivary juices inducing a hypertrophy of root cells, causing typical galls and swellings.

Roots and stem base

498 Several white or brown masses can be observed on this root, which is also brown and altered. These masses are in fact live females or cysts of G. *tabacum*.

499 After the female has died, its body hardens and becomes dark, thus becoming a cyst full of eggs that does not take long to break away from the root.

500 This live female is light-colored and attached to the root by its head.

501 The cuticle of a cyst which has a brown colour.

502 A young larvae with a stylus is emerging from an egg.

Globodera tabacum (cyst nematodes)

Cyst nematodes

Some species of endoparasitic nematodes of the *Globodera* genus have been reported throughout the world on tomato (*Table 30*, p. 269). These nematodes do not appear to be very common on this host but can occasionally be harmful.

Their presence on the roots is visible through numerous reddish brown necrotic lesions scattered with brown to white masses (cyst nematodes), whose size varies between species (Photos **498** and **499**). They are actually the females of these nematodes, full of eggs and hypertrophied, which remain connected to the roots by their heads. These females are white when they are alive (Photo **500**). They eventually die, are increasingly a brown colour (linked to their parched and leathery cuticle) and give rise to a cyst that will protect the eggs (Photos **499** and **501**). These cysts are clearly visible when the roots are carefully washed. They allow the easy identification of these rather special nematodes.

In addition to root symptoms, the affected plants, which often occur in groups, can have reduced growth depending on the soil population level.

Cyst nematodes (Figure 43) do not generally pose problems for tomato crops. Significant damage to KNVF type rootstocks (see p. 659) has been reported in south-west France, where *Globodera tabacum sensu stricto* has been present for very many years.

As with root-knot nematodes, cyst nematodes are not the only pests observed on the roots and they are often accompanied by fungi such as *R. solani* and *C. coccodes*.

Roots and stem base

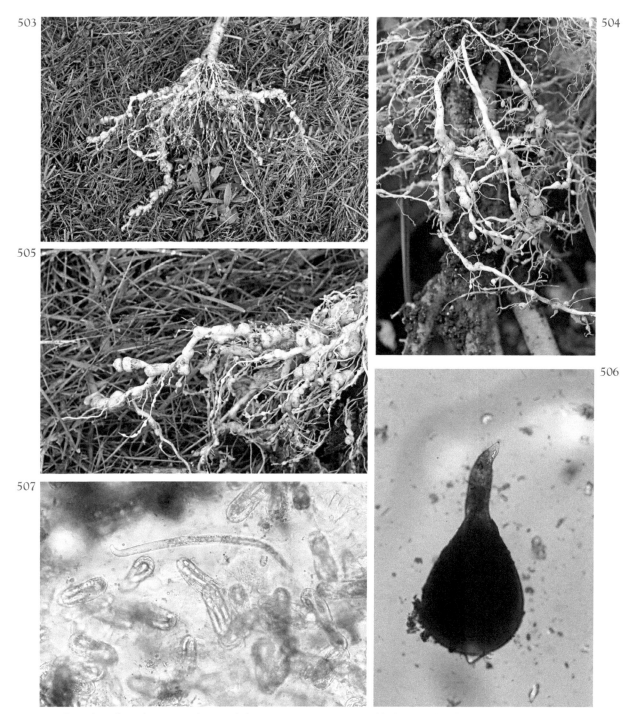

503 Many galls and swellings have caused major deformities on the root system, whose colour varies from dirty white to brown.

505 When the soil population level is high, it possible to see strings of swellings and galls along the roots.

507 Near the roots, some mature eggs can be seen containing young larva with a highly visible stylus.

504 Galls, first pearly white, vary in size and affect both rootlets and larger diameter roots.

506 The galls contain one to several swollen adult females. After detaching from a gall the stylus of the female is sometimes visible.

Meloidogyne **spp. (root-knot nematode)**

Meloidogyne spp.

Root-knot nematodes (see Figure **43**) affect many vegetable crops used in rotation with tomato (salad crops, eggplant, melon) so it is not surprising to find them in many crops especially those grown in the soil.

The root symptoms caused by these nematodes – galls and swellings (Photos **503–505**) – are very characteristic and easily identifiable. To confirm diagnosis, on cutting the galls in two one to several round, whitish masses can be seen easily, corresponding to swollen adult females (Photo **506**). These root distortions heavily disrupt plant function, the more so if the soil population level is high. Also, they can cause a reduction in the growth of plants, wilting and/or yellowing, and eventually the decline of the plant.

Several species of *Meloidogyne* have been reported on tomato (*Table 30*, p. 269): M. *incognita*, M. *javanica*, M. *arenaria*, M. *hapla*.) Galls are smaller and rounder with M. *hapla* than those of other species.

Several species are prevalent on tomato. In general, nematodes, like other soil-borne organisms, are found in over-cropped soils where susceptible plants, especially tomatoes have been repeatedly grown. In such soils it is not surprising to find gall nematodes in association with R. *solani*, C. *coccodes*, and P. *lycopersici*.

Note that root-knot nematodes have been implicated with reduced resistance to some diseases such as Fusarium wilt, for example. In addition to the rootstocks (see p. 658), some tomato varieties are resistant to several species of *Meloidogyne*. This resistance is not absolute, and a few galls can still be found on the roots of resistant plants.

Finally, note that the cessation of use of methyl bromide and the use of less efficient alternative methods largely explains the rise of soil-borne diseases and pests, especially gall nematodes.

Roots and stem base

273

508 Young tumours, whitish to cream, with a smooth surface and rather characteristic, are being formed on this root.

510 Tumours can occur by the stem base. Their size and irregular shape is fairly typical.

509 When aging the tumours grow larger, they become brown and irregular in shape. Tissues may become corky on the surface and sometimes spongy.

511 In rare cases, tumours may have a very large size and measure several centimetres in diameter.

Agrobacterium tumefaciens (crown gall)

Agrobacterium tumefaciens

This worldwide bacterium is responsible for tumours (crown gall) on the stem base of many cultivated or noncultivated, woody or herbaceous plants (fruit, ornamental). Although the tomato is sensitive to this bacterium and is used as a bioassay to test or confirm the pathogenicity of bacteria suspected of being A. *tumefaciens*, attacks on tomatoes are rare.

The symptoms caused by this bacterium have only been observed on rare occasions, mainly in glasshouses. Two types of attacks have been noted, on the root system and on the stem base of tomato. The whole root system of plants grown soil-less on peat can be affected. In this situation, initially small and pearly galls develop gradually over the entire root system (Photos **508** and **509**). On tomato plants grown soil-less in rockwool large tumours are visible at the foot of some plants (Photos **510** and **511**).

Unlike many other root pests/pathogens, there is no particular structure visible in the gall that confirms the presence of this bacterium in root cells. The effects of this disease do not seem serious, whatever the nature of the attack; however, during severe attacks, tumours on the roots and stem base can still disrupt plant development, reducing growth, and inducing leaf chlorosis.

Spongospora subterranea

This aquatic fungus, now classified as Protozoa, is best known as the cause of potato powdery scab, but is also able to attack many other Solanaceae, cultivated or not. Its symptoms on tomato roots are rarely reported. However, the tomato is susceptible to this obligate parasite; firstly it produces white pustules on the roots that rapidly evolve into tumours. These tumours gradually turn brown and become corky on the surface (Photo **512**).

Although sometimes very spectacular, these root tumours do not seem to have a significant impact on plant development. On tomato, the pathogen grows poorly and does not perpetuate. Mature zoosporangia can still be observed in some cortical cells and in rootlets (Photo **513**), thereby confirming the identification of S. *subterranea*.

512 Two prominent tumors are visible on these roots. Brown, warty, and superficially corky, they can surround the root.

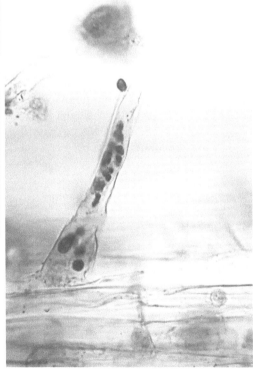

513 Inside the rootlet, a zoosporangium is easily identifiable.

Spongospora subterranea (Spongospora gall)

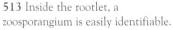

Agrobacterium radiobacter biovar[1]

Root mat is a relatively old disease, first described in the 1970s in the UK, and associated with a strain of *Agrobacterium* biovar 1 in glasshouse cucumber cultures in soil. It has only been observed on tomatoes in France since 1997, mainly in soil-less culture where the number of affected plants ranged between 1 and 50%. *A. radiobacter* strains carrying an 'Ri' (root-inducing) plasmid were responsible. The root symptoms are induced by the transfer and expression of this plasmid into the tomato cell's genome. More recent studies have shown that the plasmid 'Ri' is also present in other bacterial species associated with root mat and from other genera, including *Ochrobactum*, *Rhizobium*, and *Sinorhizobium*. These species may play a significant epidemiological role in the disease. Note that the root mat is now reported in several other European countries such as the Netherlands.

Affected plants show rather linear root proliferation at the surface of rockwool cubes and packs (Photo **514**), leading to increased vegetative growth of tomatoes at the expense of fruiting. Note that fruit size can be reduced. In addition, the density of roots makes them much more susceptible to attack by fungi present in soil-less cultures, especially *Pythium* spp. and *Phytophthora* spp.

514 A proliferation of roots in soil-less substrates characterizes root mat.

The root proliferation is caused by strains of *A. radiobacter* biovar 1 harbouring a fragment of circular DNA, a root inducing cucumopine plasmid 'Ri'. This bacterium, as in the case of *A. tumefaciens*, acts as a vector and transmits the plasmid which becomes inserted into the plant cell's genome. Subsequently root production is completely disrupted and affected plants produce roots in an uncontrolled way. The incubation period of this disease is rather long, approximately 4–8 weeks. Isolation and identification of this bacterium is difficult and requires a specialized laboratory (the plasmid can be detected by polymerase chain reaction, PCR).

---

[1]Biovars 1 of *Agrobacterium tumefaciens*, *A. rhizogenes*, and *A. radiobacter* may have been merged into a single taxon: *A. tumefaciens* (see Description 23).

It is difficult to control this disease in soil-less tomato crops. Indeed, there is no approved product to achieve effective treatment during cultivation. Only several hygiene measures can be used in the nursery and during cultivation to limit the persistence and proliferation of this bacterium:

– Clean and disinfect nurseries and shelters with a disinfectant. The equipment used in glasshouses should not be stored on the ground or in dirty environments between crops as it can become contaminated.

– Disinfection of the soil is often not effective because the bacteria easily survive in it and can quickly re-colonize;

– The blocks/slabs/cubes should not be in contact with the soil and the drippers need to be well clear;

– Ground covering plastic should not be contaminated by soil/dust on its upper surface when it is replaced.

Attempts to steam disinfect the blocks/slabs were effective in killing the bacteria but not the infectious plasmid. Keep in mind that this method of control is not advocated in countries affected by root mat.

Note that a quite similar condition, called thick root syndrome whose origin is unknown, is reported notably in the Netherlands, mainly in cucumber, pepper, and tomato nurseries. It results in excessive growth of roots which also tend to curl unevenly and to have a glassy appearance. Their diameter may be considerable (at least ten times larger than that of a normal root), and they eventually decompose. Note that they are much more susceptible to attack by *Pythium* spp.

Root malformations

The cultivation of some soils, in poor conditions, leads to the formation of an impervious pan. The latter, located at the bottom of the depth of cultivation (plough pan) forms a barrier difficult to penetrate by both water and roots. In this context, the tomato plant's root system may be largely near to the surface. Furthermore, during excessive irrigation or rainfall, this waterproof pan may induce waterlogging, causing the browning of number of roots and a limited plant growth, accompanied by yellowing and sometimes drying. Comparable symptoms may occur on seedlings which are kept too long in pots and have become pot bound (Photo **515**).

515 The roots of this tomato plant have been constricted and have grown into a knot which does not allow the tomato plant to establish satisfactorily in the soil after planting.

Root malformation

516 Several flower stalks have emerged near a tomato plant, indicating the presence of this parasitic plant.

517 On the flower heads, pale blue flowers are now clearly visible.

518 After exposing the root system, *Orobanche* sp. can be seen on a portion of the root.

Orobanche **spp. (broomrape)**

Orobanche spp.

The broomrapes are parasitic plants without roots or chlorophyll, whose nutrition is totally dependent on their hosts. Over 150 species have been identified. They attack many crops, cultivated or not, especially in the families Solanaceae, Fabaceae, and Cucurbitaceae, on which they can cause significant damage. The most harmful species are mainly located in the Mediterranean, south-west Asia, and some other regions where the climate is similar to the Mediterranean type (Western Australia, California).

In France, *Orobanche ramosa* L., syn. *Phelipaea ramosa* CA Meyer, a species with the widest host range, is increasing, notably on hemp, rapeseed, nightshade, and tobacco. It does not currently appear to pose major problems for tomato crops in open fields. Note that several other species of *Orobanche* have been reported on tomato – *O. aegyptiaca* Pers. and *O. cernua* Loefl.

The appearance of white, yellow, or purple tufts at the base of plants characterizes broomrape parasitism (Photos **516** and **517**). These tufts correspond to the floral stems that may have a height of several tens of centimetres. These stems bear flowers, pale blue to blue-purple in the case of *O. ramosa*, which will produce capsules containing 300–500 seeds each (seed measuring 0.2–0.4 mm long).

Broomrape parasitism mainly affects tomato roots, on which they set up suckers then sprawling anchorage roots, allowing them to absorb the mineral nutrients essential for growth (Photo **518**). The host plants are weakened and therefore smaller and stunted. Eventually, the fruit yields are reduced. In addition, these plants are unable to regulate their stomatal transpiration in dry periods, so they keep their host in chronic water deficit.

Broomrape seeds survive in the superficial soil layer, where they remain for many (10+) years. They can also survive and multiply in the soil through many host crops, e.g. vegetables sensitive to *O. ramosa* (eggplant, lettuce, pepper, Cucurbits, beans, peas, carrot, celery, fennel) and some weeds (*Capsella bursa-pasteuris*, *Solanum* spp., *Amaranthus* spp.). In the presence of a sufficiently moist soil and stimulated by root exudates, seeds germinate and attach to the roots of their host. Subsequently, the broomrape penetrates and set up a sucker in connection with the xylem of the host, allowing it to secure the required carbon, mineral, and water needs, to the detriment of the host plants. Broomrape stem growth begins at first underground. It does not take long to emerge and form the flower stalk, which flowers; tens of thousands of seeds are produced per broomrape plant. These can be spread by irrigation water or as a result of flooding, by wind, animals, farming tools, and contaminated seed. The fight against these very polyphagous plants is often very difficult because of their enormous potential for survival and dissemination. The following measures should be implemented:

– Pull out or destroy the weeds before they flower;

– At the start of an outbreak, treat broomrape and their hosts with a herbicide such as glyphosate, to prevent plants from producing seeds that will return to the ground;

– Maximize the fight against certain weeds (Brassicas) that can be substitutes for the tomato and perpetuate these parasite plants in the absence of more susceptible hosts;

– In some countries, soil disinfection with a fumigant is considered (methyl bromide). Solarization would destroy part of the stock of seeds in the first few soil centimetres (see p. 487);

– Deep burial of seeds is sometimes recommended because it distances them from future tomato roots that will only be infected later. This practice is controversial because it contributes to their conservation;

– The increase of nitrogen fertilization reduces infestation by reducing seed germination and attachment to their hosts. Ammonium nitrogen is more inhibitory than the nitrate forms;

– Prevent contamination of new fields by contaminated farm equipment. For this, infested fields should be worked last and equipment washed afterwards.

For example, note that a monogenic, dominant, resistance to *O. aegyptiaca* controlled by the 'Ora' gene has been developed in tomato in Egypt.

Roots and stem base

519

520

521

522

519 Moist and black lesion. Nearby roots are brown and altered.
Phytophthora nicotianae

520 Old canker is partially covered by mycelium and brownish small structures: sclerotia.
Sclerotium rolfsii

521 A moist lesion or rot surrounds the buried stem of this tomato plant. In some places, the cortical tissues have disappeared.
Botrytis cinerea

522 A black rot, very moist, causes gradual cortex decomposition of the lower part of the stem of seedlings.
Rhizoctonia solani

Examples of cankers symptoms on the stem base of the tomato

VARIOUS SYMPTOMS AT THE STEM BASE AND LOWER STEM

Possible causes

- *Alternata* f. sp. *lycopersici* (see p. 309)
- *Botrytis cinerea*: Description 2
- *Didymella lycopersici*: Description 3
- *Fusarium oxysporum* f. sp. *radicis-lycopersici*: Description 11
- *Fusarium solani*: Description 12
- *Phytophthora* spp.: Description 12
- *Rhizoctonia solani*: Description 14
- *Sclerotinia minor*: Description 15

- *Sclerotinia sclerotiorum*: Description 15
- *Sclerotium rolfsii*: Description 16
- *Thanatephorus cucumeris*: Description 14
- *Agrobacterium tumefaciens* (see p. 275): Description 23
- Corky stem base and tap root
- Excess salinity
- Various phytotoxicities

Additional information for diagnosis

As suggested in the preceding sections, the stem base can be affected by several micro-organisms (especially fungi) that can attack both the stem base and roots (see p. 243). Many affect young seedlings in nurseries (causing damping-off in particular), and the plants after planting (see p. 251).

Note that the stem base, the part of the stem located at ground level, is especially vulnerable to water stress and to some 'specialized' fungi such as

Sclerotium rolfsii. Thus symptoms of varying severity, which when severe, may encircle the stem, characterize a stem base problem. In many cases, the sap flow will be disrupted, causing yellowing and/or reversible wilting.

Note that it is not always easy to identify the cause of stem base symptoms, and it is often appropriate to take a sample to a specialized laboratory in order to get an accurate identification.

Roots and stem base

523 This very etiolated tomato plant has a black lesion above the rockwool cube.

524 A closer look shows that this lesion is several centimetres long and is moist and dark, girdling and constricting the stem.

525 When removed from the rockwool the roots are seen to be brown and rotten to varying degree.

Phytophthora nicotianae attack on a young plant
(Phytophthora crown and root rot)

Phytophthora spp.

Several species of *Phytophthora* have been reported worldwide on the stem base of tomato seedlings and mature plants: *P. erythroseptica*, *P. citricola*, *P. capsici*, *P. cryptogea*, *P. nicotianae*. This last species is the most frequently described and is particularly damaging in nurseries, and in soil after planting. Note that attacks have also been observed on tomato rootstocks (intra- or interspecific) used as an alternative to methyl bromide (Photo **527** and p. 658).

Whatever the species of *Phytophthora* attack, the symptoms observed on tomato are frequently the same: a blackish lesion with rather diffuse contours, rather moist, gradually encircling the stem base (Phytophthora crown and root rot) (Photos **523** and **526**); on young plants, the stem may be constricted in the affected area (Photo **524**). A longitudinal section of the stem shows that the canker underlying vessels is brown, and is even more visible when the cortical tissues disintegrate and locally disappear (Photo **528**). The appearance of fructifications on the decomposed tissues confirms the presence of *Phytophthora* (Photo **530**). Often, before colonizing the stem base, *Phytophthora* has already affected the roots, causing moist brown lesions and generalized rot (Photos **525** and **528**, and p. 241).

Finally, remember that some species also induce fruit symptoms, particularly *P. nicotianae*. These are concentric brown rings (buckeye rot): (Photo **529** and p. 374).

Fusarium oxysporum f. sp. radicis-lycopersici

Before attacking the stem base, FORL affects the roots on which it creates numerous brown lesions (Fusarium crown and root rot) (see p. 247). Subsequently, it reaches the base of the stem on which a well defined brown, moist lesion gradually develops (Photo **531**). It is slightly depressed and may extend over several tens of centimetres (Photo **532**). It should be noted that the stem base canker extends upwards, giving the lesion a flame shape.

Vascular tissues are also affected: they turn brown in the lateral roots and tap root but also for several centimetres above the base of the plant (Photo **534**).

Cankers may be partially covered with a characteristic pink salmon mucous (Photo **533**). It consists of many sporodochia producing countless *Fusarium* micro- and macroconidia (Photo **535**).

It is obvious that in the presence of such symptoms plants may wilt and even completely wither away.

Fusarium solani is also reported as the cause of stem base rots on various Solanaceous vegetables (eggplant, pepper, tomato), mainly in tropical production conditions. On the damaged tissues it produces its conidia form and its teleomorph (red perithecia) (*Haematonectria haematococca* [Berk. & Broome] Samuels & Rossman [1999]).

It seems to act in combination with other pests/pathogens and/or on plants that have already suffered water stress.

Roots and stem base

526 On this old plant, the roots near the stem base are brown and rotten. The rot has gradually reached the stem base that has taken on a blackish hue.

528 The cortical tissues are also thicker, sometimes corky and rotted. They have broken down locally, revealing a brownish vascular system.

527 *P. nicotianae* also attacks rootstocks; a moist brownish lesion with a diffuse outline is clearly visible. The injury does not appear to extend to the graft – an eggplant in this case.

529 Note that *P. nicotianae* induces irregular brownish rings on fruits, with looped edges and concentric patterns reminiscent of a target.

530 Elongated and mature sporangia in the damaged tissues are seen with the aid of a microscope.

Phytophthora nicotianae attack on a young plant (Phytophthora crown and root rot)

531 A wet, brown lesion is visible at the foot of this soil-less-grown tomato.

533 In wet conditions, small mucoid pink salmon spore masses (sporulating sporodochia) can cover the central part of the cankers. The cankers are well defined and slightly concave.

535 This sporodochium consists of several grouped conidiophores at the end of which macroconidia are produced.

532 The lesion may girdle the stem at the stem base and extend several centimetres, affecting only a portion of the stem at its highest point.

534 A longitudinal section at the stem base shows that penetration has taken place through several roots, and that some vessels are now brown.

Roots and stem base

Fusarium oxysporum f. sp. *radicis-lycopersici* (Fusarium crown and root rot)

536 A moist, blackish lesion, with a diffuse leading edge surrounds the stem of this soil-less-grown tomato.

537 The brown colour of underlying vessels confirms that they are also affected.

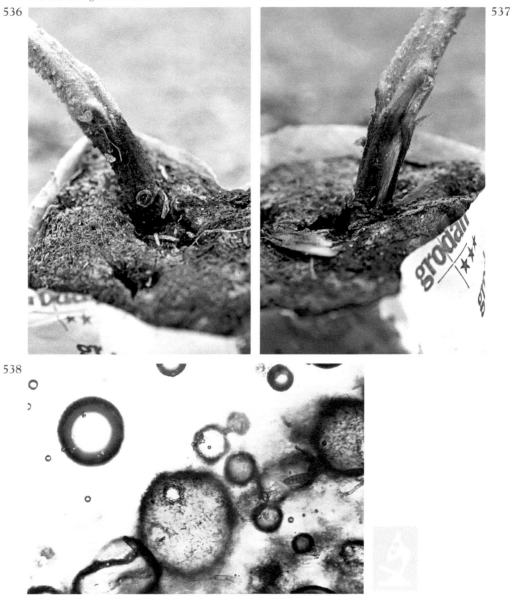

536

537

538

538 The brown, globose, and ostiolate pycnidia are visible in tissues.

Didymella lycopersici (Didymella foot rot)

Didymella lycopersici

This fungus which affects red pepper and egg-plant can be widespread on tomatoes both in the field and under protection. Although generally uncommon on this crop, it can occasionally be harmful particularly when conditions are favourable. Once present in a crop, it is difficult to eliminate.

The stem is the most commonly affect plant part. The pathogen infects through leaf removal wounds (see p. 311). It also affects the stem base, causing moist brown to black lesions on the cortex gradually encircling the base of the diseased plants (Photo **536**), which explains the name given, Didymella foot rot. Attacks at this point of the stem may have resulted from penetration through the cotyledonary scars. The underlying vascular tissue have a brownish tinge (Photo **537**).

At this point, the lower leaves may yellow, wilt, and become necrotic and the plants die. Globular structures, brownish to black, are formed in the damaged tissues, the pycnidia of the anamorph of this fungus (Photo **538**). They should not be confused with the dark brown glandular trichomes present on the stems of tomatoes.

Note that *D. lycopersici* causes symptoms on the leaves (see p. 181) and fruits of tomato (see p. 397).

Sclerotium rolfsii

Highly polyphagous and rather dependent on hot and humid conditions, this basidiomycete fungus is particularly aggressive on the stem base of tomatoes (southern blight). When acute, a moist brown to black lesion appears and surrounds the stem at soil level, then extends gradually over several centimetres to reach the foot of the affected plants (Photos **539–541**). If weather conditions are favourable, a strong and thick white mycelium quickly takes over the basal lesion and also reaches the part of the lesion above the ground (Photo **543**). In this mycelium, rounded sclerotia, at first white then beige to brown, are formed (Photo **542**). They resemble seeds of Brassicaceae and are 1–3 mm long. The mycelium also invades the root system making it completely rot (Photo **544**). Affected plants develop a reversible wilt at first, which increases in intensity very rapidly. They die soon after.

Fruit in contact with the ground have a yellowish lesion, then a wet rot causing a collapse of tissues. The wrinkled skin eventually cracks. Mycelium and sclerotia quickly cover the lesions (Photo **545**). Like fruits, leaves touching the ground may be attacked, developing large moist spots.

The mycelium of this fungus is able to grow actively on the soil surface and in this way the pathogen moves from one plant to another (Photo **547**). Its growth is saprophytic and it uses vegetable debris as its nutrient base. This mycelium is characterized by the presence of anastomosis loops (Photo **546**).

Sclerotium rolfsii is relatively rare on tomatoes, especially those growing in cooler climates. However, it is particularly common and harmful in many vegetable crops in tropical areas.

Roots and stem base

539 In the centre of withered vegetation, the affected stem base and some rotten fruit are readily visible. The lesion at the base of the plant is moist and brown.

540 A strong and thick white mycelium covers the affected stem portion visible above ground.

543 The mycelium grows on the stem, even if the stem is buried.

541 The canker brightens with age and becomes a beige to light brown colour. Mycelium and structures comparable to Brassica seeds are seen

542 The visible structures are actually circular sclerotia 1–3 mm in diameter. They are formed on the mycelium and/ or damaged tissues. Their colour changes from white to beige then to dark brown.

Sclerotium rolfsii (southern blight)

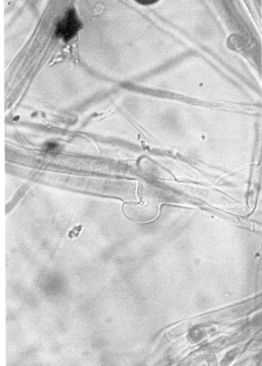

Roots and stem base

544 The fungus also invades the roots which turn brown and rot. Its mycelium is visible in places.

546 The mycelium is large and clamp connections can be seen which are characteristic of *S. rolfsii*.

545 *Sclerotium rolfsii* has partially colonized the fruit. A moist and concave lesion is clearly visible, the skin is wrinkled and broken. Mycelium and sclerotia colonize the affected fruit.

547 By growing over the soil surface the pathogen (white mycelium) has transferred from the diseased plant (right) to a healthy plant (left).

Sclerotium rolfsii (**southern blight**)

The behaviour of these fungi on tomatoes is somewhat similar (see Symptoms and irregularities on external or internal stems). They are opportunistic and penetrate their host through wounds and senescent tissues. Their symptoms are rather localized on the aerial parts of tomato, but damage to the stem base can sometimes be found.

Botrytis cinerea causes beige cankers located at the stem base or close to it (Botrytis basal stem canker, Photo **548**). It easily colonizes senescent cotyledons on seedlings planted too deep. Subsequently, it reaches the stem and gradually encircles it. It can also occur after planting causing stem base canker. Grey mould is often found on damaged tissues.

Two species of *Sclerotinia* can attack tomato, *S. sclerotiorum* and *S. minor*. Although much more common on other members of the Solanaceae, the former is sometimes responsible for moist lesionss, brown to dark brown, sometimes covered with a cottony white felting and fairly regular large black sclerotia (Photo **549**). *S. minor* produces a less abundant mycelium and smaller and irregularly-shaped sclerotia. They rarely affect the stem base (Sclerotinia basal stem canker).

Note that plants affected by these three fungi, whose stem is encircled by a stem base wound, eventually wilt and wither. The fungi are normally easy to identify as they typically produce their fructifications on the damaged tissues. In the absence of these, samples (damaged portions of stems) should be placed in a plastic bag containing damp paper such as paper towel. Under these conditions, *Botrytis* fructifications or mycelium and sclerotia will soon appear. The fructifications of other fungi mentioned in this section (*Fusarium oxysporum* f. sp. *radicislycopersici, Sclerotium rolfsii*) will also develop in these damp conditions.

548 A beige to light brown, rather dry canker is clearly visible on the stem near the peat mound. **Botrytis cinerea**

549 A thick white mycelium covers the roots and crown of this tomato. A large black sclerotium has formed on the mycelium surface. **Sclerotinia sclerotiorum**

Botrytis cinerea, Sclerotinia sclerotiorum
(Botrytis and Sclerotinia basal stem canker)

Table 31 Characteristics of the main fungi causing symptoms on the stem base of tomatoes grown in soil or soil-less[1]

Pathogen	Frequency on stem base	Identifying structures on and in the roots	Other features
Botrytis cinerea (Botrytis basal stem canker) (Description 2)	Common.	Grey mould consisting of numerous branched conidiophores bearing powdery conidia (Photo **319**).	Its attacks are mostly found in the nursery and after planting. Affects all types of nurseries.
Didymella lycopersici (Didymella foot rot) (Description 3)	Uncommon.	Globular ostiolated pycnidia immersed in damaged tissues (Photo **538**).	Limited in its distribution and now less common.
Fusarium oxysporum f. sp. *radicis-lycopersici* (Fusarium crown and root rot) (Description 11)	Nowadays uncommon.	Sporodochia producing numerous conidia often multicellular and curved (Photo **535**).	Some varieties and rootstocks are resistant. The systematic use in glasshouses of resistant varieties has dramatically reduced the incidence of this fungus.
Phytophthora nicotianae (Phytophthora crown and root rot) (Description 12)	Less common.	Sporangia visible in damaged tissues (Photo **530**).	Occurs mainly during propagation and soon after planting. Affects mainly extensive nurseries; the rootstocks are severely affected. Restricted to warm climates.
Rhizoctonia solani (Rhizoctonia basal stem rot) (Description 14)	Common.	Cellular mycelium, hyaline to brown, with a constriction at its branches (Photo **551**).	Present mainly in the nursery and after planting. Extensive nurseries are more affected.
Sclerotinia sclerotiorum (Sclerotinia basal stem canker) (Description 15)	Less common.	White mycelium covering the damaged tissues, presence of large black sclerotia (Photo **549**).	Occurs especially after planting and during cultivation.
Sclerotium rolfsii (southern blight) (Description 16)	Rare in Europe, frequent in tropical areas.	Abundant white mycelium with smooth and spherical sclerotia, browning gradually (Photo **542**).	Can attack after planting at all stages of development.

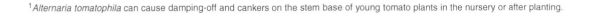

Roots and stem base

---

[1] *Alternaria tomatophila* can cause damping-off and cankers on the stem base of young tomato plants in the nursery or after planting.

550 A brownish canker, rather dry, well-demarcated, surrounds the base of the tomato plants.
Rhizoctonia solani

552 A white mycelial sheath partially covers the black symptoms of the lower stem.
Thanatephorus cucumeris

551 A brown mycelium growing on the surface of injured tissues.
Rhizoctonia solani

553 A mycelial felt is the hymenium on which basidia and basidiospores form.
Thanatephorus cucumeris

Rhizoctonia solani, Thanatephorus cucumeris
(Rhizoctonia basal stem canker)

Rhizoctonia solani, Thanatephorus cucumeris

R. *solani* is present in many soils and in some substrates, and is responsible for damping-off and root damage (see p. 245) and is able to attack the stem base of the tomato (Rhizoctonia basal stem canker). On the stem base, it causes lesions that either appear after planting or later, particularly on plants produced under adverse conditions. Lesions are first moist, brownish, sometimes reddish brown and they are well defined. Eventually these cankers may encircle the stem base and become rather dry in appearance (Photo **550**). Often, this damage also extends to the top of the root system.

Note that the brown mycelium characteristic of this soil fungus is easily observable on the affected tissues, using a binocular microscope (Photo **551**).

The perfect form of R. *solan* (*Thanatephorus cucumeris*: basidiomycete) is occasionally seen on tomatoes. The perfect stage is accompanied by a moist blackish lesion on the stem base (Photo **552**), covered with a thick whitish hymenium (Photo **553**) (see also pp. 312 and 313). *Rhizoctonia solani* and its teleomorph may also affect the stem, leaves, and fruits of tomato.

Roots and stem base

554 A corky split is visible at the foot of this tomato.

556 This stem base has a large diameter partly due to a thickening of cortical tissue. The end of the tap root has disappeared due to constriction.

555 Corky tap root tissues have a yellow/orange colour. A split is forming.

557 The stem base is sometimes brown and corky while the diameter of the tap root is greatly reduced. Roots close to the latter are also slightly brown and corky.

Corky stem base and tap root (corky crown)

Corky stem base and tap root

Corky and basal cracks of tomato stems (corky crown) is occasionally seen in some cultures. These symptoms and their development are similar to those already described in other Solanaceous plants, such as peppers (capsicum basal necrosis) and tobacco (tobacco corky stem base), but the cause of the problem remains unknown. The symptoms in particular include:

– a slight chlorosis and a reversible plant wilting during the hottest periods of the day, accompanied by a slower growth;

– necrosis and corkiness of the stem base (whose diameter is sometimes increased) and tap root with cracks (Photos **554–557**). These are partly due to loss of elasticity of the cortical tissues which, after corkiness, become less flexible to the stem growth. Note that these cracks may allow for opportunistic organisms to develop and cause browning of deep tissues and the most superficial vessels. The weakening of the stem base can also cause plant collapse;

– tap root may be constricted, secondary roots are necrotic and slightly corky.

As with other Solanaceae, the origin of this disease does not appear to be parasitic. It could be related to repeated local asphyxiation at the propagation stage or after planting, causing superficial plant damage that induce the development of corky scar tissues in underground parts. It is encouraged by the use of agricultural practices that result in water saturation. Excessive salinity may also be an encouraging factor in its appearance.

Note that stem base necrosis have been observed in hydroponic (NFT) tomatoes in the UK, due to the accumulation and crystallization of soluble salts (phosphates and nitrogen) near the plant's stem base. The presence of excess salt at the base of the stem induces tissue burns at first, then a strangled stem base and reduced water flow. Wilting follows. The tomato is more sensitive to excess salts in soil than in soil-less culture. In the latter case, the development of its root system is slower.

Various phytotoxicities

Excessive levels of herbicides in some layers of the soil may lead to abnormal root development. Growth is inhibited, which gives the stems a: swollen, dented, sometimes fluted and superficially corky appearance. Many roots, which have stopped developing, are also visible. These largely explain the characteristic symptoms (Photo **558**).

Several herbicides are known to cause such deformations of the taproot and the underground stem such as pendimethalin and trifluralin. Further information on phytotoxicities (chemical injuries) is given on pp. 77 and 131.

558 On these plants which were planted too deeply, the stems in the soil are swollen, or even fluted. The tissues are brownish and slightly corky.
Herbicide damage

Roots and stem base

Internal and external symptoms of the stem

Symptoms studied

- Cankers on the stem (often starting from pruning wounds)
- Other stem symptoms (browning, bronzing, cracks, and growth of adventitious roots)
- Symptoms of vessels and/or pith (yellowing, browning)

Stem

559 A canker with concentric patterns developed from a pruning wound.
Botrytis cinerea

560 A significant portion of the stem of this plant shows a marked browning.
Pectobacterium sp.

561 The separation of the external stem tissues shows the vessels with a brown colour, characteristic of vascular diseases.
Clavibacter michiganensis subsp. *michiganensis*

562 The pith of the stem is glassy and dark to blackish depending on position.
Pseudomonas corrugata

Examples of symptoms occurring in and on the stem of the tomato

Possible causes

- *Alternaria alternata* f. sp. *lycopersici*
- *Alternaria tomatophila*
- *Botrytis cinerea*
- *Botryosporium* sp.
- *Didymella lycopersici*
- *Fusarium oxysporum* f. sp. *lycopersici*
- *Fusarium oxysporum* f. sp. *radicis-lycopersici*
- *Phoma destructiva*
- *Phytophthora infestans*
- *Sclerotinia minor*
- *Sclerotinia sclerotiorum*
- *Thanatephorus cucumeris*
- *Verticillium albo-atrum*
- *Verticillium dahliae*
- *Clavibacter michiganensis* subsp. *michiganensis*
- *Pectobacterium carotovorum* subsp. *carotovorum*
- Other *Pectobacterium* (*P. carotovora* subsp. *atrosepticum*, *P. chrysanthemi*)
- *Pseudomonas corrugata*
- *Pseudomonas mediterranea*
- Other *Pseudomonas* (*P. cichorii*, *P. fluorescens* biotype I, *P. viridiflava*)
- *Ralstonia solanacearum*
- Various viruses
- *Aculops lycopersici*
- Lightning
- Hail
- Physiological growth of aerial adventitious roots
- Various phytotoxicities
- Distortion of the stem

Diagnosis – sometimes very difficult

Several nonparasitic and parasitic diseases are able to cause a variety of symptoms on and in the stems of tomato. Thus, all parts of the stem are likely to be affected: bark, endoderm (i.e. cortical tissues), vessels, and pith. Note that several parasitic diseases do not affect the vessels at first, but do later. The sap transport is eventually disrupted, causing wilting, yellowing, and drying of leaflets, leaves, and plants.

Note that the examination of an affected stem (which can sometimes be several metres long) should be performed very carefully, and over its entire length. In addition, to verify the status of vessels and pith, longitudinal and transverse cuts should be made at different levels: near the stem base, near the apex, and at several intermediate levels. Note that the so called 'vascular' diseases which especially affect tomato vessels, are the cause of similar symptoms in the stem leading to possible confusion.

Finally, many air-borne tomato diseases are likely to cause symptoms on the stem. Generally, they are accompanied by other symptoms on the other plant parts and these are often more characteristic. Therefore, the sections on these plant parts should be consulted – this section only deals with diseases that specifically affect the stem.

Stem

563 A pruning stub has allowed the establishment of *Botrytis cinerea*.

564 *Didymella bryoniae* entered through a pruning wound to penetrate the stem. The lesions continues to spread in spite of the fungicidal paste applied to the stem.

565 The leaflets, colonized by *Sclerotinia sclerotiorum*, are in contact with the stem, which has enabled the pathogen to colonize the stem affecting a significant portion of it. Note that resting structures (sclerotia) form on the surface; they help with the identification of this opportunistic pathogen.

Examples of cankers on tomato stem

CANKERS ON THE STEM (OFTEN STARTING FROM PRUNING WOUNDS)

Possible causes

- *Alternata alternata* f. sp. *lycopersici*: Description 1
- *Alternaria tomatophila*: Description 1
- *Botryosporium* sp.
- *Botrytis cinerea*: Description 2
- *Didymella lycopersici*: Description 3
- *Phoma destructiva*

- *Sclerotinia minor*: Description 15
- *Sclerotinia sclerotiorum*: Description 15
- *Thanatephorus cucumeris*: Description 14
- *Clavibacter michiganensis* subsp. *michiganensis* (see p. 325)

Additional information for diagnosis

The fungi *Botrytis cinerea*, *Sclerotinia sclerotiorum*, *Sclerotinia minor*, and *Didymella lycopersici* are responsible for cankerous symptoms on tomato stems, and all appear similar. Sometimes they coexist on the same lesion. They are opportunistic parasitic micro-organisms who often take advantage of special conditions to settle on plants and occasionally cause serious damage:

– a prolonged period of damp and wet weather, excessive irrigation;
– free standing water on the leaves, particularly along the leaves edges;
– plants with numerous injuries such as pruning and leafing wounds;
– the presence of senescent tissues, especially the older lower leaves in contact with the ground;
– very vigorous plants with succulent tissues.

These fungi readily colonize wounds and senescent tissues and these entry points or nutrient bases enable them to penetrate and establish. They degrade the invaded tissues by enzyme action. Generally they gradually surround the stem and eventually reach the vessels. Wilting and desiccation can then be seen.

These fungi are easily identifiable, especially *Botrytis cinerea* and the two species of *Sclerotinia* (*Table 32*), thanks to the production of spores and sclerotia which they produce quickly. To encourage the appearance of these, some samples should be placed in an airtight container or plastic bag containing paper towels soaked with water. It will provide the necessary humidity for the formation of these structures.

Note that these pathogens produce other symptoms on the other aerial parts of tomato, particularly on leaves and fruit. The sub-section 'Brown and beige discoloration, often with necrosis in advanced stages' should be consulted p. 173.

Stem

Table 32 Diagnostic criteria for the identification of the main fungi responsible for stem lesions on tomato

	Botrytis cinerea	*Sclerotinia* spp.	*Didymella lycopersici*	*Alternaria tomatophila*
Frequency in crop	Very common under protection, less common in field-grown.	Less common in field-grown/under protection.	Uncommon, in field-grown/under protection.	Common, mainly in field-grown.
Plant symptom location	Randomly on the stem, from wounds; on all other aerial parts.	Randomly on the stem, from wounds; on all other aerial parts.	Randomly on the stem, from wounds; on all other aerial parts.	Randomly on the stem, not from pruning wounds; on all other aerial parts.
Period onset of symptoms	All year.	Spring, summer.	Spring, summer.	Spring, summer.
Progression rate of symptoms in humid conditions	+++	+++	+	++
Presence of characteristic structures on lesions	Grey mould, consists of many conidiophores and conidia (Photo **576**).	White cottony mycelium, carrying sclerotia of varying size depending on the species of *Sclerotinia* (Photo **579**).	Black globular structures: pycnidia (Photo **588**).	Conidiophores and multicellular club-shaped conidia, with a rather long appendix (Figure **45**).

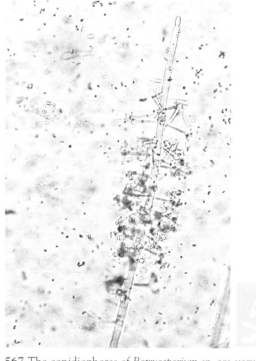

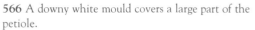

566 A downy white mould covers a large part of the petiole.

567 The conidiophores of *Botryosporium* sp. are very long (5 mm) and hyaline. Many conidia form on the swellings carried by short branches.

Botryosporium sp. (Botryosporium mould)

Botryosporium sp.

It is not uncommon to observe the presence of white mould (Photo **566**) on senescent tissues of tomatoes grown under protection (near the stubs from pruning or leaf stripping wounds, or decomposing leaflets). The mould is actually made up of numerous long aerial conidiophores (Photo **567** and Figure **44**) belonging to a fungus that appears to be a simple saprophytic colonizer of senescent tissues, causing no harm to the tomato.

Note however that some species of *Botryosporium* may be pathogenic on various plants. Two species, *B. pulchrum* Cad. and *B. longibrachiatum* (Oud.) Mayor, have been reported on tobacco, where they affect quality by developing into the leaves during drying. In addition, *B. longibrachiatum* may have been associated in Brazil with the death of pepper shoots. *B. pulchrum* has also been described as pathogenic on beet leaves and pelargonium.

Botrytis cinerea

This fungus is responsible for grey mould and occurs frequently on tomato as well as on many other vegetable crops. Often, wet and cold weather accompany its development. Although it is found in open fields, it is under protection that it damages stems and is particulary serious. It has significant saprophytic potential allowing it to colonize senescent or diseased plant tissues, and on tomato occurs notably on:

– Dehiscent cotyledons in the nursery, eventually resulting in the development of lesions surrounding the stem (Photo **568**). Cankers appear at the time of planting.
– Injured leaflets, young or old, chlorotic and senescent (Photos **572** and **574** and see p. 181);
– Senescent floral parts (especially the petals) still in place or fallen on other parts of plants (see p. 181);
– Various natural wounds or those related to plant cultivation (leaf trimming, budding control).

Senescent tissue and wounds when in a moist environment enable this fungus to colonize these rapidly, and especially those of the stem. These quickly become moist and take on a brown shade. They are eventually necrotic and dry out, their colour then ranging from beige to dark brown. Eventually, lesions or cankers are visible on the stems, mainly localized near pruning and budding wounds (Photos **569** and **570**). They gradually spread, sometimes over several centimetres (Photos **571** and **573**), and eventually surround the stem. In many cases, the distal end of the plant wilts and desiccates. A cut in the stem shows that the vessels and sometimes the pith are discoloured (Photo **570**).

The wounds are usually covered with a characteristic greyish to beige mould (Photos **575** and **576**). This consists of numerous conidiophores of the fungus and these produce innumerable conidia.

Additional photos of symptoms caused by *B. cinerea* on other parts of the tomato plant are found on pp. 243, 397, and 411.

Stem

Figure **44** Drawing of *Botryosporium* sp. fructifications.

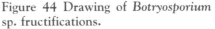

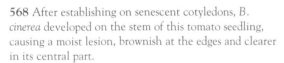

568 After establishing on senescent cotyledons, *B. cinerea* developed on the stem of this tomato seedling, causing a moist lesion, brownish at the edges and clearer in its central part.

570 A longitudinal section of a mature canker reveals symptoms of the vessels and locally of the pith, which is moist and chocolate brown.

569 A moist, brown lesion is emerging on the stem and the petiole. It is almost completely covered by a grey mould.

571 Once in the stem, *B. cinerea* continues to develop. Eventually the cankers can surround the stem for several centimetres.

Botrytis cinerea
(grey mould)

572 After the leaflets have been invaded, *B. cinerea* progressively colonizes the petiole of this leaf.

574 *B. cinerea* spreads quickly to the whole leaf and eventually reaches the stem on which it produces a canker.

573 Once the stem is girdled, the sap flow is disrupted, and the distal ends of the plant wilts and withers.

575 On all affected parts, *Botrytis cinerea* forms a characteristic grey to beige mould.

576 The conidiophores of *B. cinerea* are 'tree-like' and produce hyaline ovoid spores.

Botrytis cinerea
(grey mould)

577 In open field and in flat ground crops, the fungus rots all organs near the ground: leaves, stems, petioles, and fruit.

579 Sclerotia of irregular size may form on lesions on the surface of the stem. Note that they melanize gradually, which explains their black colour.

578 A moist lesion develops from a leaf removal wound. It is beige and is partially covered by a white cottony mycelium.

580 In some situations, a longitudinal section of the stem is necessary to reveal sclerotia located in the pith.

Sclerotinia sclerotiorum
(white mould)

Sclerotinia sclerotiorum and *Sclerotinia minor*

Both species of *Sclerotinia* are widely distributed throughout the world and affect many species including the tomato (white mould). *Sclerotinia sclerotiorum* is much more common on tomato than *S. minor* both in the field (staked or unstaked) and under protection

S. minor is much more restricted, mainly reported in the field on unstaked plants.

These two fungi cause quite similar symptoms on affected tomatoes and for this reason are considered together. As with *Botrytis cinerea*, they are opportunistic parasites that readily colonize wounds and senescent plant tissues.

It is therefore not surprising that they are a severe problem in unstaked field crops, in wet soil conditions, and at a time when well developed plants have vegetation that covers the stems and fruits. All plant parts near or in contact with the soil can be affected by a soft and moist rot (Photo **577**). Lesions on stem tissues are moist and have a beige hue and a fairly typical milky appearance. They spread and eventually surround the stem. A white cottony mycelium eventually grows on the damaged tissues, as well as black masses of varying size which are the sclerotia. These structures allow the differentiation between the two *Sclerotinia* species attacking tomato:
– clusters of small irregular black sclerotia in the case of *S. minor*;
– some large black sclerotia, rather elongated, are typical of *S. sclerotiorum*.

On this type of crop, a focus of rotting plants often occurs within the crop canopy. The pathogen spreads by contact between diseased and healthy organs. The leaflets, like the fruits, are also affected by a moist and soft rot responsible for the disintegration of the former and the liquefaction of the latter (see p. 399). As before, mycelium and sclerotia may be present on the affected organs.

In protected cultivation, only *S. sclerotiorum* is found. Often, only stem cankers are seen, which result from infection by sexually produced airborne ascospores. This is an extremely rare mode of reproduction in *S. minor*, which probably explains why it is not found on stems of protected crops. The stem lesions start, as for *Botrytis cinerea*, from senescent tissues or from various wounds. Remember that pruning wounds constitute favoured points of entry and infection results in moist and soft tissue development. The lesions are, as in field crops, beige with a white cottony mycelium gradually covering them (Photo **578**). Sometimes orange coloured droplets ooze from the lesions. Masses of fungal tissue, initially white then gradually melanized and of hard consistency, are formed either on the stem (Photo **579**) or inside it (Photo **580**): these are the sclerotia of the fungus, and they ensure its survival in the soil.

Note that the perfect form of *S. sclerotiorum* is sometimes visible on the surface of a damp soil. It is characterized by the formation of small 'trumpet-shaped' apothecia (see Description 15), which are the source of ascospores responsible for aerial contamination.

Stem

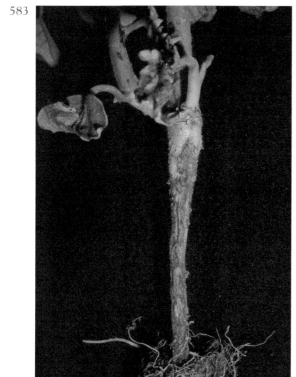

581 Several brown necrotic lesions are visible on the stems and petioles of tomato. They are often lighter in the centre and have concentric patterns.
Alternaria tomatophila

583 A dark brown to black very extensive necrotic lesion has surrounded this plant.
Alternata **f. sp. lycopersici**

582 Inter-vein irregularly-shaped necrotic spots partially cover this leaflet.
Alternata **f. sp. lycopersici**

584 This young plant with a stem canker has slightly reduced growth. Its foliage is chlorotic, dotted with numerous necrotic lesions.
Alternata **f. sp. lycopersici**

Alternaria tomatophila and Alternaria alternata f. sp. lycopersici

Both *Alternaria* species cause quite similar lesions on the stem of the tomato which can lead to confusion. For this reason the descriptions of stem symptoms are grouped together.

A. tomatophila is the cause of dark brown to black lesions on the flower stalks as well as on the petioles and stems. These lesions are rather elongated, and show concentric rings which are brighter in the centre (Photo **581**). They can surround the stem, causing the wilting of the branch distal part (see also p. 173).

Cankers on stem (and/or stem base) caused by *A. alternata* f. sp. *lycopersici* are also elongated, but rather close to the ground. They have a brown to dark beige colour and a dry consistency. Concentric patterns are also visible on their surface. The underlying vessels located on either side of the canker may have brown streaks. These lesions gradually spread and eventually surround the stem, killing the plants (Photo **583**). Note that a toxin spreads from the cankers and causes leaf chlorosis and the widespread development of inter-vein necrotic spots (Photos **582** and **584**).

Lesions also form on the fruit.

Figure **45** should allow identification of these two *Alternaria* species, showing their different conidia.

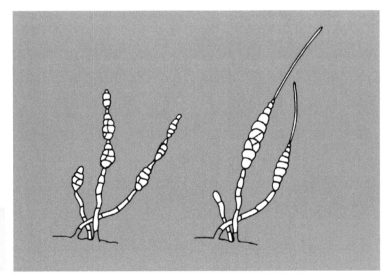

Figure 45 Drawing of fructifications of *Alternaria alternata* f. sp. *lycopersici* (left) and *Alternaria tomatophila* (right).

Stem

309

585 A moist and dark brown symptom has developed from a pruning wound. In addition to surrounding the stem, it now reaches a leaf petiole.

587 Despite the application of a fungicidal paste on a canker, the latter eventually disrupt the sap flows, which resulted in wilting of the plant.

586 Tiny globular black dot structures on the diseased tissues are characteristic of parasitism by *D. lycopersici*.

588 These structures are in fact the pycnidia of the anamorph of this fungus (*Phoma lycopersici*).

Didymella lycopersici
(Didymella stem rot)

Didymella lycopersici

This fungus has a more restricted world distribution. It is known to be responsible in glasshouses for stem cankers (Didymella stem rot), located on the crown and pruning wounds. In fact, it can attack all plant organs, especially in the field and following significant rainfall or irrigation. *D. lycopersici* causes moist dark brown stem cankers that often develop from de-leafing and other wounds (Photo **585**). These lesions gradually spread and eventually surround the stem and/or petioles, thus disrupting the sap flow. Ultimately, it is not surprising to see yellowing and wilting of the leaflets and plant parts below the lesions (Photo **587**).

Brownish tiny globular structures (pycnidia) dot the damaged tissues (Photos **586** and **588**). Note that they are more easily distinguishable with a magnifying glass than with the naked eye. These fungal structures produce bicellular hyaline conidia, fairly typical of the *Didymella* genus.

Note that lesions are also visible on the stem base (see p. 287) and on the leaflets and fruit (see pp. 184 and 397), although less frequent on the latter two. They are often darker than those caused by *Botrytis cinerea* and *Sclerotinia* spp.

Thanatephorus cucumeris

During 2004, an outbreak of rot affected dense crops in processing tomato plantations in south-western France (Photo **589**). All aerial parts were decomposed. The affected tissues suddenly rotted following moist lesion development on the leaves. Moist and brown symptoms gradually appeared on stems and eventually surrounded them (Photo **590**). A dense white mycelium covered them. The tissues of the stem cortex were eventually completely rotten and crumbled (Photo **591**).

A microscopic observation of the mycelium showed that it was actually a hymenium, on which basidia formed, structures which produce sexual basidiospores in basidiomycetes (Photo **592**).

The fruits were also affected. They had two types of lesions on the parts in contact with the ground (Photo **593**):
– a circular canker (Photo **594**), sometimes superficially corky, revealing brown concentric patterns. Superficial brown hyphae of *Rhizoctonia solani* could be seen on the lesions and/or healthy tissues nearby (Photos **596** and **597**);

– moist and soft rots covered by the same dense white felting, already noted on other aerial parts of the tomato.

This disease, previously unknown on tomato, was the result of the simultaneous manifestation in these crops of *Rhizoctonia solani* and its teleomorph *Thanatephorus cucumeris* (bottom rot). It was therefore not surprising to see the characteristic brown mycelium of *R. solani* and the whitish hymenium bearing basidia characteristic of *T. cucumeris* (Photo **595**) on the same plant. *R. solani* is well known on tomato, responsible in particular for damping-off (see p. 243), lesions on roots and stem base (see p. 245), and fruit rots (see p. 411).

The symptoms observed occasionally in the south west of France are fairly original and seem attributable in part to *T. cucumeris*. This basidiomycete is known on other plant species to be responsible for damage related to aerial contamination, initiated by its basidiospores. So, for example, on, tobacco, it is responsible for leaf spots called 'target spots'.

Stem

589 A rot outbreak has developed in the centre of the vegetation. All aerial organs seem affected.

590 Wet and brownish lesions affect sections of petiole and stem. A dense and regular mycelial felting covers them.

591 The stem portions in the centre of the rot outbreak have a completely decomposed cortex. A large white mycelial felting covers the ground nearby.

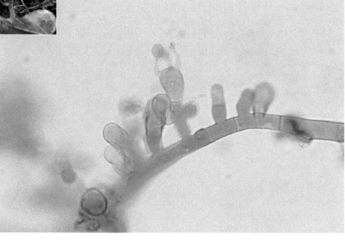

592 Rare basidia, often bearing four basidiospores, can be observed on the white hymenium partially covering the affected organs and the soil.

Thanatephorus cucumeris

593

594

595

596

597

593 Several fruit in contact with the soil are affected by a darkish rot. Some diseased fruits are slightly concave.

594 Lesions more conventional of an attack by *R. solani* were observed on this unripe fruit. Lesions are rather limited, brown to beige, they have some concentric patterns.
T. cucumeris

595 This totally rotten fruit is entirely covered with a velvety whitish hymenium of *T. cucumeris*.
T. cucumeris

596 A brown mycelial network covers the surface of some unripe fruits.
T. cucumeris

597 Light microscopy shows that the brown mycelium visible on some fruits display the characteristics of that of *R. solani*: cell walls and a constriction located at the 45° branches.
R. solani

Stem

Thanatephorus cucumeris **and** *Rhizoctonia solani*

598 A diffuse brown to black lesion is clearly visible on the stem and several translucent droplets exude from it.
Pseudomonas corrugata

600 This stem has a bronze green colour over its entire length.
Aculops lycopersici

599 Both stems have splits and numerous preliminary aerial roots.
Pseudomonas corrugata

Examples of other stem symptoms

OTHER STEM SYMPTOMS (BROWNING, BRONZING, CRACKS, AND GROWTH OF ADVENTITIOUS ROOTS)

Causes possibles

- *Phytophthora infestans* (see p. 185)
- *Pectobacterium carotovorum* subsp. *carotovorum*: Description 26
- Other *Pectobacterium* (*P. carotovorum* subsp. *atrosepticum, P. chrysanthemi*): Description 25
- *Pseudomonas corrugata*: Description 26
- Other *Pseudomonas* (*P. cichoric, P. fluorescens* biotype I, *P. viridiflava*)

- Various viruses
- *Aculops lycopersici* (see p. 147)
- Hail (see p. 81)
- Physiological growth of aerial adventitious roots
- Various phytotoxicities

Additional information for diagnosis

Several bacteria can cause extended browning of the stem of tomato: *Pectobacterium* and *Pseudomonas corrugata* (responsible for black pith). In fact in many cases, these bacteria colonize and change the inside of the stem before attacking the external tissues (p. 341).

Note that the symptoms they cause outside of the stem are often quite similar, which can be confusing. Moreover, their development is also promoted by the same conditions:

– plants grown under protection;
– often very vigorous plants, with a fleshy and thick stem and lush vegetation;
– high humidity in the crop or in the glasshouse;
– the presence of free water on the plants;
– cloudy weather preceding the appearance of the symptoms.

Pectobacterium spp.

Several species of *Pectobacterium* (bacterial stem rot) have been reported on tomato (*P. carotovorum* subsp. *atrosepticum, P. chrysanthemi*), but *P. carotovorum* subsp. *carotovorum* is the most common. This bacterium affects many vegetables including chili, eggplant, lettuce, and carrot. It attacks mainly the stems and fruits of tomato. In Europe it infrequently occurs under protection and sometimes on processing tomato crops in open field but it does not represent a threat to this production.

Stem

P. carotovorum subsp. *carotovorum* causes symptoms within the stem, often resulting from wounds. The pith on both sides of the penetration zone quickly becomes brown, sticky and liquifies to varying degrees. The stem eventually becomes hollow (see also p. 341). The cortex may also be affected: the browning may extend over several centimetres, while the affected tissues are of moist appearance and gradually decompose. Severely affected plants eventually wither and dry up (Photos **601** and **602**). Smelly, liquefying rots may occur on fruits, in the field and during storage.

Pseudomonas corrugata

The bacterium, responsible for tomato pith necrosis, does not seem to affect all the world production areas. It is present in North America (US, Canada) and in several European countries (UK, Netherlands, Germany, France). Sometimes observed in open field, it mainly affects crops in glasshouses. Fairly recent work has shown that 'P. corrugata' may be a mixture of two very similar and long confused bacteria: *P. corrugata* and *P. mediterranea*. The latter species has been identified in France among strains supposed to belong to the species *P. corrugata*. It has also been identified in Italy, Spain, Portugal, and Turkey.

Diseased plants have their growth inhibited. Young leaves gradually become chlorotic and sometimes wilt (Photo **603**). Elongated lesions, brown to black and irregularly shaped are visible on the surface of the stem (Photo **604**), and sometimes the petioles. In some situations the bacterium exudates and appears as beads on the surface of the leaflets (Photo **605**). It should be noted that some stems become brittle. A longitudinal stem section reveals that the pith is affected. It first takes a translucent appearance and a pale brown colour, then gradually darkens (Photo **606**); the stem can also be hollow and with several pith cavities which can alternate with still intact parts. Vascular tissues may have a brownish colour (see p. 341).

601 Several wilted and withered plants show a stem surrounded by several centimetres of a dark brown to black discoloration.

602 The initially moist cortex browns quickly while the stem takes on a soft consistency.

Pectobacterium **spp. (bacterial stem rot)**

603 The apex of this plant is reduced and the young leaflets are chlorotic.

604 On this rather wide and slightly flattened stem, several blackish irregular lesions are spreading.

605 In some lesions, blackish and still moist, translucent bacterial exudates can be observed when humidity is very high.

606 A longitudinal section of the stem shows that the pith is altered.

Pseudomonas corrugata
(tomato pith necrosis)

Stem

607 Adventitious roots are being formed on the stem. The stem is bumpy longitudinally to varying degree.

608 The formation of adventitious roots can be quite sudden, and they occur inside large longitudinal splits which develop on the stem.

609 These stems, formerly affected by pith necrosis, have developed numerous adventitious roots which cover fairly extensive areas.

Pseudomonas corrugata
(tomato pith necrosis)

Note that in the case of pith necrosis, the stem may also be affected by other symptoms which reflect the 'malfunction' of the plant, for instance, the growth of adventitious aerial roots (Photos **607–609**):
– these roots can remain in a undeveloped state, the stem showing 'bumpy' areas (Photo **607**);
– some longitudinal splits can occur, with many roots emerging from them (Photo **608**);
– many roots can cover the stem over a substantial length (Photo **610**), the latter also having a longitudinal split.

This disease can cause the total destruction of certain plants, but some plants may recover: weakly affected plants may recover if the conditions of production improve (sunny weather, so more light in the glasshouses, lower relative humidity). In some cases, the disease may even go unnoticed.

Pseudomonas fluorescens is very closely related to *P. cichorii*, but differs in its pathogenicity notably on lettuce. It has been described in Canada as responsible for brown lesions on stem. These grow at the nodes, the adjacent rachis, and internodes. The cortex and pith may also deteriorate. Unlike with previous bacterial diseases, plants do not seem to wilt and wither. This bacterium, which is considered opportunist, may attack plants in the spring grown in soil or soil-less conditions, but in particular those that are stressed as a result of inadequate nutrition or excessive humidity.

Other causes

A number of other biotic, but also abiotic factors, can cause brown lesions and the growth of roots on the surface of the tomato stems.

Thus, in addition to symptoms described on p. 107, several viruses can also cause longitudinal brown and necrotic lesions on stems (streaks). These range in length from a few centimetres upwards (Photo **610**). They can be located at the apex of plants, on stems, and on petioles. In the case of early infection, whole plants become necrotic and dry up. Among the viruses most often associated with such symptoms are Alfalfa mosaic virus (AMV) and Cucumber mosaic virus (CMV).

610 A brown necrotic lesion has appeared on one side of the apical part of the stem of this plant.
Afalfa mosaic virus (AMV)

Stem

 **Various viruses**

613

611 In addition to having completely dried leaves, this tomato plant has a stem with superficial irregular brownish patches.
Phytotoxicity due to a herbicide

613 This portion of stem was superficially burnt after the application of an incorrect concentration of a chemical treatment, or the treatment may have been applied in adverse conditions. The cortex is almost entirely necrotic and has a beige colour.
Phytotoxicity

612 This stem has some wide light green patches where adventitious roots emerge.
Phytotoxicity due to a herbicide

614 The appearance of adventitious roots on certain portions of the stem is not always of parasitic origin; some varieties are more prone to this physiological condition.
Physiological growth of aerial roots

Some nonparasitic diseases

Remember that spots or more extensive brown patches (Photo **611**) occasionally occur following the use of certain herbicides absorbed by the roots. Some herbicides may also induce yellowish patches on stems, together with the localized growth of adventitious roots (Photo **612**). Aerial spray treatments can also cause superficial burns of the stem (Photo **613**) resulting in multiple extensive beige injuries.

More complete information and photos showing these types of conditions are presented on pp. 77 and 131.

Aerial roots may appear on the tomato stem, as already mentioned for pith necrosis. Often, first many small isolated or aligned protrusions appear (Photo **614**). These then split giving rise to a root that may remain very short (preliminary root) or reach several centimetres. In some cases, the roots appear much more suddenly, stems burst, and the cortical tissues break off. This phenomenon can occur at all levels of the stem but is rarer near the apex.

This phenomenon is not specific to pith necrosis, but may be associated with various tomato diseases, parasitic or not. It is often an indicator of poor plant functioning. In fact, when plants are too vigorous or when their water and nutrients supply is disrupted (in the case of changes to the roots or stem), they often produce adventitious roots. Note also that some tomato varieties placed in specific production conditions (excessive humidity, lack of water) can produce adventitious roots (Photo **614**). These symptoms do not appear to affect plant development. Varieties that readily express these symptoms should not be grown.

Finally, the various tomato parts colonized by the *Aculops lycopersici* mite tend to turn a fairly typical bronze colour, especially the stem, peduncles, and petioles (Photo **615**). The leaflets and fruits are also affected. Page 147 should be consulted for more details on this pest.

615 The stem in the background, the peduncles and sepals of the fruits have a bronze tint characteristic of *Aculops lycopersici* damage.

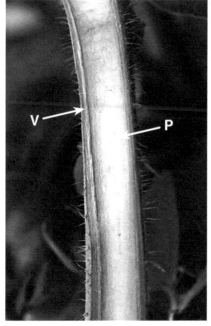

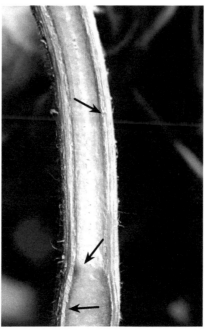

616 Longitudinal section of a healthy stem. Vessels (V) and pith (P) can be clearly distinguished and have a whitish colour.

617 The portion of the pith near the vessels is slightly yellowish.
Clavibacter michiganensis **subsp.** *michiganensis*

618 After careful removal of the stem cortex, a marked browning of the vessels and the presence of darker streaks in places can be seen.
Fusarium oxysporum **f. sp.** *lycopersici*

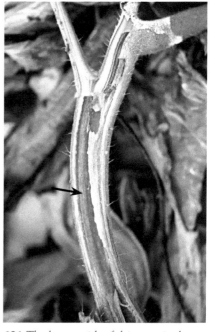

619 The area located along the vascular system has taken a brown colour and appears more moist.
Pseudomonas corrugata

620 In this stem, the pith has browned and hollowed out near the vessels which are also yellow.
Clavibacter michiganensis **subsp.** *michiganensis*

621 The brown pith of this stem is almost completely liquefied. Some remains only close to the vascular system.
Fusarium oxysporum **f. sp.** *lycopersici*

Examples of internal stem symptoms

SYMPTOMS OF VESSELS AND/OR PITH (YELLOWING, BROWNING)

Possible causes

- *Fusarium oxysporum* f. sp. *lycopersici*: Description 19
- *Fusarium oxysporum* f. sp. *radicis-lycopersici*: Description 11
- *Verticillium dahliae* or *V. albo-atrum*: Description 20
- *Clavibacter michiganensis* subsp. *michiganensis*: Description 24

- *Pectobacterium carotovora* subsp. *carotovora*: Description 25
- Other *Pectobacterium* (*P. carotovorum* subsp. *atrosepticum*, *P. chrysanthemi*): Description 25
- *Pseudomonas corrugata*: Description 26
- Other *Pseudomonas* (*P. cichorii*, *P. fluorescens* biotype I, *P. viridiflava*)
- *Ralstonia solanacearum*: Description 27
- Lightning

Additional information for diagnosis

Among the many diseases attacking the tomato, a few cause symptoms within the stem. Thus, various symptoms can be observed in the pith, stem vessels, and tissues constituting the cortex. Some of these symptoms are illustrated by Photos **616–621**, and are compared with a healthy stem.

Some micro-organisms are responsible for these internal stem symptoms and these affect the vessels (vascular micro-organisms). As a consequence, the plants produces gummy substances in the vessels, which can cause other symptoms such as wilting, yellowing, and drying out of leaves, often partial or unilateral.

Note that when the vessels are colonized, they become yellow and/or brown. Adjacent tissues such as the cortex or pith may be affected, and thus some symptoms may begin to occur on the outside of the stem:
– longitudinal yellowing on one side of the stem;
– a brown or longitudinal necrotic lesion on one side of the stem;

– the browning of a significant portion of the stem.

To identify these vascular diseases, the plants should be very carefully observed for signs mentioned in the following pages. These diseases cause quite comparable symptoms, called 'generalists': wilting, yellowing, and vascular browning.

To avoid diagnostic confusion, other more characteristic symptoms should be identified and/or a specialist laboratory approached to analyze a plant by making vascular isolation on artificial medium.

Some diseases, parasitic or otherwise, affecting the roots or stem base may be the cause of vascular browning in the lower part of the stem which then spreads further up.

If none of the conditions detailed in this section are identified, the sections corresponding to these plant parts should be consulted.

Stem

Table 33 Main differentiation criteria of common micro-organisms affecting the different internal tissues of the tomato stem (vessels, pith, cortex)

	Ralstonia solanacearum	Clavibacter michiganensis subsp. michiganensis	Fusarium oxysporum f. sp. lycopersici	Verticillium dahliae	Pseudomonas corrugata
Frequency in crop	Very common in the tropics and sub-tropics.	Quite common, in field crops and under protection.	Uncommon in Europe.	Uncommon or insignificant in Europe.	Quite common, especially under protection.
Location of first vascular browning	Lower part of stem.	Anywhere in the stem, at the pruning wounds.	Lower part of stem.	Lower part of stem.	Top and middle of stem.
Intensity of vascular browning[a]	+++	++	+++	+	+ to +/–
Potential browning of the pith and cortical tissues[a]	+	+	–	–	+++
Location of leaf yellowing, wilting and drying	Apex then quickly the entire plant. Always sudden wilting.	Random leaves.	Lower leaves first.	Lower leaves first.	Young leaves.
Presence of symptoms on other parts	Milky exudate leaking from vessels when stem portion is placed in water.	Small cankerous lesions on stem, petioles, leaflets, and mainly fruits.	None.	None.	Large brown lesions on the stem (Photo 604).
Diseased plant distribution in crop	Isolated or in group(s).	Isolated and in lines.	In group(s).	In group(s).	Isolated or in group(s).
Existence of resistant varieties[b]	+	–	+++	+++	–

[a] –: No symptom; +/– : hardly noticeable symptoms; +: minimal symptoms; ++: visible and extensive symptoms; +++: highly visible and extensive symptoms.

[b] –: No resistant varieties available; +: resistant varieties still rare; ++: resistant varieties available; +++: numerous resistant varieties available.

Ralstonia solanacearum

This bacterium is responsible for bacterial wilt. and is particularly feared in warm production zones, primarily tropical to sub-tropical, both in field crops and under protection. It is polyphagous and better known by its synonym *Pseudomonas solanacearum*. In protected crops it occurs in more northern production areas, notably in some European tomato glasshouses after having developed on potatoes.

Symptoms may occur very early on tomato seedlings. Young leaves wilt (Photos **622** and **623**), and the leaflets tend to curve upward (epinasty, see Photo **624**). Brownish inter-vein necrotic lesions are occasionally seen on some leaflets (Photo **625**). Vascular tissues turn yellow and brown quite early (Photos **626** and **627**).

Comparable symptoms are observed on adult plants. Again, sudden wilting and vascular browning characterize this bacterial disease (Photos **628** and **629**). The pith and cortex of the stem display large brown lesions in conditions particularly favourable to the development of *R. solanacearum*.

A relatively simple to implement test is often recommended to differentiate this bacterial disease from vascular diseases caused by fungi. It involves cutting the stem of an infected plant transversely and plunging it into water. If a milky trail consisting of countless bacteria exudes from the vessels within minutes, it is indeed a bacterial disease.

Clavibacter michiganensis subsp. michiganensis

This bacterium is transmitted by seeds during propagation and is particularly devastating. It is reported in many production areas of the world, both in field and protected crops growing in soil or in soil-less crops.

In the glasshouse, bacterial canker often appears early in the crop. Conspicuous inter-vein spots appear on the leaflets which quickly become necrotic (Photos **630** and **631**). Often they are followed by leaf wilting (Photos **632** and **633**), the leaflets tend to roll down in this case. Eventually, the seedlings or the mature plants may wilt and fully desiccate. In some situations, the leaf wilting is partial and accompanied by a localized yellowing (Photo **634** and **635**). On the leaves, only some leaflets on one side may be affected (Photo **636**). These symptoms are,

in this case, more typical of a vascular disease. Weather conditions greatly influence the expression of symptoms of bacterial blight. Also, it can go undetected for a long time, appear suddenly and evolve rapidly, or stagnate and allow the plants to grow more or less normally. Symptoms occur on leaflets located either at the base, top, or mid-height of the plants.

In field crops leaf symptoms are slightly different: the wilting and drying out of leaflets is mainly along the leaf edges. Brown lesions of varying widths in the shape of an inverted 'V' often accompanied by a yellow halo can be found on some leaflets (Photo **637**). They spread to affect the whole leaf, which eventually dries up completely. Many affected leaves and plants may eventually die (Photo **643**).

Stem

622 Bacterial wilt just beginning to emerge on the tomato seedling. Only one leaf shows a marked wilting.

623 The blight is now well established on the seedling, which has completely withered and will soon dry up.

624 When wilting, the leaflets tend to curve upward (epinasty).

625 Some inter-vein browning is occasionally seen on some leaflets.

Ralstonia solanacearum
(bacterial wilt)

626 A longitudinal section in the lower part of the stem shows that a vascular disease is present. The vessels are yellow and starting to brown in places.

628 These adult plants, like seedlings affected by *Ralstonia solanacearum*, eventually completely wilt.

627 Some more localized vascular browning can also be detected in the upper part of the stem.

629 The vessels are also affected. Here, clusters of very dark vessels are seen. The pith is moist and brown in places.

Ralstonia solanacearum
(bacterial wilt)

Stem

630 Several bright inter-vein patches are developing on this slightly wilted leaflet.

631 These patches gradually become necrotic and beige.

632 This adult plant shows inter-vein necrotic patches on the leaflets of several middle leaves. It is also slightly wilted.

633 The bacterial canker is now present on the completely wilted plant. In addition, many leaflets are dried up.

Clavibacter michiganensis subsp. *michiganensis*
(bacterial canker)

634

635

636

637

634 This leaflet shows early wilting affecting only one side of the leaf. Leaf tissues have a pale to brownish colour in the centre with chlorosis on the periphery.

636 Unilateral distribution of wilted and withered leaflets, characteristic of vascular disease.

635 On this plant, many leaflets have some partial yellowing, sometimes accompanied by peripheral drying causing the leaf to roll.

637 On this leaflet, from an open field plant, two necrotic lesions in a 'V' shape can be seen. They are located on the leaf edges and have a rather marked yellow halo.

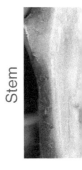

Stem

Clavibacter michiganensis subsp. *michiganensis*
(bacterial canker)

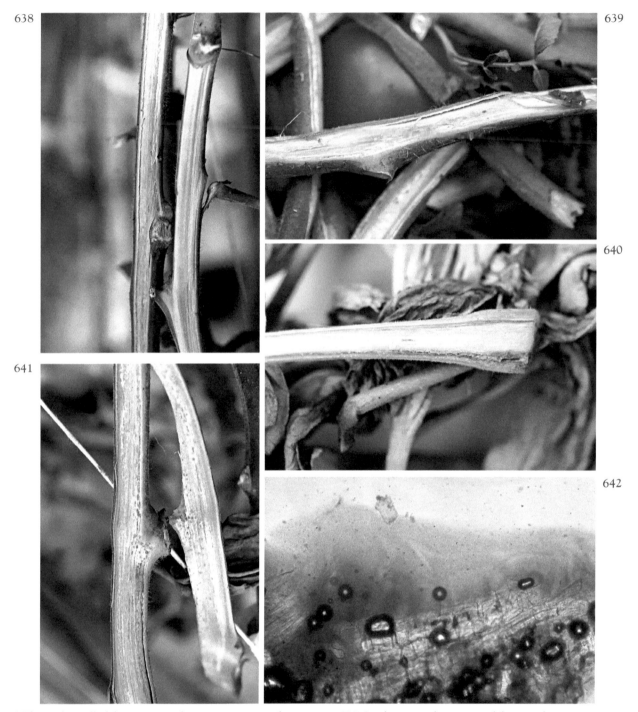

638 The peeling off of the cortex on these two stems reveals a yellowing or even a slightly diffuse browning of all vascular tissues on the diseased stem (on the left). The right stem is healthy.

641 A longitudinal section through the petiole reveals a slight vascular browning on one side and a more marked change of the vessels and adjacent tissues on the other, especially the pith which has a reddish brown colour.

639 On this stem where some of the vascular system is visible, an area shows some irregular brownish patches.

640 The removed cortex reveals many small areas of damaged tissues (replaced by small cavities) and a yellowing or browning of the vessels.

642 The microscopic observation of a longitudinal cut made in the stem near the vessels shows mucus consisting of countless bacteria.

Clavibacter michiganensis subsp. *michiganensis* (bacterial canker)

In both glasshouses or in the open fields, the distribution of diseased plants is often the same: in rows, shorter or longer (Photo **643**). This distribution is related to the characteristic mode of transmission of C. *michiganensis* subsp. *michiganensis* that occurs between plants during cultivation operations, especially during the removal of axillary shoots.

Leaf wilting as previously described, is not always sufficient to diagnose this bacterial disease, some other more specific symptoms must be present on the diseased plants such as:
– affected vessels seen after making a longitudinal or transverse stem cut. Thus creamy lines, white to yellow, and gradually turning brown, are visible along the vessels and on the cortical tissues and pith nearby (Photos **638–641**). Depending on circumstances, the internal stem symptoms can be very difficult to detect and require cutting the entire length of the stem or, conversely, are particularly marked and accompanied by vas-

cular brown streaks. A microscopic observation of histological sections of the vascular zone of the stem reveals the presence of bacterial mucus (Photo **642**);
– small lesions of a few millimetres in diameter, at first whitish, and quickly becoming cankerous, appear on the tomato aerial parts. These small superficial cankers, which slowly become necrotic, are rare on the leaflets (see p. 154), but are more frequent on the stem (Photo **644**), petioles (Photo **645**), and fruit. On the latter, they often have a distinctive appearance: small flat white circular spots on the periphery and whose centre is raised, sometimes broken, beige to tan. These characteristics give them the appearance of a bird's eye (bird's eye spot, see Photos **648** and **650**). Also, adventitious roots may grow on the stem.

Only the combined observations of foliar symptoms and more specific lesions will help to identify this disease with certainty.

643 Several fully dried plants are visible in this supported open field crop. Note that they are distributed on the same line, which is characteristic of an established attack of *Clavibacter michiganensis* subsp. *michiganensis*.

Stem

644 Tiny beige cankerous lesions are visible on the stem, on either side of the supporting twine.

645 Several contiguous microsplits resulted in the cankerous corky symptoms visible on the peduncle.

646 Under certain conditions, bacteria exude from the leaf. This is the case for this leaflet on which a milky droplet formed from the main vein.

Clavibacter michiganensis subsp. *michiganensis* (bacterial canker)

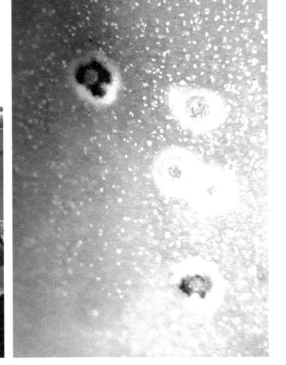

647 Several whitish to necrotic spots dot this unripe fruit. They sometimes split in their centres and have a haloed oily ring.

649 When the conditions are very humid, cankerous lesions may be more substantial. This fruit has some small cankerous brown spots and a split with necrotic edges.

648 The lesions on this unripe fruit are older. They resemble a 'bird's eye'.

650 The bird's eye spots are circular and white; their cankered centre eventually becomes necrotic and tan to brown.

Stem

Clavibacter michiganensis subsp. *michiganensis*
(bacterial canker)

651 This young plant with stopped growth, has almost completely withered.
Fusarium oxysporum f. sp. *lycopersici* (Fusarium wilt)

652 On this young plant, the lowest leaflets are the first to yellow and wither.
Fusarium oxysporum f. sp. *lycopersici* (Fusarium wilt)

Fusarium oxysporum f. sp. *lycopersici*

This fungus is widely distributed throughout the world and was once much feared, but is not currently a threat in areas of production where all the varieties are resistant. In fact, all varieties currently grown in intensive systems are resistant to *Fusarium*.

Fusarium wilt can affect very young plants as well as adult plants. On the first, it causes a growth slowdown, wilting, and yellowing of the lower leaves (Photos **651** and **652**), vascular tissues are very brown. Often, the seedlings eventually wither and die.

On mature plants, once again, the first symptoms affect the old leaves Some of the leaflets show a sector of partial chlorosis (Photos **653** and **654**). Subsequently, the yellowing spreads throughout the leaves and extends to other leaflets located on the same side of the plant. Petioles and stem also show a longitudinal yellowing (Photo **656**), which intensifies and gives rise to a necrotic lesion affecting leaves (Photo **655**). In addition to yellowing, the leaves wilt for much of the day. Petioles and stem also show a longitudinal yellowing (Photo **656**) which gets stronger and gives rise to a necrotic lesion affecting one side of the stem over several centimetres (Photos **657–659**). In some cases, adventitious roots may appear. A cut made in the stem shows a very dark brown colour in the vessels (Photo **660**). The pith does not seem affected.

Gradually as it develops, the disease causes a more pronounced wilting of plants, which eventually completely desiccate.

Note that these symptoms may occur on resistant varieties. In this case, several hypotheses are possible:
– *Fusarium* is truly involved, a new race may have developed, overcoming the genetic resistance expressed by the variety (quickly contact a diagnostic laboratory, this type of observation may be of interest);
– *Fusarium* is present but could not develop and colonize the plant because its root system was affected. In fact, during severe nematodes attacks or in the case of root asphyxia, we repeatedly found attacks by *F. oxysporum* f. sp. *lycopersici* on plants normally resistant;
– a few seeds of a *Fusarium* susceptible variety have been mixed with the seeds used;
– some symptoms have been confused and the diagnosis is incorrect; be very careful when repeating the observations.

If there is any doubt, samples should be sent to a specialized laboratory that will conduct invaluable microbiological isolations.

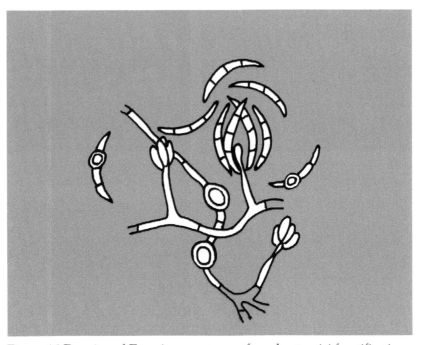

Figure 46 Drawing of *Fusarium oxysporum* f. sp. *lycopersici* fructifications.

Stem

653 A section of this leaflet is chlorotic. The yellowing seems to spread from some veins, some of which are slightly bronzed.

655 The yellowing reaches several leaflets on the same side of the leaf. Note that several veins are still green.

654 The chlorosis is now apparent, and its unilateral nature is typical of vascular disease.

656 The yellowing gradually affects several petioles on one side of the leaf rachis.

Fusarium oxysporum f. sp. *lycopersici*
(Fusarium wilt)

657

658

659

660

657 A discrete longitudinal yellowing is developing on part of the stem.

658 A fairly strong yellow band is now in place. Cortical tissues are becoming necrotic in the centre.

659 The cortical tissues are necrotic and have a brownish tinge. They are slightly depressed and give the longitudinal stem lesion a cankerous appearance.

660 After exposure, the dark brown colour of the vessels can be seen.

Stem

Fusarium oxysporum f. sp. *lycopersici*
(Fusarium wilt)

661 A localized 'V'-shaped inter-vein yellowing is gradually developing on the leaflet, which is also withered.

663 Several leaflets distributed on only one side of this leaf are affected. This distribution is characteristic of a vascular wilt.

662 The yellowing increases and the leaf tissues eventually become necrotic. They then become a brownish colour.

664 Numerous leaflets distributed on different leaves have the same 'V'-shaped necrotic lesions with a chlorotic border.

Verticillium dahliae
(**Verticillium wilt**)

Verticillium dahliae, Verticillium albo-atrum

These two fungi are responsible for Verticillium wilt on several vegetables, including tomatoes. The disease is reported in many production areas of the world and can cause severe damage. It affects both field and protected crops As with *Fusarium*, there are now many resistant varieties. The development of this disease in plants appears to be slowed or even inhibited when temperatures rise above 30°C, which may explain the apparent reversibility of the disease. A rapid increase was observed a few years ago, when some seed companies considered that they could do without resistance for soil-less tomato crops. Damage occurred quickly on new varieties lacking the gene 'Ve', a reminder that this fungus is still present on many farms, especially in the soil-less crop substrates. Race 2 has begun to appear on tomato rootstocks (intra- or interspecific) and more occasionally on some resistant soil-less

Figure 47 Fruiting of *Verticillium dahliae*.

grown tomato varieties Like many other vascular diseases, attacks of Verticillium wilt on tomato result in plants wilting during the hottest times of the day. The leaflets quickly develop a yellowing of the leaves, which takes the form of a 'V' (Photo **661**). Gradually, the leaf tissues become necrotic and dries up in the discolored central part (Photo **662**). As the disease progresses, several leaflets and leaves show similar symptoms (Photos **663** and **664**).

A longitudinal section through the stem shows that the vessels are shades of brown (Photos **665** and **666**); the vascular browning observed is less severe than that caused by *Fusarium oxysporum* f. sp. *lycopersici*. The exterior of the stem shows no particular symptoms.

As in the case of *Fusarium*, a specialist laboratory should be consulted if there is any doubt, as much confusion in diagnosis is possible.

665 The vessels of the stem on the right are pale brown in contrast with the healthy ones on the left stem.

666 A cut in the vessels shows that these are invaded by fungi including *Verticillium dahliae*.

Stem

339

667 The pith is glassy and gradually turns brown from the outside to the centre of the stem. The latter will eventually be completely hollow.
Pectobacterium carotovorum subsp. *carotovorum*

668 A longitudinal section of the stem shows the pith with a blackish colour near the vessels while the centre is becoming hollow in some places.
Pseudomonas corrugata

Pectobacterium carotovorum subsp. *carotovorum*,
Pseudomonas corrugata (tomato pith necrosis)

Pectobacterium spp.

Several species of *Pectobacterium* have been reported on tomato (bacterial stem rot), particularly *P. carotovora* subsp. *carotovora*. This bacterium is the cause of internal symptoms in the stem which affects mainly the pith in the first place. The pith becomes glassy in appearance, then brown, gradually breaking down, and eventually the stem becomes hollow (Photo **667**). In some situations, the stem cortex can also be affected, displaying a local brown to black colour. For more information, see p. 315.

Pseudomonas corrugata

This bacterium causes tomato pith necrosis. As the disease name suggests, it affects the pith of the stem. The pith is first translucent and a pale brown colour, then gradually darkens. The stem can also be hollow. Several cavities may be formed which can alternate with intact parts of the pith (Photo **668**). In such cases it is referred to as a 'stack of plates pith'. Vascular tissues may have a brownish colour. In addition, the apex is often arrested and young leaves are chlorotic. Irregularly-shaped, brown to black elongated lesions are visible on the surface of the stem and sometimes petioles. A proliferation of aerial adventitious roots may also occur (see also p. 316).

In addition, other bacteria have occasionally been associated with symptoms of the pith: *Pectobacterium carotovorum* subsp. *atrosepticum*, *P. chrysanthemi*, *Pseudomonas fluorescens* biotype I, *P. cichorii*, and more recently *P. mediterranea*.

Fusarium oxysporum f. sp. *radicis-lycopersici*

Unlike F. *oxysporum* f. sp. *lycopersici*, FORL (Fusarium crown and root rot) affects the root system of tomato, where it causes browning of many roots that eventually rot (Photo **669**). Typically, sections of xylem located in the central root and in some large roots show brown discoloration as do the stem vessels. Thus, a brown streak, a few millimetres wide, may extend over several centimetres. Subsequently, browning affects a larger portion of vessels and can be observed more than 30 cm above the ground or substrate (Photo **670**).

These various alterations cause the wilting of leaflets and leaves at the apex and/or yellowing of the lower leaves of plants. These foliar symptoms often occur at the approach of harvest, and on hot days at a time when the plants begin to be laden with fruit.

Other symptoms are present on diseased plants, especially stem base cankers. For more information on this disease, pp. 247 and 283 should be consulted.

Stem

669 Brown rotting roots and vessel browning often characterizes *F. oxysporum* f. sp. *radicis-lycopersici* attack on tomato roots.

670 The stem vessels may turn brown over several centimetres, a symptom which may be confused with that of a vascular disease.

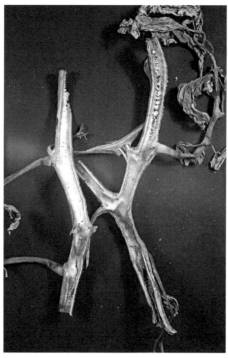

671 The pith of this lightning struck stem section has retracted in places and its colour is white to pale brown.

Lightning

Lightning

After a storm, producers can sometimes see in their tomato crops symptoms caused by lightning: large outbreaks of plants which have suddenly wilted and subsequently became necrotic and then died.

The stem of these plants is also affected and shows occasional superficial brown symptoms.

The pith may remain white or may brown slightly. Frequently the pith, shrinks and appears like a of multiple stack of discs ('pile of plates pith', see Photo **671**).

Plants in the centre of the affected area are frequently more damaged than plants at the periphery, and some may be totally dried up.

Stem

Fruit symptoms

Symptoms studied

- Fruit distortions (cracks, splits, punctures)
- Fruit discoloration (external and internal, mottling, yellowing, browning)
- Fruit spots and other symptoms
  – Small spots
  – Spots and other symptoms with a specific location (exposed side, peduncular or stylar scars)
  – Spots and other symptoms, gradually evolving into rot
  – Spots in rings, circles, and various patterns

672 This deformed fruit shows growths at the stylar end.
Physiological abnormality

674 Several small brown spots are scattered on these fruits.
Pseudomonas syringae* pv. *tomato

673 A highly contrasted mottling is visible in these two fruits.
Unknown virus

675 These two shrivelled fruits gradually liquefied as a result of colonization by two fungal rotting agents.
Geotrichum candidum* and *Rhizopus stolonifer

Examples of symptoms on fruit

Possible causes

- Many fungi
  - *Alternaria* spp. (*A. alternata*, *A. alternata* f. sp. *lycopersici*, *A. tomatophila*)
  - *Aspergillus* spp.
  - *Aureobasidium pullulans*
  - *Botrytis cinerea*
  - *Colletotrichum* spp. (*C. coccodes*, *C. dematium*, *C. gloeosporioides*)
  - *Corynespora cassiicola*
  - *Didymella lycopersici*
  - *Fusarium* spp.
  - *Geotrichum candidum*
  - *Mucor* spp.
  - *Penicillium* spp.
  - *Phoma* spp.
  - *Phytophthora infestans*
  - *Phytophthora nicotianae*
  - *Pleospora herbarum*
  - *Pythium* spp. (*P. aphanidermatum*)
  - *Rhizoctonia solani*
  - *Rhizopus* spp.
  - *Sclerotinia minor*
  - *Sclerotinia sclerotiorum*
  - *Sclerotium rolfsii*
  - *Stemphylium botryosum*
  - *Thanatephorus cucumeris*
  - *Trichothecium roseum*
  - *Ulocladium consortiale*
- Various bacteria
  - *Bacillus* spp.
  - *Clavibacter michiganensis* subsp. *michiganensis*
  - *Lactobacillus* spp.
  - *Leuconostoc* spp.
  - *Pectobacterium* spp. (*Erwinia* spp.)
  - *Pseudomonas fluorescens*
  - *Pseudomonas syringae* pv. *tomato*
  - *Xanthomonas* spp.
- Various viruses
  - Alfalfa mosaic virus (AMV)
  - Cucumber mosaic virus (CMV)
  - Tomato mosaic virus (ToMV)
  - Eggplant mottled dwarf virus (EMDV)
  - Pepino mosaic virus (PepMV)
  - Tobacco etch virus (TEV)
  - Tomato aspermy virus (TAV)
  - Tomato infectious chlorosis virus (TICV)
  - Tomato mosaic virus (ToMV)
  - Tomato spotted wilt virus (TSWV)
  - Tomato torrado Virus (ToTV)
  - Tomato yellow leaf curl virus (TYLCV)
- Other pests (*Aculops lycopersici* [see p. 147], *Bemisia tabaci* [see p. 213], moth caterpillars, various birds, bugs, thrips)
- Nonparasitic diseases
  - Genetic abnormalities
  - Silvering
  - Various injuries (vibrator damage)
  - Sunburn
  - Internal browning
  - Corky peduncular scar
  - Corky stylar scar
  - Yellow stem base
  - Green stem base
  - Nutritional disorders (deficiencies or toxicities) (see p.125)
  - Growth splits
  - 'Zipper' fruit
  - Puffiness
  - Cordiform fruit
  - Mucronate fruit
  - Pointed fruit
  - Sooty mould
  - Hail damage
  - Blotchy ripening
  - Microcracks
  - Gold specks
  - Apical necrosis (blossom-end rot)
  - Grey wall
  - Golden points
  - Low tissue temperatures
  - Internal white fibrous tissue
  - Fruit pox
  - Zebra stripe

Easy diagnosis

Most symptoms observed on fruit are quite characteristic, so the cause should, in most cases, be easily identified. Note that certain diseases, whether parasitic or not, do not cause symptoms on fruit. Other cause symptoms both on fruit and one or more other parts of the plant. It is often useful to examine other structures in order to confirm the diagnosis.

It can be difficult to identify causes on fruit, particularly those where spots or similar symptoms are the only ones seen. Sometimes their very limited size, a particular location such as the peduncular scar, or the presence of patterns such as rings are characteristic of particular causes. It is obvious that this approach is restrictive. It is useful to consult several sections of the book at once to increase the chances of identifying the problem on fruit.

676

676, 677 How to identify the main parts of a fruit.
S: sepal
P: peduncle or peduncular area
St: stylar area
C: columella
T: placental tissue
Pe: pericarp

677

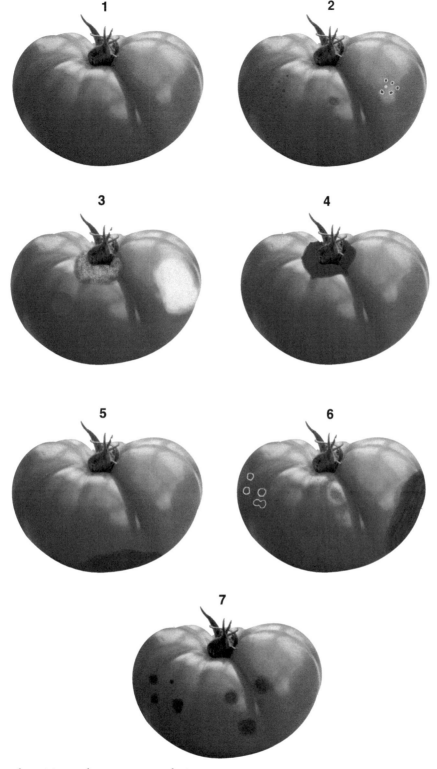

Figure 48 Types and positions of symptoms on fruits.

1 Healthy fruit.
2 Small spots on fruit.
3 Larger spots on fruit.
4 Change localized at the peduncular scar (axil of calyx).
5 Change localized at the stylar scar (end of fruit).
6 Fruit with rings, circles, patches, etc.
7 Spots with well-defined edge (left) or diffuse edge (right).

678 This large green fruit, in addition to being deformed, has several marked sides.
Puffiness

680 Several radial and concentric cracks affect this fruit.
Growth cracks

679 In addition to a corky scar, this fruit has several stylar growths.
Poor fruit set

681 A large corky scar is visible at the stylar end of fruit.
Catface

Examples of deformed fruits

FRUIT DISTORTIONS (CRACKS, SPLITS, PUNCTURES)

Possible causes

- Various viruses
  - Eggplant mottled dwarf virus (EMDV)
  - Tomato aspermy virus (TAV)
  - Tomato infectious chlorosis virus (TICV)
  - Tomato yellow leaf curl virus (TYLCV)
  - Tomato spotted wilt virus (TSWV)
- Caterpillars of moths
- Various birds
- Nonparasitic diseases
  - Various injuries (vibrator damage)
  - Corky peduncular scar
  - Corky stylar scar
  - Growth splits
  - Puffiness
  - 'Zippering'
  - Cordiform fruit
  - Mucronate fruit
  - Pointed fruit
  - Fruit pox
  - Microcracks

Additional information for diagnosis

In this section, a number of causes of symptoms are described, especially nonparasitic, linked to agricultural and environmental factors which affect fruit appearance.

Among these the climatic conditions are dominant and in particular temperature. For many years intensive tomato cultivation has been started early in the year, in the 'wrong' season, when the climatic conditions are unfavourable for flowering, and in particular fruit set. In these circumstances and depending on the variety, the crops (in cold/insufficiently heated glasshouses or in open fields) will produce deformed fruit on their first trusses. They may be irregular in shape, have corky scars sometimes at the stylar end or in other locations, have teat-like outgrowths, be heavily ribbed, pointed, or hollow (see below).

To help fruit set in these adverse conditions, growth substances are sometimes still used. If misapplied, they produce phytotoxicity affecting plants and fruits which may then be deformed. If these symptoms are recognized, the following questions need to be asked: Were the prescribed doses used? Was treatment repeated on the same flowers? Was the product applied to the leaves and terminal buds? Was the application done in good conditions (not too low temperatures)?

Also, note that the use of such products may lead to leaf deformation and reduction of the leaf surface (see p. 131).

The use of herbicides, on or near to the crop can also cause fruit deformation and other symptoms as can some genetic abnormalities.

Puffiness

Symptoms   Deformed, ribbed, angular fruits, flat in the 'equatorial' zone (Photo **682**). Several sides can be seen whose number varies according to the locular cavities present. The fruits inside can be hollow. The locular cavities are partially empty, with little gel and few seeds (Photo **683**). The fruits are lighter and hollow.

The fruits on the latest trusses are the most affected.

Cause   **Nonparasitic disease**

Disruption of flowering, pollination and fruit set and seed development in relation to several factors:

– short days, insufficient light intensity and temperatures too low;

– inappropriate nutrition; too rich in nitrogen and phosphate or too low in potassium;

– conductivity too low;

– the use of fruit setting hormones.

Vigorous varieties, with fleshy fruits, are more sensitive to this physiological condition, especially in early production.

Control methods   Optimize environmental conditions during the flowering period of the first flowers of early crops.

Climatic conditions

– Maintain large day/night temperature differences, especially with vigorous plants.
– Avoid keeping night time temperatures too low.
– Ensure optimum humidity under protection to encourage plant transpiration and pollen release. Avoid excessive humidity.

Fertigation

– Control irrigation to prevent excesses, particularly in periods of low evapotranspiration.
– Avoid excess nitrogen, especially nitrates, and provide a balanced potassium and phosphate level.

The plant and its care

– Grow less sensitive varieties.
– Reduce the planting density if the plants are vegetative.
– Restrict the first trusses to four or five fruits.
– Make sure the bees are sufficiently active.
– Master the use of fruit setting hormones.

683 A cross-section shows that the locular cavities are largely hollowed. There is only a little gel and very few seeds left.

682 This very deformed fruit is heavily ribbed and angular.

Lemon-shaped fruits

Symptoms Fruits showing an elongated stylar end giving them a lemon-shaped appearance (Photos **684** and **685**). This sharp protuberance is called a mucronate appendix.

Cause **Nonparasitic disease**

The absence of falling petals and stamens during early ovary enlargement is the cause of this stylar extension.

The persistence of the floral parts at the end of the fruits may be influenced by climatic conditions that limit their drying or ovarian development (excessive humidity, low temperatures).

Note that the use of hormones may also contribute to the emergence of mucronate appendices and that all cultivars do not show the same sensitivity.

Control methods Optimize plant conditions during the flowering period of the first flowers in early crops.

Climatic conditions
– Avoid low night temperatures.
– Ensure optimum humidity, especially not excessive, in the glasshouses.

The plant and its care
– Grow low sensitivity varieties.
– Avoid plants that are too vegetative.
– Use hormone fruit setting sprays correctly.

Under favourable conditions for mucronate appendices, 'pointed fruit' may also develop. Too low temperatures at night, and perhaps in the day, may disrupt the development of ovary locular cavities: the placenta does not grow and there are a limited number of seeds, which explains the distortion of some locular cavities and the appearance of heart-shaped fruits. This disorder may also be a problem in warm production zones during periods of very low humidity.

685 At mature stage, this stylar extension is still visible on the fruits, making them difficult to market.

684 Many of these unripe fruits have a pointed stylar extension. Dried flower parts persist at their end.

Catface

Symptoms Irregularly deformed fruits, sometimes with protrusions (Photos **688** and **689**), but mostly with spectacular corky scars of varying sizes in the stylar area (Photos **686** and **687**). Thus, large irregular scars, heavily corky, sometimes covering a large proportion of the fruits, are present at the end of the fruits. Cavities sometimes revealing residual seeds are also visible.

The disease mostly affects early field and protected crops which have had adverse weather conditions during flowering and fruit set.

Cause **Nonparasitic disease**

Disruption of flowering and then fruit set, related to poor flower formation and defective pollen quality.

Several factors influence the expression of this condition:

– the nature of the variety, the multilocular large-fruited cultivars being most susceptible;
– weather conditions at flowering time, especially during the 3 weeks before anthesis: too cold or too hot and poor light intensity;
– effects of some herbicides, including 2,4-D (Photo **690**);
– plant vigour and excessive nitrogenous fertilizers also exacerbate catface.

Control methods Optimize growing conditions during the flowering period of the first trusses.

Climatic conditions

– Avoid keeping night time temperatures too low.
– Ensure optimum humidity in protected crops. Excessively high or low can be favourable.

Fertigation

Do not use nitrogen fertilizers excessively.

The plant and its care

– Grow more resistant varieties: the old cultivars and/or large fruited with many locules appear more sensitive.
– Balance vegetative growth of plants and flowering.
– Remove deformed fruits early in order to regulate the trusses.

686 This unripe fruit has a fairly pronounced stylar scar. It is brown and corky.

687 The scar is sometimes spectacular; this is the case on this ripe fruit which has a fairly irregular corky area at its end. 'Holes' are visible in places.

688 Inclement weather conditions may affect fruit set; some may subsequently have out-growths at the stylar end.
Teated fruits

689 The growths are sometimes more at the peduncular end.
Teated fruits

690 Following the use of hormonal herbicide, several fruits are particularly deformed by the presence of several corky scars spread over the entire fruit surface.
Phytotoxicity

Zippering[a]

Symptoms Thin brown necrotic and corky scars encircle the fruit over their length and/or width (Photos **691** and **692**). They can connect the stem scar to the stylar scar or sometimes form a perpendicular linear scar, giving them the appearance of a zipper, hence the origin of the name.

The scars are also the cause of localized splits on some fruit. The corky tissues has lost its elasticity and is not able to expand as the fruit grows resulting in the symptoms. Seeds and placental tissues are visible (Photo **693**).

Cause **Nonparasitic disease**

This is a disruption of flowering and fruit set linked to temperatures being too low. Thus, the anthers remain attached to the ovary wall during the formation of new fruit, leading to the appearance of fine corky scars.

Varieties differ in sensitivity to this physiological condition.

Control methods Once the damage is present, it is often too late to intervene. This problem is highly dependent on the variety grown. Optimizing environmental conditions during the flowering period of the first trusses is recommended.

Climatic conditions

– Avoid keeping night time temperatures too low.
– Ensure optimum humidity in the shelters avoiding extremes.

The plant and its care

– Grow less susceptible varieties.
– Remove deformed fruits early in order to regulate the trusses.

[a] A symptom like zippering and called 'spider track' is reported in the US in particular. Of unknown cause, this problem is characterized by a greater number of black necrotic scars, mainly located in the peduncular area of the fruit. These are not connected to the peduncular scar, but arranged in a radial direction in relation to the latter. Streaks of dark green to yellow, depending on the stage of fruit ripening, 'radiate' from the peduncular scar. The tomato varieties do not all have the same sensitivity to this condition, large-fruited cultivars are more sensitive.

691 A thin corky scar surrounds this ripening fruit.

692 The scar extends from stylar area to peduncular area. A brownish and partially corky split can be seen.

693 On this ripe fruit, a wide irregular split with a corky edge occurred on the corky scar. Seeds and placental tissues are clearly visible.

Miscellaneous viruses

Viruses are the cause of various symptoms on fruits, in addition to causing symptoms on leaflets and stems. *Table 34* presents several viruses disrupting fruit growth and responsible for fruit deformation. This list is not exhaustive, but serves as a reminder of viruses when considering the cause of such symptoms. Photos **694** and **695** illustrate the cankerous and corky lesions on some fruit affected by a few rare viruses.

In most cases of viral infections, other symptoms are visible on tomato leaves and stems (see other sections of the book).

Table 34 Examples of viruses causing tomato fruit deformation and reduced size

Viruses	Symptoms	Vectors
Eggplant mottled dwarf virus (EMDV) Description 45	Wrinkled, puffy, and poorly coloured fruits.	Transmitted by leafhoppers: *Agallia* spp.
Serrano golden mosaic virus (SGMV) Description 42	Deformed fruits.	Transmitted by whiteflies *(Bemisia tabaci)*, in the semi-persistent and persistent manner. Not transmitted by seeds.
Tomato aspermy virus (TAV) Description 38	Greatly reduced fruiting; the few fruits that are formed are small and deformed. Very poor or no seed production.	Transmitted by 22 aphid species including *Myzus persicea* in the nonpersistent manner. Transmitted by seed in the *weed Stellaria media.*
Tomato infectious chlorosis virus (TICV) Description 40	The fruits are small and limited in number.	Transmitted by whiteflies (*Trialeurodes vaporariorum*, but not *Bemisia tabaci*), in the semi-persistent manner.
Tomato yellow leaf curl virus (TYLCV) Description 41	Large flowers drop, very few and small fruits.	Transmitted by whiteflies (*Bemisia tabaci*, mainly B biotype, *B. argentifolii*), in the persistent and circulative manner.
Tomato yellow mosaic virus (ToYMV) Description 42	Very few and very small fruits.	Transmitted by whiteflies (*Bemisia tabaci*, mainly B biotype, *B. argentifolii*), in the persistent manner.

695 Several unripe fruits, somewhat deformed, have large open necrotic and corky wounds. Dark green patterns are also observable.
Unknown virus

694 The unripe fruit is locally surrounded and therefore deformed by a tortuous necrotic and corky lesion, caused by Tomato spotted wilt virus (TSWV, see pp. 163 and 220).

Growth cracks

Symptoms Splits, cracks of varying depth and size, occurring mostly in the peduncular area and sometimes in the stylar area of the fruits. These cracks may be longitudinal, radial (radial cracking, see Photos **696** and **697**) and even concentric (concentric cracking, see Photo **698**). The earlier they occur the more damaging they are.

They mainly affect crops growing in soil both under protection and in the field. Somtimes soil-less crops are affected under protection, and also especially crops harvested at advanced stages of maturity.

Cause **Nonparasitic disease**

Excessive water flow into the plant and fruits; the elasticity of the skin of the latter is not sufficient to compensate for their sudden growth. They eventually split, and burst. This condition is often seen when environmental conditions change substantially altering the rate of plant growth. This often occurs after periods of hot weather and high light intensities.

Control methods Control the factors influencing water flow in plants.

Irrigation

– Avoid irrigation in spurts.
– Favour steady irrigation.

Climatic conditions

Be wary when alternation of cold and overcast weather periods are followed by sunny, hot and dry periods which are subsequently followed by high humidity.

Fertigation

– Prevent sudden falls in soluble salts in the nutrient solution.
– Avoid using too much nitrogen fertilizer and significant low levels of potassium.

The plant and its care

– Maintain an optimal balance of foliage/fruit to avoid plants with too little or too much vigour and succulent tissue.
– The sunlight exposed fruits are the most affected. Maintain the leaf cover by protected them from air-borne pathogens and pests.
– Furrow irrigated crops are more vulnerable.
– The large-fruited cultivars seem more sensitive.

696 Large deep longitudinal cracks, corky at the periphery in unripe fruits.

698 Some radial and concentric cracks, of various sizes, partially cover the peduncular end.

697 The skin of this ripening fruit recently split at the stylar end.

Fruit pox, tomato pox

Symptoms

Small whitish lesions, slightly elongated to oval, appearing on the skin in the peduncular area of the fruits of tomato. These spread slightly and become necrotic, the epidermis eventually splits superficially (Photo **699**).

This disease often has a restricted distribution in crops affecting a few plants and a limited number of fruits.

Cause

Nonparasitic disease

Several factors seem to influence the appearance of this condition:

– high temperatures causing a too rapid growth of the leaves and fruits;
– limited leaf cover for the fruit.

Control methods

Once the damage has occurred, it is often too late to intervene. This disease is highly dependent on the cultivar.

Climatic conditions

– Avoid excessive temperature rises in the glasshouses by ventilating and providing some shade.

Fertigation

– Prevent sudden falls in soluble salts levels in the nutrient solution.
– Do not use too much nitrogen fertilizer and avoid significant low levels of potassium.

The plant and its care

– Grow nonsensitive varieties.
– Promote optimal foliage to best protect the fruits.
– Abandon sensitive varieties for the next crop.
– Breeders must be careful when selecting new varieties that they do not use a parent that is susceptible to this genetic disorder.

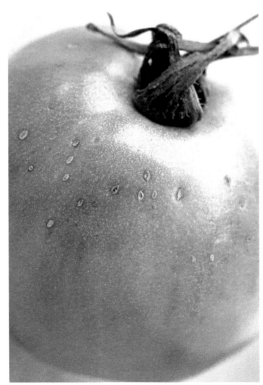

699 Several small whitish lesions, slightly necrotic, are visible on the peduncular area of this unripe fruit. The tissues of the skin split open, giving the lesions the appearance of small cracks.

Fruit

Russeting

Symptoms Superficially corky cuticle with a greyish colour, revealing many concentric cracks several millimetres long (Photo **700**). The fruit areas which are particularly exposed to treatments are the most affected.

Cause **Nonparasitic disease**

Superficial burns of the skin (and therefore loss of elasticity) resulting from the application of a mixture of inappropriate pesticides or used in poor conditions.

Control methods Unfortunately, the damage observed is irreversible, It will be necessary to be particularly vigilant during the next pesticide treatment.

Phytotoxicities can cause other symptoms on fruit (see pp. 355 and 379).

See pp. 77 and 131 for more information on these nonparasitic diseases.

700 This superficially corky fruit is covered by many fine radial and concentric cracks.

Note that in the literature, several other nonparasitic diseases are reported to cause microcracks in the peduncular area of unripe tomato fruit.

Among these are fruit russeting, also called 'shoulder checking' or 'weather checking'. This disease is due to the presence of water (due to irrigation, rain, dew) on the surface of the fruit for too long a period when mornings are still cold.

A somewhat comparable condition is described as 'rain check', characterized by the development of tiny concentric splits in the peduncular zone, sometimes combining to form larger splits. These local microcracks give the fruit surface a rough touch and change their colour development: they are green when mature or gradually blacken in the affected area. The causes of this disease are not yet known. As with fruit russeting, the most exposed green fruit are the most affected.

Moreover, the phenomenon also occurs as a result of rainfall after a period of drought. Note that varietal differences are found. Cultivars with vegetation covering much of the fruits are less sensitive. In addition to using tolerant varieties, all agricultural measures for maintaining the leaf cover as well as protecting it from pests would be a wise policy to follow for the control this disease.

Finally, another condition, with an unknown cause,' black shoulder', causes spots, grey streaks in the peduncular area of green fruit. These lesions spread, later becoming dark grey to black in colour. Microcracks present in fruits allow the development of *Alternaria alternata*, which results in their decay. The emergence of this disease appears to be associated with the same conditions as those influencing the development of rain check. Varietal differences exist.

Physiological russeting

Symptoms Onset of many superficial skin microcracks, a few tenths of a millimetre to 1–2 mm long, mostly localized at the peduncular end of fruit (Photos **701** and **702**). The tissues involved are greyish and dull.

The microcracks occur in protected crops during fruit ripening.

Cause **Nonparasitic disease**.
The lower cuticles elasticity at maturation does not compensate enough for fruit growth.

Control methods Optimize the factors affecting the fruit growth.

Climatic conditions

– Avoid too high day-night temperature contrasts.
– Maintain optimum average temperatures preventing uneven growth.
– Beware of cloudy weather and poor light intensity in glasshouses with thermal screens.

Fertigation

– Avoid excessive irrigation.
– Do not irrigate during periods of low plant activity (from end of day until early next day).
– Maintain a high level of soluble salts in the root zone.

The plant and its care

– Maintain an optimal foliage/fruit load balance.
– Maintain an adequate fruit load at end of the season by not stopping too early.
– Varietal differences in sensitivity can be observed.

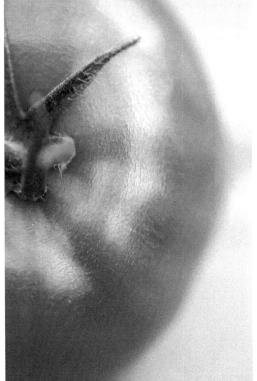

702 Scores of tiny concentric cracks can be observed in the peduncular area.

701 The skin at the peduncular end of the fruit has a greyish colour and a dull appearance.

703

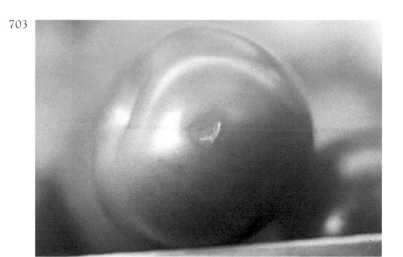

704

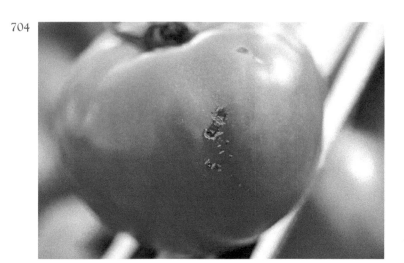

706

705

703 This fruit shows recent mechanical damage which has not yet healed.
Accidental injury

704 Various mechanical injuries are visible on the fruit. The presence of corky tissue shows that they are healing and also that the injury occurred some time ago.
Accidental injury

705 One of these three fruits has suffered the pecks of a bird.

706 A circular and regular hole is present on two of these fruits. They were made by a caterpillar that was able to penetrate them.

Various injuries

During their development, harvest or storage, the fruits may receive some localized injuries. These accidents occur throughout cultural operations: pruning, trellising, truss pruning, staking, and harvesting. During this last phase, the fruits may be bruised as a result of falling on the ground or when moved from one crate to another.

Thereafter, the fruits will bear traces of healed damage (Photos **703** and **704**), perforations, bruises, squashing, which may be the cause of their downgrading or disposal as waste.

Some injuries can be linked to vibrator damage in countries where this tool is still used to assist pollination. Thus, when the flower trusses are shaken, the operator may sometimes hit some fruit with this tool. Such damage causes wounds of varying depths.

Birds

Green or ripe tomatoes are very tempting food for many bird species. During feeding, the many pecks cause lesions of varying depths in the fruit (Photo **705**). The latter are entry routes for many micro-organisms, notably various moulds, which can colonize the fruits and cause rot.

Bird damage can be considerable in production areas where they congregate.

Moth catepillars

The caterpillars of several moths can cause fruit damage (Photo **706**). This can be considerable if nothing is done to prevent the outbreak of these pests. More information on this can be found on p. 83.

Fruit

708 All of these fruits are streaked with dark green longitudinal patterns.
Zebra stripe

707 The fruits of this truss are mottled. Unripe patches alternate with ripe patches.
Tomato mottle

709 Large brown rotting areas with a looped edge partially cover these unripe fruits.
Phytophthora nicotianae (ground mildew)

Examples of fruit discoloration

FRUIT DISCOLORATION (EXTERNAL AND INTERNAL MOTTLING, YELLOWING, BROWNING)

Possible causes

- *Phytophthora infestans*: Description 7
- *Phytophthora nicotianae*: Description 12
- Other fungi inducing flesh browning (*Alternaria* spp., *Colletotrichum* spp., *Corynespora cassiicola*, *Phoma destructiva*, *Pleospora herbarum* [*Stemphylium* sp.]) (See p. 389)
- Various viruses
- Eggplant mottled dwarf virus (EMDV)
- Pepino mosaic virus (PepMV)
- Tobacco etch virus (TEV)
- Tomato mosaic virus (ToMV)
- Silvering

- Internal browning
- Yellow neck
- Green neck
- Blotchy ripening
- Microcracks
- Grey wall
- Golden specks (see p. 379)
- Internal white fibrous tissue
- Zebra stripe
- *Bemisia tabaci* (see also p. 213)
- *Aculops lycopersici* (see p. 147)

710 These fruits all display a discreet mottling: areas of still chlorotic tissues alternate with areas of normally coloured and mature tissues.
Pepino mosaic virus (PepMV)

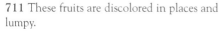

711 These fruits are discolored in places and lumpy.
Eggplant mottled dwarf virus (EMDV)

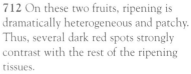

712 On these two fruits, ripening is dramatically heterogeneous and patchy. Thus, several dark red spots strongly contrast with the rest of the ripening tissues.
Unknown virus

713 The peduncular area of the fruit has not ripened evenly, and some unripe green spots (blotchy ripening) remain.
Tomato mosaic virus (ToMV)

Additional information for diagnosis

Various viruses

Many viruses are capable of causing fruit discoloration(Photos **710–713**) which appear as mosaics, mottling, discolored patches of various shapes, diffuse browning, or marked necrotic areas(see p. 373). *Table 35* lists several viruses that cause such discoloration on tomato fruit.

It is often very difficult to identify accurately a virus from fruit symptoms alone. Look for other symptoms in other parts of the plant. If any are found, refer to the corresponding sections in the book to confirm the diagnosis. Otherwise, if there is any doubt, consult a specialized laboratory that will perform specific tests for viruses (serological or molecular test in particular).

Table 35 Examples of viruses causing tomato fruit discoloration

Virus	Symptoms	Vectors
Colombian datura virus (CDV) Description 38	Poorly coloured fruits.	Transmitted by aphids (*Myzus persicae*), in the nonpersistent manner.
Eggplant mosaic virus (EMV) Description 45	Poorly coloured fruits, mottled to spotted.	Transmitted by beetles (*Epithrix* sp.). Unknown transmission methods.
Eggplant mottled dwarf virus (EMDV) Description 45	Poorly coloured fruits, wrinkled to puffy (Photo **711**).	This virus may be transmitted by *Agallia* leafhoppers.
Pepino mosaic virus (PepMV) Description 29	Sepals superficially and locally changed, mottled fruits (Photo **710**).	Mechanically transmitted by contact during agricultural operations by pollinating bees and by seeds.
Peru tomato virus (PTV) Description 38	Mottled fruits.	Transmitted by aphids (*Myzus persicae*) in the nonpersistent manner.
Tobacco etch virus (TEV) Description 37	Mottled and small fruits when infected early.	Transmitted by various aphids in the nonpersistent manner (*Myzus persicae*, *Aphis gossypii*, *A. craccivora*, *A. fabae*, *Macrosiphum euphorbiae*).
Tomato mosaic virus (ToMV) Description 31	Chlorotic rings visible on green or ripe fruits. Incomplete maturation (Photo **713**). Internal brown necrotic lesions (see p. 368).	Transmitted by contact and by seeds.

Physiological mottling (blotchy ripening)

Symptoms Green to yellow patches of various sizes, appearing on the mature fruits (Photo **714**), with a firm consistency and making them unsuitable for marketing. In other words, partial ripening of fruit in blotches (blotchy ripening) occur mainly in the peduncular area. A tranverse section of the fruits can confirm their partial ripening (Photo **715**), and the browning of some pericarp vessels in the areas that remain green (internal browning).

It mainly affects early unheated crops or late autumn protected crops. It is reported worldwide and is sometimes described under other names[a]. Similar symptoms are associated with attacks of ToMV (see p. 99).

Cause **Nonparasitic disease, not well known**

Several factors appear to influence the disease:

– excess nitrogen, and calcium potassium deficiency;
– periods of reduced light, alternating periods of clouds and sunshine;
– low levels of soluble salts in the soil or in the nutrient solution, combined with excess water and lower temperatures;
– heavy soils.

There are large differences in sensitivity between varieties.

Control methods Better management of the parameters influencing its appearance is sufficient.

Plant care

– Grow nonsensitive varieties, old cultivars being rather more sensitive.
– Avoid too vegetative plants.
– Thin out the leaves to improve the trusses' exposure to light.

Climatic conditions

Avoid maintaining too low temperatures, especially at night.

Fertigation

– Master the watering and maintain high levels of soluble salts.
– Limit nitrogen intake and increase potassium fertilization.

[a] A condition is reported in the US as the 'gray wall', in open field and under protection.
It could be the same or a variant of the blotchy ripening, like another disease described in New Zealand and named 'Cloud'. It occurs on unripe fruits prior to harvest, sometimes later. The presence of dark brown to black necrosis in the pericarp tissues characterizes this disease. The vessels are particularly affected. The affected fruits have a grey appearance, and their size is slightly reduced. It is during the late appearance of the disease that symptoms similar to those of blotchy ripening are observed: large unripe green and/or yellow areas, of firm consistency and located in places where internal necrosis has occurred. The conditions influencing the expression of this condition are the same as those described above for blotchy ripening.

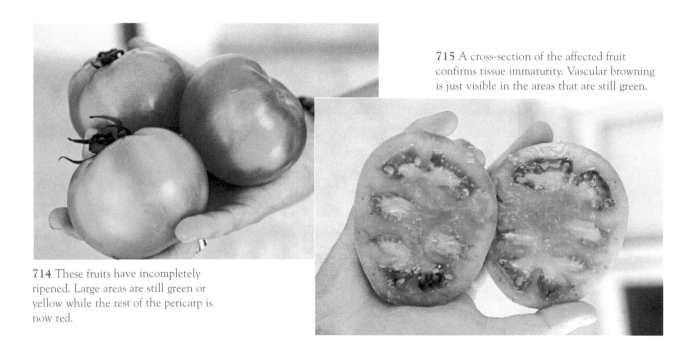

715 A cross-section of the affected fruit confirms tissue immaturity. Vascular browning is just visible in the areas that are still green.

714 These fruits have incompletely ripened. Large areas are still green or yellow while the rest of the pericarp is now red.

Green or yellow shoulder, greenback

Symptoms Darker colour of peduncular end of unripe fruits. At maturity, this area remains green or gradually turns and stays yellow. Eventually, the shoulders of the fruits have a green (Photo **716**) or yellow (Photo **717**) colour, quite characteristic and of variable intensity, either uniform or in irregular patches. The pericarp of the affected areas is of hard consistency and sometimes of a white colour.

Cause **Nonparasitic disease**

During fruit ripening, in the affected areas chlorophyll is not or incompletely degraded.

Several factors appear to influence the occurence of this physiological disease:

– high temperatures when fruit is ripening;
– plants with reduced foliage or excessive leaf loss exposing the fruits to sun radiation;
– unbalanced use of fertilizers, especially low in potassium.

In these conditions, different tomato varieties do not show the same sensitivity. For example, 'uniform colour' fruited cultivars are resistant. On the other hand, those of 'long shelf life' type quite readily show this defect.

Control methods Better management of the parameters influencing its appearance is sufficient.

Plant care

– Grow resistant varieties.
– Avoid excessive de-leafing in summer, especially for less vigorous varieties.
– Ensure good plant protection to prevent damage associated with various defoliating pests.

Climatic conditions

Apply shading in good time.

Fertigation

Ensure proper potassium fertilization.

717 In the case of yellow shoulder, the pericarp tissues in the peduncular area remain yellow, even when the fruit is ripe.

716 The peduncular area of the fruit to the right has remained uniformly green/orange.
Green shoulder

Fruit

718 These fruits, affected by silvering, are deformed. They show incomplete maturation and some line-patterns from the stylar scar to the peduncular scar.

720 A large segment of this fruit has an orange colour and is dotted with clusters of green irregular tissue.
Silvering

719 Green and ripe fruits are striped lengthwise with irregular dark green streaks.
Zebra stripe

721 In the fruit cut in half, a white fibrous mass is partially filling the pericarp.
Internal white tissue

722 Several fruits of this truss are discolored.
Bemisia tabaci

Examples of fruit discoloration caused by some genetic disorders and insects

Many mutations are known in tomato, modifying the expression of some of its genes. Their consequences may be clearly visible on the plants or go unnoticed, essentially altering physiological processes. They correspond to rare events that geneticists sometimes use to create new varieties. Unfortunately, they may also be the cause of defects affecting tomato production, the symptoms affecting various plant parts especially fruit. In all cases we are dealing with 'genetic' conditions. Among these diseases, at least two cause partial discoloration, sometimes leading to diagnostic confusion (Photos **718–720**).

Silvering

This nonparasitic disease, in addition to the symptoms it causes on leaflets and stems, is the source of partial fruit discoloration. The sometimes deformed fruits show some green areas (Photo **718**), sometimes silvery streaks, becoming pale yellow when ripe. Sometimes more important and well defined pericarp sectors are affected. Maturation is incomplete, and the fruits are interspersed with green patches, irregularly shaped and arranged on the whole fruit surface (Photo **720**).

Leaf symptoms are caused by abnormal development of the leaf palisade mesophyll tissues occurring at a very early stage of development. The origin of those observed on fruits is uncertain. They are probably interrelated with chlorophyll synthesis. Page 143 should be consulted for more information on this nonparasitic disease and the factors that influence its appearance. Varieties are sometimes listed as resistant to silvering, with the symbols 'wi' and 'si', according to the seed catalogues.

Zebra stripe

This nonparasitic curiosity is characterized by the presence on unripe fruit of a series of spots of limited size and dark green colour arranged in a particular way: in-line, from the peduncular scar to the stylar scar (Photo **719**). When numerous and confluent, these spots form bands between the two ends of the fruit, literally like zebra stripes. Most of the time, these changes disappear during fruit ripening. Note that this symptom gives affected tomatoes the appearance of the fruits produced by the 'Green Zebra' variety. This disease, probably genetic in origin, occurs under certain still poorly known environmental conditions. It is genetically related to tomato fruit pox and gold flecks.

Internal white tissue

This nonparasitic condition is characterized by the presence of white fibrous structures, sometimes grouped, inside the tomato fruit, in the pericarp, but especially in the outer wall and occasionally in the placental area (Photo **721**).

This condition is linked to too low potassium fertilization and to high temperatures. Other factors are involved, some linked to various stresses, the presence of sweet potato whitefly, some possibly genetic (some varieties more readily exhibit the phenomenon).

Also, note that two insects in particular are responsible for fruit discoloration. *Bemisia tabaci* induces a patchy fruit discoloration (Photo **722**). More detailed information on this insect can be found on p. 213. The numerous feeding bites of *Aculops lycopersici*, tomato russet mites, are the cause of imperfect fruit colour. These can be smaller and present superficially corky, brownish patches, of various sizes (see p. 149).

723 All the fruits on this plant show brown to black lesions, small and very well defined on the green fruits, rather diffuse and large on the nearly ripe fruits.
Alfalfa mosaic virus (AMV)

725 This fruit, affected by Tomato mosaic virus (ToMV), has irregular brown lesions and sometimes internal ones. It is also bumpy.

727 Severe necrotic brownish lesions deform these three fruits.
Cucumber mosaic virus (CMV)

724 On this plant, the frequency of affected fruits is high, they are covered with large diffuse brown lesions.
Tomato mosaic virus (ToMV)

726 A cut in the fruit shows that the sub-epidermal tissues and vascular vessels are brown and corky.
Tomato mosaic virus (ToMV)

Table 36 Examples of viruses causing browning and brown necrotic lesions on tomato fruits

Virus	Symptoms	Vectors
Alfalfa mosaic virus (AMV) Description 33	Fruit with necrotic brown lesions, marked, fruits can also be bumpy (Photo **723**).	Transmitted by many aphid species in the nonpersistent manner *(Myzus persicae, Aphis fabae, A. gossypii, A. craccivora)*.
Tomato mosaic virus (ToMV)[a] Description 31	Chlorotic rings are present on green and mature fruit (Photos **724** and **725**). Internal brown necrotic lesions are also observed (*internal browning*, see Photo **726**).	Transmission by contact and by seeds.
Cucumber mosaic virus (CMV) Description 35	Severe necrotic brown, round and raised areas, cover large parts of the fruit (Photo **727**).	Transmitted by many aphid species in the nonpersistent manner *(Myzus persicae, A. gossypii, A. craccivora, A. fabae, Acyrthosiphon pisum)*.
Tomato spotted wilt virus (TSWV) Description 43	Large patches, rings, arcs partially cover the fruits (Photo **728**). These can be locally discolored, bronzed and deformed.	Transmitted by at least nine thrip species in the persistent manner (including *Frankliniella occidentalis, F. fusca, F. schultzei, Thrips palmi, T. tabaci, Scirtothrips dorsalis*).

[a] Note that in the case of Tomato mosaic virus (ToMV) browning and brown necrosis may be found in and on fruit (*internal browning*) of resistant tomato varieties. These symptoms may be related to a breakdown of the resistance that may occur if such varieties are placed near infected sensitive plants or following periods of high temperatures and high luminosities. These symptoms are quite similar to those caused by tomato mottling which is a physiological disorder (see p. 368). Moreover the Tomato Torrado virus (ToTV) causes brown necrotic lesions on fruit (see p. 225).

728 These fruits are almost entirely covered by brown rings of varying widths, some of them are also deformed.
Tomato spotted wilt virus (TSWV)

729 Numerous small concave black lesions are scattered across the surface of these fruits.
Unknown virus

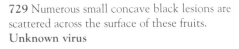

Air-borne and soil-borne mildews: *Phytophthora infestans* and *P. nicotianae*

Two mildews are likely to produce brown lesions on tomato fruits. Firstly, the classic mildew of this Solanaceous plant, caused by *P. infestans*, also called 'aerial blight' because it attacks all aerial parts of this plant. Early affected fruits develop very characteristic brown blotches and are often very bumpy (Photos **730** and **731**). On such affected fruits the mottling has a poorly defined edge and spreads rather slowly. If attacks occur later, the mottled patches are more homogeneous and may be distributed in looped concentric circles. A whitish down is occasionally visible on the surface of the lesions. Further information is available on p. 185.

Soil-borne mildew, caused by *P. nicotianae*, occurs mainly on the root system, stem base, and fruits in contact with the ground. It should be noted that fruit located high up on plants can sometimes be affected as a result of heavy rain or heavy spray irrigation: the soil splashes ensuring dispersal of the *Phytophthora* sporangia. Brown spots occur on fruits in contact with the ground (Photos **732–734**). These gradually spread in the form of diffuse and concentric brown bands, progressively invading the fruit (buckeye rot, see Photo **734**). When the humidity is very high, the mycelium of *P. nicotianae* can superficially develop in the form of a fluffy white felting (Photo **735**). Other symptoms caused by *Phytophthora* on tomatoes can be found on p. 283.

Note that particularly affected fruits in contact with the ground may also be colonized by secondary invaders such as *Fusarium* spp. or *Geotrichum candidum* that add to the rot. The parasitic combination thus formed will often cause the decomposition and complete liquefaction of fruit (Photo **736**). Note that several other Phytophthora have been reported on tomato fruits: *P. mexicana*, *P. capsici*, *P. drechsleri* (see p. 241).

731 On this fruit, the browning is more marked and diffuse. It is referred to as 'brown mottle' and it is characteristic of the damage of *P. infestans*. Note that a sparse white down has formed in some places.

730 This young immature fruit is almost completely brown. Its surface is slightly bumpy. Note that many flowers are completely necrotic.

Phytophthora infestans
(downy mildew)

732 A recent lesion, brown and diffuse, grows on one side of this unripe fruit.

733 These two fruits are widely covered by one to several vaguely target-shaped lesions consisting of concentric bands, dark brown, diffuse, and scalloped. Note that the fruit surface remains smooth and firm.

734 On this ripening fruit, we find the same type of damage, namely the juxtaposition of brownish concentric bands. In this case, the browning of the tissues is much less pronounced than on immature fruit.

735 A fluffy white felting (mycelium) partially covers the completely blackened and rotten young fruit.

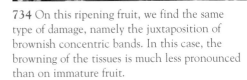

736 Several fruit rotted by *P. nicotianae* have been colonized by various secondary invaders that magnify the damage.

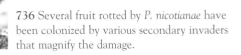

FRUIT SPOTS AND OTHER SYMPTOMS

- Small spots
- Spots and other symptoms with a specific location (exposed side, peduncular or stylar scars)
- Spots and other symptoms gradually evolving into rot
- Spots in rings, circles, and various patterns

Possible causes

- Many fungi
- *Alternaria* spp. (*A. alternata*, *A. alternata* f. sp. *lycopersici* [Description 9], *A. tomatophila* [Description 1])
- *Aspergillus* spp.
- *Aureobasidium pullulans*
- *Botrytis cinerea*: Description 2
- *Didymella lycopersici*
- *Colletotrichum* spp. (*C. coccodes*, *C. dematium*, *C. gloeosporioides*): Description 10
- *Corynespora cassiicola*: Description 9
- *Fusarium* spp.
- *Geotrichum candidum*
- *Mucor* spp.
- *Penicillium* spp.
- *Phoma* spp.
- *Phytophthora infestans*: Description 7
- *Phytophthora nicotianae*: Description 12
- *Pleospora herbarum*
- *Pythium* spp. (*P. aphanidermatum*): Description 12
- *Rhizoctonia solani*: Description 14
- *Rhizopus* spp.
- *Sclerotinia minor*: Description 15
- *Sclerotinia sclerotiorum*: Description 15
- *Sclerotium rolfsii*: Description 16
- *Stemphylium botryosum*
- *Thanatephorus cucumeris*: Description 14
- *Trichothecium roseum* *Ulocladium consortiale*
- Various bacteria
- *Bacillus* spp.
- *Clavibacter michiganensis* subsp. *michiganensis*: Description 24
- *Lactobacillus* spp.
- *Leuconostoc* spp.
- *Pectobacterium* spp. (*Erwinia* spp.): Description 25
- *Pseudomonas fluorescens*
- *Pseudomonas syringae* pv. *tomato*: Description 21
- *Xanthomonas* spp.: Description 22
- Various viruses
- Bugs
- Thrips
- Sunburn
- Corky stylar scar
- Corky peduncular scar
- Hail damage
- Golden speck
- Blossom-end rot
- Golden points
- Cold temperatures

Additional information for diagnosis

This section describes various changes occurring in tomato fruits during cultivation, and for some, after harvest during storage and marketing. They can all be ascribed to spots that vary in size according to their origin. They may be parasitic or nonparasitic.

The various spots and lesions considered are schematically illustrated in Figure **48**, p. 349. It is difficult to classify them into well differentiated fields. Indeed, they may have several common characteristics and evolve over time, making their classification difficult. Yet these characteristics are essential for making a diagnosis, so have still been used by integrating them sparingly throughout this section.

Fruit

Golden speck

Symptoms   Presence of numerous tiny yellowish spots mainly at the peduncular end of the fruit (Photos **737** and **738**). These microlesions give the fruit a dull appearance and reduce their shelf life.

Cause   **Nonparasitic disease**

Formation of calcium oxalate crystals in tissues correlated with a high calcium concentration in fruits.

All the parameters influencing the migration of calcium in the fruits may be implicated, including:
– climatic conditions, with high humidity levels by day and night, accompanied or not by low light intensity and low temperatures;
– fertigation applying too much calcium, phosphates, or nitrates or too little potassium or chlorine.
Very vigorous plants are more sensitive to this physiological condition.

Control methods   Ensure optimal growing conditionst during the flowering period of the first flowers of early crops.

Climatic conditions

– Ensure optimum humidity in the glasshouses, so as to promote plant transpiration (mulching, setting up dehumidification if humidity is excessive).
– Avoid too low night time temperatures.

Fertigation

– Use high level of soluble salts and maintain significant levels of magnesium and low levels of phosphates.
– Avoid watering too early in the morning and too late in the afternoon.

The plant and its care

– Grow nonsensitive varieties.
– Conduct regular leaf thinning out.

737 The peduncular area is dotted with many tiny yellowish microlesions giving it a dull appearance.

738 These microlesions, side by side and less than a millimetre in diameter, are due to calcium oxalate accumulation in the tissues.

■ Small spots

Gold fleck

This disease is the cause of small irregular dark green specks, randomly arranged on the surface of unripe fruits. When the fruits ripen the spots are then a yellow colour. The cause(s) of gold fleck are somewhat controversial. Some authors think it has a genetic origin, with the 'GDF' gene being responsible. Others report that this problem would be partly solved through insecticides use and could be due to the effects of insects, notably thrips. Whatever the cause, varietal differences exist.

Chemical injuries

As a result of pesticide applications on tomato, in particular under protection, the fruits can be superficially injured in the days following treatment. Small dark green lesions appear covering parts of the most exposed fruits (Photo **739**). To cause this symptom the product(s) sprayed must diffuse through the cuticle and locally affect the underlying tissues. In some cases, microcracks may also be observed (see p. 360).

See also pp. 77 and 131 for more information on chemical injuries occurring on tomato.

739 Following the application of a miticide, these fruits are mottled with small translucent to dark green lesions on the most exposed parts.
Phytotoxicity (chemical injuries)

740 An adult bug walks on this unripe fruit.

742 The damage seen on these two fruits is typical of that caused by bugs on ripe fruits: tiny bites with chlorosis with internal changes in the fruit tissue. Spongy white areas appear internally in the pericarp.

743 The effects of bug bites are very dramatic on this fruit. Large dark lesions in the shape of a diffuse chlorotic ring are visible around the position of each bite.

741 The ripe fruit reveals numerous irregular lesions with chlorosis. *Nezara viridula* larva can be seen in the affected area.

744 Several necrotic bites with a diffuse whitish halo varying in size are visible on the very young fruit. They result from attack by *Frankliniella occidentalis*.

Bugs

Several bugs, often polyphagous and attacking various crops and weeds, are likely to be harmful on tomato, for example *Lygocoris pabulinus* (Linnaeus), *Lygus* spp. *Nesidiocoris tenuis* (Reuter), *Nezara viridula* (Linnaeus). These insects belong to the Hemiptera order, Heteroptera suborder, Miridae family (for the first three) and Pentatomidae (the last). The characteristics of one example will be described, the soybean plant green bug, *Nezara viridula*. It is widely distributed around the world and occurs on various crops, vegetables or other.

Note that the *Nesidiocoris tenuis* species was deliberately introduced in France for biological control to fight whiteflies and thrips.

• Nature of the damage

This piercing-sucking insect, like many other bugs, is likely to cause symptoms, mainly on leaves and fruit (Photos **741–743**). The apical leaves wilt. On young fruit, the feeding punctures cause tiny dots around which the underlying tissues is lighter than the rest of the fruit. This is best seen by cutting the affected fruits. On ripe fruit, the spots are wider, white to dark yellow. They result from the effects of enzymes released during feeding punctures which diffuse and subsequently lead to the white and spongy appearance of degraded tissues which remain firm when ripe (cloudy spot). The bugs could be vectors of bacteria and yeasts presumed responsible for fruit symptoms once introduced.

• Biology

Nezara viridula goes through several developmental stages: egg, five larval stages, adult. The duration of the cycle varies with temperature, approximately 3 weeks at 30°C to 2 months at 20°C.

Forms of survival and/or alternative hosts: adults overwinter in the building structures, such as behind the tunnels hoops. This bug lives as a parasite of many plants such as soybeans, rice, various vegetables (eggplant, cucumber, tomato, peppers, beans), as well as weeds.

Developmental stages (Figure **49**): the eggs (1), off-white, are shaped like small kegs grouped in a honeycomb on the underside of the leaves. Once the eggs hatched, dark red to black with white spotted (2) larvae gradually disperse on the plants. Five instars follow, from the newly hatched larva which is barely 1 mm long to the aged larva 1 cm long (3), the last stage before adult (4) (Photo **741**). The latter measures 12–16 mm long and 8 mm wide. It is light green in summer and becomes purplish in autumn and winter. It has two pairs of wings, the first pair, named hemi-elytra are thick and hardened in their upper part.

Distribution in culture: crawlers and flying adults ensure the dissemination of these insects.

Favourable conditions for development: the restriction of the use of broad-spectrum insecticides, due to the development of integrated pest management (IPM), seems to have contributed to the emergence of these bugs indoors.

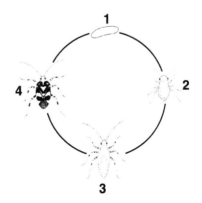

Figure 49 Stages of development of bugs.

745 Several small black spots, similar to fly-specks, are scattered on these young unripe tomatoes.
Pseudomonas syringae pv. *tomato*

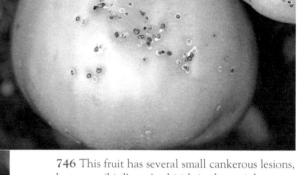

746 This fruit has several small cankerous lesions, known as 'bird's eye', whitish in the periphery; their necrotic centre is cankerous and brown ochre.
Clavibacter michiganensis **subsp.** *michiganensis*

747 Four pustular and corky spots, oily at the periphery, are clearly visible on the still unripe fruit.
Xanthomonas sp.

748 These fruits have locally brown and corky lesions caused by hailstones
Hail injury

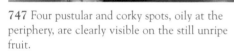

749 Damage caused by hail is more substantial on this fruit. Large brown, circular, diffuse, spots, which show cracks have resulted from hailstones.
Hail injuries

Various bacteria

Many bacteria are capable of attacking tomato fruits (see *Table 37*). They can be divided into two broad categories:

– the bacteria that cause spots of limited size on fruit, which are *Pseudomonas syringae* pv. *tomato* (Photo **745**), *Clavibacter michiganensis* subsp. *michiganensis* (Photo **746**), and *Xanthomonas* spp. (Photo **747**). Their symptoms have been widely described in the diagnosis part of this book; see p. 155 for more information on these and other characteristics of these bacteria;

– the bacteria that cause lesions developing into soft rot on fruit, detailed in the section 'Extensive spots gradually evolving into rot', p. 407.

Table 37 Incidence of the main bacterial diseases of tomato fruits at various crop stages and post harvest

Bacteria	Fruit development stages and incidence				Development potential of the symptom
	Green	Nearly mature	Ripe	Post harvest	
Pseudomonas syringae pv. *tomato*	++	+	+/–		+/–
Xanthomonas spp.	++	+	+/–		+/–
Clavibacter michiganensis subsp. *michiganensis*	+	+/–	+/–		+/–
Pectobacterium spp. (*Erwinia* spp.) (*P. carotovorum* subsp. *carotovorum*, *P. chrysanthemi*, *P. carotovorum* subsp. *atrosepticum*)		+/–	++	++	+++
Pseudomonas fluorescens				+/–	+
Leuconostoc spp.				+/–	+
Bacillus spp.				+/–	+

– : None, +/–: low, +: medium, ++: high; +++: considerable.

Hail damage

Following heavy rains accompanied by hail it is not uncommon to find substantial damage in tomato crops (hail injuries). The fruits are particularly vulnerable to hail, which, when they are hit, cause splits at the points of impact. Brown lesions appear later (Photos **748** and **749**), which can be corky and reminiscent of the corky spots caused by *Xanthomonas* spp.

The leaves may also be affected. They are holed and shredded (see p. 81). The identification of this nonparasitic problem is easy because of the relationship of cause and effect between the presence of a thunderstorm and the onset of damage.

Note that following such damage, many microorganisms described in this section, can colonize the wounds caused by hail damage to gain entry into the fruit and subsequently induce rot.

Fruit

750 All these fruits have been subjected to sunlight on their most exposed side to varying degrees. White lesions are slightly wrinkled and yellowish on the periphery. **Sunscald, sun burn**

751 The area of the fruit in contact with a heating pipe has become whitish and translucent caused by the effects of heat.

752 The peduncular area of these two fruits has extensive and irregular lesions. Besides being slightly concave, they show browning under the cuticle. **Chilling injury**

Spots and other symptoms with a specific location (exposed side, peduncular or stylar scars)

Before reviewing the main fungi and the few bacteria that cause various rather extensive spots evolving into fruit rot on tomato (see *Table 38*, p. 388), some nonparasitic diseases that have a specific location on the fruits will be described: peduncular or stylar ends, the most exposed face.

The symptoms studied reflect damage due to solar radiation, low temperatures, and water stress. In any event, it must be remembered that some of these abiotic diseases may allow the entry of other micro-organisms and secondary invaders.

Sunburn damage

On unripe or exposed fruit, during very hot weather periods, spots sometimes appear on the side in direct sunlight (sunscald, sun burn). These lesions are irregular, with a whitish centre, and can be surrounded by a yellow halo (Photo **750**). They are slightly sunken, and their surface is wrinkled with a dry papery texture. They are often superficial and sometimes invaded by various opportunistic fungi, especially *Alternaria alternata*. In this case, a brown to black mould gradually covers the lesions.

This nonparasitic disease occurs more in field crops but can also be seen under protection. It mostly affects varieties with sparse foliage or crops with reduced foliage; in the latter case, the reduction of foliage may have various causes:

– severe attack of one or more air-borne diseases, especially fungal (*Alternaria*, *Stemphylium*, mildew, *Septoria* spp.) or bacterial;

– foliar damage resulting from the attack of pests or due to physiological diseases (e.g. physiological leaf roll);
– too much leaf removal;
– inadequate shade under protection during particularly hot and sunny periods;
– turning of plants at the first harvest of field-grown unstaked crops, etc.

The unripe fruits of field crops are particularly vulnerable. Their pericarp undergoes irreparable damage from solar radiation as a result of the locally raised temperatures of tissue. Controlling these factors helps prevent sunburn. In some crops the fruits may be covered by straw for protection.

It should be noted that similar symptoms can be observed in glasshouses on fruits accidentally in contact with heating pipes (Photo **751**).

Low temperature

Field tomatoes, produced in late autumn and suffering the effects of early autumn cold, sometimes show different symptoms (chilling injury) mainly localized around the peduncle. These may be of various types. In most cases, the lesions are recessed and of various sizes. They are also diffuse with an irregular brown colour in the centre (Photo **752**). Other symptoms are irregular ripening and early softening. Note that similar damage can occur during tomato cold storage.

These symptoms are due to the effects of temperatures below 12°C, especially at night, and the presence of water around the peduncular area. Cuticular and epidermal tissues are affected to a greater or lesser extent. Subsequently, they can be secondarily colonized by various opportunistic organisms, especially *Alternaria* spp.

Fruit

753 Several wet lesions rapidly becoming necrotic have spread at the end of these unripe fruits.
Blossom-end rot

756 These three fruits have stylar scars of varying length, width and corkiness. On one of them, the scar has opened exposing the inside of the fruit.
Stylar scar corky (catface)

754 Both fruits show a marked discoloration around their stylar scar. It is dark brown to black, well defined, and slightly concave.
Blossom-end rot

755 A cut in this tomato reveals the browning and blackening of some seeds and a portion of the placenta in the distal part of the mature fruit.
Internal necrosis (blossom-end rot)

757 This large still unripe fruit has a large corky peduncular scar.
Corky peduncular scar

Blossom-end rot

Symptoms   Development of small, moist lesions at the end of fruit at or near the pistil scar, rather diffuse at first, gradually turning brown and spreading (Photo **753**). Ultimately, a large brownish black lesion, concave and well defined ,with a rather dry consistency, affects the blossom-end of one or more fruit (Photo **754**). This symptom is nicknamed 'black bottom' by the producers.

Mainly internal browning of some seeds and a portion of the placenta located at the blossom-end of fruit (Photo **755**). Sometimes this internal injury takes the appearance of a corky fibrous mass. This internal necrosis corresponds to a form of classical apical necrosis.

This physiological disorder is commonly observed in all types of crops, particularly in those irrigated by furrow, by flood, or when home gardeners use the garden hose. It can appear at any stage of fruit development, but it mostly occurs when they have reached one-third or one-half of their maximum size. Tomatoes affected by blossom-end rot are often the first formed, which mature more quickly. Note that this lesion may be colonized by various micro-organisms, secondary invaders, responsible for the rot described in this section.

Origin   **Nonparasitic disease**

This condition is linked to a lack of calcium in the distal part of the fruit resulting from a lack of absorption of this element by the roots or its inadequate transportation via the sap in the xylem.

Several parameters may explain these two situations:

– a real deficiency in calcium or an antagonism of this element with other elements in the soil or nutrient solution (NH_4^+, NO_3^-, Mg^{++});

– high salinity induced by insufficient irrigation or high electrical conductivity of the nutrient solution, limiting the absorption of calcium;

– heavy transpiration;

– too rapid growth of plants and fruits;

– a naturally limited root system or resulting from the development of lesions of biotic origin (root pathogens/pests) or abiotic origin (poorly prepared soil, root damage caused by cultivation, roots, waterlogging), thus reducing water and calcium absorption (see p. 227);

– insufficient or poorly time distributed irrigation causing too much fluctuation in soil moisture.

In addition to these factors, blossom-end rot is particularly evident during and after periods of hot and dry weather.

Control methods   Optimize plant growth.

Climatic conditions

– Ensure optimal humidity in shelters and minimize plant transpiration (whitening or shading the roofs).

– Avoid subjecting the plants to dry and hot wind.

Fertigation

– Ensure balanced nutrition (avoid excess nitrogen in particular) and optimal calcium intake.

– In soil-less culture enrich the nutrient solution in $PO_4H_2^-$ and Cl^-, both of which promote calcium absorption.

– In soil culture, maintain adequate phosphorus levels, especially for planting, and a soil pH 6.5–6.8.

– Avoid excessive salinity.

– Foliar applications of anhydrous calcium chloride are recommended in the US.

The plant and its care

– Grow less sensitive varieties.

– Regularly removel leaves to maintain a good balance with the fruit load.

– Mulch the soil to maintain a more constant moisture.

– Do not damage the roots during cultivation.

Cat face

This physiological disorder (catface) is quite common in tomato crops. It mainly affects the first fruits of early crops, in the field or under protection, which are subjected to adverse climatic conditions (low temperatures) during blossoming and fruit formation. This is reflected by the presence of corky scars (Photo **756**), sometimes with protrusions appearing at the stylar scar. See p. 354 for more information about this disorder.

Corky peduncular scar

The large fruited varieties of tomato (multilocular with few seeds per loculus and large central columella) often have large corky peduncular scars (Photo **757**). These must be accepted as a problem associated with this type of cultivars.

Many rots can start from the peduncular and stylar scars (see p. 389).

Fruit

Table 38 Incidence and/or development potential of the main fungi causing rot or other symptoms on tomato fruit during production and post harvest

Fungi	Fruit development stages and incidence				Speed of symptom development
	Green	Nearly mature	Ripe	Post harvest	
Acremonium recifei				+/–	+/–
Alternaria tomatophila	+ +	+/–		+/–	+
Alternaria alternata f. sp. lycopersici	+	+/–		+/–	+/–
Alternaria spp. (A. alternata, A. tomato, A. arborescens)		+/–	+++	++	++
Aspergillus spp. (A. niger, A. flavus, A. fumigatus, A. ochraceus, A. tamarii)			+/–	+	+
Botryodiplodia theobromae			+/–	+/–	+
Botrytis cinerea	+	++	+++	+++	++
Cladosporium tenuissimum, C. oxysporum			+/–	+/–	+/– (Niger, India)
Cochliobolus spicifer				+/–	+/–
Coleophoma empetri			+/–	+/–	+/– (India)
Colletotrichum spp. (C. coccodes, C. gloeosporioides, C. dematium)		+	+++	+	+
Corynespora cassiicola	+	+	+		+
Cristulariella moricola[a]			+		+ (US)
Curvularia spp. (C. lunata, C. trifoli)[b]			+/–	+	+/– (India)
Didymella lycopersici (P. lycopersici)	+	+	+		++
Eremothecium coryli			+/–		+ (California)
Exserohilum rostratum				+/–	+/– (Niger)
Fusarium spp. (F. chlamydosporum, F. equiseti, F. moniliforme, F. nivale, F. oxysporum, F. roseum, F. semitectum, F. solani, F. verticillioides, F. xylarioides			+	+	++
Geotrichum candidum	+/–	+	+++	+++	++
Gilbertella persicaria var. 'indica'			+/–	+/–	+/– (India)
Helminthosporium fulvum			+/–	+	+
Helminthosporium spp. (H. carposaprum, H. fulvum)			+/–	+/–	+/– (Panama)
Mucor spp. (M. hiemalis, M. circinelloides)			+	+	++
Myrothecium spp. (M. roridum, M. carmichaelii)			+/–	+/–	+/– (India)
Penicillium spp. (P. expansum, P. citrinum, P. notatum...)		+/–	+	+	+
Pestalotiopsis sp.			+/–	+/–	+/– (India)
Phoma spp. (P. lycopersici, P. destructiva, P. sorghina)	+	+	+	+	+
Phytophthora infestans	+++	+	+/–		++
Phytophthora nicotianae	+++	++	+/–		++
Pythium spp. (P. aphanidermatum, P. ultimum...)	+	+	+		++
Rhizoctonia solani	++	++	+		++
Rhizopus spp. (R. stolonifer, R. oryzae, R. rhizopodiformis)	+/–	+/–	+++	+++	+++
Sclerotinia minor	+/–	+/–	+/–		++
Sclerotinia sclerotiorum	+	+	+		++
Sclerotium rolfsii	+	++	+++	+	+++
Stemphylium botryosum			+/–	+	++
Syncephalastrum racemosum			+/–	+/–	+ (India)
Trichoderma koningii				+/–	+ (India)
Trichothecium roseum		+/–			+ (Czech Republic)
Ulocladium consortiale			+/–	+	+

–: None; +/–: low; +: medium; ++: high; +++: considerable.

[a] Cristulariella moricola: responsible for zoned spots on leaves and fruit rots in New York state, US. This fungus occurs on trees (Acer negundo, Juglans nigra) near affected tomato crops. The tomato crops provide the primary inoculum. Very rare.

[b] Curvularia lunata var. 'aeria' and Curvularia trifolii may be responsible for leaf spots and fruit rots in India.

■ Spots and other symptoms gradually evolving into rot

Numerous micro-organisms are capable of attacking tomato fruits and causing spots which evolve at varying speed into rots. The main fungi and some characteristics that identify them are listed in *Table* 38. According to their parasitic potential on tomato, they can be subdivided into two main groups:
– opportunistic fungi affecting mainly fruit, detailed in this section;
– 'true' pathogenic fungi also affecting other tomato plant parts, which have already been described in other sections of the book. These micro-organisms will therefore only be mentioned here (see *Table 39*, p. 397, and the corresponding descriptions).

The first group members are for the most part ubiquitous fungi, ever present in the tomato environment. They often affect fruits in contact with the ground and/or which have lost their structure because of being over-ripe and/or with various wounds aiding penetration (growth cracks, cold damage etc, see Figure **50**). Once present they develop and induce lesions or rots which can be very spectacular. In all cases they produce many spores that enable their spread via air currents, splashing water, and various vectors (workers, animals, insects). These fungi are capable of causing damage both during cropping, as well as after harvest. Wherever the crop is grown the same fungi are responsible: *Alternaria alternata*, *Rhizopus stolonifer*, *Mucor* spp., *Geotrichum candidum*, *Penicillium* spp., *Aspergillus* spp. etc.

A summary of the control methods for these fungi is on p. 521.

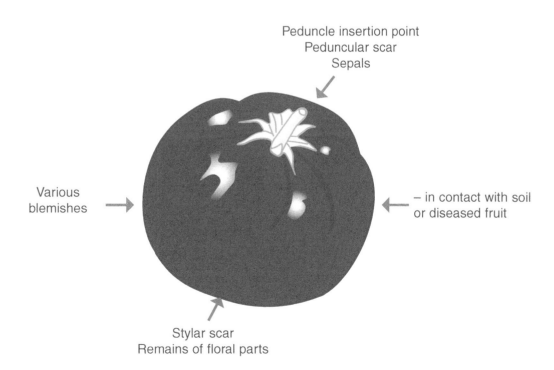

Peduncle insertion point
Peduncular scar
Sepals

Various blemishes

– in contact with soil or diseased fruit

Stylar scar
Remains of floral parts

Figure 50 Main entry points or nutrient bases allowing pathogenic or opportunist micro-organisms to penetrate and affect the fruits.

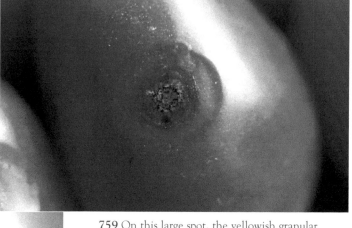

Colletotrichum coccodes (anthracnose)

758 Many yellowish and slightly sunken spots are visible on the fruit. Their centre gradually turns brown.

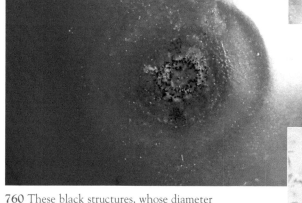

759 On this large spot, the yellowish granular texture of the tissues below the cuticle and the presence of tiny black structures can be seen.

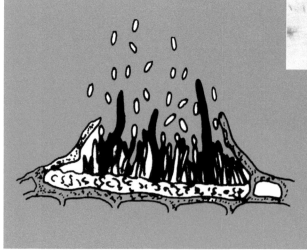

760 These black structures, whose diameter is less than 1 mm, are the microsclerotia of *Colletotrichum coccodes*. They are buried in the tissues beneath the epidermis.

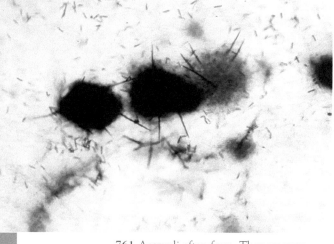

761 Acervuli often form. They are very distinctive as they have erect black setae and unicellular hyaline conidia.

Figure 51 Drawing of *Colletotrichum coccodes* fructification.

Colletotrichum spp. (anthracnose)

Several Colletotrichum species may be associated with tomato anthracnose. *C. coccodes*, which is the most common species on fruits, has been previously described in roots, responsible in particular for cortical rot (see p. 261). Other species are also found on tomatoes with anthracnose symptoms: *C. gloeosporioides* (Penz) Penz. & Sacc., *C. dematium* (Pers) Grove. Anthracnose causes major losses in many production areas. It mainly damages ripe fruits in the field, and sometimes after harvesting. This parasitic disease is common, both in the processing crops of mechanically harvested field tomatoes and in the gardens of amateurs.

The first symptoms mainly appear on ripe fruit in the form of small brown lesions that change into randomized circular slightly recessed moist patches (Photo **758**). These lesions spread, gradually deepening and darkening. The underlying flesh is of a lighter colour with a grainy texture (Photo **759**). The centre of advanced lesions is brownish and black dots appear, which are the microsclerotia produced by these fungi (Photo **760**). The cuticle of the fruit remains intact and, in damp climatic conditions, may be covered with small masses of salmon-coloured mucoid spore masses. Note that several spots present on the fruit may merge and cause extensive rot.

The rather slow development of the spots and the presence of microsclerotia and acervuli on them (Photo **761** and Figure **51**) allow easy identification of this fungus.

Alternaria alternata f. sp. *lycopersici*

Several *Alternaria* species may attack tomatoes. The best known is *A. tomatophila* that affects all aerial parts of plants in the Solanaceae (see also p. 173). Other species have a more limited parasitism, merely colonizing and rotting ripe fruits; this is the case of *A. alternata*, and to a lesser degree, *A. tomato*.

A specialized form of *A. alternata*, *A. alternata* f. sp. *lycopersici*, is reported in several countries (particularly in the US but also in Iran, Iraq, Egypt, Greece, Sardinia, India, Japan, Korea), on susceptible varieties which do not possess the incompletely dominant 'ASC' allele in the homozygous state. It causes stem cankers and specific fruit symptoms.

This *Alternaria* species is responsible for various symptoms and damage can be significant, in the field but also under protection.

The lesions occur on green fruit, often in the form of concentric rings are dark brown, concave and rather small. *A. alternata* f. sp. *lycopersici* can also cause rot covered with black mould but is not frequently associated with this type of symptom especially in comparison with opportunistic species of *Alternaria* (*A. alternata*, *A. tomato*). It sporulates in the affected tissues in the form of conidiophores bearing short chains of brown multicellular conidia.

Remember that it causes elongated stem cankers, close to the ground (see p. 309). In addition, its toxin diffusing into the vascular tissues causes secondary leaf symptoms.

762 Several slightly concave and confluent spots, are now forming in the peduncular area of this red fruit. Their centre begins to brown.
Alternaria **sp.**

764 An *Alternaria* sp. has established on the ripe fruits from the peduncular scar. Tissues gradually collapse and split while the pericarp blackens locally.

766 A spot has developed from the peduncular scar of the fruit. The tissues are slightly recessed with a yellowish colour. Tiny brown to black globular structures, concentrically arranged, are visible within the tissues.
Pleospora **sp.**

763 These spots are more advanced and much larger. Small ruptures of the cuticle occur, revealing a black mould: the fructifications of *Alternaria* sp.

765 The rot is now in place, the fruits gradually shrivel while fructifications are formed.
Alternaria **sp.**

Black mould rots

Among the fungi responsible for rots on tomato fruit (see *Table 38*, p. 388), several cause symptoms characterized by tissues taking on a black colour accompanied by a mildew covering of a similar colour. This is particularly the case during the development of several *Alternaria* species, including *A. alternata* (Fr.) Keissl.[1] (syn. *A. tenuis* Nees), *A. tomato* (Cooke) R. L. Jones, *A. subtropical* Simmons, *Stemphylium botryosum* Wallr., and *Ulocladium consortiale* (Thüm.) EG Simmons (1967). They are saprophytic and worldwide in their distribution They are constituents of the surface flora of many plant species, including the tomato and are ever present in the crops environment. They are most likely to occur on mature fruits. Their damage is reported in many countries and on all continents. They affect mainly fruits grown in unstaked field-grown crops.

Symptoms

Initially small concave moist lesions with a diffuse edge appear on the fruits (Photo **762**). They are often located close to the peduncle or in the peduncular area, called the 'shoulder' (Photos **763** and **764**). Subsequently, they gradually spread, and sub-cuticular tissues located in their centre quickly turn brown then black (Photo **759**).

Ultimately, if the humidity is high, the changed surface is covered with a dense velvety coating, dark green to black (Photos **767** and **768**), particularly near small skin slits. This mould is formed by the mycelium, conidiophores, and numerous conidia of *Alternaria alternata*, *Stemphylium botryosum*, or *Ulocladium consortiale*. The spores of these fungi have characteristic morphology (Photos **770–772**)[a].

In some cases small black globular structures form in the damaged tissues in place of the mould (Photo **776**). These symptoms are then caused by *Pleospora* (*Pleospora herbarum* [Pers.] Rabenh., syn. *Pleospora lycopersici* El Marchal & Em Marchal), teleomorph of a *Stemphylium*. These globular structures are perithecia containing several asci and ascospores (Photos **773** and **774**).

Biological cycle

These fungi are widely present on plants and in their environment. They have some saprophytic potential[b] enabling them to remain on/in the soil, on plant debris of any kind (leaflets, stems, senescent floral parts), from tomato and many other hosts, cultivated or not.

 The conidia of these opportunistic pathogens germinate on the fruits and penetrate them directly through the cuticle or through stomata, lenticels and various wounds (insect punctures, cracks, and splits, perforations, damage resulting from 'sunburn' or cold conditions, microlesions of the cuticle [**764**, **768**, and **769**]). Their mycelium invades the tissues and does not take long to produce conidiophores and conidia which are dispersed by wind and air currents and sometimes water splash. Their spread can occur by contact from diseased to healthy fruit in the field and in storage.

Their development is encouraged by wet weather and optimum temperatures between 18 and 30°C. Damaged fruits, already affected by primary pathogen/pests and/or over-ripe are particularly vulnerable. Remember that some species can produce mycotoxins.

[a] The morphology of the conidia of *Alternaria alternata* and *A. tomato* is similar and varies with the culture medium and the substrate from which they are collected. Risks of confusion are therefore possible. *A. tomato* has been associated with nail head-shaped spots on fruit (nail head spot).

[b] These fungi may occur alone or in combination with other micro-organisms responsible for soft rots. They must be regarded as constituents of the flora of rotten fruits in field crops, with the *Alternaria* being among the most damaging.

[1] *Alternaria*, Pleosporaceae, Pleosporales, Pleosporomycetidae, Dothideomycetes, Ascomycota, Fungi.

767 The peduncular area of this fruit was invaded by *Alternaria* sp. after the first frost. This area has now completely collapsed and is covered with black mould.

768 These fruits were occasionally without water. Apical necrosis occurred thereafter, quickly colonized and covered by mycelium, but mostly by the many sporing structures of *Alternaria* sp.

769 The sunburn injuries on these two fruits have been colonized by *Alternaria* sp. which is sporulating in places.

**Black mould
rots**

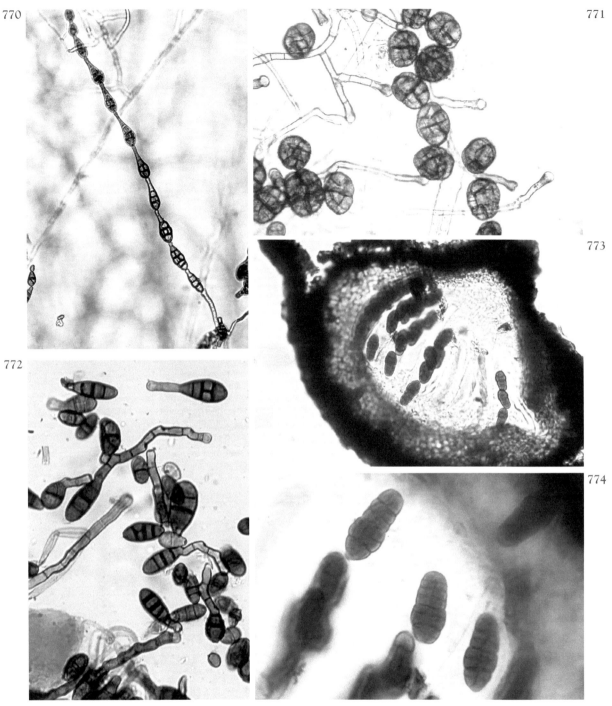

774

772

770 The spores of *Alternaria alternata* are produced in long chains at the end of conidiophores (longicatenatae). Conidia are brown, multicellular, with a relatively short appendix.

772 The conidia of *Ulocladium* sp. are quite polymorphic. Of variable size, they are formed on long brown conidiophores, simple or branched. The spores which are transversely and/or longitudinally partitioned have a golden to olive brown colour.

771 *Stemphylium botryosum*, a common saprophyte, forms brown multicellular warty conidia with rounded ends (15–24 × 24–33 µm).

773 *Pleospora* produce perithecia on the fruits that are first embedded, then superficial and without paraphysis. They contain many clearly visible asci.

774 Ascospores of *Pleospora* are multicellular, ovoid, bulbous, and muriform.

Fruit

Phoma black rot

Several *Phoma*[1] have been isolated from tomato fruit[2]. *P. lycopersici*, a synonym of *Didymella lycopersici* will not be discussed here. Some of its characteristics are listed in *Table 39* and in Description 3. Instead *P. destructiva* Plowr. (syn. *Diplodina destructiva* [Plowr.] Petr.) a deuteromycete fungus responsible for leaf spots, but mostly fruit rot (Phoma rot), in many countries (US, Europe, India, China) will be discussed.

Symptoms

Initially spots of limited size, slightly concave, appear at the peduncular scar of some fruits (Photo **775**), or near growth cracks or various wounds. They gradually spread with a fairly typical dark brown to black colour. The lesions decay further while their perimeter is clearly identified. They eventually cover a large proportion of ripe or unripe fruits if the environmental conditions are favourable. Whatever their development stage, the fruits undergo internal rot extending to the seeds. Brown to black globular structures, pycnidia, can be observed in the damaged tissues (Photo **776**, Figure **52**). They are spherical and irregularly shaped, ostiolate, and contain many hyaline and unicellular conidia, containing two lipid droplets (3.5 × 2 μm).

Note that spots can be observed on the leaves. They are primarily small, irregular and dark coloured. Subsequently, they spread and become necrotic, with visible concentric patterns, causing diagnosis confusion with *Alternaria*. Strongly affected leaves eventually turn yellow and wither. Black lesions may develop on the stem.

Biological cycle

Biological cycle

In production areas where it occurs, the fungus survives very well on plant debris in soil. It is also capable of attacking the fruits of various other hosts.

Colonization occurs through the inevitable fruit wounds, especially at the peduncular scar and growth cracks. Once in place, the mycelium of *P. destructiva* colonizes the fruits radially and in depth. It does not take long to produce pycnidia that are randomly distributed, and embedded in the tissues. These produce conidia which transmit the fungus resulting in secondary infections. Transmission occurs in wet periods, during or after rainfall or sprinkler irrigation, during splashing, or during cultural operations. Seedlings may also contribute to the spread of the fungus.

The development of *P. destructiva* is favoured by wet weather and mild temperatures. Wounded fruits, sometimes affected by primary pathogens such as *Alternaria alternata,* are more vulnerable.

775 A large particularly advanced spot has developed from the peduncular scar of the unripe fruit. The tissues are characteristically black and are now collapsed.

Phoma destructiva (Phoma rot)

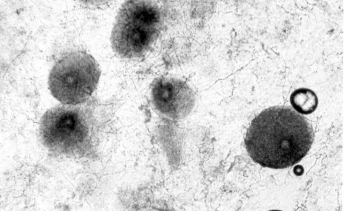

776 Several pycnidia are visible in the damaged tissues of this tomato fruit. Some have a very distinct ostiole Aggregated fungal mycelium can also be seen.

[1] Other *Phoma* have occasionally been reported on tomatoes: *P. sorghina* (Sacc.) Boerema, Dorenb. & Kesteren, *P. exigua* Sacc., *P. terrestris* H.N. Hansen.

[2] *Phoma*, Leptosphaeria, Incertae sedis, Pleosporales, Pleosporomycetidae, Dothideomycetes, Ascomycota, Fungi.

Table 39 Some characteristics of fungi causing symptoms on fruit, and also on other tomato plant parts

Fungi	Symptoms	Structures observed
Didymella lycopersici	A brownish circular lesion spreads quickly from the peduncular scar (Photo **777**). Tissues tend to slightly collapse and darken, due to the colonization of the flesh by melanized mycelium of the fungus and the formation of many fructifications (see also p. 181).	Globose and ostiolate pycnidia are visible under a dissecting microscope. They are dark to different degrees depending on their maturity and are immersed in damaged tissues (see Photo **588**).
Alternaria tomatophila	Large concave circular lesions, with a rather hard texture and sometimes a wrinkled cuticle, form at the peduncular scar of green or ripe fruit (Photo **778**). The fungus first colonizes the sepals and/or the peduncular scar and these progressively blacken. Subsequently, it reaches the fruit and colonize it. Lesions sometimes appear on the side of fruit (see also p. 173).	A dense black felting finally covers the lesions, which is the sporulation of *A. tomatophila*. Thus, many conidiophores are produced which mainly carry isolated conidia (noncatenatae). These are brown, multicellular, and club-shaped. They have a long filiform appendage (see Photo **809** and Figure **45**).
Corynespora cassiicola	On green fruits, the spots are small, light brown, with a darker edge and a rather dry consistency. On ripe fruits, large circular spots develop. They are slightly recessed, light brown, and split at the centre (Photo **779** and p. 177).	In general, many fructifications can cover the lesions, giving them a dark grey to black colour (see Figure **55**). Its long conidia are characteristically elongated, subhyaline to slightly brown, isolated or in short chains, they contain 4–20 pseudo-partitions.
Botrytis cinerea	Large spots evolving into rot form at the peduncular scar, often from the sepals (Photo **780**), or at other locations on fruit, from the remains of floral parts, wounds (Photo **781**). They are white and their consistency is soft. The skin eventually splits at the centre of the lesions (see also p. 181).	Abundant grey mould, consisting of many branched conidiophores bearing powdery conidia, covers the lesions (see Photo **576**).
Sclerotinia sclerotiorum	The fruits (Photo **782**), and also the leaflets, are affected by a wet soft rot followed by the liquefaction of the former and the disintegration of the latter (see p. 307).	A thick white mycelium forms on the surface of the fruits on which large black sclerotia are seen (Photo **579**).
Sclerotium rolfsii	Translucent lesions arise on the fruit in contact with the soil. They rapidly develop a wet and soft rot resulting in tissue collapse (Photo **783**). The wrinkled skin eventually cracks (see also p. 287).	Thick white mycelium covers the affected fruit. This mycelium is relatively characteristic as it has anastomosis loops. Moreover, it forms round sclerotia 1–3 mm in diameter. The colour of these changes from white to beige then to dark brown (see Photo **783**).

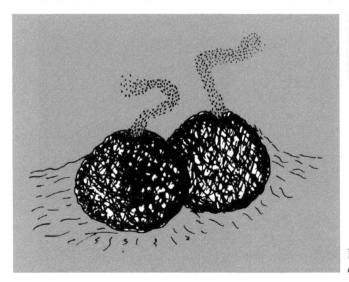

Figure 52 Drawing of *Phoma destructiva* fructifications.

777 A large wet lesion develops around the fruit peduncular scar. The collapsed tissues are a brownish to black colour due to the presence of the fungus mycelium and pycnidia, some of which are melanized.
Didymella lycopersici (Phoma lycopersici)

778 After invading the sepals and peduncle, *Alternaria tomatophila* colonizes this green fruit and causes a circular brown and recessed lesion, with discrete concentric patterns.
Alternaria tomatophila

779 Several spots are visible on the nearly ripe fruit. The smallest are light brown and darker at the periphery. The most advanced is circular, strongly recessed, brownish, radially split, with concentric patterns.
Corynespora cassiicola

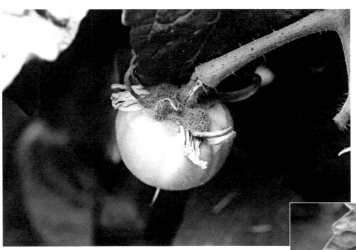

780 A large milky and soft rot has developed around the peduncular scar of the green fruit. A dense grey mould partially covers its centre.
Botrytis cinerea

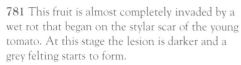

781 This fruit is almost completely invaded by a wet rot that began on the stylar scar of the young tomato. At this stage the lesion is darker and a grey felting starts to form.
Botrytis cinerea

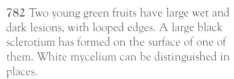

782 Two young green fruits have large wet and dark lesions, with looped edges. A large black sclerotium has formed on the surface of one of them. White mycelium can be distinguished in places.
Sclerotinia sclerotiorum

783 A soft, wet rot has developed on the ripe fruit. The tissues have collapsed and the skin is shrivelled. In addition to a little soil, some mycelium and brownish sclerotia partially cover the affected fruit.
Sclerotium rolfsii

Other fungi causing rot

Many fungi, listed in *Table 39* (p. 397), have been associated with symptoms on tomato fruits. Only those that have some significance will be discussed here.

• *Penicillium* spp.

Several *Penicillium*[1] species are reported on tomato fruits, causing damage both during cultivation and after harvest (*P. expansum* Link, *P. citrinum* Thom, *P. notatum* Westling, *P. brevicompactum* Dierckx, *P. cyaneo -fulvum* Biourge, *P. cyclopium* West., *P. puberulum* Bainier, *P. italicum* Wehmer). They are responsible for soft, moist, and dark lesions, which can develop either from the peduncular scar (Photo **784**) or elsewhere on the fruit, near a wound. The damaged tissues are gradually covered with spore-bearing structures whose colour may vary depending on the species present and the maturing of the sporing structures at the time of observation (blue, grey, green).

As an example, *P. expansum* is the species most often seen on many fruits: apple, pear, kiwi, orange, mandarin, grapefruit, strawberry, and various vegetables such as garlic or onion. Like many other *Penicillium* species, it is naturally present in soil and on the different surface of the tomato plant where it is present as mycelium or spores. The spores, very light and mass produced, are airborne but may fall on the surface of a fruit near a wound. They germinate and the mycelium grows extensively, in all directions within the flesh that gradually breaks down because of the multiple enzymes produced. The fungus does not take long to sporulate; conidiophores develop generating conidia (Photo **785**). The number of spores present on the surface of a contaminated fruit can be counted in millions. Once they reach maturity, they separate and float easily in the air for some time before settling and causing new infections. Germination is best in a moist atmosphere between 20 and 25°C and the mycelial growth stops beyond 30°C. Note that *P. expansum* only infects ripe or ripening fruit and it synthesizes two toxins, patulin and citrinin. Patulin is a mycotoxin that occurs frequently in animal feed.

• *Trichothecium roseum*

T. roseum (Pers.) Link[2] is a worldwide fungus found on living and dead plants. It has been reported on various fruits (apple, pear, plum, cherry) and vegetables (melon, cucumber). It seems more unusual on tomato but has been reported particularly on protected fruit: US (North Carolina), Czech Republic, Italy (Liguria), Japan. It has very occasionally been observed in France, also in glasshouses. It may have been isolated from lesions caused by *Phytophthora infestans*.

This fungus causes dark lesions, mottled with brown veins that appear near the peduncular scar (Photo **786**). This area, besides having a brown halo, is covered with a pale pink mould. This mould consists of the long arbuscular conidiophores of the fungus. These have bicellular smooth, hyaline conidia, which can be oblong or pyriform (Photo **787**). In general, the affected fruits may drop prematurely. *T. roseum* can synthesize trichothecene, a mycotoxin in the damaged tissues.

• *Aspergillus* spp.

Several *Aspergillus*[3] species appear to be responsible for rot on tomato fruits, mainly after harvest. *A. niger* is the most frequent and damaging. The ripe fruits show soft and dull lesions. When environmental conditions are favourable, this species forms a brown to black mould, powdery to charcoal-like, consisting of numerous conidiophores and conidial chains of this fungus. Other species have moulds of various colours. The *Aspergillus* species are widely distributed throughout the world. Several species, among approximately 185 identified, are found on plants, in many soils, on very diverse plant debris and on seed. They have biological characteristics quite similar to those of *Penicillium*: they mostly penetrate through wounds, but on some ripe fruit they are able to do so directly through the cuticle.

---

[1] Anamorph of *Eupenicillium*, Trichocomaceae, Eurotiales, Eurotiomycetidae, Eurotiomycetes, Ascomycota, Fungi.

[2] *Trichothecium*, Incertae sedis, Incertae sedis, Incertae sedis, Incertae sedis, Ascomycota, Fungi.

[3] *Aspergillus*, Trichomoceae, Eurotiales, Eurotiomycetidae, Eurotiomycetes, Ascomycota, Fungi.

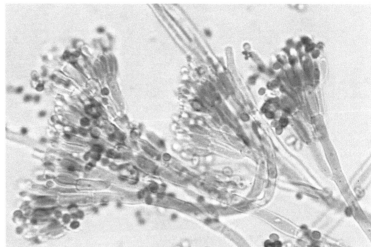

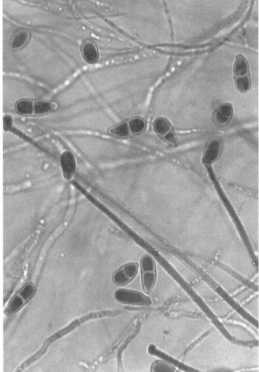

784 Some rot surrounds the peduncular scar of the ripe fruit and is the cause of the tissues collapse. A velvety grey-green coating covers it in its centre.
Penicillium sp.

786 The green fruit is almost completely rotten. It has become dark and oily in appearance and brown veins are visible in places. The tissues surrounding the peduncle are brown and some pink mould emerges from beneath the sepals.
Trichothecium roseum

785 *Penicillium* species are characterized by their brush-shaped sporing structures. These consist of branched conidiophores whose ends are not swollen, with or without differentiated sterigmata, and phialide whorls at the base of many conidial chains.

787 Several conidiophores still bear a few hyaline conidia, the oldest became detached. Most of them are pyriform, elongated, and bicellular.
Trichothecium roseum

Fruit

788 The ripe fruit is fully degraded by a wet and soft rot. The liquefied tissues have collapsed and the skin has broken and shrivelled in places. An aerial black mould has grown in the cracks.
Rhizopus stolonifer

790 The use of a microscope shows the columella, the rhizoids, and numerous brownish spores of this Mucoraceous fungus.
Rhizopus stolonifer

792 As for *Rhizopus*, *Mucor* produce spherical sporangia at the end of straight sporangiophores. A young evolving sporangium is clearly visible.
Mucor sp.

789 This mould is formed by the aseptate mycelium of *Rhizopus stolonifer*. Many black spherical sporangia on rigid sporangiophores can be seen. They are hyaline before gradually browning.

791 Generalized wet rot on this ripe fruit. The flesh has liquefied and the skin is wrinkled. A greyish aerial mould develops around the peduncle and the stylar end.
Mucor sp.

Mucoraceous rots
(Mucor and Rhizopus rot)

Rhizopus stolonifer (Ehrenb.) Vuill. (syn. *Rhizopus nigricans* Ehrenb. [Rhizopus rot])

Rhizopus, Mucoraceae, Mucorales, Incertae sedis, Zygomycetes, Zygomycota, Fungi

Symptoms

Moist and soft lesions expanding rapidly mainly on ripe fruits. The affected tissues liquefy and eventually collapse, the cuticle splits (Photo **788**). Exudation of juice may be found. The fungus mycelium invades the tissues which are covered by a whitish grey mould. This mould is formed by the mycelium, sporocystophores, and sporocysts of this Zygomycete (Photos **789** and **790**, Figure 53). Ultimately, many fructifications, 'black pinheads' form on the damaged tissues (Photo **789**).

Biology

This ubiquitous saprophytic fungus grows very easily on and in soil, on plant debris of many types, on various wet and dry substrates, and remains there for several years sometimes in different forms: spores, chlamydospores, zygospores, mycelium. It is also known to induce rot on many plant species, notably fruit and vegetables. It is thus present in the plants environment in the fields and stores, on the packaging and storage materials. It enters fruits through various wounds, but also from the stylar and peduncular scars. Once in the tissues, it invades them, quickly rotting them, and eventually sporulates on the surface with many sporangiophores containing countless black spores in sporangia. These spores are spread by wind over long distances, but also by splashing water, by workers and some 'passive vector' insects, including *Drosophila melanogaster* (Photo **793**). Contact transmission of rotten fruit to healthy fruit is also possible.

Wet weather and the high temperatures during summer are very favourable for *Rhizopus stolonifer* development. Thus, it thrives at temperatures of around 23–27°C and high humidity. It also occurs at low temperature during fruit storage. Ripe tomatoes are particularly sensitive.

In addition to *Rhizopus* spp. the Mucorales includes other genera likely to attack tomatoes. This is particularly true of the genus *Mucor* which includes some species that can grow on tomato fruits: M. *hiemalis* Wehmer, M. *circinelloides* Tiegh. The rot they cause and the mould that covers them are quite comparable to those produced by *R. stolonifer*. They are moist and soft (Photo **791**) and covered with a greyish fluffy mould (Photo **792**). Their biological characteristics are similar to those of *Rhizopus*.

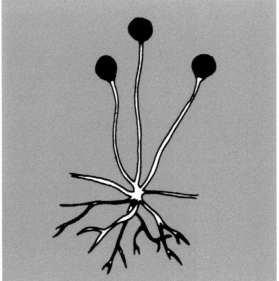

Figure 53 Drawing of *Rhizopus stolonifer* fructifications.

793 Many small flies (*Drosophila melanogaster*) dot this fruit, rotting with a Mucoraceous fungus.

794 Half of the mature fruit shows a wet rot. The liquefied tissues are collapsed, and the skin eventually split longitudinally. Note the presence of a dense whitish 'slime' on the exposed flesh.
Geotrichum candidum (sour rot)

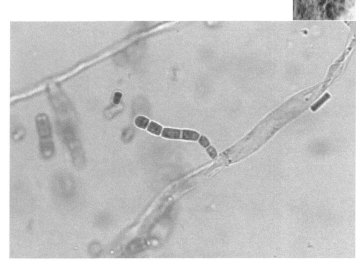

796 *Drosophila melanogaster* is an efficient vector of *Rhizopus stolonifer*, but mostly of *Geotrichum candidum*.

795 On this hyaline and septate mycelium, subglobose to cylindrical conidia form in chains. Conidia measure on average of 3–6 × 6–12 μm.
Geotrichum candidum (sour rot)

Geotrichum candidum Link[1] (sour rot)

This yeast-like[2] fungus is associated with wet fruit rots affecting mainly unstaked field-grown tomatoes for processing. It occurs more rarely on fruits from trellised crops and for the fresh market. It is a saprophyte widely distributed in the environment and is able to attack green or ripe fruit at the slightest opportunity.

Symptoms

It produces wet and oily lesions that appear on fruit wounds or on peduncular scars. These lesions enlarge rapidly on mature fruits, whose skin eventually splits, causing the exudation of juices. Locally white and dense mucus sometimes covers tissues (Photo **794**); it is the development of *G. candidum* colonies. Within the cellular mycelium are arthrospore chains which eventually unravel and are readily observable by light microscopy (Photo **795**).

It should be noted that a characteristic acid smell accompanies this rot, it reflects a relatively low pH of the fruit flesh, around 4.

Affected green fruit remain firm for longer, but they eventually develop a similar rot to ripe fruits when fully colonized.

Biology

This fungus survives in soil, on plant debris, in water tanks, on pulp contaminated material and/or material having contained rotten fruits. Penetration occurs through wounds, notably around the peduncular scar, in no case directly through the cuticle. Note that *Drosophila melanogaster*, the fruit fly, may be a spores vector of *G. candidum* depositing them on the many microwounds present on ripe fruit. Once in the tissues, this fungus grows rapidly and multiplies in large quantities. Spread may occur in various ways: through contact between diseased and healthy fruits, after water splashing, through insects in particular *D. melanogaster* (Photo **796**), through workers during cultural operations. The presence of wounded ripe fruits in a humid atmosphere and a temperature around 30°C promotes the development of *G. candidum*.

---

[1] *Geotrichum*, Incertae sedis, Saccharomycetales, Saccharomycetidae, Saccharomycetes, Ascomycota, Fungi, another species of *Geotrichum*, morphologically identical, may cause similar rot on tomato. This is *G. penicillatum* (Carmo Souza) Arx. Unlike *G. candidum*, the rot produced is not accompanied by a sour odour.

[2] *Aureobasidium pullulans* (de Bary) G. Arnaud (syn. *Pullularia pullulans* [de Bary & Löwenthal] Berkhout), a yeast-like fungus, is responsible for fruit rot in India. It is present as a saprophyte on the surface of many plants, especially fruits. It seems able to parasitize several plant species.

Fruit

797 A dark and wet rot is developing inside the mature fruit.
Pectobacterium (*Erwinia* sp.)

798 A large wet and dull lesion has spread from the peduncular scar of the fruit. A white bacterial 'slime' is locally present.

799 This fruit, already affected by *Didymella lycopersici*, has been secondarily infected by a bacterium that causes the milky 'slime' visible where the skin broke.

Bacterial soft rot[1]

The Pectobacteria (*Erwinia* spp.), which also affect the stems (see p. 315), are the most commonly found bacteria on tomato fruits. They are responsible for quickly developing wet rots, which are smelly and can sometimes be very harmful. The most aggressive is certainly *P. caro-* *tovorum* subsp. *carotovorum* and to a lesser extent, *P. chrysanthemi*. *P. carotovorum* subsp. *atrosepticum* responsible for potato blackleg has been isolated from tomato. These bacteria are fairly widespread in many production areas throughout the world (see also Description 26).

Symptoms

These bacteria are the cause of small, dark, moist lesions, starting from wounds and scars and quickly spreading on the surface and at depth (Photo **797**). Eventually, milky bacterial exudates can be observed in the damaged epidermis (Photo **798**). Tissues liquefy quickly, and bubbles of gas are sometimes reported. These symptoms can occur both during cultivation and after harvest, during storage. Note they can also infect lesions caused by other pathogens (Photo **799**).

Biology

These ubiquitous bacteria easily remain in the epiphytic state on a variety of plants, in the soil and surrounding wet areas. They seem unable to penetrate the fruit directly, they always enter through various wounds or natural openings. Contamination may occur in the field or after harvest, during handling and/or fruit washing. Once established, they multiply more or less rapidly in the tissues, depending on climatic conditions. Under favourable conditions, damage can be observed in the day or days following contamination.

Note that 'lactic' gram-positive bacteria have recently been linked with moist lesions on tomatoes harvested for the fresh market, in California and Florida, US. The symptoms caused by these bacteria have, incorrectly, been attributed to several post harvest rotting agents described in this section. In fact, they are mostly comparable to those induced by *Geotrichum candidum* (sour rot). Of firm consistency and dull and oily, they emit a rather characteristic odour of lactic acid.

In addition to being present in and on fruit, these bacteria are found throughout the tomato vegetation, but also in storage and packaging stations. Two bacterial genera have been involved: *Leuconostoc* and *Lactobacillus* and the species *Leuconostoc mesenteroides* subsp. *mesenteroides*. The latter bacterium enters the fruits through wounds and could be associated with other post harvest rotting agents.

The identification of these bacterial diseases is difficult. A specialist laboratory should be consulted if there is any doubt. The absence of fungal structures in tissues, particularly mycelium and the presence of many gram-positive bacteria should guide the diagnosis.

---

[1] Some of them are relatively rare and do not constitute a major threat to tomato production. This is particularly true of some pectinolytic strains of *Pseudomonas fluorescens* causing more slowly developing rots than those of Pectobacterium (*Erwinia* spp.) Similarly some *Bacillus* spp. occasionally cause some damage.

800 The green fruits are widely covered by raised necrotic lesions with a circular shape. **Cucumber mosaic virus (CMV)**

801 The superficial and diffuse browning affecting these green fruits is not widespread. Here and there, circular patches of tissues appear to be unaffected. **Parietaria mottle virus (PMoV)**

802 Several brown and well defined rings, variable in size, have randomly formed. **Pelargonium zonate spot virus (PZSV)**

803 Chlorotic rings of varying widths are clearly visible on both mature fruits. **Tomato spotted wilt virus (TSWV)**

Spots and marks in rings, circles

Spots in rings, circles, and various patterns

Several micro-organisms are able to produce lesions and other fruit symptoms, characterized by having more complete or partial circles or rings, isolated or interlacing with each other.

Some viruses are known to induce chlorotic brown, necrotic rings, often on green tomatoes, sometimes persisting to maturity. Photos **800–803** illustrate the diversity of their lesions on fruit.

Some of these viruses and their main characteristics are listed in *Table 39a*. Their identification based on fruit symptoms alone is very difficult. It often is necessary to observe the symptoms they cause to other parts of affected plants, which will help with the diagnosis (see pp. 59 and 97). If such symptoms are not visible, it is advisable to use a specialized laboratory.

Table 39a Main viruses causing irregular patterns and/or concentric rings on tomato fruits

Viruses	Symptoms	Transmission methods and vectors
Cucumber mosaic virus (CMV) Description 35	Severe brown necrotic lesions, circular and raised, covering large areas of affected fruit (Photo **800**).	By aphids in the nonpersistent manner. Many species are able to acquire and transmit it: *Myzus persicae, Aphis gossypii, A. craccivora, A. fabae, Acyrthosiphon pisum.*
Parietaria mottle virus (PMoV) Description 44	Lesions and chlorotic rings are present on the fruit. These lesions later turn brown (Photo **801**), becoming corky and causing deformation.	Probably through thrips.
Pelargonium zonate spot virus (PZSV) Description 47	The rings on fruit are at first chlorotic and concentric, and subsequently become necrotic (Photo **802**).	Unknown.
Tobacco streak virus (TSV) Description 44	The necrotic flowers may fall. The fruits are sometimes covered with necrotic rings.	Mechanical transmission through contaminated pollen and by thrips *(Frankliniella occidentalis, Thrips tabaci).*
Tomato bushy stunt virus (TBSV) Description 32	Rings and chlorotic patterns are visible on the fruits.	No known vector. This virus is associated with soil; it is transmitted by water and by seeds (via pollen) to a rate of 5% in tomato.
Tomato mosaic virus (ToMV) Description 31	Chlorotic rings are visible on green and mature fruits. Internal brown necrotic lesions are also observed (*internal browning*).	Transmitted through contact and seeds.
Tomato spotted wilt virus (TSWV) Description 43	Large patches, rings and arcs partially cover the fruits (Photo **803**). Fruits can be locally discolored, bronzed, and deformed.	Through thrips, in the persistent manner. Nine species are vectors: *Frankliniella occidentalis, F. fusca, F. schultzei, Thrips palmi, T. tabaci, Scirtothrips dorsalis.*

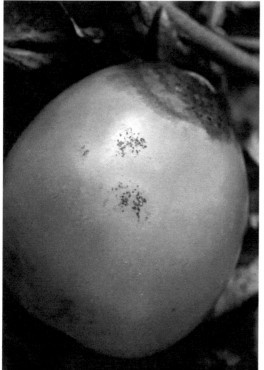

804 These three mature fruits show the diversity of lesions induced by *Rhizoctonia solani* on tomato.

806 Several brown mycelial masses (sclerotia), concentrated in places, indicate the presence of *Rhizoctonia solani* on the tomato.

805 Small beige cankerous lesions and a large spot consisting of brown concentric rings, alternating with lighter rings are visible at the same time on this fruit.
Rhizoctonia solani

807 One of these green fruits is largely covered by a brown lesion made up of diffuse dark brown concentric semicircles with a looped edge.
Phytophthora nicotianae

Rhizoctonia solani, soil fungus, attacks the fruits when they are in contact with wet soil. It causes characteristic lesions, especially on ripening or ripe tomatoes. These lesions are initially small and cankerous. As they develop, they become fairly regular and well defined circular brown bands, alternating with lighter and yellowish bands, giving them the appearance of a target (Photos **804** and **805**). Note that the centre tends to become corky and the cuticle can crack in time. The damaged tissues are rather firm, unless secondary invaders contribute to their softening and their liquefaction. Brown mycelium and mycelial clusters of the same colour (sclerotia) are frequently visible on fruit, on or near the lesions (Photo **806**). *Thanatephorus cucumeris*, the teleomorph, also induces symptoms on fruits described on p. 311.

Rhizoctonia solani also attacks other tomato parts (see p. 293).

Phytophthora nicotianae, responsible for a blight, which mainly affects the roots, the stem base (see pp. 241 and 283), but also the fruits of tomato in contact with the ground. It causes brown spots which gradually spread in the form of brown rings that are sometimes regular, diffuse, and concentric. These progressively invade the fruits (buckeye rot, see Photo **807**). When conditions are very wet, the mycelium can grow on the surface in the form of a fluffy white felting (see p. 375). Note that diagnostic confusion on fruit can sometimes occur with aerial blight, especially for nonstaked crops (for the differences between these two mildews, see p. 375).

Botrytis cinerea is well known for its ghost spots on fruits. They are illustrated in Photo **808**, which shows the presence of whitish rings, several millimetres in diameter, distributed at different locations of the fruit. These are very superficial and inconsequential, and in fact correspond to abortive infections of this aerial fungus, with the mycelium from germinated spores failing to establish in the tissues. They mainly occur on adult green fruits under protection and sometimes in open fields in wet conditions. They can disappear at maturity. Their presence on the fruits often makes them unmarketable.

More information is available on pp. 181 and 303.

808 Whitish rings with a tiny necrotic central point, are scattered on this ripening fruit.
Botrytis cinerea

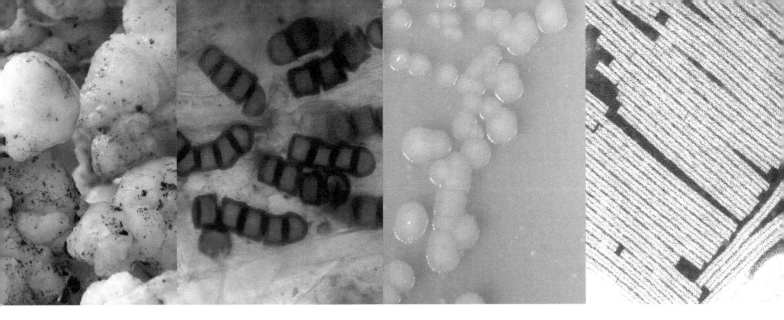

3

Principal characteristics of pathogenic agents and methods of control

FUNGI AND RELATED ORGANISMS

Phytopathogenic fungi are generally small and more complex organisms than bacteria. They are, sometimes visible to the naked eye but more often with the aid of a microscope. They are characterized by the formation of free or interlaced filaments (hyphae), all of which constitute the 'mycelium'. This can be coenocytic (without cell walls, as *Pythium* spp., *Phytophthora infestans*, and other *Phytophthora* spp.,[1] *Rhizopus stolonifer*), or with cell walls (*Alternaria tomatophila*, *Stemphylium* spp., *Botrytis cinerea*, *Pyrenochaeta lycopersici*, *Sclerotinia* spp., *Sclerotium rolfsii*, *Rhizoctonia solani*).

It should be noted that some aquatic fungi are devoid of mycelium, as for example *Olpidium brassicae*, a vector of viruses and associated with the roots of soil-less tomato cultures. They often form amoeboid or plasmodial structures, sporangia, and uni- or bi-flagellate zoospores.

On tomato, fungi also produce specialized structures, very resistant to adverse elements and perfectly adapted to their preservation. The most common are chlamydospores, thick-walled spores, for example *Thielaviopsis basicola*, *Fusarium oxysporum* f. sp. *lycopersici*, and *Fusarium oxysporum* f. sp *radicis-lycopersici*, *Pythium* spp. *Pythium*, like the *Phytophthora*, survives in the form of oospores which result from sexual reproduction. More visible, of variable size and equally effective for survival are sclerotia; these are dense mycelium masses, and are characteristic of certain fungi on and/or in affected plant tissues: *Sclerotinia sclerotiorum*, *Sclerotinia minor*, *Sclerotium rolfsii*, *Botrytis cinerea*, *Rhizoctonia solani*.

During their distribution, fungi come into contact with the different tomato plant parts. Fairly quickly, spores germinate and produce an attachment structure called an 'appressorium' which also serves to assist in the penetration of the host. Some fungi prefer younger tissues, others senescent organs. They also exhibit some specificity with respect to the part of the plant that is attacked. It is easy to distinguish between the fungi that mainly affect tomato roots and/or stem base (*Olpidium brassicae*, *Pythium* spp., some *Phytophthora* spp., *Fusarium oxysporum* f. sp. *radicis-lycopersici*, *Pyrenochaeta lycopersici*, *Thielaviopsis basicola*) or vessels (*F. oxysporum* f. sp. *lycopersici*, *Verticillium dahliae*.), while others are limited solely or mainly to aerial organs (leaves, stems, fruits, flowers) (*Phytophthora infestans*, *Leveillula taurica*, *Oidium neolycopersici*, *Stemphylium* spp., *Alternaria tomatophila*).

Penetration into tomato tissues can be done in different ways:
– directly through the cuticle and plant cell wall (*O. neolycopersici*, *Stemphylium* spp., etc.);
– through natural openings such as stomata (*P. infestans* in part, *L. taurica*);
– through wounds or senescent tissues always present on plant organs: the old leaves, dead flowers, pruning wounds (*B. cinerea*, *Sclerotinia sclerotiorum*, *S. minor*), breakage of epidermal hairs, sites where adventitious roots develop, nematode damage (*F. oxysporum* f. sp. *lycopersici*, *T. basicola*).

Once in their host, fungi put in place various parasitic processes that allow them to establish and invade, corresponding to the infection phase. Up to this point, they live on their reserves, but then begin to collect various substances from their host. Obtaining nutritious substances can be done without causing damage to the affected plant cells or, on the contrary, can lead very quickly to their destruction. In the first case, the fungi are completely dependent on living cells, and it is for this reason they cause minimal damage; they are called obligate parasites of fungi (*L. taurica*, *O. neolycopersici*). For example, in one of the tomato powdery mildews, *O. neolycopersici*, invasion of the host is very limited. Its mycelium remains outside the cells, but it has

---

[1] There are about 800 species of parasitic or saprophytic Oomycetes which have for a long time been classified among Phycomycetes or 'Lower fungi' (Eumycota). This classification was revised several years ago because the ultrastructure of these micro-organisms, their biochemistry, and molecular sequences indicated that they belonged to the Chromista, which includes mainly algae (green and brown), and diatoms. Currently, according to bibliographical sources, they can be associated either with the kingdom of Chromista (*Index Fungorum*) or to the kingdom of Stramenopila (*Tree of Life*).

particular structures, the haustoria (suckers), enabling it to take its nutrients from the cells without destroying them, at least initially. In the second case, fungi have significant saprophytic potential and the survival of their host is not essential for their own survival. These facultative parasites destroy the plant cells quickly, thanks to pectolytic enzymes (*B. cinerea*, *S. sclerotiorum*, *Sclerotium rolfsii*) and, occasionally, toxins causing necrosis.

In both situations, after having colonized the tissues of their host (invasion phase) and if weather conditions are favourable, the fungi form specialized structures that allow them to produce many spores (sporulation). These structures take different names according to their origin: asexual vegetative multiplication, termed the'anamorph form' (sporocystophores, conidiophores, pycnidia, acervuli) or sexual reproduction, known as the 'teleomorph form' (oospores, zygospores, cleistothecia, perithecia, apothecia). Frequently seen inside or on the surface of damaged tissues, the varied forms can identify the fungi, and have long been a very important factor in the fungi botanical classification. Some of them are important in survival, but their role is to produce spores (zoospores, sporangia, conidia, ascospores, basidiospores) to ensure their dissemination.

The incubation period is the time required between the penetration of the parasitic fungus in the plant and the first symptoms of the disease. The latency period also starts when the first infection takes place, but it ends when sporulation occurs.

The spores are then disseminated in several ways:
– wind or air currents that carry them directly or through soil dust from contaminated soil. Release occurs most frequently passively, depending on climatic conditions and especially the ambient humidity (*P. infestans*, *L. taurica*, *O. neolycopersici*, *B. cinerea*);
– by water carrying the fungal propagules or passively during streaming on and in the soil (*Pythium* spp., *Phytophthora* spp.), more actively following leaching, splashes occurring during rain or overhead irrigation (*P. infestans*, *Didymella lycopersici*, *Stemphylium* spp., etc.). Some less evolved fungi are specifically adapted to aquatic life. They have spores, referred to as 'zoospores', which are equipped with flagella to enable them to move easily in wet conditions (*Olpidium brassicae*, many species of *Pythium* and *Phytophthora*).

Both modes of transportation frequently act together. For example, it is not uncommon for small water droplets to release some conidia of a fungal plant pathogen present on the leaves of a diseased plant and for them to be carried away by the wind before being deposited a few metres away on healthy plant(s).

On the other hand, workers and their tools during farming operations and the technicians during their successive visits to producers, can easily spread the different propagules of fungi. These can also be dispersed by different vectors such as aerial insects or seeds (*Alternaria tomatophila*, *D. lycopersici*, *Septoria lycopersici*). These dissemination methods are quite exceptional among pathogenic fungi in tomato.

The development of fungi is influenced by climatic conditions. Temperature may not be the most important factor, but like other plants, fungi require minimum temperatures for growth. During the winter, temperatures are often too low, and fungi reduce or cease their activity. It restarts as soon as temperatures rise and stay within a fairly wide temperature range, between a few degrees and 30°C and more. The temperature may intervene indirectly by weakening the plants (some frozen organs), or by amplifying some of the symptoms (increased wilting intensity in the case of root or vascular diseases).

The overriding factor is certainly the humidity, which is affected by various factors such as rain, irrigation, dew, fog. It is essential to almost all stages of development of aerial fungi, notably spore germination and germ tube penetration into the host. The germination of spores frequently requires a relative humidity above 95%, and in some cases, the presence of free water on the leaves. Spore formation, longevity, and their release are also heavily conditioned by humidity.

The vast majority of fungi like moist environments, whether on or around the foliage or the roots. It is now well accepted that plants in waterlogged soils are very prone to attack by soil fungi such as *T. basicola*, and many species of *Pythium* and *Phytophthora*. The roots of plants growing in these asphyxiating conditions are much more sensitive.

Some rare fungi prefer drier environments, at least at certain times of their cycle.

Thus, the wind contributes to the release and dissemination of spores. It can however, sometimes have a negative effect on the development of a fungal infection by reducing the ambient humidity. The effects of light are generally not significant, but low light can lead to the plants withering, making them more susceptible to certain fungi such as *Botrytis cinerea* in protected nurseries.

Soil pH and plant nutrition also have some influence on the development of certain fungi. Their effects are not always easy to identify.

⚠️ Note to the reader

To enable readers to benefit from the most effective control measures, various methods of control, including the use of fungicides, are listed. Some of these recommendations have originated in particular countries and may not have universal application. Recommendations may need to be modified depending on the particular country and the local legislation on pesticides and the use of other products. It is therefore vital that readers check their local regulations before selecting a fungicide, making absolutely certain that the product selected is approved for use in their country. These remarks are also valid for all other pesticides mentioned in this book and also for those products that are organic or based on micro-organisms or natural substances.

Note that enhanced pesticide degradation and pesticide resistance may affect the efficacy of some products. For instance, decreased efficacy of metam-sodium and dazomet has been found in some soils in Israel due to an accelerated degradation of methyl isothiocyanate. Similarly, failure to control *Botrytis cinerea* with some products worldwide may be due to the occurrence of fungicide-resistant isolates. Similar situations may occur with other pathogens and products.

■ Air-borne fungi

Alternaria tomatophila E.G. Simmons (2000)

syn. *Alternaria solani* Sorauer

Alternaria early blight

(Anamorph of *Lewia*, Pleosporaceae, Pleosporales, Pleosporomycetidae, Dothideomycetes, Ascomycota, Fungi)

Principal characteristics

Alternaria solani, reported for several decades on the Solanaceae, has long been described as affecting tomato, eggplant, potato, and several weeds of this botanical family.

In fact, colonization by *Alternaria* spp. on these plants is much more complex. Thus, several species of *Alternaria* are found on more than 60 members of the Solanaceae. Moreover, the name '*Alternaria solani*' should be reconsidered, at least on tomato. Indeed, it appears that, on this plant, another species, morphologically very similar to *A. solani*, called '*Alternaria tomatophila*'[1] is prevalent. In addition, two phenotypes exist within this species, distinguishable by the appearance of their colonies on agar in Petri dishes: a clear phenotype, more aggressive on tomato, and a dark one.

A. solani may be the pathogenic agent of potatoes, and *A. beringelae* Simmons may colonize eggplant.

It should be noted that *A. subcylindrica* Simmons & Roberts was occasionally observed on cherry tomato leaves, and *A. cretica* Simmons & Vakalounakis was observed in Greece associated with classic *Alternaria* leaf lesions. Note that *A. subtropica* Simmons can cause spots on fruit.

Among the other *Alternaria* sp. attacking tomato, are described *A. alternata* f. sp. *lycopersici* (see Description 9), *A. alternata*, and *A. tomato* (Cooke) Jones (see p. 393).

• **Frequency and extent of damage**

Among tomato air-borne diseases, Alternaria blight is certainly one of the most common and most prevalent in the world: it is found on every continent, wherever these tomatoes are cultivated. It is found in many climatic zones, in tropical, sub-tropical, and temperate production areas. The presence of dew in semi-arid regions allows its development. It mainly affects field crops, and sometimes those in cold glasshouses. It is also often present in many gardens of amateurs. Its damage can be substantial if wet weather persists and/or if no method of protection is practised. It sometimes leads to significant defoliation causing not only a reduction in yields, but also many lesions on fruits due to the sun effects on them as they are less protected by the foliage (see p. 385).

In some countries this disease is common in field crops and is found occasionally in crops in plastic tunnels, but never in heated glasshouse crops.

• **Main symptoms**

A. tomatophila attacks all aerial parts of the tomato plant at all stages of growth.

It affects young seedlings, before and after transplanting, in wet and cold extensive nurseries. In addition to seedling blight, the fungus is the cause of sometimes extensive

[1] Simmons E.G., 2000. *Alternaria* Themes and Variations (244–286): Species on Solanaceae, *Mycotaxon*, **75**, 1–115.

black symptoms on the stem, located near the stem base or higher up. Having surrounded the stem, the lesions can quickly lead to the plants drying and to their death. Attacks at this stage of plant development are due to seed and/or nursery soil contamination.

On mature plants, it mainly causes leaf spots mostly developing on lower and mature leaves and then progressing to the apex. These spots are initially dark green, then quickly become brown to black. They have a rounded shape, sometimes angular when delineated by the veins. Their diameter is often about 10 mm, and can reach several centimetres when climatic conditions are particularly favourable and/or when the spots coalesce. These have the characteristic of presenting discrete concentric patterns, giving them a target appearance. They are located throughout the leaf, including at its periphery. In the latter case, they do not always show concentric rings. A bright yellow halo (due to the effects of a toxin) is sometimes seen around these spots that eventually become necrotic. When they are numerous and with very favourable climatic conditions, entire leaflets may turn yellow and dry out, sometimes leading to significant defoliation of plants.

Similar spots are also observed on stems, petioles, and peduncles; they have a blackish colour and often have a more elongated shape. These lesions may surround the affected organ, causing its distal part to die.

Well defined concave spots appear on fruits near the peduncular scar and sepals. The latter, initially colonized, are often affected. The surface of the affected areas of fruits will shrivel and may be covered with a fairly characteristic black smooth mould, with concentric patterns, as on the leaves. Green and ripe fruits can be affected, the over-ripe ones are little or not affected. Note that contamination can occur through wounds such as growth cracks. Lesions on fruits can cause them to fall in large quantities. Heavy attacks of blight can also cause the reduction of the size and number of fruits produced.

Several tomato aerial pests can cause symptoms on leaflets which can be confused with those caused by A. tomatophila, notably Cristulariella moricola (Hino) Redhead (syn. C. pyramidalis Waterman & Marshall), a rare fungus on tomato causing large concentric spots (zonate leaf spots).

(See Photos **296**, **299–308**, **581**, **778**.)

• Biology, epidemiology

Survival, inoculum sources: A. tomatophila can survive for several years on the surface of tomato seeds, in the soil and on plant debris, as melanized mycelium, its conidia, and its chlamydospores. In some regions, it may also be able to sustain itself from one season to another on other Solanaceous plants such as potato, eggplant, pepper, black nightshade (Solanum nigrum), S. carolinense, S. pseudocapsicum.

Penetration and invasion: once the multicellular spores come into contact with the tomato cells, germination can occur in 2 hours in water at temperatures between 6 and 34°C. The fungus enters the tissues either directly through the cuticle or through stomata or various wounds. It rapidly invades the tissues, and lesions begin to be visible 2–3 days after the initial infection.

Sporulation and dissemination: on the colonized tissues, if the climatic conditions are humid, A. tomatophila soon produces short conidiophores topped by long multicellular club-shaped conidia (Photo **809**). These are spread by wind, but also by rain and after spray watering. The presence of water is necessary for sporulation to occur. Seeds, workers (especially through their tools) also contribute to the spread of the pathogen. The conidia produced ensure secondary spread and several parasitic cycles can occur in the crop.

Conditions for its development: this blight is favoured by high humidity levels and temperatures between 18 and 30°C. Dews, low ongoing rainfall (5 mm) or sprinkler irrigation are sufficient for its development but these must be repeated for the disease to develop rapidly. Stressed or poorly manured plants or plants

⚠ It may be appropriate to reconsider the potential hosts of A. tomatophila that are not necessarily those of A. solani. It is possible that these two species have different host spectra.

laden with fruit are more sensitive. The disease never develops very rapidly, but progressively becomes more pronounced over time, as and when the plants age. It becomes severe in late season. Under favourable conditions, the symptoms are quite noticeable in 5–7 days.

Control methods[1]

• During cultivation

Once the first symptoms are observed during cultivation, the plants should be treated as quickly as possible using one of the following fungicides: mancozeb, maneb, chlorothalonil, iprodione, difenoconazole, cymoxanil + famoxadone, azoxystrobin, thiophanate methyl, famoxadone, pyrachlostrobine. Applications must be repeated after heavy rain and sprinkler irrigation. Note that these promote the spread of the disease, therefore must be conducted during the day at a time that will allow plants to dry and not to stay moistened for too long.

It is also necessary to avoid plant stress, and to provide them with balanced fertilizer, especially nitrogen.

Wherever possible, the maximum amount of plant debris (leaves, stems, and especially fruit) should be removed during and at the end of culture, and destroyed. Some authors advise deeply burying debris.

• Subsequent crop

During the next crop and if the crop is in the same location as the previous year, soil disinfection or preventive applications of fungicides must be applied. Healthy seed must be used. In case of doubt, uncoated seeds can be treated in hot water or with chlorothalonil, iprodione, or thiram.

Long enough crop rotation, of 3–4 years, involving small grain cereals, maize etc, should be considered in the most severely affected areas. It will be necessary to eliminate weeds that can be hosts.

In addition, one should avoid establishing a culture close to already affected tomato crops, or near other sensitive crops such as potato, eggplant. Unbalanced fertilizers may promote blight, particularly excess or lack of nitrogen. Those well supplied with potassium may have the opposite effect.

In some countries, bed soil disinfection with steam or with a fumigant is recommended.

Seedling quality will be monitored prior to use. It is also recommended to have the crop at a density that allows proper ventilation of the vegetation, good drying after rainfall or sprinkler irrigation, and to produce vigorous plants. Avoid planting in waterlogged areas. It is better to irrigate the plants by drip rather than spray. As far as possible the tomato plants must be staked or grown on frames so that the foliage and fruit remain off the ground; ground mulching has the same effect by forming a mechanical barrier. The material used for staking should be disinfected between each crop.

Fungicide treatments, as preventative measures, should be applied with the previously mentioned active ingredients. It is worth noting the existence of several decision making systems which can be used to determine the periods of applications of fungicides, improve their efficiency, and reduce their number. These prediction models, now well known, are NJ-Fast, Fast-CU, and especially Tom-Cast. The advantage of these tools is well established particularly where the principles of sustainable tomato protection are being applied.

Sodium bicarbonate sprayed on tomatoes may reduce the germination of *A. tomatophila* conidia and help to control early blight. Other unconventional products have been used to help control with varying degrees of suc-

[1] See note to the reader p. 417.

cess. These include, various plant extracts, vegetable oils (*Acacia concinna*, *Bassia latifolia* [syn. *Madhuca longifolia* var. '*Latifolia*'], garlic, *Azadirachta indica*) which may limit the growth of this fungus.

Several potential antagonists are reported in the literature: some bacteria (*Pseudomonas fluorescens*, *Bacillus* spp.) as well as some fungi (*Trichoderma polysporum*, *Chaetomium globosum*); their performances has not proved to be significant in situations of natural epidemics.

Sources of resistance were found in several wild species, *Lycopersicon hirsutum* and especially *L. pimpinellifolium*. They have been used in breeding programmes and have resulted in lines and cultivars with a polygenic partial resistance to *Alternaria* and also relatively late maturity.

Lower sensitivity of the stem and stem base to Alternaria blight has been reported in other wild species without having been used effectively. Moreover in certain crop contexts in India, a number of tomato varieties ('Ace', 'Flora-Dade') are less susceptible to this disease.

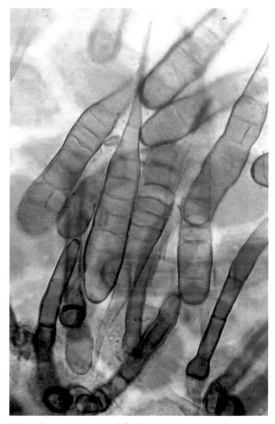

809 *Alternaria tomatophila* has a septate mycelium which melanizes progressively with age. It produces short septate and brown conidiophores on which only a single conidium often forms. Conidia are brown, multicellular and very elongated. They have a long hyaline appendage (beak), sometimes forked and longer than the body of the spore, which measures between 120 and 300 µm in length (from base to tip of beak).

810 Grown on a malt-agar, this fungus produces a fast growing colony that takes on a reddish brown hue as does the underlying agar.

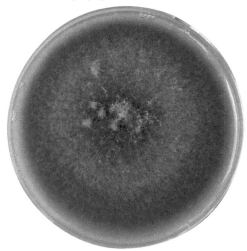

Botrytis cinerea Pers. (1794)

Botrytis grey mould

(Anamorph of *Botryotinia fuckeliana*
[De Bary] Whetzel, Sclerotiniaceae,
Helotiales, Leotiomycetidae, Leotiomycetes, Ascomycota,
Fungi)

Principal characteristics

The biological variability of this fungus is still unknown. It should be noted that strains resistant to several fungicides have been reported in many countries.

Recent molecular biology studies have revealed a genetic variability within the B. *cinerea* species. This fungus may actually be a species combination which can be divided into two groups. Group I, or 'pseudo-cinerea' sub-population, is distinguished by its resistance to the fenhexamid fungicide and has one of the two alleles of the '*Bc-hch*' gene of vegetative incompatibility. Group II consists of vacuma and transposa strains. Transposa sub-population has in its genome two active transposons, Boty and Flipper, while these two transposons are absent or inactive among vacuma and pseudo-cinerea sub-populations.

Note that some specialization has been demonstrated in Canada from 91 strains of B. *cinerea* from tomato, primarily based on their glasshouse origin, and not on their geographical origin or their isolation host.

• Frequency and extent of damage

This fungus, ubiquitous and highly polyphagous, is observed on tomato in virtually all areas of production in the world, where it can locally cause significant damage. It can be rampant in open fields as well as in protected crops, but it is particularly feared in this latter context, especially for the cankers it causes on the stems. These inevitably surround the stem and cause the withering of distal parts often affecting many plants. In some glasshouses, up to 25% of plants may be affected. In these circumstances, it becomes difficult to manage the plants' irrigation in slabs where plants are missing. Rots on fruits are also very damaging, in protected and open field crops, during cultivation and post harvest (during storage, transportation and marketing).

B. *cinerea* is rife in open field and under protection, but it is in the latter case that the attacks are often the most severe as the environment there is generally more humid. In addition to climatic characteristics, culture duration in these production systems is often very long, sometimes almost the entire year. Under these conditions, plants have many leafing and budding wounds, very conducive to the development of this opportunistic fungus. In addition, it easily colonizes weakened or senescent tissues (constituting nutrient bases), which are present in such crops. They allow it to be sustained on the plants and to multiply in this environment.

• Main symptoms

All aerial parts of tomato can be affected at any stage of its development.

B. *cinerea* can occur in the nursery, leading to gaps at the germination stage or to damping-off.

Senescent cotyledons often allow it to colonize the young stems; stem cankers can develop and then affect the development of young plants in the weeks after planting. The leaflets often display circular and wet spots at first. Beige to light brown, they also have a wizened appearance and reveal concentric arcs. They evolve rapidly and cause the drying out of large sectors of the leaf. This may lead to rot, which then spreads and eventually damages entire leaves, with gradual necrosis and tissues collapsing.

Once the petioles are reached, the rot will develop on the stem.

When the stem is affected, whether on seedling as stated above or on an older plant, one or more cankerous lesions are seen essentially starting from pruning and budding wounds. These cankers are wet at first and tend to dry with age. With well defined contours, they are a beige to brown colour; they eventually surround the stems and may be several centimetres in length. Subsequently, the distal leaves turn yellow, wilt, wither, and die.

Senescent petals are particularly vulnerable. They allow *B. cinerea* to settle on the inflorescences and to rot them, but also to use them as nutrient base in the first instance, to ensure subsequent infections, notably on leaflets and fruit.

The green fruits, rarely the mature tomatoes, show thin whitish rings, 2–10 mm in diameter, encircling a small central necrotic lesion. These rings, known as 'ghost spots', are yellowish on ripe fruit. They are due to a reaction of the very young fruits to aborted infection, after spores have germinated and the germ filaments have penetrated into the tissues.

Soft and wet rot may appear at the fruit peduncular scar and to a lesser degree at the end. The lesions are circular and whitish, pale brown to grey; often tissues eventually collapse. Fruits can also be mummified.

Whatever the affected organs, dying tissues are covered with a very characteristic dense grey mould, consisting of conidiophores and conidia of the fungus. *B. cinerea* can produce black sclerotia 2–5 mm in diameter, rarely visible on lesions.

(See Photos **239, 298, 314–322, 521, 548, 559, 563, 568–576, 780, 781, 808**.)

• **Biology, epidemiology**

Survival, inoculum sources: the fungus is sometimes found on seeds. It is able to remain in the soil on diverse plant debris and in many forms: conidia (Photo **811**), mycelium, or sclerotia (Photo **812**). They persist in soil for several years. The saprophytic potential of *B. cinerea* allows it to survive easily on organic matter. It is polyphagous and able to attack and colonize hundreds of cultivated plants or weeds and these contribute to its survival and constitute potential sources of inoculum, as is the case with this fungus in most vegetable crops. On these plants it may produce sclerotia as well as sporulating abundantly. The primary infections are thus often aerial and in this case involve conidia which are very easily wind transported. These spores germinate in a few hours (5 hours at 20°C) on wet leaves and/or in the presence of an ambient humidity of at least 95%. Conidia germination is adversely affected by temperatures over 30°C.

Penetration and invasion: once the germ tube has formed, it enters the tissues and leads to the mycelium destroying the cell walls and their contents. Penetration occurs either directly through the cuticle, or from various wounds, especially on the stem through side-shooting and leafing wounds. Infections occur after approximately 15 hours. *B. cinerea* spores also germinate and establish on trusses stalks and wounds which have resulted from their removal. It can also invade all senescent, necrotic and/or dead tissues such as petals, necrotic sepals, old leaflets. It sometimes colonizes tissues already affected by other pathogens or pests. It rapidly spreads to tissues and causes rots in a few days due to hydrolysis of proteins.

Sporulation and dissemination: on all its hosts as on plant debris, it produces mycelium and numerous long, branched conidiophores. At their ends ovoid to spherical conidia develop that ensure the dissemination of *B. cinerea* (Photo **811**). Sporulation may begin 3 days after the initial contamination. Dissemination occurs primarily through wind and air currents and to a lesser degree through rain and splashing water. Farm workers during cultural operations are also a means of dissemination. The mycelium extends in the affected tissue to the point between obviously diseased and healthy tissues. *B. cinerea* may eventually produce on damaged tissues small sclerotia that also allow its survival. Under favourable conditions, the duration of a cycle is quite short, about 4 days.

Conditions encouraging development: as with many air-borne fungi, it develops well in humid environments. Relative humidity around 95% and temperatures between 17 and 23°C are very conducive conditions. These parameters are found especially under protection, but also in the field during rainy periods or after sprinkler irrigation. Etiolated or too vigorous plants are particularly vulnerable. This happens when plastic tunnels are closed to conserve heat: plants then have a tendency to grow rapidly. Their vegetation becomes quite lush with succulent tissues which are very sensitive to *B. cinerea*. Note that attacks of this fungus are particularly severe in the earliest crops, in new glasshouses, often better sealed and equipped with a thermal screen, and polythene tunnels. In the latter, the type and quality of the plastic cover can influence the development of certain diseases, especially *B. cinerea*. Indeed, less damage was observed in EVA compared to PVC and polyethylene. Under EVA, it seems that light transmission is better, and a lower humidity is present with less formation of water droplets on the walls. Finally, covers sometimes used to protect plants from insects cause a rise in humidity, aggravating the damage.

Note that some varieties are more susceptible than others to *Botrytis*.

Control methods[1]

• During cultivation

Control of *B. cinerea* is still difficult in many countries of the world, particularly under protection. This situation has several explanations:
– A climate favourable to the fungus development;
– Particularly susceptible plants, with plants more succulent and tender than in the open field and numerous wounds caused by pruning and side-shooting;
– A special ability of this fungus to adapt rapidly to fungicides used for *Botrytis* control, and a too limited choice of active ingredients to allow alternating programmes.
– Few available fungicides with different modes of action (some new ones not yet approved and others withdrawn from the market following the recent EU legislation on pesticides).

When symptoms of *B. cinerea* occur in the crop and if a protective spray programme has not been established, a registered fungicide should be applied. Several active ingredients are used around the world, alone or in combinations, to control *B. cinerea*. We have included the large majority in *Table 40*.

As shown in *Table 40* (overleaf), the fungus is resistant to several families of fungicides. The genetic variability of *B. cinerea* has enabled it to adapt to many fungicides that were used against it. For example, dicarboximides, very powerful in the early days of their use, quickly led to the selection of resistant strains (nonpersistent), which are now very common in many

[1] See note to the reader p. 417.

Table 40 Main fungicides used worldwide to control *Botrytis cinerea* on tomato

Mode of action	Chemical family	Active ingredient
Multisite (respiratory processes and cellular energy production)	Dithiocarbamates	Thiram or TMTD
Cell division and microtubules	Phenylcarbamate	Diethofencarb
Sterol biosynthesis	Benzimidazoles[r]	Carbendazim, benomyl
	Hydroxyanilides[r]	Fenhexamid
Carbohydrate and polyols metabolism	Dicarboximides[r]	Iprodione
		Procymidone, vinclozolin
Amino acid and protein biosynthesis	Anilinopyrimidines[r]	Pyrimethanil
Multisite (respiratory processes and cellular energy production)	Sulfonamides	Tolyfluanid

[r] Chemical family mainly concerned with the phenomena of resistance.

crops. Note that currently there are still in the field strains tolerant to fungicides of the benzimidazole family, used at one time against *B. cinerea*. Unfortunately, a similar situation has been seen during the intensive use of almost all the latest anti-*Botrytis* fungicides.

Therefore, it is strongly recommended to alternate chemical groups to avoid the development of resistant strains.

Treatment can be applied following leaf removal to protect the pruning wounds. Also, when a canker develops at a wound site, the lesion may be cut out to rid the stem of diseased tissues and a thick fungicide mixture applied with care to cover the wound.

Similar prophylactic measures involving chemi-cal control will also be required in nurseries and also in field crops

Plastic tunnels must be ventilated to the maximum, to reduce the ambient humidity and, in particular, to avoid the presence of free water on plants. In periods of overcast and damp weather, tunnel ventilation and heating will ensure humidity control: achieving progressively a temperature of about 17°C at sunrise will prevent condensation on the plants. The use of aeration should be done sparingly, however, when outside air temperature is below 12°C.

Pruning should be done carefully: removing side-shoots, leaves, and stems of trusses flush with the stems keeping them as smooth as possible. The early removal of sideshoots will avoid large wounds. It should be kept in mind that the presence of long snags greatly increases the risk of disease. The pruning will be done in the morning because at that time, tissues are more turgid and are easier to cut and the wounds have more time to dry during the day. It is not advisable to prune during overcast and humid periods. Note that stem wounds, although apparently dry, remain susceptible to *B. cinerea* for some time.

Young cankers will be removed to prevent further colonization by this fungus. Wounds should be protected by a fungicide painted or stems protected by preventive spray treatments particularly directed at the lower part of the stem. The thinning out of the lower foliage promotes fruit ripening and simplifies the harvest, helping to improve the microclimate by improving air circulation in the crop.

Under protection, irrigation will only take place when the plants are able to evaporate.

Water retention in tomato leads to more vegetative plants with waterlogged and succulent tissues, and therefore more susceptible to *B. cinerea*. This type of lush plant also induces a more humid climate, and therefore more favourable to the fungus. In the field, sprinkler irrigation preferably applied early in the morning – and never at night – will allow plants to dry as quickly as possible.

Avoid any stress to the plants, leading to growth spurts. Nitrogen fertilization should be controlled. It should be well balanced so that it does not cause very susceptible succulent tissues, nor too low, causing chlorotic leaves which are ideal nutrient bases for *B. cinerea*.

Conventionally, plant pathologists advise the elimination of plant debris during cropping, particularly dying plants with stem canker(s) and rotten fruits on which *B. cinerea* sporulates.

It is also desirable to have the same requirements for removed leaves and sideshoots. In practice these are often left on glasshouse soil without necessarily increasing *Botrytis* damage. If this cultivation method is chosen, soil should be kept dry so the leaves do not rot. This option is only valid under protection. In all cases, ripe fruits should be harvested as quickly as possible. At the end of the crop, all debris must be eliminated from the glasshouses and open field plots as soon as possible. In the latter context, it will avoid debris being subsequently buried in the soil and the fungus remaining in it. Deep ploughing will facilitate the decomposition of the few remains.

• **Subsequent crop**

Obviously, the measures recommended during cultivation will be systematically applied during the next crop. They may be supplemented by the following operations.

Where propagation is done annually in the same place and/or in the same glasshouse, it will be necessary to implement the measures recommended on p. 541.

B. cinerea is polyphagous and saprophytic and the fact that the inoculum can come from the crop environment makes crop rotations ineffective. In the field, the ground for future crops should be well prepared and drained to prevent water puddles which induce *B. cinerea* attacks on leaves in contact with the ground. The lines of planting should be directed where possible in the direction of prevailing winds so that the leaf canopy is well ventilated.

Avoid setting up the culture with too high densities and damaging the plants. Under protection, it will be desirable to provide a crawl space, to replace the plastic that may be used to cover the soil, and to disinfect the glasshouse carefully (see p. 436).

Although the nature of the substrate does not have a direct role in the development of *B. cinerea*, it must be remembered that the organic substrates lead to more vegetative plants and therefore greater susceptibility to *Botrytis*.

On fertilization, it has been demonstrated that replacement of nitrates by chlorides reduces very significantly the damage by *B. cinerea* in soil-less culture in glasshouses. Furthermore, increased calcium in tomato tissues greatly reduces its sensitivity to this fungus.

In the field, a localized drip irrigation system should be used, to avoid wetting the vegetation as is the case when spray irrigation is used.

Preventive fungicide treatments are often required under protection at certain times of the year to avoid the development of the fungus. These can be applied during the first de-leafing operation preferably on the same day. Note that it appears that the lower stem is more sensitive to attack by *B. cinerea*. For this reason the first few centimetres of stem are particularly vulnerable and should be protected.

Other diseases and pests should be controlled because they are the cause of wounds and tissue necrosis conducive to *B. cinerea*. Once vegetation becomes significant, vigilance is necessary, especially on overcast days.

Sources of resistance have been identified in different countries among the wild relatives of tomato, notably *Solanum lycopersicoides*, *Lycopersicon peruvianum*, and *L. hirsutum*. The latter species appears to be of particular interest to breeders.

A model based on weather forecasts and named 'Botman' is being developed in Israel. This model, coupled with the use of a biological agent, *Trichoderma harzianum*, allows the control of *B. cinerea* with weekly fungicide applications.

A number of alternative and/or original methods have been or are being tested in several countries. Thus, composting extracts have been used in the UK; sprayed on the plants, they may reduce the damage caused by *B. cinerea* and increase yields. Antioxidants have also been employed in Israel to limit the development of this fungus. Essential oils of lemon have also been tested.

Bacteria and several fungal antagonists have been evaluated *in vitro* or *in vivo* to control this fungus, including *Streptomyces* spp., *S. griseoviridis*, *Bacillus licheniformis*, *Candida guilliermondii*, *Cryptococcus albidus*, *Rhodotorula glutinis*, *Aureobasidium pullulans*, *Gliocladium virens*, *G. catenulatum*, *Trichoderma harzianum*, *Ulocladium atrum*, *Chaetomium globosum*, *Clonostachys rosae*, and *Rhodosporidium diobovatum*.

Please note that biological preparations based on *Bacillus subtilis* and an extract of *Reynoutria sachalinensis* showed some effectiveness in controlling *B. cinerea* in protected crops in the US. The combination of yeast (*Cryptococcus laurentii*) and baking soda reduces the effects of this fungus after harvest. Finally, note that the fungus *Microdochium dimerum* can successfully limit the development of *B. cinerea*, in particular on the pruning wounds of plants grown under protection.

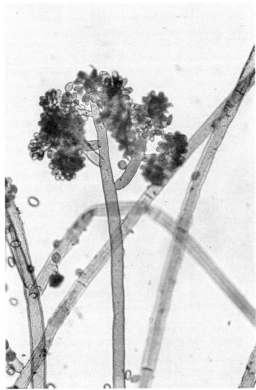

811 *Botrytis cinerea* produces strong and long conidiophores, irregularly branched, that melanize progressively from the base. Conidia are formed on conidiophores at the end of sterigmata. They are unicellular, ovoid to elliptical and hyaline to slightly pigmented (6–18 × 4–11 mm).

812 Its sclerotia are irregular in size and are rather flat.

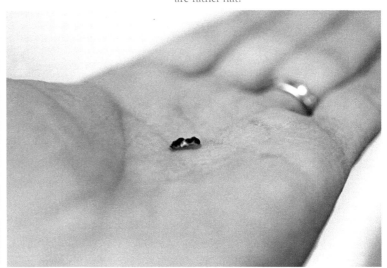

813 On artificial media, *Botrytis cinerea* mycelial colonies have a variable appearance depending on the strains. Some only form sclerotia and/or numerous conidiophores, others are sterile. In this case, the strain grown on malt-agar medium has strongly sporulated. The characteristic grey mould of the fungus is clearly visible.

Didymella lycopersici (Fuckel) Rehm (1881)

Syn.: *Phoma lycopersici* (Plowr.) Jacz., 1898.

(Incertae sedis, Pleosporales, Pleosporomycetidae, Dothideomycetes, Ascomycota, Fungi)

Didymella stem canker and fruit rot

Principal characteristics

Several Phoma species have been described on the Solanaceae in the world. On tomato both *P. glomerata* (Corda) Wollennw. & Hochapf. and *P. exigua* Sacc. 1879 occur, although both are polyphagous and also affect plants in other families. Two other species, *P. destructiva* Plowr. and *P. lycopersici*, also affect *Lycopersicon esculentum*. They are more specialized affecting only one or a few hosts in the Solanaceae. They are morphologically very similar and their respective parasitic specificities debatable. These two species and the *exigua* species can be differentiated by their morphological characteristics, notably those of their colonies on potato dextrose agar (PDA).

• Frequency and extent of damage

Given the problems in identifying *Phoma* species, it is quite difficult to know precisely the impact of *D. lycopersici* in the world. It has been reported in many countries, particularly in Europe (UK, Germany, France, Netherlands, Italy, Spain, Norway), where it can attack field and protected crops, in soil and soil-less systems. It is also described in New Zealand, India, Africa, and particularly in Morocco, in several American countries (Mexico, Canada, US), and Asia.

The severity of *D. lycopersici* infection in crops varies widely from country to country depending on the production system. Often sporadic, it can be particularly severe where the disease is well established and can cause significant damage to tomato fruits and stems.

• Main symptoms

D. lycopersici is best known for the damage it causes on the stems and to a lesser extent, the fruits of tomatoes.

Cankers, with slightly recessed tissues, may appear at different stem levels: on the lower part, at soil surface level, or just below (blackfoot), or higher up at leafing and sideshoot wounds. Cankers are moist and dark brown. The epidermis and cortex gradually decompose and xylem tissues turn brown. These cankers spread and gradually surround the stem and/or petioles, thus disrupting the sap flow. Ultimately, yellowing, wilting, and drying of the leaflets affected by stem or petiole lesions occurs. When a canker has surrounded the stem base, the whole plant may die.

Symptoms also develop on the fruits, often at the peduncular scar; they are wet, black, and develop quickly. Ultimately, large spots with concentric rings cover the fruit. These may gradually mummify or fall. Symptoms on fruits also appear during storage and marketing.

On the leaflets, brown and wet spots appear, mainly localized at the periphery of the leaf, and develop quickly, sometimes covering large areas. Concentric brown arc patterns are clearly visible on the damaged tissues. These spots, often with chlorosis at the edges, eventually become necrotic and dry up. Decomposed tissues may fall out. The flowers may also be attacked and destroyed.

On all affected organs, tiny brown to black globular structures (distinguishable to the naked eye but more easily so with a magnifying glass) dot the damaged tissues. They are abundant on fruits, more sparse on stems, and rare on leaflets.

(See Photos **323–327**, **434**, **536–538**, **564**, **585–588**, **777**.)

• Biology, epidemiology

Survival, inoculum sources: D. *lycopersici* survives between tomato crops on plant debris and organic matter in the soil. In the soil, survival is increased in the presence of moisture, organic matter, and relatively low temperatures. It is found on seeds, as mycelium and pycnidia. Survival on the Solanaceae is quite controversial: eggplant, potato, pepper, and black nightshade show some sensitivity after artificial inoculation, which would make these plants alternative hosts ensuring this fungus survives. It is also found in the plant environment in glasshouses and/or on the equipment (poles, stakes, pots) used for cultivation which has been in contact with diseased plants. Pycnidia are mostly the source of primary infection through conidia; the role of perithecia and ascospores seems much more limited.

Penetration and invasion: the first infections occur after a wet period, directly through the cuticle, through stomata or wounds. The fungus then quickly invades the tissues.

Sporulation and dissemination: two types of sporing structures are formed in the tissues, both brownish to black. The first and most frequent, are sub-epidermal ostiolate pycnidia (anamorph of the fungus, see Photo **814**) and are the source of colourless conidia, mostly unicellular but also bicellular (Photo **815**). Perithecia of the teleomorph are rarely formed in nature (diameter 120–210 µm). These give rise to hyaline and bicellular ascospores ($5.5–6.5 \times 16–18\,\mu m$). The dissemination of this fungus occurs primarily through its conidia which are dispersed by splashing during rain or overhead irrigation. They are also dispersed through the tools, hands, and clothes of the workers during cultural operations. When perithecia are formed, ascospores are spread by wind. Seeds may also contribute to survival and dissemination of the fungus.

Conditions encouraging its development: humidity is the factor that most influences the development of D. *lycopersici*. Indeed, the disease is particularly damaging as a result of rain, prolonged morning dew, and spraying. The stagnant water on the leaves stimulates the germination of spores and the penetration of germ tubes occurs over the temperature range of 13– 29°C, the optimum being around 20°C. It appears to be most virulent on old plants and those that have received reduced nitrogen or potassium fertilizers. Its growth is substantially reduced at 30°C.

Control methods[1]

• During cultivation

When the first fruit spots or cankers are seen affecting a limited number of plants, it may be wise to remove the latter carefully by putting them in a plastic bag prior to removal from the crop.

During cultivation, fungicide treatments can be applied with the following products: mancozeb, chlorothalonil, hexaconazole, difenoconazole, carbendazim associated with iprodione, and procymidone. Strains tolerant to benomyl and carbendazim are reported in the literature.

Sprinkler irrigations that promote germination, development, and sporulation of D. lycopersici should be avoided.

Plant debris should be removed from plots and destroyed. They should never be buried in the soil.

• Subsequent crop

French seeds and seedlings are normally free of contamination. In countries where health quality is suspect, be careful to use seeds free of contamination.

It will be necessary to disinfect nurseries carefully where seedlings are produced, especially when previous crops on the site have been chronically affected by D. lycopersici.

When possible (especially in open field), crop rotations over several years should be considered. One should of course avoid alternating tomato with other Solanaceous crops, or leave weeds from the Solanaceae on the site and surrounding area, particularly black nightshade which should be fairly routinely destroyed.

Contaminated soils may be subject to disinfection. The method used will vary by country and crop types: steam, fumigants (dazomet, metam-sodium), fungicides (maneb, thiophanate-methyl), solarization (see p. 487). The latter method, used especially in Sicily and Morocco, seems particularly effective in significantly reducing the amount of inoculum in the soil surface. Note that rootstocks from inter-specific crossing are resistant to D. lycopersici. Grafting may be of interest if the attacks are mainly located at the stem base.

The glasshouse and equipment that has been in contact with diseased plants should be thoroughly disinfected (bleach, fumigants, various products, see p. 436). Stakes may be solarized. They will be covered with a transparent polyethylene film and exposed in this way to summer solar radiation. The high temperatures obtained under the film will destroy a high proportion of the contaminating inoculum, the fungus being totally eliminated at more than 50°C.

Treatment with the fungicides recommended above can be made preventively, especially after pruning in wet periods. Pruning cuts can be protected with a fungicide mixture in water or mineral oil. Liprodione and fenarimol are useful for this purpose.

Sprinkler irrigation should be completed at times of the day that allows rapid drying of the vegetation. However, drip irrigation is preferable.

If no varieties currently marketed are resistant to this fungus, Lycopersicon hirsutum has a high resistance to D. lycopersici. For this reason inter-specific hybrids between KNVF-type rootstock and tomato are resistant to this fungus. The resistance may be dominant and controlled by more than one gene.

Note that some attempts at biological control of D. lycopersici by Trichoderma harzianum have been attempted with varying degrees of success.

---

[1] See note to the reader p. 417.

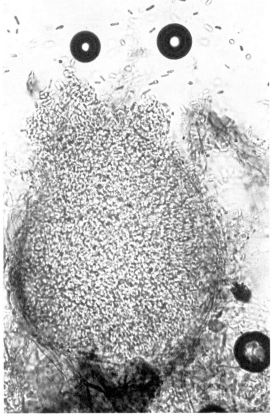

814 This young pycnidia of *Didymella lycopersici*, subglobose to elongated (100–270 μm in diameter), will gradually melanize. Numerous conidia are visible around it. This fungus can form perithecia (120–150 × 100 μm) with bicellular ascospores (16–18 × 5–6.5 μm).

815 The pycnospores are smooth and hyaline, oblong to ellipsoid or obovoid, uni- or bicellular. Their dimensions vary (3–11 × 2–4.5 μm).

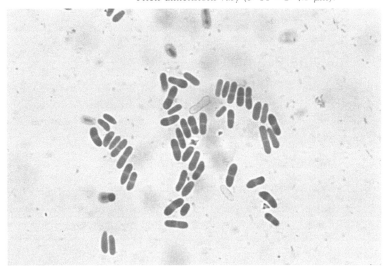

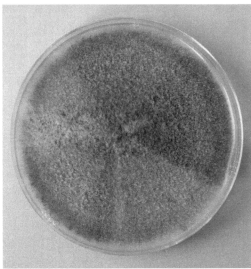

816 This colony of *Didymella lycopersici*, grown on malt-agar, has a greyish to blackish colour depending on the isolate. It has not yet developed specific fruiting bodies.

Mycovellosiella fulva (Cooke) Arx 1983

Cladosporium leaf mould

Syn.: *Fulvia fulva* Cooke Ciferri (1954), *Cladosporium fulvum* Cooke.

(Anamorph of *Mycosphaerella*, Mycosphaerellaceae, Capnodiales, Dothideomycetidae, Dothideomycetes, Ascomycota, Fungi)

Principal characteristics

This fungus was described by Cooke in 1883. Best known in the field under the name '*Cladosporium*' it is widespread on tomatoes, sometimes severely affecting the leaflets. Its sexual form (teleomorph) has never been observed.

The interaction between this fungus and tomato, governed by the gene for gene relationship, has been studied for many years and is now a very well used biological model in plant pathology. The selection of tomato varieties resistant to *Cladosporium* has been very actively developed in recent decades, but has suffered numerous setbacks in the wake of the emergence of virulent strains that can bypass resistance specific genes ('*Cf*') introduced in some cultivars. Several culitvars are now present in the field, with different virulence profiles.

Note that another *Cladosporium*, *C. oxysporum* Berk. & MA Curtis, has been described in a glasshouse in upstate New York. It may be responsible for irregular, angular, and dark brown lesions on older tomato leaves.

The spots are of variable size (1–5 mm), and are surrounded by a yellow halo. Unlike *M. fulva*, *C. oxysporum* seems to sporulate on the upper side of the leaf at the centre of the lesions. Note that in India this fungus is also known to cause rot on fruit post harvest. It is also responsible for leaf spots on pepper.

• Frequency and extent of damage

M. fulva is globally widespread, especially in humid production areas and is very specific to tomato. It mainly affects crops in poorly ventilated glasshouses (soil and soil-less), but also field crops of wet and warm areas. Its damage can be considerable in tunnels, especially if susceptible varieties are cultivated. The destruction of foliage resulting from attack may cause significant yield losses.

• Main symptoms

M. fulva is a parasitic fungus almost entirely attacking leaves. It causes light green to pale yellow spots, with diffuse edges, especially on the leaflets of lower leaves. At first a whitish then purplish-brown to olive fungal growth gradually covers the spots on the underside of the leaf. Eventually, the tissues located at the centre of the spots turn brown, become necrotic and dry while the leaves curl. In very favourable conditions, *M. fulva* also sporulates on the upper side of the leaf. The disease later spreads to reach the upper parts of plants while older leaves eventually wither and sometimes fall. The stem may also be affected.

The flowers are rarely attacked, but when they are, they die before fruit set. Irregular lesions develop from time to time on green or mature fruit; these are black with a diffuse edge, and give rise to peduncular rot.

(See Photos **14, 238, 338, 340–344, 373.**)

• Biology, epidemiology

Survival, inoculum sources: mycelium, sclerotia, and conidia of M. *fulva* can survive on and in soil and also on the glasshouses walls. Its saprophytic potential allows it also to remain on plant debris. Very resistant to desiccation, its conidia can survive over a year in a glasshouse in the absence of susceptible host. This fungus can also perpetuate on seeds. Note that it has been described in Brazil on *Carica papaya*, a host capable of sustaining it.

Penetration and invasion: by getting into contact with the leaves, conidia will germinate if a film of water is present or if the humidity is above 85%. M. *fulva* enters leaves through stomata. Infection takes place in 24–48 hours in poorly ventilated glasshouses. The mycelium invades the mesophyll intercellular spaces. The incubation period is long: it usually lasts 10–15 days, and sporulation occurs within hours.

Sporulation and dissemination: once produced in great numbers on the underside of the leaflets, conidia (Photos **817** and **818**) are easily spread by wind, air currents in glasshouses, splashing water, tools, and workers' clothing. Some insects also ensure the spread of this fungus. Prolonged periods of high humidity increase the amount of sporulation.

Conditions encouraging development: in Europe, M. *fulva* is typically a parasitic fungus of protected crops where high humidities prevail. It is mostly active in autumn, early winter, and spring. In these glasshouses, day/night climatic contrasts encourage the fungus, as well as overcast weather with low light. It is sometimes found in open fields, shaded by hedges and when conditions are wet. It particularly likes temperatures of about 20–25°C and humid environments, and its activity is limited below 11°C. In tropical areas, it occurs mainly during 'fresh', periods when air humidity is high. Excessive nitrogen fertilizers also promote *Cladosporium*.

Control methods[1]

• During cultivation

As soon as the first spots are detected, the glasshouses must be ventilated as much as possible to reduce the humidity of the ambient air below 85% and to avoid the presence of free water on the leaflets. It is also necessary to heat the glasshouses at certain times to maintain night temperatures higher than those outdoors. Defoliation at the base of the plant removes the first leaves attacked and increases ventilation of the lower parts of the plant.

Treatments can be applied with the following fungicides: maneb, mancozeb, chlorothalonil, thiophanate-methyl, azoxystrobin, benomyl, triforine, fenarimol, hexaconazole, and cyproconazole. In the areas at risk, they must be applied carefully and each week. The plants must be well covered with the fungicide spray, especially the underside of lower leaves. Strains resistant to certain fungicides have been detected in several countries. For example, cross-resistance to flusilazole, myclobutanil, and triadimefon has been identified in M. *fulva* in China.

Plant debris must be removed fairly quickly during cultivation following various cultural operations. After cultivation once the plants have been uprooted they must be destroyed.

[1] See note to the reader p. 417.

Table 41 Examples of commercial chemicals for disinfection of structures (glasshouses and tunnels) and equipment (trade names as used in France)

Product	Active ingredients[a]	Dosage
Agroxyde II	Acetic acid 194.4 g/l + hydrogen peroxide 157.5 g/l + peracetic acid 58.6 g/l	0.75%
Alca Chlorine 0147	Sodium hypochlorite 348 g/l	2%
Arvo HDL	Didecyldimethyl ammonium 9% + glutaraldehyde 20%	1%
Bactesam	Didecyldimethyl ammonium chloride 45 g/l	1%
Best-Top	Didecyldimethyl ammonium chloride 32.1 g/l + formaldehyde 131.6 g/l + glutaraldehyde 133.7 g/l	1%
Halamid-R	N-sodium-n-chloro-para-toluene sulfonamide 100%	25 l/100 m^2
Menno FloradeS	Benzoic acid 9%	1%
Phenoseptyl Pov	Orthophenylphenol 250 g/l	0.5%
Sicacid	Lauryl dimethyl benzyl ammonium bromide 100 g/l + glutaraldehyde 40 g/l	without
Virkon	Sulfamic acid 5% + potassium monopersulfate 22.5% + malic acid 10 %	1 l/hl
Zal Perax SU 380	Peracetic acid 55 g/l + hydrogen peroxide 220 g/l	1%

[a] Many other commercial products are available, often using the same active ingredients. Before making a choice, product(s) should be checked for current registration in the relevant country.

• **Subsequent crop**

Before planting the next crop, the structures and walls of the glasshouse must be disinfected in order to destroy any spores of M. *fulva* present. Various chemicals can be used. In some countries, 2–5% formaldehyde is traditionally used, sprayed at high pressure against the walls. It is also used in fumigation at 0.9 l of commercial solution (38% formaldehyde) per 100 m^3. Potassium permanganate (at a rate of 360 g for this amount of formaldehyde per volume unit) is often added as an oxidizing agent. Temperatures during the disinfection must be at least 10°C and relative humidity should lie between 50 and 80%. The glasshouse should be left closed for at least 24 hours, and ventilated well for at least a day before planting. Formaldehyde is sometimes applied by spraying at a rate of 3500 l/ha. A solution of bleach at 4–7% of the commercial solution at 48°Chl.[1] can also be sprayed.

Other products are commonly recommended for disinfecting the glasshouses walls and structures. The most representative are listed in *Table 41*.

Heat is sometimes used between two crops to destroy the inoculum residing in glasshouses. For this, they are kept closed, with no ventilation, to ensure a temperature of about 57°C for at least 6 hours.

[1] The chlormetric system of measuring active chlorine uses degrees: 1% active chlorine converts to 3.16 chlormetric degrees of chlorine.

Sometimes contaminated seeds may be soaked in water at 50°C for 25 minutes. Avoid excessive nitrogen fertilizer applications and plants too vegetative with waterlogged tissues.

High planting densities should be avoided to aid good ventilation in the crop and to avoid a dense leafy canopy and shading. In addition, all sanitary and agricultural practices as previously described, must be implemented.

Preventive fungicide treatments are recommended above, during periods of risk, especially in conditions of high humidities. They are best applied in the morning so that plants dry in the day. Furthermore, sprinkler irrigation should be avoided. Note that a system for decision support Greenman developed to reduce the use of pesticides under protection, is being validated on tomato for the control of *Botrytis cinerea* and M. *fulva*.

There are many varieties resistant to *Cladosporium*. They are the result of intense selection. The resistances have been investigated in several wild species: *Lycopersicon cheesmanii*, *L. chilense*, *L. hirsutum*, *L. pennellii*, *L. peruvianum*, and *L. pimpinellifolium*. Among the genes used are 'Cf-2', 'Cf-4', 'Cf-5', 'Cf-6', 'Cf-8', 'Cf-9', and 'Cf-11'. Unfortunately, more than a dozen strains of M. *fulva* can overcome one or more resistance genes present in cultivated varieties. In Europe, many hybrids grown under protection have resistance genes which protect against most of the races of M. *fulva*. They are listed as 'C5' in the catalogues. This genetic combination sometimes breakdown in practice by a relatively uncommon race. This is particularly true in the Mediterranean region.

Note that different biopesticides have been tried to control M. *fulva*: *Trichoderma harzianum*, *Hansfordia pulvinata*, and *Bacillus subtilis*.

817

819

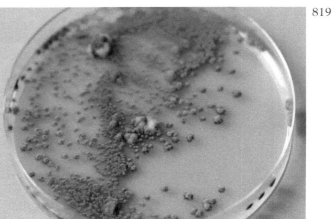

818

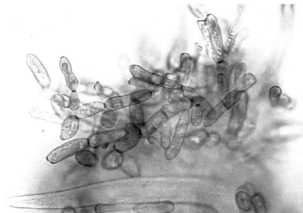

817 Several conidiophores are visible, light brown with cell walls. On one of them, a conidium is still in place. They measure 57–125 µm long and 1.5–7 µm wide.

818 *Mycovellosiella fulva* conidia are cylindrical to elliptical, smooth, sometimes slightly curved and of brownish colour. Their dimensions are variable, from 12–47 µm long and 4–10 µm wide. Uni-or bicellular, they are formed in chains.

819 This slow-growing fungus produces small dark green or purple colonies. Grey and sterile mutant areas are sometimes visible. Some strains are also rather grey. On the reverse side, as in many species, *Cladosporium* is black.

Leveillula taurica (Lev.) G. Arnaud (1921)

Syn.: *Oidiopsis sicula* Scalia, *Oidiopsis taurica* (Lev.) Salmon

(Erysiphaceae, Erysiphales, Leotiomycetidae, Leotiomycetes, Ascomycota, Fungi)

Powdery mildew (white mould)

Principal characteristics

The name *Leveillula taurica* may cover an association of species that are capable of attacking a large number of crops and weeds: more than 1000 species belonging to 74 botanical families have been recorded as hosts. This obligate parasitic fungus is well known on the Solanaceae, notably on tomato, pepper, and eggplant. Its peculiar biology, and in particular its parasitic process on these hosts, brings it closer to *Alternaria tomatophila* than to other classic ectoparasite powdery mildews.

• Frequency and extent of damage

Although found in hot and dry zones, tropical to sub-tropical, this fungus is actually more widely distributed in the world because of the breadth of its thermal and water requirements.

It can affect field and protected crops. It sometimes causes considerable damage, greatly reducing the photosynthetic capacity of plants and thus their yield. Production decreases of several tens of percent have been reported. Impact on fruit quality has been also observed, particularly related to fruit overexposure to the sun and sunburn.

This fungus can be present without affecting production. However, it is occasionally severe in some crops in glasshouses where epidemics have not been detected in time or have been misdiagnosed.

• Main symptoms

At first pale green spots gradually appear on the upper side of the lower leaves. Rounded to angular in shape when their edges are delimited by the veins of the leaf, they inevitably turn yellow over time. On the lower side, the presence of a discrete white down can be seen on the spots, dotting tissues that are yellowish brown but with few other symptoms. The fungus can also sporulate on the upper surface of the leaflets in wet weather conditions.

Eventually, the spots become entirely necrotic and lighter brown in the centre. Concentric patterns are also visible. The spots may coalesce and cause yellowing of the leaf and complete death of leaflets and leaves, which however remain on the plant. Some fruits, no longer protected under the leaf canopy, suffer sunburn (see p. 385).

No symptoms are visible on other parts of the plant, including the stems, petioles, and fruits.

(See Photos **146, 339, 348–355, 375.**)

• Biology, epidemiology

Survival, inoculum sources: as previously suggested, *L. taurica* is mainly active in summer. This obligate parasite is capable of surviving as mycelium on living hosts: crops (peppers, eggplant, artichoke, cucumber, several *Allium* spp. [onion, garlic, leek], cotton) and weeds (*Sonchus asper*, *Physalis* spp., *Chenopodium ambrosioides*, *Oxalis cernua*, *Urtica urens*) that contribute to its survival and propagation. It should be noted that more than 1000 different hosts have been listed. This information is relative though, if as suggested previously, *L. taurica* is actually an aggregate species. For the same reason, some potential hosts described may not be significant in relation to the disease on tomatoes.

This fungus has a perfect form producing cleistothecia measuring from 140–250 µm in diameter. They are rarely observed in nature, their role in the parasitic cycle of the fungus is therefore rather small.

Penetration and invasion: fungus conidia on the leaf germinate relatively well at humidity above 40%. Subsequently, mycelial hyphae invade the leaf tissues after entering directly into the leaf through the cuticle or through stomata, progressing between mesophyll cells. The fungus is thus protected from desiccation, ultraviolet rays, and leaching. It is also less exposed to fungicide treatment.

Sporulation and dissemination: after 20 days, long conidiophores emerge from stomata present on the underside of the leaf and produce spearheaded terminal conidia (Photos 820 and 821). These spores will ensure the dissemination of *L. taurica*, often through wind and air currents in the glasshouses. Workers may also disseminate it during cultural operations. In some countries, infected plants may help to introduce and spread the pathogen.

Conditions encouraging development: temperatures around 26°C and relative humidity of 70–80% help its development, although infection can still occur at temperatures between 10 and 33°C with relative humidity being higher or lower. The combination of warm dry days with cool and humid nights promotes this disease. The presence of dew on leaves encourages this disease.

Control methods[1]

• During cultivation

As soon as any early symptoms are detected, the first leaves attacked should be removed, taking care to put them directly in a plastic bag to avoid spreading the spores. These leaves are then taken away from the crop and destroyed.

The following fungicides can be used for treatment: sulfur, myclobutanil, bupirimate, azoxystrobin, hexaconazole, triforine, triadimefon, propiconazole, fenarimol, and pyrazophos. Applications need to be done with care to ensure full coverage of the lower leaves and the leaf underside. Alternations of chemical groups with different action modes should be used.

A number of hygiene measures should also be followed throughout the crop:

– limit the presence of visitors in the glasshouse to reduce the risk of dissemination;

– defoliate the base of plants. This will eliminate the first leaves attacked and promote ventilation of the lower parts of plants and penetration of applications;

– control the climatic conditions in the glasshouse to reduce humidity and promote air circulation. Avoid water condensation on the foliage.

Plant debris and discarded plants should be fairly quickly removed and destroyed from plots during cultivation (after the various cultural operations).

• Subsequent crop

In the field, long crop rotations may be beneficial. When the seedlings have been produced in other production areas which may already be affected, it will be necessary to control their quality and/or to use a fungicide on arrival for safety.

Under protection, disinfection of structures and walls is recommended to destroy the spores present. Products identical to those

---

[1] See note to the reader p. 417.

used against *Mycovellosiella fulva* may be used. Subsequently, the crops should be carefully monitored to detect the earliest symptoms of this powdery mildew. Indeed the earlier the detection of this disease the greater the chance of effective control by the protection methods used. It is particularly difficult to eradicate an outbreak of *L. taurica* once it is established.

All the preventive measures outlined above should be implemented. In addition, crops or weeds which may harbour the fungus should be removed from the glasshouses and their immediate environment.

The resistance gene 'Lev' to *L. taurica* is dominant and located on chromosome 12; it has been introduced by back-crossing of the wild species *Lycopersicon chilense* with cultivated tomato. This resistance is now available in several varieties.

Several antagonistic micro-organisms have been tested and found to be effective to varying degress against *L. taurica*, including *Trichoderma harzianum* and *Ampelomyces quisqualis*, as have plant extract (*Reynoutria sachalinensis*) and solutions of potassium dihydrogen phosphate or potassium bicarbonate sprayed on leaves.

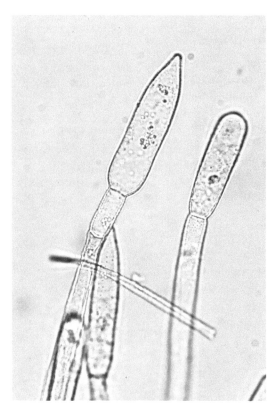

820 The conidiophores of *Leveillula taurica*, an obligate fungus which cannot be grown on artificial media, are long and sometimes branched. They emerge through the stomata and bear single conidia or in very short chains. Note that this fungus rarely forms cleistothecia.

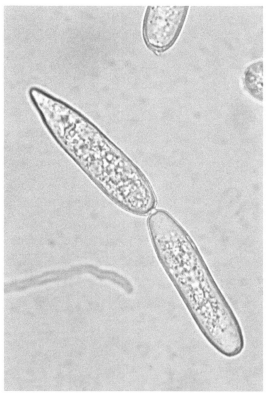

821 The first conidium is more or less lanceolate and pointed at its end. The second is more or less ellipsoidal to cylindrical. Their measurements vary: 30–80 × 12–22 µm.

Leveillula taurica

Oidium neolycopersici L. Kiss (2001)

Powdery mildew (white mould)

(Anamorph of Erysiphe, Erysiphaceae, Erysiphales, Leotiomycetidae, Leotiomycetes, Ascomycota, Fungi)

Principal characteristics

The identification of this powdery mildew, whose sexual form has never been observed, has resulted in much confusion. Thus, several other names have incorrectly been attributed to it: *Erysiphe orontii* Cast., *E. polygoni* D.C., *E. cichoracearum* DC, perhaps *Oidium lycopersicum* Cooke & Massee Emend. Noordeloos. Some studies, notably on samples from several herbaria, have shown that in many cases this disease caused by *Oidium neolycopersici* (sub-genus *pseudoidium*) has affected tomatoes for some time. It is therefore not new, and its presence in Asia dates back for at least 50 years.

Different species of powdery mildew are prevalent in Australia, especially *Oidium lycopersici* Cooke & Massee, Grevillea (sub-genus *reticuloidium*). It causes symptoms on tomato comparable with those due to *O. neolycopersici*. It is known to have been present in Australia since 1888. It may be close to *O. longipes* Noordel. & Loer., affecting the eggplant, but also *Cyphomandra* sp., tobacco, and petunia.

The world status of species of powdery mildews associated with tomatoes is probably not yet definitively clarified. Differences are known to occur in the pathogenicity of *O. neolycopersici*, both in terms of host range and virulence. Moreover, a recent study on tomato in Australia has highlighted a species suspected of being *Golovinomyces biocellatus* (Ehrenb.) VP Heluta (1988), previously known mainly on the Lamiaceae.

• Frequency and extent of damage

Tomato powdery mildew caused by *O. neolycopersici* is quite different from *Leveillula taurica*, and has been a cause for concern in many countries around the world, including in Europe, Africa, North and South Americas, and Asia. In Europe, it appeared in the early 1980s in the Netherlands, UK (1987), Belgium, Switzerland (1987), Germany (1987–1988), Czechoslovakia (1988), and Italy (1989), and then spread to the Mediterranean basin, in Crete (1990). It was found in the US and Canada during the 1990s.

This powdery mildew spreads rapidly even in countries where it is is recorded for the first time, and can cause considerable damage if nothing is done to control it. It is found both in glasshouses and field crops.

It was introduced into various European countries in the mid 1980s through young tomato plants imported from the Netherlands. Subsequently, it spread to all areas of production, attacking protected crops of tomato almost all year round.

It is now feared by glasshouse growers who are often forced to use an intensive preventive chemical control strategy at certain times of the year.

• Main symptoms

O. neolycopersici causes spots on leaves which are fairly typical of powdery mildew. These are powdery and white and mostly cover the upper side of tomato leaflets. This white felting is actually composed of a mycelial network colonizing the leaf surface, topped by numerous conidiophores producing isolated hyaline conidia (Photo **822**) or sometimes in pseudochains of four to six spores when relative humidity is high.

Locally, spots become chlorotic, brown, and eventually necrotic. During severe attacks, the whole leaf can be covered by the mycelial network and some leaflets turn yellow and become entirely necrotic.

Comparable spots can be seen on the stem. Fruits are not affected.

(See Photos **240**, **372**, **376–382**.)

• Biology, epidemiology

The epidemiology of this obligate parasitic fungus is poorly known.

Survival, inoculum sources: like many powdery mildews, it can perpetuate from one year to another in several ways:
– through alternative hosts, crops, or weeds. Its potential host range is quite broad; it seems to differ from one country to another, which suggests that this species contains specialized forms. Artificial inoculations have shown its aggressiveness on different plants. It establishes with difficulty on melon, cucumber, courgettes, peas, and *Solanum dulcamara*; and it develops more readily on eggplant, potato, *Solanum albicans*, *S. acaule*, and *S. mochiquense*, but does not persist. However, tobacco is very sensitive and could be an alternative host. *Datura stramonium*, *S. capsicoides*, *S. jamaicensis*, *S. laciniatum*, *S. lycopersicoides*, *S. ptycanthum*, and *Petunia hybrida* are also rather susceptible. Moreover, given the cycle of tomato production in glasshouses, *O. neolycopersici* is able to survive on this plant in its mycelial form, from one year to another;

– globose cleistothecia (perithecia without ostiole) are the structures of its sexual reproduction. But as noted earlier, the teleomorph of this fungus has never been observed, so should not play an important role in its life cycle.

Penetration and invasion: conidia of *O. neoly-copersici* certainly cause primary infection. Once in contact with the host, they germinate quickly in a few hours if environmental conditions are right. They form an appressorium and penetrate directly into epidermal cells, developing haustoria. These enable the fungus to obtain the necessary elements for its growth.

Sporulation and dissemination: within a few days after infection, short conidiophores arise on secondary hyphae formed on the leaf surface and produce conidia (Photo **822**). Sporulation is abundant. Conidia are rather fragile and live only a few days or even hours, even when conditions are favourable. They are very light and can be easily carried and spread by wind and incidentally by rain or splashing during sprinkler irrigation.

Conditions encouraging development: *O. neolycopersici* prefers humidity levels at or below 80% moisture; above this level its development is gradually reduced. Therefore, excessive humidity would reduce the severity of the disease. A high nitrogen content of leaf tissues reduces their resistance to this powdery mildew. This fungus is particularly well suited to warm and humid environments.

Control methods[1]

• During cultivation

At the beginning of a localized attack of mildew, fungicide treatments should be done with one or more of the following products: sulfur, myclobutanil, bupirimate, azoxystrobin, hexaconazole, triforine, triadimefon, propiconazole, fenarimol, and pyrazophos. Remember that treatment is only effective if applied in time, at the right dose, with sufficient volume of spray and with equipment suitable for spraying tomatoes. Once an attack is established it is often difficult to control the development of an epidemic in the crop.

Among the few prophylactic measures able to improve control of this fungus are:
– removal of plant debris from the plots and their surroundings, or they must be buried quickly and deeply into the soil;
– the destruction in the crop and its surroundings of weeds which can serve as plant 'bridges' for the fungus.

Resistance has been found in the wild species *Lycopersicon hirsutum*. This is controlled by an incompletely dominant gene named 'Oi-1', located on chromosome 6. A gene 'Oi-3', genetically indistinguishable from *Oi-1* is

[1] See note to the reader p. 417.

located between the same two genetic markers on the same chromosome. Other wild species of *Lycopersicon* have proved resistant: *L. pennellii* (three dominant additive genes), *L. cheesmanii* and *L. parviflorum* (polygenic resistance) (differences in resistance between accessions), *L.chilense* (one partially dominant gene), *L. peruvianum* (one dominant gene, '*Oi-4*'). Note that resistance to powdery mildew is often associated with a reduction of mycelial growth and sporulation of the fungus.

• Subsequent crop

During the next crop, a number of additional preventive measures can be implemented to reduce the risk of occurrence of powdery mildew.

In the field, it is desirable to practise crop rotation. It does not need to be long. The future crop should be located in a very airy and sunny place. Use a balanced fertilizer for the plants and eliminate known hosts of *O. neolycopersici* in the vicinity of crops.

In glasshouses, an interval between crops is required. Surfaces should be disinfected with one of several disinfectants for this purpose (see Section 4). Note that maintaining a humid climate could slow the progression of the disease. However, this measure carries some risks of encouraging other air-borne fungi in tomato.

In areas of production and/or glasshouses where mildew is a constant threat every year, preventive treatments can be done with the products mentioned above.

Foliar application of several salts such as $CaCl_2$, $Ca(NO_3)_2$ and K_2HPO_4, may limit the number of leaf spots on tomatoes, reducing them to levels comparable to those achieved with the application of sulphur.

Genetic resistances to *O. neolycopersici* is reported in the literature and has been exploited by breeders.

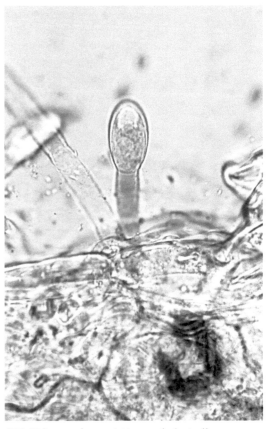

822 *Oidium neolycopersici* is morphologically characterized by the frequent formation of a single conidium at the end of its conidiophores measuring 22–46 × 10–20 µm.

Phytophthora infestans (Mont.) de Bary (1876)

Late blight, downy mildew

(Incertae sedis, Peronosporales, Peronosporomycetidae, Oomycetes, Oomycota, Chromista)

Principal characteristics

Late blight of potato and tomato, caused by *P. infestans*, is a disease that has marked the history of mankind. Its first outbreaks in Europe during the 1840s, were the cause of dramatic famines and the deaths of more than one million people (particularly in Ireland), and contributed to mass emigration of the Irish to the US.

The spread of this disease, from its native area (the Andean[1] area, long contested at the expense of Mexico), took place in several stages from the 1840s. The first spread may have allowed the US-1 lineage to enter the US, and subsequently Europe, probably via infected potato seeds. After colonizing Europe, late blight spread to all world countries producing potato and tomato. Note that this line did not seem particularly aggressive on tomato in the 1980s. Indeed, outbreaks occurring in this crop were often of low severity, and blight was thought to have virtually disappeared from tomatoes.

Subsequently, the increase in international trade from the late twentieth century onwards may be the cause of the distribution of one or more aggressive strains, all from Mexico, which was considered to be the source of the genetic diversity of *P. infestans*. Thus, several strains[2] have progressively been detected in the US from the late 1970s. Some of them were also found in other countries in the Americas, but also in Europe and in many other countries. They have gradually replaced the US-1 lineage.

During this period, the sexuality of *P. infestans* has been studied. It is heterothallic and its sexual reproduction requires the presence of strains belonging to two groups of complementary compatibility groups: A1 and A2. Until the late 1970s, the A1 sexual type was predominant in many areas of production (US-1 lineage is type A1), except in Mexico where both A1 and A2 co-existed with a ratio of 50/50%. New strains[3] of A1 type, but especially of A2 type, have notably been introduced in Europe in the early 1980s. They are now found in many areas of potato production in the world, but with a ratio often unfavourable to the A2 sexual type.

These sexual types are now present in various countries including in Europe. The introduction of these strains has strongly affected the incidence of blight in many tomato areas of production. They risk increasing the genetic

---

[1] Analysis of mitochondrial and nuclear loci of *P. infestans* can support the hypothesis that this pathogen originates from South America. An ancestral population would have diverged into several lineages in the South American Andes in association with various *Solanum* spp. Two of these divergent lines would have led to existing haplotypes of *P. infestans* able to infect potato, tomato, and some wild species of *Solanum*.

[2] There are about 800 parasitic or saprophytic species of Oomycetes which for a long time have been classified in Phycomycetes or 'Lower fungi' (Eumycota). This classification was revised several years ago because the ultrastructure of these micro-organisms, their biochemistry, and molecular sequences indicated that they belonged to the Chromista, which includes mainly algae (green and brown) and diatoms. Currently, according to the different bibliographical sources, they can be associated either with the kingdom of Chromista (Index Fungorum) or to the kingdom of stramenopiles (Tree of Life).

[3] Several other strains were isolated in the US, especially in recent decades:
– the US-6 clonal lineage, infecting both tomato and potato (1979);
– two clonal lines identified in 1992 in the US and Canada, US-7 and US-8, both of A2 sexual type, metalaxyl-resistant, and pathogens on both Solanaceous hosts;
– a recent US-17 isolate, sexual type A1, showing a higher aggressiveness on tomato comparable to the US-8, and metalaxyl resistant.

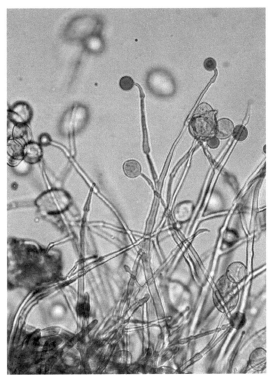

823 This chromist forms well differentiated sporangiophores with swellings.

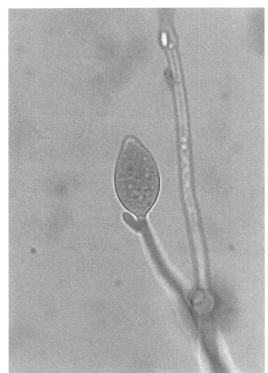

824 The sporangia are first terminal then lateral, following the growth of the sporangiophore. They are hyaline and measure 21–38 × 12–23 µm.

variability of *P. infestans*, and thus increasing its aggressiveness and virulence, but also facilitating its adaptation to fungicides and its survival through oospore formation, constituting a permanent inoculum in fields.

This oomycete has a rather narrow host range and limited to the Solanaceae.

• Frequency and extent of damage

Blight causes severe damage in many production areas of the world. It is particularly dangerous and destructive in areas with a humid climate, and its damage can be very important during poorly controlled sudden epidemics. It is not unusual in this case to observe completely devastated crops in which the number of marketable fruits is very limited. Blight mainly affects tomatoes grown in open fields, but it can be seen in glasshouses with a poorly controlled climate. Like many agents of mildew, *P. infestans* is most evident in the production areas that experience prolonged periods of moisture (rain, sprinkler irrigation, mist, dew) and mild weather. As the name 'late blight' suggests, it is

rather late in the season. It is often a constant threat and requires preventive fungicide treatments.

Some countries have climates that are particularly favourable for the development of *P. infestans*. Since the 1990s and the arrival of A2 strains in particular, its impact on tomato crops has risen sharply and it is now common and dreaded by producers of field tomatoes and home gardeners alike, especially during the summer production period. Its incidence is lower in protected crops. The diversity of strains isolated from tomato is greater than that observed from potatoes; A2 strain ratio is also higher.

• Main symptoms

Blight can infect all aerial parts of tomato. It is initially characterized by the development of wet spots or patches, on the leaflets. These affected tissues are a light green to brownish green colour. Significant portions of the leaves are eventually affected and quickly turn brown and necrotic. These spots are frequently

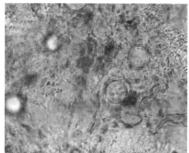

825 In damaged tissues, there are sometimes some brown oospores 28–32 μm in diameter.

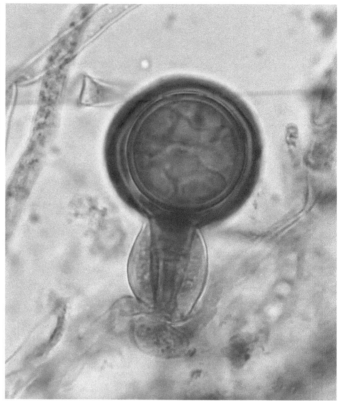

826 To obtain oogonia and oospores *in vitro* requires two strains belonging to complementary vegetative compatibility groups, as *Phytophthora infestans* is heterothallic.

surrounded by a poorly defined edge of pale tissues, on which, at the underside of the leaf, a discreet and elusive white down is sometimes seen. The white down consists of sporangiophores and sporangia (sporocystophores and sporocysts) of *P. infestans*.

When conditions are particularly favourable, symptom progression on the leaflets is rapid. Leaves, branches, or even complete plants eventually become necrotic and dry up entirely.

Brown cankerous lesions, of various sizes and irregularly shaped, are visible on the stems and petioles, and often surround them. Comparable browning can be observed on the floral trusses and cause the fall of many flowers.

The fruits infected at an early stage have a very characteristic brown mottling. They are often bumpy. In this case, the mottling progression is rather slow, with an irregular edge. If attacks occur later, the patches are more homogeneously mottled and are often divided into concentric circles with looped edges. They

may be confused with those caused by other *Phytophthora* sp., especially *P. nicotianae*.

The same whitish down as that observed on the leaflets is occasionally visible on the surface of fruits. They remain relatively firm regardless of the earliness of the attack. In some cases, secondary micro-organisms can invade the damaged tissues and lead to various soft rots (see p. 374). Note

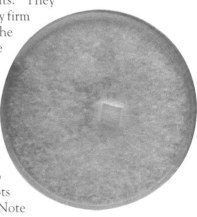

827 On pea medium, *Phytophthora infestans* develops quite well and produces a white downy mycelial colony, fairly typical of *Phytophthora*.

that some seed from diseased fruits can result in the infection of future seedlings.

Remember that diseased plants are distributed in groups in the crop; these can enlarge rapidly if the weather is mild.

(See Photos **6, 7, 297, 328–337, 730, 731**.)

• Biology, epidemiology

Survival, inoculum sources: the mode of survival of tomato strains of *P. infestans* is sometimes difficult to understand, especially in countries where frost occurs in the winter. It seems to survive in different ways depending on the areas of production:

– via oospores resulting from sexual reproduction; these are formed in diseased and necrotic tissues and occur later in the soil, along with plant debris (Photos **825** and **826**). As stated previously, for over a decade new strains from potato and tomato of both sexual types A1 and A2 have been quite commonly found in several countries in North America and Europe. Type 2 is frequently found on tomato but not so much on potato. Thus, *P. infestans* must now be able to complete its full sexual cycle on tomato and form oospores which can survive in the soil for several months or even years. These survival structures can be found on the leaflets, fruits, and occasionally on the seeds. Moreover, sexual reproduction may be the origin of new genotypes with new biological properties that may adversely affect control methods, especially chemical and genetic;

– in the form of mycelium inside diseased potato tubers remaining in the soil. When infected tubers germinate, they give rise to infected shoots on which sporulation quickly occurs. The importance of this inoculum is difficult to estimate. Indeed, the strains of *Phytophthora infestans* isolated from tomato in France are quite different from those found on potato, confirming specialized parasitic strains. Tomato strains appear to be more diversified and are capable of infecting potato. The possibility, albeit limited, for a shift towards host change cannot be excluded, especially in areas of production where both tomatoes and potatoes are grown;

– on various other alternative hosts, surrounding crop or wild plants, on which this pathogen sporulates to a greater or lesser extent; many Solanums (eggplant, red pepper, potato, pepino, black nightshade, *Solanum incanum*, *S. indicum*) *Datura stramonium*, *D. metel*, several species of *Hypomea*, *Lycium hamilifolium*, *Nicotiana glauca*, *Petunia* sp., *Physalis angulata*, blueberry (Cameroon). On these hosts, *Phytophthora infestans* produces fructifications consisting of long sporangiophores bearing sporangia. Zoospores from the latter are often the primary infection source for the tomato. Moreover, the presence of tomato crops almost all year round in some areas of production allows the pathogen to survive easily from year to year. Regrowth, natural regeneration, also ensure the survival of *P. infestans*.

Penetration and invasion: as noted above, once on the leaf, sporangia release flagellated zoospores. The optimum temperature for release is about 13°C. These zoospores, once settled, grow a germ tube that enters the leaf through stomata but also sometimes directly through the cuticle and epidermal cells. Sporangia may also give rise directly to a germ filament of mycelium. The infection takes place in 3–4 hours. Leaf tissues are subsequently quickly invaded by the noncellular mycelium (the optimal growth temperature is 23°C) whose activity gradually disrupts the colonized tissues.

If weather conditions are favourable, the first spots appear between 4 and 7 days after the initial infection.

Sporulation and dissemination: once within the host, *P. infestans* produces sporangiophores through stomata, sometimes directly through the epidermis. These produce numerous lemon-shaped sporangia, several thousand per spot (24 000/cm²) (Photos **823** and **824**). This production requires the presence of high humidity (relative humidity of at least 90%) and temperatures between 3 and 26°C. Sporangia are easily dispersed by wind and rain, sometimes over long distances (several hundreds of metres) and reach healthy new plants, resulting in secondary infection. Zoospores can also perform the same function, but more locally and mainly in the aqueous phase.

The spread of the disease sometimes occurs through contaminated plants. In some countries, some farmers have specialized in producing large quantities of seedlings which they sell in different production areas of the world. Seedling contamination sometimes goes unnoticed and infected plants are then shipped to distant producers, thus contributing to the spread of the pathogen and the early development of the disease in crops. If seed contamination by the presence of oospores is confirmed such seeds could ensure a spread of the disease over long distances.

Conditions encouraging its development: this pathogen is extremely influenced by climatic conditions. It grows at temperatures between 3 and 25°C. Its sporulation is optimal between 16 and 22°C. It needs high relative humidity, above 90%. Cool nights and moderately warm days with high humidity encourage its development. In contrast, dry conditions and temperatures around 30°C inhibit it. Rainy periods, sprinkler irrigation, and dew are also very favourable to epidemics of blight. Only 2 hours of water on the leaves are needed to start infection. Production of sporangia is abundant at 18°C, but it is zero at 28°C. Oospores are formed in large numbers between 8 and 15°C; their production requires the presence of moisture and high humidity at all times.

Control methods[1]

• During cultivation

Blight develops very quickly so it is necessary to react promptly when the very first symptoms are detected, especially if no preventive treatment has been applied. An antiblight fungicide treatment should be done immediately.

Several active ingredients are used around the world, alone or in combination, to control *P. infestans*. A list of the large majority of these is presented in *Table 42*.

Note that for multisite fungicides to be relatively effective they must be applied preventively and every week, especially in field crops. Despite limited effectiveness over time, they have still have the advantage of being versatile enough and not being affected by fungicide resistance.

This is not the case of some singlesite fungicides (such as those of the anilides family), which are rarely used alone, but in association with multi-site fungicides to reduce the risk of the development of resistance. Curative treatments are only moderately efficient and may, promote the risk of the development of resistance to fungicides. In addition, it is strongly recommended to alternate fungicides with different modes of action. With some products, no more than two to three applications per season should be given, once the disease has occurred.

In the nursery, the first outbreak(s) can be contained by treating plants with a more concentrated dose of a fungicide and by eliminating the affected seedlings. The marketing of the seedlings produced in such conditions will only be possible if the attack of blight has been totally contained and by warning farmers of the risks.

[1] See note to the reader p. 417.

Table 42 Main fungicides used worldwide to control tomato late blight

Mode of action	Chemical family	Active ingredient
Multisites (respiratory processes and cellular energy production)	Dithiocarbamates	Mancozeb
		Maneb
		Metiram zinc
	Phtalimides	Folpel
	Copper salts	Copper hydroxide
		Copper oxychloride
		Cuprous oxide
		Copper sulfate
	Sulfonamides	Tolyfluanid
	Chloronitriles	Chlorothalonil
Fatty acids: unknown target	Carbamates	Propamocarb HCl
Cell wall formation	Acetamides	Cymoxanil
	Cinnalates	Dimethomorph
RNA biosynthesis	Anilides[r]	Benalaxyl
		Mefenoxam
Mitochondrial complex III: QiI	Cyanoïmidazoles	Cyazofamid
Mitochondrial complex III: QoI	Strobilurin[r]	Azoxystrobin
		Pyrachlostrobine
	Oxazolidinediones	Famoxadone
	Imidazolinones	Fenamidone

([r]) Chemical family particularly concerned with the phenomena of resistance.

In addition to fungicide treatments, a number of prophylactic measures should be applied.

Ventilate glasshouses as much as possible to reduce moisture. Avoid sprinkler irrigation or misting. If it cannot be avoided, overhead irrigation should not be done in the late afternoon and evening, but rather at an earlier and relatively warm time so that the plants have time to dry before nightfall.

In field crops, the irrigation recommendations are the same as those suggested in nurseries. It may be necessary to heat up the nurseries to reduce humidity. Everything must be done to avoid the presence of a water film on the plants. At the end of crop, all plant debris must be quickly removed. Residues can be buried deep in the ground to facilitate their rapid decomposition.

• **Subsequent crop**

The next propagation areas will be set up in a sunny spot, never in a shady and damp place. If in the same glasshouse as the previous year, it is advisable to use the hygiene measures and disinfection procedures recommended on p. 541.

The quality of the plants should be checked on arrival, especially if they come from nurseries in countries or areas of production with temperate climates. If there is any doubt, a fungicide application should be performed as insurance.

Crop rotations of at least 3 years are recommended. Avoid new plantations near tomato or other Solanaceous crops that might already be affected. Avoid planting tomatoes in poorly drained fields with heavy water retention and in soils too rich in organic matter. The fertilization must be balanced and by no means excessive. Where it is possible to reduce planting densities, this may increase the aeration of crops and also reduce the humidity within the crop canopy. If possible, planting mounds and/or rows positioned in the direction of prevailing winds will promote aeration of the vegetation to the maximum. Sensitive weeds need to be eliminated from crops and their surroundings.

Mulching, and to a lesser extent staking, contribute to reduce the development of blight. In addition, in extensive crops, corn production with inter-row tomatoes reduced both the incidence of blight and alternaria in El Salvador. It is the same with sesame and soya in Uganda. In addition, plants supplied with organic manure rather than chemical fertilizer may be less sensitive to blight. This is linked to nitrogen metabolism.

A number of other health measures, on their own but mostly when used in combination, can reduce the development of late blight epidemics:
– removal at regular intervals of attacked leaves (this will interfere with the proper growth of plants);
– the use of plastic covers to protect plants from the weather;
– reducing the density of plants.

Given the rapid development of blight and the risks it poses to the crop, preventive fungicide treatments are essential during propagation and during cropping. The products used and the rates adopted will be chosen according to local farming practices. The interval between two treatments should not exceed 7–12 days depending on the active ingredients. Fungicides with different modes of action should be rotated to limit the selection of resistant strains.

It should be noted that several prediction models or decision support systems have been developed, especially in temperate zones, to better time fungicide applications, improve their efficiency, and reduce their number. These include the Hyre system (based on temperature and rainfall), Wallin system (based on temperature and humidity), but mostly Blitecast, which integrates the two systems in a computer program. The advantage of these tools is proven. Other systems are reported in the literature: Phymet in Italy, MacHardy system, developed in New Hampshire in the US, and Maschio and Sampaio systems developed in Brazil in particular.

Also, note that experiments have been conducted in the US to detect blight by plant spectral analysis, in order to manage the use of fungicides better. In countries where such services exist, heed the agricultural warnings issued fairly regularly by the plants protection office or other organizations.

Several genes, all from *Lycopersicon pimpinellifolium* were used to control *P. infestans* on tomato but their effectiveness has been questioned by various breeders:
– 'Ph-1 (on chromosome 7) whose resistance was quickly bypassed by race T-1;

– 'Ph-2 (on chromosome 10) and from the accession WVA 700. It is now available in F1 hybrid varieties and confers partial resistance. Virulent strains towards this gene are also cited in the literature;

– 'Ph-3' (present on chromosome 10) obtained from accession L3708. This gene is capable of controlling more strains.

The differentiation of races of *P. infestans* can only be done on tomato leaves and stems. Fruits are susceptible to all strains. T0 race occurs mainly on tomato varieties lacking resistance genes. T1 race is more aggressive than race T0. Note that research done in China involving five tomato genotypes (TS19, Ts33, WVA 700, LA1033, L3708) has allowed us to distinguish eight different races: T0, T1, T1-2, T1-2-3, T1-2-3-4, T1-2-4, T1-4, and T3.

Varieties with partial resistance to late blight are available in several countries, notably in France. While representing a real asset for tomato growing, the plant material must be used in conjunction with other control methods, especially with a complementary chemical control. This will reduce the risk of emergence of races as described above, and thus contribute to the sustainability of these resistances. For similar reasons, never mix cultures of resistant and sensitive varieties, and avoid the proximity of sensitive varieties.

A number of micro-organisms and plant extracts (rosemary, lavender, thyme, fennel) have been tested against *P. infestans*. Some of them have shown some efficacy *in vitro*. Unfortunately, used in field conditions they have been ineffective against the aggressiveness of the late blight pathogen. Soils antagonistic to this pathogen have also been described in Mexico in the Toluca valley. This property is conferred by the biological activities of several micro-organisms (*Pseudomonas* sp., *Burkholderia* sp., *Trichoderma* sp.).

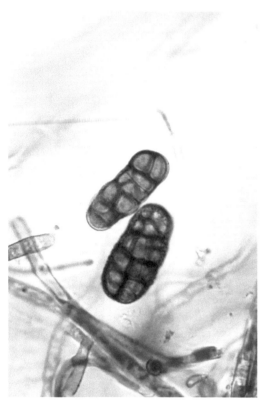

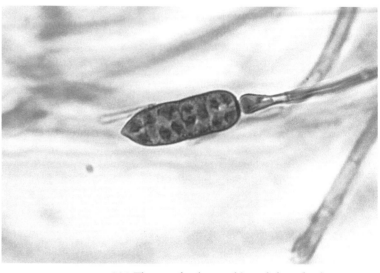

828 *Stemphylium vesicarium* has conidia with rounded ends, with echinulate walls, and two to four divisions. They have a brown colour, a rectangular shape, and measure 26–44 × 12–20 µm. The perfect form of the fungus, *Pleospora allii* (Rabenh.) Ces. & De Not., occurs *in vitro* on artificial medium.

829 The conidiophores of *Stemphylium floridanum*, as those of S. *solani*, are septate, rigid, of variable length and melanized, with a swollen end. Muriform conidia are formed.

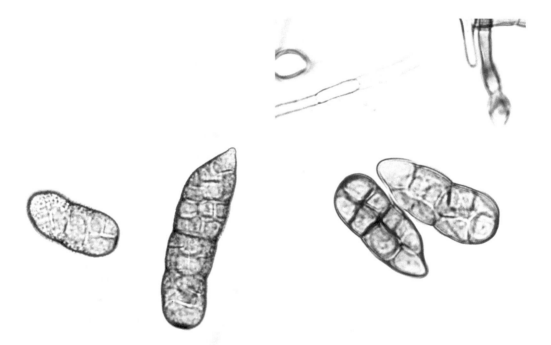

830 The conidia of *Stemphylium floridanum* in addition to being multicellular have a pointed apex and two to five constrictions, often three. They are also quite long as they measure 44–72 × 12–20 µm. Their wall is warty. The sexual form of this fungus has not been observed.

831 The conidia of *Stemphylium solani* are also muriform, brown, and have a pointed apex (sometimes eccentric) and a warty wall. They show fewer constrictions (one to two) and are shorter (36–52 × 14–20 µm) than those of S. *floridanum*. This species does not have a sexual form.

Stemphylium spp.

S. solani G.F. Weber (1930)

S. lycopersici (Enjoji) W. Yamam. (1960) (syn. *S. floridanum* C.I. Hannon & G.F. Weber [1955])

S. botryosum f. sp. lycopersici Rotem, Y. Cohen & I. Wahl (1966) (anamorph of *Pleospora herbarum* [Pers.] Rabenh. [1854])

S. vesicarium (Wallr.) E.G. Simmons (1969) (syn. *Pleospora allii* [Rabenh] Ces. & De Not.)

S. botryosum Sacc. (1886)

(Pleosporaceae, Pleosporales, Pleosporomycetidae, Dothideomycetes, Ascomycota, Fungi)

Stemphylium blight (grey leaf spot)

Principal characteristics

Stemphylium blight is a foliar disease of tomato with symptoms classically associated with three different species of *Stemphylium*: *S. solani*, *S. lycopersici*, and *S.botryosum* f. sp. *lycopersici*. Some identifications made in the 1980s, especially in the Mediterranean area, excluded *S. botryosum* f. sp. *lycopersici*. *S. vesicarium* has been demonstrated to be pathogenic on tomato. These four *Stemphylium* spp. are controlled by the same resistance gene in *Lycopersicon esculentum*.

• Frequency and extent of damage

The disease is globally distributed but is particularly serious in humid tropical and subtropical production areas. It is described in the Americas: US, Canada, Mexico, Brazil, Colombia, Peru, and it is very serious in Cuba. It affects crops in many African countries (Ivory Coast, Gambia, Tanzania, Tunisia, Morocco) and is less severe in Europe (Germany, Italy, Spain). Stemphylium blight is also reported in Asia (China, Japan, Indonesia, Thailand), and Oceania (Australia). It is found both in field and protected crops.

The species involved depend on production areas: *S. solani* and *S. floridanum* are mainly associated with the tropics, while *S. botryosum* f. sp. *lycopersici* and *S. vesicarium* occur in more northern areas. It is however possible that several species can occur on tomatoes within the same country.

In hot and humid conditions, these fungi cause severe damage on sensitive varieties, resulting in significant defoliation.

• Main symptoms

Stemphylium blight is a disease which mainly attacks the leaflets, rarely the petioles and the stem, and, to our knowledge, never the fruits.

Tiny brown spots, slightly angular, sometimes discreetly haloed with a chlorotic border, can be seen on all the tomato leaflets. Their size is often limited to a few millimetres, except in particularly wet conditions where they expand and join together. They then may cover large portions of the leaf. Generally, the spots gradually lighten and eventually become greyish in colour as they become necrotic and dry up. Fructifications of *Stemphylium* spp. (brown masses corresponding to muriform conidia) are visible under a binocular microscope on both sides of the leaf. The damaged tissues are of dull appearance and sometimes split. In particularly severe attacks, the spots are numerous and as they develop and join together induce significant leaf death, often preceded by yellowing and eventually leaf fall. A few small and longitudinal, brown lesions are sometimes observed on the petioles or the stem. In some countries severe attacks can occur on seedlings in nurseries, accompanied by yellowing and defoliation. Note that S. floridanum has been described in the 1970s, as responsible for black lesions on flowers and later on fruits.

(See Photos **242, 260–269**.)

• Biology, epidemiology

Survival, inoculum sources: *Stemphylium* sp. have saprophytic potential which enables them to survive from one season to the next, in and on the soil, on tomato plant debris of all kinds (leaflets, stems, senescent floral parts), and on many alternative hosts, cultivated or not. *S. solani* and, to a lesser extent, *S. floridanum* is capable of attacking many cultivated Solanaceous plants (eggplant, pepper) and wild ones (*Solanum carolinense*, *S. lycocarpum*) that enables it to survive and/or ensures the production of primary inoculum. Note that the strains of *S. solani* affecting cotton are aggressive on tomato, potato, and blue lupin in Brazil (cotton strains are nevertheless genetically different from tomato strains). *S. floridanum* is described on *Allium*, safflower, gladiolus, and chrysanthemum. It is possible that these different host can serve as inoculum reservoirs. *S. vesicarium* and *S. botryosum* f. sp. *lycopersici* also produce perithecia and the ascospores enable them to survive from one season to the next.

Penetration and invasion: the conidia of these fungi germinate on the leaflets and penetrate directly through the cuticle or through various wounds. The mycelium then rapidly invades the tissues, and spots are visible 5 days after the initial infection.

Sporulation and dissemination: once the mycelium is established, conidiophores and

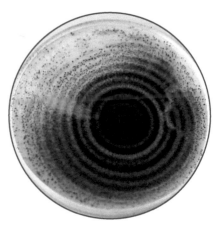

832 Mycelial growth of *Stemphylium vesicarium* is fast on malt-agar medium. Septate mycelium, hyaline at first, gradually melanizes and readily forms several conidiophores with conidia. One can also observe the development of microstromas, the beginnings of black globular structures corresponding to the perithecia of the sexual form.

conidia are formed on the underside of the leaf (Photos **828** and **831**). These are mainly dispersed by wind and air currents, but also by splashing water resulting from rainfall or sprinkler irrigation. Remember that the seedlings can transmit the disease if infection takes place in the nursery.

Conditions encouraging development: the development of *Stemphylium* spp. includes wet weather, especially the presence of water on plants (rain, dew, spraying, condensation in glasshouses) and by high temperatures. Their temperature optima are between 23 and 27°C, depending on the species.

Control methods[1]

• During cultivation

Once the first symptoms are observed during cultivation, the plants need to be treated as soon as possible using one of the following fungicides: mancozeb, maneb, chlorothalonil, anilazine, iprodione, difenoconazole, azoxystrobin, thiophanate-methyl. The application must be repeated after heavy rainfall and sprinkler irrigation. Spraying may help the spread of the disease, so it will need to take place in the late morning, never late in the afternoon or at night, to avoid the plants remaining moist all night. Plants should not be stressed and must receive a balanced fertilizer, especially in relation to nitrogen.

Wherever possible, the maximum amount of plant debris (leaves, stems) should be removed during and after cultivation.

• Subsequent crop

If the plants are propagated in the same place as the previous year, the soil should be disinfected and/or preventive applications of fungicides used. In some countries, be wary of the seedlings quality as they can transmit the disease.

Eliminate weeds that can serve as hosts. In addition, avoid setting up a crop near other sensitive crops, such as peppers and eggplant, or in areas where they have been produced in recent years.

It may be advantageous to use lower density planting that allows good ventilation of the vegetation and drying out after rainfall or sprinkler irrigation. This also gives more vigorous plants. It is best to irrigate the plants by drip irrigation rather than by sprinkling.

Fungicide should be applied preventively with the previously reported active ingredients.

High-level genetic resistance to S. solani was demonstrated many years ago. Sourced from Lycopersicon pimpinellifolium, it is controlled by the dominant gene 'Sm' on chromosome 11. It is actually effective against all species of Stemphylium attacking tomatoes. This resistance is particularly stable as, so far, it has never been overcome.

---

[1] See note to the reader p. 417.

◼ Main other fungi attacking the leaves

Alternaria alternata (Fr.) Kessler (1912) f. sp. *lycopersici* Grogan

Alternaria stem canker

(Anamorph of *Lewia*, Pleosporaceae, Pleosporales, Pleosporomycetidae, Dothideomycetes, Ascomycota, Fungi)

Corynespora cassiicola (Berk. & M.A. Curtis) C. I. Wei (1950)

Corynesporiose (target spot)

Anamorph of *Corynesporasca*, Corynesporascaceae, Pleosporales, Pleosporomycetidae, Dothideomycetes, Ascomycota, Fungi)

Distribution and damage

The specialized form of this *Alternaria* has been reported in several countries including the US, Iran, Iraq, Egypt, Greece, Sardinia, India, Japan, and Korea. It attacks sensitive varieties which do not have the incompletely dominant allele '*Asc*' in the homozygous state. It is not currently a major threat to tomato crops, except perhaps in the few countries where sensitive varieties are still cultivated, in open fields or under protection.

This fungus, sometimes confused with *Helminthosporium*, is fairly widespread in the world. It is a problem on tomato in many humid tropical and sub-tropical countries of America (US, Brazil, Mexico), Europe (Romania), Africa (Nigeria), India, Asia (Japan, Taiwan) and the Caribbean (Cuba, Puerto Rico, Haiti, Trinidad).

It is very polyphagous and is reported on more than 145 plant genera belonging to at least 53 families. Some strains have nevertheless certain host specificity. On tomato, this is partly due to the production of a phyto-toxin.

Symptoms

This *Alternaria* mostly causes elongated cankers on stems (Alternaria stem canker), close to the ground. They have concentric patterns and are brown to dark beige, with a dry consistency. The underlying vessels show some localized browning. These cankers even-tually surround the stem, killing the plant. By colonizing the stem, the fungus also produces a toxin that spreads through the plant and causes chlorosis of the leaves and the appearance of many inter-vein spots, necrotic brown to black and angular.

Small concave and dark brown lesions develop on green fruit. They often have concentric rings. Black mould rots are sometimes found on fruit (see also p. 393).

In all cases, *A. alternata* f. sp. *lycopersici* sporulates on the affected tissues, in the form of short conidiophores bearing chains of brown multicellular conidia (Figure **54**, p. 395).

(See also Photos **582–584**, and Figure **45**.)

Small wet lesions appear on the upper side of the leaf. They grow radially and are sometimes locally limited by a vein. They are circular and can reach 2 cm in diameter. They are also surrounded by a highly visible yellow halo. Concentric patterns form, similar to those caused by *Alternaria solani*.

Brown longitudinal lesions also appear on the stem and petioles, and sometimes completely surround these organs, causing the leaves and leaflets to desiccate.

On young fruits, the lesions are a dry consistency and are small, light brown with a darker edge. They are wider and circular on ripe fruits. They are slightly recessed and eventually turn brown and split in the centre. Fungus fructifications can dot the lesions, sometimes giving them a slight dark grey to black tinge (Figure **55**, Photo **402**).

(See also Figures **30** and **31** and Photo **779**.)

Elements of biology

Alternata f. sp. *lycopersici* is a fungus widely distributed on plants and their environment. It is saprophytic and readily survives on and in the soil, on the most varied plant debris, either from tomato or from other plants, cultivated or not.

Its conidia germinate and the germ tubes penetrate the plant directly or through various injuries. Its mycelium invades the tissues and soon produces conidiophores and conidia (Figure **54**), which are dispersed by wind and air currents, and sometimes by splashing water.

C. cassiicola is extremely polyphagous, and attacks a variety of crops (peppers, eggplant, tobacco, melon, cucumber, some beans, hydrangea, soybeans, rubber, sesame, cotton) or weeds (*Commelina benghalensis*, *Verronia cinerea*, *Aspilia africana*, *Lepistemon* sp.) which ensures the propagation of this fungus and also act as reservoirs. It easily survives on plant debris for over 2 years. Infections occurs during wet periods, via the stomata or directly through the cuticle. Once in the tissues, the fungus colonizes fast and sporulates on the damaged tissues, forming characteristic elongated conidia, slightly sub-hyaline to brown, isolated or in short chains. These have four to 20 pseudo-partitions (Figure **55**). They are scattered during the morning by the wind,

Its development is encouraged by humid climatic conditions, the presence of water on the plants (rain, dew, sprinkler irrigation), and temperatures of about 25°C.

rain, and splashes. They result in secondary infection.

This fungus is encouraged by heavy rain, long periods of humidity, and by temperatures of 24–31°C, optimal around 28°C.

Control methods[1]

Several prophylactic methods recommended to control the other tomato *Alternaria* spp. can be applied to limit the development of *A. alternata* f. sp. *lycopersici* (see Description 1, p. 419).

In general, chemical control is not successful against this pathogen, only the use of resistant varieties provides effective and sustainable tomato protection.

Almost all varieties are resistant to *Alternaria* due to the '*Asc*' gene. Note that there are some claims of resistance to *A. alternata* f. sp. *lycopersici* in the seed catalogues that wrongly suggest a resistance to *A. tomatophila*.

Long crop rotation are required, avoiding the use of any sensitive crops.

Avoid setting up new crops close to sensitive or affected crops.

Remove the diseased old leaves at harvest time and do not, under any circumstances, leave them on the soil.

Fungicides can be applied preventively during risk periods, or as soon as the very first symptoms are detected. The following active ingredients are reported in the literature for their efficacy against *C. cassiicola*: chlorothalonil, mancozeb, maneb, copper, dichlofluanid, iprodione. Unlike cucumbers, no resistant tomato varieties are currently available. However, a few lines and cultivars (*Lycopersicon esculentum*, and *L. pimpinellifolium*) are very resistant as a result of varietal screening.

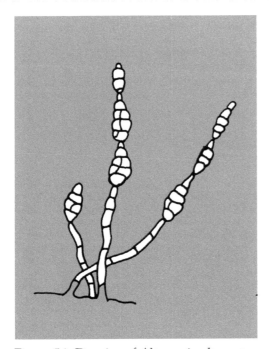

Figure 54 Drawing of *Alternaria alternata* f. sp. *lycopersici*.

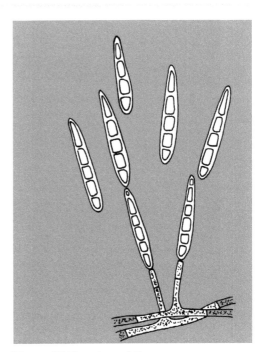

Figure 55 Drawing of *Corynespora cassiicola*.

[1] See note to the reader p. 417.

457

Pseudocercospora fuligena (Roldan) Deighton (1976)
Cercospora leaf spot (black leaf mould)

(*Mycosphaerella*, Mycosphaerellaceae, Capnodiales, Dothideomycetidae, Dothideomycetes, Ascomycota, Fungi)

Septoria lycopersici Speg. (1881)
Septoria leaf spot

(*Mycosphaerella*, Mycosphaerellaceae, Capnodiales, Dothideomycetidae, Dothideomycetes, Ascomycota, Fungi)

Distribution and damage

This fungus is found in crops with high humidity and poor ventilation in several tropical production areas around the world. It is reported in Asia (including Japan and China), India, Africa, and Central America. It is responsible for symptoms that could be described as 'black cladosporium' (see the damage associated with *Mycovellosiella fulva*).

P. fuligena is not prevalent in Europe.

Note that *Cercospora diffusa* has been isolated from tomato in Cuba.

S. lycopersici is reported in virtually every continent and in many countries in open field and protected crops. It is present in Europe, particularly in several East European countries, Germany and Italy. In prolonged wet conditions, it is likely to cause major damage and significant destruction of the foliage and, subsequently, sunburn of fruit.

Two races were isolated in the US many years ago.

Symptoms

Chlorotic spots appear on the upper side of the leaf, similar to those induced by *Mycovellosiella fulva*. First pale yellow, they have a poorly delimited outline but spread and eventually are covered by a sporing down, first white then gradually becoming black as spores are formed. Eventually some large spots covered with a velvety matt-black coating are seen on the leaf, and are visible on both the underside as well as upperside (see Figure **32**). Its conidiophores are brown, segmented, and measure 3.5–5 × 25–70 µm. They carry sub-hyaline conidia with multiple partitions and measure 3.5–5 × 15–120 µm.

The development of the spots and their eventual fusion can lead to necrosis and desiccation of a variable number of leaflets and leaves that remain in place on the plants.

The fungus may also colonize the petioles, the stem, and the tomato fruit peduncle at all stages of development.

(See Figure **32** and Photo **356**.)

The symptoms of Septoria leaf spot are initially sometimes confused with those of Alternaria. Thus, small circular spots, moist at first, appear on the leaf of the old leaves. They spread slightly, become brown and are limited to a diameter of 2–6 mm. A chlorotic halo surrounds them and the centre becomes necrotic and is tan to a whitish colour.

Its tiny brown to black pycnidia are visible in the leaf tissues (Figure **56**). They generate filiform and multi-compartmentalized hyaline conidia, whose average dimensions are 67 × 3.2 µm. In times of high humidity, the disease spreads to the young leaflets and some spots may coalesce, resulting in entire sectors of the leaf being affected and leaf drop.

Similar symptoms but less extensive and darker in colour, are described on other tomato aerial parts; however, their description in the literature is rather vague.

(See Figures **28** and **29** and Photo **309**.)

Elements of biology

P. fuligena survives on diseased plants and plant debris in its mycelial form or through its conidia which can survive for more than 18 months. It can affect some weeds such as *Solanum nigrum* so may remain in or near crops. The same is true for other Solanaceous plants such as *Solanum* and *Capsicum*. Spread is fast in humid conditions, through the stomata. Relative humidity of at least 85% is required for the spores to germinate in 5 hours. The germ tubes penetrate through the stomata. Once in place in the tissues, the fungus sporulates abundantly, producing a velvety black conidial coating. The conidia are then dispersed by wind over long distances, and by splashing water from rain or overhead irrigation. They are also dispersed by the workers and their tools.

The fungal pathogen develops in the presence of water on the plants during periods of prolonged humidity (rain, dew, fog). Its optimum temperature is around 27°C.

This fungus survives in the soil on plant debris. It also survives on a number of Solanaceous hosts, cultivated or not, including *Solanum nigrum*, *S. carolinense*, *Physalis subglabrata* and *Datura stramonium*. It also survives on the equipment used for cultivation, notably on the stakes and on contaminated seeds. It develops during wet periods in the presence of free water on the leaves. Conidia present on the surface of the leaf germinate and the mycelium penetrates through the stomata and then rapidly invades the tissues. Symptoms are already visible after 6 days, and *S. lycopersici* sporulates 14 days after infection at temperatures between 15 and 27°C. Spores formed in the pycnidia (Figure **56**) are dispersed by water splashing from rainfall or sprinkler irrigation Some insects transmit the spores from one plant to another, as does the movement of workers and equipment in wet vegetation. Seed contamination could contribute to the survival and dissemination of the disease over long distances.

Temperatures between 20 and 25°C and extended periods of high humidity are very favourable conditions for its development.

Control methods[1]

Use crop rotations. Plants should be staked and planting density reduced. Avoid planting near other sensitive members of the Solanaceae (peppers, eggplant) and eliminate black nightshade from crops and their surroundings. As soon as the first symptoms are detected fungicide treatments need to be applied. Chlorothalonil is effective against *Pseudocercospora* spp.

Avoid sprinkler irrigation, or carry it out during the day to allow the vegetation to dry quickly.

Tying and leaf stripping can improve the ventilation of the crop and canopy, making the plants less vulnerable to the disease.

Differences in sensitivity between varieties exist.

Some sources of resistance have been identified, notably from *Lycopersicon esculentum* and *L. hirsutum*.

This fungus easily survives on plant debris, which will need to be removed during and after cultivation.

Crop rotations of at least 1–2 years are required. Future crops should be grown in disease-free plots which are well drained and with a reduced plant density.

Seeds must be from production areas free from the disease. Care should be taken to remove weeds that may harbour the fungus.

It is best to stake the plants so the foliage is not in contact with the soil, and to irrigate plants in the course of the day so that they dry quickly. Mulching the soil also contributes to a reduction of the disease. Do not work in the plots where the vegetation is wet.

In badly affected countries, preventive or protective treatments may be carried out at regular intervals with fungicides: copper, maneb, mancozeb, thiophanate-methyl, benomyl, chlorothalonil, tolylfluanid. In many countries where antimildew/*Alternaria*/*Oidium* treatments are used these may also control Septoria leaf spot.

Biofungicides likely to limit the development of Septoria leaf spot include *Cryptococcus laurentii* and *Pseudomonas putida*.

Resistant varieties are reported in the literature. Also, most wild species of *Lycopersicon* have been shown to be resistant to Septoria leaf spot in artificial inoculations tests.

Plant debris should be buried deep after harvest.

Figure 56 Drawing of *Septoria lycopersici* fructifications.

[1] See note to the reader p. 417.

459

833 A young acervulus is forming on a pseudo-sclerotium of *Colletotrichum coccodes*. A few hyaline conidiophores sometimes carrying hyaline and cylindrical conidia are visible. The size of conidia is slightly variable (15–20 × 3–4 μm).

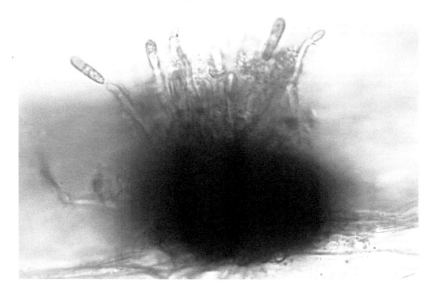

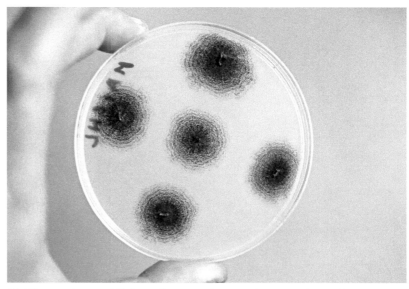

834 Young colonies of *Colletotrichum coccodes* on malt-agar medium. They are sclerotic and some black, pink areas are producing acervuli.

Soil fungi

Colletotrichum coccodes (Wallr.) S. Hughes (1958)

Anthracnose, black dot, root rot

(Anamorph of Glomerella, Phyllachoraceae, Phyllachorales, Incertae sedis, Sordariomycetes, Ascomycota, Fungi)

Principal characteristics

This fungus has two synonyms: *C. atramentarium* (Berk. & Br) and *C. Taube phomoides* (Sacc.) Chester. Polyphagous, it is reported on every continent and in many countries that produce tomatoes and potatoes.

Isolates of *C. coccodes* appear to have quite variable pathogenicity, growth rates, and sclerotia size. They primarily cause root and fruit damage on tomato. Other species of *Colletotrichum* are reported on the fruits, notably *C. gloeosporioides* (Penz.) Penz. & Sacc. in Penz. and *C. dematium* (Pers.) Grove.

• Frequency and extent of damage

C. coccodes is very common in all types of cultures, in field and protected crops, in soil and soil-less systems. Commercial tomato growers especially fear its attacks on fruits, that mostly appear on ripe and over-ripe fruits. This condition is considered the most serious threat to mature fruits in processing tomato production in several production areas of the US. In these areas and without any control methods, the proportions of rotten fruit can reach 70%.

Damage to roots by this fungus are often underestimated. Long considered a secondary pathogen associated in particular with attacks of *Pyrenochaeta lycopersici*, it must now be regarded as a serious tomato fungal pathogen, especially in intensive soil-grown systems.

It results in damage to a variable extent in soil and soil-less cultures. It is now developing on the KNVF type rootstocks used as an alternative to methyl bromide. This is notably the case in Italy and France, where it threatens the choice of this crop. In some countries, it acts in combination with other soil-borne pathogens/pests attacking the roots, such as *Rhizoctonia solani*, *Phytophthora nicotianae*, several species of *Meloidogyne*, and *Globodera tabacum*.

• Main symptoms

This fungus seems able to attack almost every part of the tomato: roots, leaves, stems, and fruits. It is on fruits and roots that its symptoms are most frequently seen.

On fruits, symptoms of anthracnose appear at maturity in the form of tiny light brown lesions, which evolve gradually into small circular spots, moist and dull. As they expand, they become concave and their centres gradually turn brown. The underlying flesh is discolored with a grainy texture. Concentric rings are sometimes visible as well as dark brown to black structures: the microsclerotia of *C. coccodes*. One can also observe acervuli with black hairs (setae). In humid conditions these acervuli produce masses of salmon-coloured spores in a mucous, visible on the surface of the lesions that remain smooth and intact. Often, numerous spots develop on the same fruit. As they merge, they cover large portions of the fruit and lead to rot.

On the roots, the fungus is responsible for brown lesions. Fine roots are few, if any, as a cortical brown rot has destroyed them. The cortex of the main roots displays dark brown lesions, that can be corky. Rotting of the base of the stem has sometimes been reported. Sometimes the cortex splits revealing the central vascular cylinder. In hydroponics, the attacked roots are weak, discolored, and partially decomposed. Root rot and leaf wilts are also attributed to this fungus.

In addition to the associations mentioned above, *C. coccodes* is found with *R. solani* in the US on stem base cankers in processing tomato crops. On the stem, it seems to act mainly as a secondary invader because its development is rather limited. On the roots, the combination of these two fungi can lead to more severe effects.

On potato, when in combination wirth *Verticillium dahliae* it leads to much more severe wilt than that observed when the plants are infected with only one of these pathogens. It is also a component of the micro-organism complex causing brown root rot (brown root rot complex), which also includes *Fusarium oxysporum* f. sp. *radicis-lycopersici*, *F. solani*, *Pyrenochaeta lycopersici*, *Pythium debaryanum*, and *Rhizoctonia solani*.

Note that in some heavily contaminated soils, *C. coccodes* is able to destroy germinating seeds and tomato seedlings. This is a very rare occurrence now. In addition, it can produce small, circular, brown spots on leaves, surrounded by a yellow halo.

(See Photos **459, 477, 487–491, 758–761.**)

• Biology, epidemiology

Survival, inoculum sources: although a poor competitor *C. coccodes* readily survives in the soil, on plant debris or through its sclerotia that enable it to remain there for several months or even years. It is also able to colonize at least 58 plant species belonging to at least 19 botanical families, primarily vegetables belonging to the Solanacea family (peppers, eggplant, potatoes) and to cucurbits (watermelon) that can act as alternative hosts or contribute to the increase in soil inoculum. It has been successfully inoculated to strawberries and onions. Many weeds can be colonized: *Solanum capsicastrum*, *S. dulcamara*, *S. nigrum*, *Abutilon theophrasti*, *Amaranthus retroflexus*, *Chenopodium album*, *Convolvulus arvensis*, *Capsella bursa-pastoris*. It has also been isolated from the roots of various plants without any symptoms: cabbage, lettuce, watercress, mustard, chrysanthemum. It has been isolated from water used for soil-less crops and from water storage tanks in the open.

Penetration and invasion: the sclerotia may produce mycelium or acervuli, in which conidia are formed. The latter germinate on the surface of green and mature fruits and adhere by means of appressoria. From these structures, the penetration is made directly through the cuticle. It can also be achieved through various wounds present on mature tomatoes.

The penetration method in the roots is similar as some appressoria form when in contact. Root colonization increases with age. In addition, the sclerotia, then the mycelium can also cause infections by coming into contact with fruits and roots. Once in place, the mycelium colonizes the tissues under the cuticle or cortex tissues aided by the production of an extracellular protease. The infection can also remain dormant which is referred to as 'latent infection'. This type of situation can occur on the green fruits, on the roots, and even on the stem. Latency on green fruits disappears when they mature or if they are exposed to low temperatures.

Sporulation and dissemination: on the fruits or, to a lesser extent, on the roots, *C. coccodes* produces intra-or sub-epidermal acervuli which are its asexual means of reproduction (Photo **833**). These structures, 200–300 μm in diameter, produce cylindrical hyaline and nonpartitioned conidia (16–24 × 3–7.5 μm) which are immersed in a gelatinous matrix that protects them from desiccation. These conidia ensure part of the fungus dissemination, which is carried out by water splashes from rain and overhead irrigation and the movements of workers, animals, and equipment in the crops. Flowing water and soil particles transported by tillage equipment also contribute to its spread. It is also easily spread by the nutrient solution in soil-less cultures, and even more so if it is recycled. Microsclerotia, often less than 1 mm in size, are also formed in the tissues; they contribute to preserve and to disseminate *C. coccodes*.

Conditions encouraging development: the fungus is able to develop over a broad range of temperatures, although germination of its conidia is optimal at 22°C. At this temperature, the appressoria form after 6 hours. Germination does not takes place at 7°C and is strongly slowed down at 10°C and above 35°C.

Once its conidia have germinated, the development of anthracnose is greatly influenced by climatic conditions. Attacks can occur at temperatures between 15 and 30°C; they are much more limited at 10°C, and no damage occurs above 38°C. Sporulation on the fruit is optimal at 28°C. High humidity and the presence of liquid water are necessary for spore germination: rainy periods and sprinkler irrigation encourage anthracnose. Free water on the fruits for several hours, linked to rainfall and length and frequency of sprinkler irrigation, greatly increases its impact on the crop.

Tomato or KNVF rootstock type monoculture on the same soil increases the rate of inoculum, and therefore an aggravation of root symptoms. Moist soils or excessive salinity predispose plants to attack by *C. coccodes*.

Control methods[1]

• **During cultivation**

Root attacks

By the time root system symptoms are observed it is too late to control this disease. Indeed, there is no effective way to eliminate the parasite on the roots of plants.

To keep *Pyrenochaeta lycopersici* plants alive:
– earth up the plants to encourage adventitious roots that can supplement the main roots affected. In soil-less culture (of peat or peat + pozzolan) and during severe attacks, peat should be applied to the stem base to allow additional rooting. Sawdust is sometimes used;
– damp the plants during the hottest parts of the day to avoid excessive and uncompensated evaporation, which leads to wilting, desiccation, and plants death;
– closely monitor irrigation. Indeed, if the plants wilt, it is not necessarily related to a lack of water, but rather to root damage caused by *C. coccodes*. In some cases, producers tend to increase irrigation in response to wilting, which increases root lesion from asphyxiation.

In addition, carefully remove and destroy diseased plants and their root systems. Also remove rotten fruits at all times but especially at the end of the crop in order to prevent them being buried in the ground and thus adding to the inoculum of *C. coccodes* already present.

Attacks on fruit

The fruits, which are sensitive to anthracnose, should be harvested as soon as possible after

maturity. Sprinkler irrigation must be well controlled by avoiding excesses and the rapid drying of the vegetation should be encouraged.

Treatment for plants can include the following fungicides: chlorothalonil, difenoconazole, mancozeb, maneb, anilazine, copper, thiophanate-methyl, from the time of fruit formation to harvest time. Renew applications after heavy rain and sprinkler irrigation.

• **Subsequent crop**

Root attacks

If plant propagation is to take place each year in the same glasshouse, it is then necessary to implement the hygiene measures recommended on p. 541. A healthy substrate (disinfected) should be used and avoid laying the blocks directly on the soil. They often become contaminated by contact with it, especially if it has not been disinfected.

Crop rotations are only significant if done preventively, before the soil becomes heavily contaminated. They should last at least 3–4 years. Of course, the other plants used in the rotation should not be sensitive. In addition, weeds which may harbour the fungus must be removed from the plots. The soil must drain well and be well worked: good sub-soiling allows root access to new layers.

Highly contaminated soils must be disinfected. Several fumigants can be used (dazomet, metam-sodium). Be aware that these fumigants are partially effective, some are problematic and have a number of drawbacks, including:

[1] See note to the reader p. 417.

– the destruction of natural micro-organisms antagonistic to pathogens;

– the increased sensitivity of soils, potting soils and disinfected substrates as a result of the elimination of antogonists.

Soil disinfection by steam is also possible. In addition, in sunny regions, solar disinfection of the soil (solarization or pasteurization) may be done. It involves covering very well prepared and moistened soil with a polyethylene film 35–50 µm thick. It is held in place at least for 1 month at a very sunny time of the year. This is a cost-effective and efficient method for control of C. coccodes.

Note that in some countries, treatment of the soil before planting has been proposed, with the fungicide solutions carbendazim or thiophanate-methyl.

In hydroponics, bags, pots, trays, etc. must be disinfected or replaced. Only limited information on disinfection of substrates is available. Steam treatment is effective if adequate temperatures are reached. Metam-sodium is not suitable for substrates made from organic materials, but it is to be used for all mineral substrates (sand, perlite, pozzolana). A dry substrate requires injection treatment repeated several times, followed by leaching it thoroughly. Formaldehyde has been tested successfully to disinfect a mixture of peat and pozzolana (by soaking the substrate with a 3% solution of commercial formalin).

Apart from soil disinfection, the attacks of C. coccodes may be delayed by setting up the seedlings in warmed up soil and/or by increasing the volume of healthy substrate used for the blocks. It is best to mulch the ground and stake the plants to reduce attacks on fruits.

If the irrigation water is contaminated, an alternative source of water supply must be used.

To our knowledge, no resistance to root attack by C. coccodes has been described. Grafting is not a feasible alternative because the KNVF type rootstock are susceptible to C. coccodes (see p. 659), and its extensive use leads to an inexorable increase in root damage.

Genetic resistance to anthracnose on fruits has been identified and is likely to be polygenic and partially dominant. Nevertheless, the level of protection achieved is not sufficient and must be supplemented by chemical protection, especially in crops where the disease occurs frequently and is serious.

The use of predictive models, developed for early blight, do not appear to aid the timing of fungicide applications.

Note that some antagonistic fungi and bacteria to C. coccodes such as Coniothyrium minutans, Trichoderma harzianum, and Streptomyces griseoviridis have been evaluated in vitro or in the field to control this disease.

Fusarium oxysporum f. sp. *radicis-lycopersici* Jarvis & Shoemaker (1979)

(Anamorph of *Gibberella*, Nectriaceae, Hypocreales, Hypocreomycetidae, Sordariomycetes, Ascomycota, Fungi)

FORL (Fusarium crown and root rot)

Principal characteristics

Nine vegetative compatibility groups (VCG) have been detected in *F. oxysporum* f. sp. *radicis-lycopersici* (0090–0099). Groups 0090, 0091, and 0094 are quite cosmopolitan, the other six have a more limited distribution. For example, five groups are established in Italy, and are also present in Israel: VCG 0090, VCG 0091, VCG 0092, VCG 0093, and VCG 0096. Two groups, VCG 0090/sub-group II and VCG 0091/sub-group I, have been reported in Turkey. The first one was also identified in Cyprus. In northwestern European countries (Belgium, Netherlands and UK), the group VCG 0094, divided into three sub-groups, predominate, except in France where VCG 0090 III sub-group was dominant (in addition to VCG 0090 sub-groups I and II, to VCG 0091 sub-groups I and II, and to VCG 0094 sub-groups I and II), as in Israel. The group VCG 0094 is a founder population in Europe resulting from inter-continental migration of some isolates from Palm Beach County, US. It is also from this county that *F. oxysporum* f. sp. *radicis-lycopersici* has migrated to other states, such as Florida. Two new groups, VCG 0098 and VCG 0099, were detected in the state, but with low frequency. Note that there is also a VCG 0097.

These different groups and sub-groups observed in the same production areas demonstrate the high level of genetic diversity expressed by this fungal pathogen of tomato. In addition, there are also differences in aggressiveness between strains.

Note that other *Fusarium* sp. have been associated with tomato root and stem base damage, especially *F. solani* (see end of this description).

• Frequency and extent of damage

This fusarium root rot was first described in Japan in 1969 (with yield losses of up to 40%) and later in California in 1971 and in other states of the US (yields reduced up to 50%). It is now present in many production areas around the world (Canada, Mexico, Korea, Europe, Mediterranean) and can occur in the field and in glasshouses both in soil and soil-less crops. It developed in Europe during the 1980s (Netherlands, Sweden, Belgium, UK, Germany, Spain) and in the Mediterranean region (Israel, Greece). It has caused considerable damage in soil-less crops, on various substrates (peat, rockwool), especially during the colder time of the year.

In a number of countries it was first recorded in the 1980s, following the use of imported peat. *F. oxysporum* f. sp. *radicis-lycopersici* later quickly spread to many glasshouses, causing considerable losses, especially in soil-less crops in winter.

Fortunately, the development of varieties resistant to this aggressive fungus has effectively solved this disease problem in many countries. The arrival on the market of new varieties resistant to Tomato yellow leaf curl virus (TYLCV; see Description 41) but sensitive to *F. oxysporum* f. sp. *radicis-lycopersici* showed that this resistance could not be ignored as damaging root *Fusarium* attacks reoccurred. This confirms that the fungus is permanently present on many farms.

F. oxysporum f. sp. *radicis-lycopersici* was recently reported in several countries of the Mediterranean

465

where it is effects are variable (Crete, Turkey, Malta, Tunisia), and in Slovakia.

• Main symptoms

F. oxysporum f. sp. *radicis-lycopersici* can attack tomato seedlings and cause their death. The disease occurs mainly near to harvest time when the plants are laden with fruits. In contrast to vascular diseases, some wilting, not always severe, appears on leaflets and leaves at the top of the plant, where the stem is also reduced in thickness. Depending on the plant, wilt can initially be reversible overnight, and its impact can vary depending on weather conditions. Wilting can also be sudden, rapidly evolving towards necrosis and drying of the leaves and leaflets, and leading to plant death. Some authors also report the onset of leaf yellowing on the periphery of the leaf of older leaves. These are followed by necrosis of the petioles and leaf drop. Some early affected plants have stunted growth.

Regardless of the wilting severity, the primary symptoms are to be found on the roots and stem base of plants. Many lesions appear on the first ones, on the cortex and on the vascular cylinder. They are reddish brown and moist, changing quickly into rot. The smaller the diameter of the roots, the more they rot and decompose rapidly. This is particularly the case with some substrates in soil-less crops where affected roots are particularly abundant.

A canker frequently develops on the stem base. It is initially dark brown, and the diseased tissues can be recessed. The canker usually develops longitudinally on one side of the stem, taking the form of a flame that can sometimes reach 30 cm above the stem base. The central part of the canker is a salmon pink colour with a rather mucous appearance, due to fungal sporodochia forming.

It should be noted that the vascular system also has some symptoms, although this is not a vascular disease. In general, the central cylinder of large roots shows some pronounced browning. The same is true for the vascular tissues at the stem base and those located on either side. The browning can be very pronounced and spreads within the stem over several tens of centimetres above the stem base. Adventitious roots may develop on the stem.

The fruits of the diseased plants are often flaccid and dull.

(See Photos **442–446, 458, 531–535, 669, 670.**)

• Biology, epidemiology

Survival, inoculum sources: the fungus is able to survive in the soil through crop residue harbouring some mycelium, micro-and macro conidia, and chlamydospores with thick and strong walls. It is also found in the dust of the glasshouses. *F. oxysporum* f. sp. *radicis-lycopersici* has saprophytic life qualities that enable it to colonize and survive on various organic compounds, and to remain in soil-less cultures, even with resistant varieties.

It is likely to grow on many plants belonging to different botanical families, without them always displaying symptoms: peppers, eggplant, beans, peas, beans, melons, beets, and spinach. Several weeds are host without any symptoms: *Chenopodium album, Solanum nigrum, Panicum fasciculatum, Trifolium repens, T. pratense, Schinus terebinthifolius, Mollugo verticillata, Gnaphalium* sp., *Stellaria media, Spergulum arvense, Rumex crispus, Plantago lanceolata, Amaranthus retroflexus, Scoparia* sp., *Capsella bursa-pastoris,* and *Polygonum* convulvulus. For example, the fungus multiplies on a weed *Tamaris nilotica* common in tomato crops in Israel. It is both found in the roots and on inflorescences of this plant which ensures its survival in the soil. Contaminated seeds are a means of distribution.

The pathogen is present in glasshouses, particularly in the irrigation and drainage systems as well as in the nutrient solution recycling system. The water storage tanks are often contaminated. Water is a significant source of inoculum, as is the nutrient solution later on. It has been shown that the fungus can survive more than 52 weeks in a nutrient solution stored at room temperature.

F. oxysporum f. sp. *radicis-lycopersici* is likely to infect the seeds of tomato. For example, seed infection rates of 0.01–0.1% were found on plants with infected stems and fruits. There is evidence that the fungus can persist for more than 12 weeks on stored seeds. Disinfection of seeds with bleach does not seem to be entirely effective against this fungus.

Penetration and invasion: after germination of its chlamydospores, *Fusarium* penetrates the roots of tomatoes through root tip natural wounds (the point of emission of secondary roots) or through accidental ones. Direct penetration in epidermal cells is possible. Recent work has shown that the fungus mycelium can contact the root hairs, intermingling with them and eventually establishing on the rootlets. The preferential site of infection of the root surface appears to be in the grooves at the junction of epidermal cells. The fungus invades the tissues and eventually colonizes the entire root system.

Sporulation and dissemination: the fungus produces numerous micro- and macroconidia or chlamydospores on and in the colonized tissues, be they located on the roots or on the stem. Its dissemination may occur on shoes and farm equipment by means of contaminated soil and also through plant debris (tillage equipment, boxes, stakes). Soil dust containing various spores is easily dispersed by air currents, and by splashing water.

Note that air contamination is possible through the micro-and macroconidia that are carried by air currents. These spores can be deposited either on aerial vegetation, or on the soil or on soil-less substrates. Injuries seem to encourage foliar contamination. It has been shown that the fungus can spread over a radius of 4 m around an infected plant, in particular by root contact between plants.

The spread of *F. oxysporum* f. sp. *radicis-lycopersici* is also possible by irrigation water, or by the nutrient solution that has been contaminated during tray storage, or when it is recycled from contaminated substrates. Indeed, the fungus spreads rapidly in soil-less culture systems, notably in those using a recycled solution. It is found 1 week after artificial inoculation, in both the drainage water and the nutrient solution in recycling systems.

Infected seedlings without symptoms and contaminated stakes, contribute to its spread.

Substrates and composts can also be contaminated, and seed is another potential source.

Some insects of the genus *Bradysia* could play a role as a vector and promote the dissemination of this fungus. The sciarid fly also ensures the dissemination of FORL. It also influences its parasitism due to the injuries they cause to the roots.

Conditions encouraging development: *F. oxysporum* f. sp. *radicis-lycopersici* seems to prefer rather low temperatures. Its optimum temperature is between 18 and 20°C and the most serious attacks take place at temperatures between 10 and 20°C.

Inoculations of tomato seedlings showed that *Fusarium* rot could be influenced by certain micro-and macronutrients. The severity of the disease is increased for example in the presence of ammonia nitrogen, but also NaH_2PO_4, $MnSO_4$, $ZnSO_4 \cdot 7H_2O$, and to be reduced with $Ca(NO_3)_2 \cdot 4H_2O$, $CuSO_4 \cdot H_2O$. In addition, the root disease is particularly severe in acid or water saturated soils. Excessive salinity increase its incidence and symptom severity.

In Israel, it was shown that the use of salt water increased the severity of the disease. The same is so in the presence of small amounts of $CaCO_3$.

Note that *F. oxysporum* f. sp. *radicis-lycopersici* colonizes or rapidly recolonizes recently disinfected soil and substrates. In addition, plants that have suffered water or heat stress are more sensitive.

The fight against *F. oxysporum* f. sp. *radicis-lycopersici* is not easy because the fungus survives in soils and substrates for a very long time, and it recolonizes them very quickly when they have been disinfected. Therefore, for the control of this soil fungus to be effective it should involve a complete set of measures and methods to eliminate it or to limit its development.

467

Control methods[1]

• During cultivation

No control method is good enough to control this disease once it is observed during cropping. In order to keep plants alive as long as possible, they must be treated quickly with a registered fungicide, by applying at the base of the plants or through the drip irrigation system. Several active ingredients have been shown to be effective to varying degrees: hymexazole, thiophanate-methyl, carbendazim, benomyl (beware of the risk of phytotoxicity especially in case of crops on rockwool, see p. 138), and prochloraz. The effectiveness of these treatments is not always high. It is important to know that benzimidazoles resistant strains of *F. oxysporum* f. sp. *radicis-lycopersici* have been detected in cultures where these products have been used repeatedly.

In soil cultivation, mounding plants allows the growth of adventitious roots that will supplement the diseased roots. In hydroponics, peat can be applied to the stem base of the plants to encourage the growth of new roots. In such crops the first fruits must be harvested quickly to reduce the stress on the plants. Plants should not be irrigated with water that is too cold.

During cultivation, the first diseased plants should be removed with care. To do this, they are placed in a plastic bag to avoid contact with other healthy plants. They must be destroyed as soon as possible. If an outbreak is present in the crop, affected plants can be tagged, quarantined, and worked separately. Carts and harvest containers from unaffected areas must not be used or move across this area.

Visitors should not be allowed in the affected areas. Foot baths should be installed at every entrance and the disinfectant solution renewed frequently (see p. 540).

It is quite common to bury crop residues in the soil after harvest. Buried plant tissues are abundantly colonized by *F. oxysporum* f. sp. *radicis-lycopersici* and produce numerous chlamydospores. The removal of plants with their root systems limits this phenomenon and reduces the amount of inoculum left in the plots. Plant debris should not be stacked and stored near plots or glasshouses where tomatoes will be grown later. Indeed, they are important sources of inoculum that should be destroyed as soon as possible. Otherwise, wind, the workers, and some insects will introduce *F. oxysporum* f. sp. *radicis-lycopersici* into crops. If it cannot be removed, the pile of debris must then be covered with a plastic film to form a mechanical barrier.

• Subsequent crop

Production of tomatoes should be avoided in already affected plots. Crop rotations with nonhost plants (like corn or similar plants) will help prevent the onset of the disease or limit its impact. To be effective, rotations must be long enough. Avoid eggplant, peppers, and some of the vegetables mentioned above in the rotation with tomatoes. Lettuce, which is not sensitive, can be used.

In protected crops, in particular soil-less ones, the whole farm should be disinfected in order to get rid of the maximum amount of inoculum. Therefore the surface of the internal structures of the glasshouses should be disinfected. Formalin, an excellent surface disinfectant that acts very quickly when mixed with steam, is sometimes used. Other products shown on p. 436 can be used. Bags, cubes of substrate, and other materials that may have been contaminated should be removed. If they are to be re-used, they must be disinfected.

Soil disinfection has only a short term and unsatisfactory effect. Indeed, like all *Fusarium* diseases, FORL quickly recolonizes the disinfected soils. In addition to steam, several fumigants can be used: chloropicrin, metam-sodium, and dazomet. The nature of the soil and its organic matter content influence the effectiveness of these fumigants. Cyanamide calcium can reduce the soil inoculum. The combination of metam-sodium and formaldehyde has

[1] See note to the reader p. 417.

been successful in Israel. The effectiveness of the disinfection of soil-less substrates is often variable. It is better to remove them and start with fresh substrate. The floor of the glasshouse must be concrete or covered with a plastic film which is replaced as soon as tears appear. This will isolate the crop from the soil and avoid aerial contamination via soil dust in particular. For the same reasons, water tanks and troughs of nutrient solution should be covered.

Solarization may also be used effectively. It may sometimes be partial in its efficacy, since *F. oxysporum* f. sp. *radicis-lycopersici* is less heat sensitive than some other soil-borne pathogens such as *Pythium* spp. or *Sclerotium rolfsii*. Therefore, solarization is often associated with other biological control methods (such as antagonistic fungi like *Trichoderma harzianum*) or chemical ones (fumigants at low doses of dazomet, metam-sodium).

Ozone, used for ionization of water and associated with *Trichoderma* sp., reduces attacks of *F. oxysporum* f. sp. *radicis-lycopersici*. Some composts also limit its effects.

Chamaecyparis obtusa and *Cryptomeria japonica* fibre bark based substrates slow its development. The incorporation of whole plants of rice, soybean, or lettuce in the soil reduces the incidence of FORL in tomato.

It is essential to use healthy seedlings, produced in a healthy substrate, and the blocks should not have been laid on the ground. Nurseries should not be located near fields or in glasshouses where crops have been affected by the disease. Aerial contamination of seedlings is always possible. Avoid over-watering the seedlings and injuring them when planting in the field. Trays and boxes re-used to hold the seedlings must be disinfected. Be particularly vigilant about their cleanliness. Those from farms affected by FORL should not be used. The seedlings should be raised in heated soil; plastic mulch should only be put in place after planting.

The tools used for working the soil in the contaminated areas must be thoroughly cleaned before being used in other healthy ones. The same must apply to tractor wheels. Thorough rinsing with water will often be sufficient to get rid of the infested soil.

Be wary of the sanitary quality of water used for preparing the nutrient solution and/or for plant irrigation, especially if it comes from an irrigation canal, stream, or tank that may have been contaminated.

If the nutrient solution is recycled, it can be disinfected. Several methods can be used: chlorination, iodination, ozonation, biofiltration, UV radiation, TiO_2 photo-catalysis. These methods are effective against a range of fungi, including the *Fusarium* sp., but some are not totally satisfactory. Indeed, some of them are partially effective and may be disadvantageous for the cropping system (interference with mineral nutrition, particularly iron). The drip irrigation pipes must be cleaned or replaced. The deposits are removed with an acid solution, and the disinfected system rinsed with water. In glasshouses where the nutrient solution is recycled, the measures taken are even more important. The system must be disinfected several times to make sure the *F. oxysporum* f. sp. *radicis-lycopersici*[2] is completely removed.

Avoid using ammonia nitrogen, and keep the soil pH at 6–7. The stakes should be disinfected.

Because of the possibility of aerial contamination via wounds, foliage protection must be considered. Also the possible sporulation of the fungus on the stem must not be ignored as this is the likely origin of the air inoculum.

The most effective method to control Fusarium root rot is to use resistant varieties. A 'Frl' dominant gene, derived from *Lycopersicon peruvianum* and located on chromosome 9, gives a high level of resistance. The close association of this gene with 'Tm-2' (resistance gene to tobacco mosaic virus) has aided the selection of resistant varieties. Several tomato rootstock of KNVFFr type are now resistant to this *Fusarium*. They allow the cultivation of susceptible varietal types, in soil-less culture and in infected soil.

Several bacteria interfer with Fusarium root rot, either directly or through the tomato:

---

[2] Note that *Fusarium oxysporum* is able to multiply in the drippers of soil-less culture irrigation systems and to clog them during cropping, disrupting irrigation. In fact, these fungi develop from organic matter (chelate, seaweed, root debris) accumulated in the drippers, with mycelial masses eventually sealing them. These are clearly visible the drippers are examined. Disinfection of the irrigation system with bleach or hydrogen peroxide can solve the problem.

Pseudomonas fluorescens (whose efficiency is influenced by zinc), *P. chlororaphis*, *P. putida* (induces acquired systemic resistance to *F. oxysporum* f. sp. *radicis-lycopersici* in tomato), *Bacillus megaterium* and *Burkholderia cepaciae* (controlling *F. oxysporum* f. sp. *radicis-lyco-persici* in combination with an addition of carbendazim).

Moreover, several fungi have been effective in controlling the development of *F. oxysporum* f. sp. *radicis-lycopersici*, and therefore the expression of this root disease: *Trichoderma harzianum*, nonpathogenic *Fusarium* spp. (avirulent strains of *Fusarium oxysporum* and *F. solani*), a hypovirulent and binucleate strain of *Rhizoctonia solani*, *Pythium oligandrum*, and *Gliocladium roseum* (alone or in combination with benomyl). In addition mycorrhizal fungi of the *Glomus* genus (*Glomus intraradices*) reduce the severity of the symptoms of Fusarium root rot on tomato.

Finally, note that several volatile oils and non-volatile substances adversely affect the development of *F. oxysporum* f. sp. *radicis-lycopersici* on tomato.

The cultural and morphological characteristics of *F. oxysporum* f. sp. *radicis-lycopersici* are not very different from those of *F. oxysporum* f. sp. *lycopersici*. Photos **867** and **868** show this fungus in a Petri dish and under a light microscope.

7 *Fusarium solani* (Mart.) Sacc. *(Haematonectria haematococca* [Berk. & Broome] Samuels & Rossman, *in* Rossman, Samuels, Rogerson & Lowen [1999])

This fungus is some what polyphagous and capable of causing symptoms on most parts of the tomato plant:
– damping-off;
– foot rot and root lesions often reddish brown (Australia, Egypt, Chile, Brazil, Argentina, California);
– fruit rot in the field and during storage.

Note that its association with *Meloidogyne incognita* and *M. javanica* on tomato roots has led to increased damage by these root-knot nematodes and to a greater reduction in plant growth. Moreover, parasitic synergy with *Rhizoctonia solani*, on the stem base and the proximal portions of roots, is also reported on tomato.

This fungus is able to attack many plants belonging to different botanical families, notably several vegetables: beans, peas, lentils, melon. Many of the methods suggested to control *Fusarium oxysporum* f. sp. *radicis-lycopersici* can be used against this pathogen.

Pythium spp.
Phytophthora spp.

Damping-off, foot and root rot, buckeye rot

(*Pythium*, Pythiaceae, Pythiales, Saprolegniomycetidae, Oomycetes, Oomycota, Chromista)

(*Phytophthora* sp., Incertae sedis, Peronosporales, Peronosporomycetidae, Oomycetes[1], Oomycota, Chromista)

Main characteristics

The many species of *Pythium* and *Phytophthora* that are pathogens of tomato are covered in the same description (except *Phytophthora infestans*, see Description 7) for at least three reasons: their biology and the symptoms they cause are quite similar (see *Table 43*, p. 474), their development is often influenced by the same conditions, and the protection methods implemented to control them are identical. Some of their morphological characteristics are listed in *Tables 43* and *44*, p. 478.

• Frequency and extent of damage

The Pythiaceae are present in all soils throughout the world and wherever tomatoes are produced. The vast majority of known species are polyphagous, notably on cultivated plants.

They cause damage on tomatoes throughout its production cycle, both on seedlings in nurseries and on adult plants in the field, both in soil and soil-less systems. Symptoms can be found on the roots, the stem base, the lower leaves, and even the fruit. With the development of soil-less crops, these aquatic micro-organisms have thrived; they are a limiting factor for production on many farms.

The sometimes intensive use of a few rare anti-Pythiaceous fungicides (especially metalaxyl) has led to the emergence of resistance in a number of countries in several species of *Pythium* and *Phytophthora*.

Damping-off caused by *Pythium* spp. is most frequently observed in extensive and/or poorly managed nurseries. It is mainly in soil-less crops that the damage caused by these fungi is sometimes spectacular. The diversity of the Pythiaceae was studied in the early 2000s and many species have been identified in soil-less systems. *Pythium* spp. are in the majority, *Phytophthora* spp. are not so common. Note that up to four different Pythiaceous fungi can be found on the roots of tomato and/or in the nutrient solution on the same farm. The study of their pathogenicity has shown considerable variation:

– some species are not pathogenic, or only mildly so (*Pythium torulosum*, *P. vexans*, *P. sylvaticum*, *P. dissotocum*);

– others have medium aggressiveness (*P. ultimun*, group F *Pythium*, *P. intermedium*);

– finally, several species are particularly aggressive (*P. irregulare*, *P. salpingophorum*, *P. myriotylum*, *P. aphanidermatum*).

---

[1] There are about 800 species of saprophytic or parasitic oomycetes which for a long time have been classified as Phycomycetes or 'lower fungi' (Eumycetes). This classification was revised a few years ago because the ultrastructure of their organisms, their biochemistry, and molecular sequences indicated that they belonged to the Chromista, which includes mainly algae (green and brown) and diatoms. Currently, according to bibliographic sources, they can be associated with either the Chromista (*Index Fungorum*) or the Stramenopila (*Tree of Life*).

P. aphanidermatum, a ubiquitous species on the roots of soil-less-grown cucumber plants is rarely harmful on tomato. When harmful, it is often on farms where cucumbers are also produced. It seems much more damaging in soil-less systems, especially in some countries (Canada, South Africa, Japan).

Among the *Phytophthora* sp., *P. nicotianae* is the most important. It is responsible for attacks on seedlings in nurseries and soon after planting in field crops as well as in underheated glasshouses. In hydroponics, it does not cause severe damage. Its accidental introduction into this type of culture, through contaminated seedlings, did not result in significant damage. This does not seem to be the case in other countries where effects are reported in soil-less crops, particularly in Germany. This *Phytophthora* is more known for the serious damage it causes on the fruits of field crops for the canning industry in particular. Its species name is somewhat controversial and many synonyms exist. It should be noted that several isolates have been reported worldwide, differing in their aggressiveness and virulence to a diversified host ranges. The tomato is rather sensitive to a high proportion of these isolates.

Of the Pythiaceae, only isolates of *Pythium* F group are widely present in the field, on almost all farms in soil-less culture in France. It should be noted that the aggressiveness on tomato of the isolates studied is highly variable. Other Pythiaceae, of either *Pythium* or *Phytophthora* spp., are found only very occasionally or in a rather limited number of crops.

• Main symptoms

Pythiaceae are likely to cause symptoms on various parts of the tomato throughout its development cycle (see *Table 43*).

On seedlings

Pythium spp. and *Rhizoctonia solani* are well known damping-off agents causing significant damage in nurseries. In addition to preventing seeds from germinating, they attack the seedlings both pre- and postemergence. Symptoms can be seen quite diverse depending on the stage of seedling development and on the species involved:
– wet and brown lesions on roots, sometimes reaching the stem base, causing them to decay;
– wet and soft tissue on the stem base which can spread over several centimetres. In this case, it can give the impression of being pinched. The affected tissues gradually turn brown.

Whatever the primary symptoms, seedlings are quick to wilt collapse, and die. They eventually break down and disappear completely (damping-off).

Several *Phytophthora* spp. can cause similar symptoms on seedlings. For example, *P. cryptogea* is responsible for brown to black lesions that develop on the stem, at or close to ground level, which eventually surround it. Subsequently, adventitious roots emerge on the top part of the stem while the roots eventually rot completely. Over time the leaflets can curl and turn yellow.

Note that with damping-off, very similar symptoms can be caused by *Rhizoctonia solani* (see p. 488).

In addition, several members of the Pythiaceae may be responsible for aerial blight on seedlings or on mature tomato plants. For example, moist and brown leaf lesions, of various sizes, or even rots have been observed in Florida, US, on plants that have been splashed by water after heavy rains. Similar lesions were also visible on the petioles and stem. *Pythium myriotylum* has been associated with this damage.

On adult plants

Pythium spp. are frequently isolated on either apparently healthy or affected tomato roots. Their presence on a root system is not always associated with damage. Indeed, *Pythium* spp., according to the species, show very different pathogenicity on tomato roots. In addition, the expression of their pathogenicity is often dependent on complex interactions between the development of the host (the tomato in this case), the environment, and the *Pythium* spp. present. Several species are nevertheless likely to cause root browning (Pythium root rot, wilt), and decay sometimes located in the root tip, and disappearance of rootlets and small diameter roots. The cortex of the roots and crown rot locally. Vascular tissues in the affected areas turn brown. These symptoms are sometimes found in soil-grown crops (especially in humid tropical and sub-tropical areas), more frequently in soil-less crops (in temperate and northern production areas).

Note that fairly sudden attacks that caused significant plants death occurs in the tropics, as a result of flooding in tomato crops. These were linked to *P. aphanidermatum* attacks in soils at high temperatures, above 30°C. It has also been observed that tomato roots can tolerate the latter species in presence of low temperatures. That same species of *Pythium* is responsible for root and stem base rot in the summer on soil-grown tomatoes in Italy.

Various *Phytophthora* spp. are also reported on tomato roots and are responsible for root and stem base rot (*Phytophthora* foot and root rot). The symptoms induced are quite similar to those caused by *Pythium* spp. but are more severe.

For example, *Phytophthora cryptogea* is very damaging in soil-less culture; it causes brown to black root and stem base rot. Other species, *P. nicotianae* (syn. *P. parasitica*), *P. erythroseptica*, *P. capsici*, cause comparable root damage in soil and/or soil-less crops. *P. capsici* is particularly rife in the US in open field tomato crops in rotation with cucurbits. *P. arecae*, now classified as *P. palmivora*, was reported to be responsible for root rots and stem base cankers in the Netherlands many years ago.

Root attacks of *Pythium* spp., as those by *Phytophthora* spp. (in soil-less crops in particular) strongly disturb the functioning of the root system, notably reducing the absorption of water and nutrients to a variable extent. They are also accompanied by leaf yellowing and wilting which is reversible. The fruits of some trusses may also have blossom-end rot symptoms (blossom-end rot, see p. 387). In some situations, the plants show poor growth and a reduced fruit size. The appearance of foliar symptoms depends on the balance between the production by the plant of new roots and the proportion of parasitized and affected roots. Note that unlike the cucumber, tomatoes grown in soil-less systems may well tolerate significant root loss without showing any aerial symptoms.

Pythiaceae also attack fruit. The *Phytophthora* spp. are responsible for irregular, moist, greyish-green to brown lesions, which appear on the fruits in contact with the ground. They evolve quickly and create patterns in concentric rings, complete or incomplete, light brown to dark brown. These lesions are rather firm, with a smooth surface and a diffuse edge. A cottony mycelial felting may cover the lesions in wet conditions (*P. nicotianae*, *P. capsici*, *P. drechsleri*). The lesions caused by *P. mexicana* produce concentric rings less consistently.

Some species of *Pythium* are the cause of wet lesions evolving into rapidly developing rot in mature fruits in contact with the ground. They are often associated with oozing and the presence of white cottony mycelium (*P. ultimum*, *P. acanthicum*, *P. aphanidermatum*).

Whatever the plant part affected and the symptoms observed, light microscopy often shows the presence of oospores and/or chlamydospores in the damaged tissues which confirm the presence of one or more Pythiaceous fungi. Indeed, it is not uncommon to find a complex of these fungi acting in concert in soil-less crops on the roots and fruits.

• Biology, epidemiology

Survival, inoculum sources: the *Pythium* spp. and *Phytophthora* spp. are able to live in a saprophytic state on organic matter in the soil or substrate. In the latter case, root exudates, but especially the remains of dead roots, are important substrates for saprophytic development and maintenance of the fungus in the soil.

Oospores (Photo **840–846**, *Tables 43* and *44*) and chlamydospores, either wet or dry, ensure survival in the soil. For example, *Phytophthora cryptogea* can remain for at least 4 years in the absence of any host, and *Phytophthora parasitica* for more than 6 years, probably because of its chlamydospores.

Their low host specificity allows them to attack a large number of hosts, cultivated or not, which also enables their multiplication and survival. *Phytophthora nicotianae* is found on dozens of crops (turnip, onion, pepper, eggplant, potatoes, watermelon, cucumber, carrot, parsley, sweet potato, lettuce, chayote, tobacco, avocado, cotton, ornamental plants, and many different trees). *Phytophthora cryptogea* also attacks many plants, cultivated or not, from at least 23 different botanical families: ornamental plants (petunia, chrysanthemum, dahlia, carnation), vegetable crops (celery, cabbage, turnips, peppers, eggplant, potato, lettuce and chicory, cucumber, asparagus, spinach, strawberries). *Phytophthora drechsleri* is also very polyphagous, parasitizing plants belonging to more than 43 botanical families.

In soil and soil-less systems, the sources of inoculum may be varied: the substrate, a few

Table 43 Symptoms associated with the main species of *Pythium and Phytophthora* reported on tomato

Reported species	Damping-off	Root and/or stem base rot, soil and/or soil-less	Fruit rot
Pythium aphanidermatum (Edson) Fitzp. (syn. *P. butleri* Subramaniam)	+	+	+
Pythium acanthicum Drechsler			+
Pythium arrhenomanes Drechsler	+	+	+
Pythium dissotocum Drechsler		+	
Pythium group F		+	
Pythium deliens Meurs (syn. *P. indicum* Balakrishnan)	+		
Pythium inflatum Matthews		+	
Pythium intermedium de Bary		+	
Pythium irregulare Buisman (syn. *P. debaryanum* Hesse)		+	
Pythium myriotylum Drechsler	+	+	+
Pythium paroecandrum Drechsler		+	
Pythium periplocum Drechsler	+	+	
Pythium salpingophorum Drechsler		+	
Pythium segnitium B. Paul		+	
Pythium spinosum Sawada	+	+	
Pythium ultimum Trow	+	+	+
Pythium ultimum var. *'sporangiiferum'* Drechsler	+	+	
Pythium vexans de Bary		+	
Phytophthora arecae (Coleman) Pethybridge (syn. *P. palmivora* [E.J. Butler] E.J. Butler)		+	
Phytophthora capsici Leonian	+	+	+
Phytophthora cinnamomi Rands var. *'cinnamomi'*		+	
Phytophthora citricola Sawada	+		
Phytophthora citrophthora (R.E. Sm. & E.H. Sm.) Leonian			+
Phytophthora cryptogea Pethybridge & Lafferty	+	+	
Phytophthora drechsleri Tucker			+
Phytophthora erythroseptica Pethybridge			
Phytophthora mexicana Hotson & Hartge	+		+
Phytophthora nicotianae Breda de Haan (syn. *P. parasitica* Dastur)	+	+	+
Phytophthora palmivora (Butler) Butler var. *'palmivora'*			+

(See Photos 435–439, 454, 455, 519, 523–530, 709, 732–736, 807.)

seedlings, or water from various sources (ponds, dams, rivers, and so on; *Phytophthora cryptogea*), plant debris, sludge. For example, in the region of Almeria in Spain, several species of *Pythium* (*P. aphanidermatum*, *P. spinosum*) were isolated from dust settling on the glasshouses sometimes in significant concentrations, thus this dust may be a source of inoculum.

Penetration and invasion: Pythiaceus fungi directly penetrate epidermal tissues of young roots and fruits (for example in 1–3 hours at 25–30°C for *Phytophthora capsici*), but also through wounds. They quickly invade the tissues through the combined action of various cellulolytic and pectinolytic enzymes, and progress between and within cells. Depending on the species' parasitic potential, their presence in the roots can go undetected or lead to rapid browning and blackening of tissues, and thus to serious damage. When colonizing a fruit, it idoes not take long to rot. Subsequently, sporangia, oospores (that will ensure the survival of the Chromista i.e. *Pythium* and *Phytophthora* spp.) are formed within the tissues or on their surface. The temperature at which their production occurs varies from one species to another: 15–35°C for *Pythium aphanidermatum* and *P. myriotylum*, 10–20°C for *P. ultimum*.

Sporulation and dissemination: *Pythium* spp. and *Phytophthora* spp. are perfectly adapted to life in soil water and in the nutrient solution for soil-less crops. As noted above, they sporulate abundantly in and on the tissues they have invaded (roots, fruits). They particularly form sporangia (Photos **835–837** and **843–845**, Tables 43 and 44) which can either germinate directly or produce flagellated and mobile zoospores. These are easily dispersed in the nutrient solution and are attracted to root exudates. In nurseries and soil-less crops, where the densities of seedlings but also the number of roots are important, they are transmitted from plant to plant during the growth of the mycelium in the soil or in the substrate. Aerial dissemination is possible as a result of splashing occurring during sprinkler irrigation or heavy rainfall.

Some insects, especially flies, may be contaminated by and transport oospores. Thus, aerial transmission of *Pythium aphanidermatum* by the *Scatella stagnalis* fly have been reported on cucumbers; this species is sometimes used in rotation with tomatoes in protected crops.

Conditions encouraging development: the development of these organisms on tomatoes, which do not all have the same pathogenicity, can be influenced by various parameters:
– the high density of seedlings in nurseries and of roots in the slabs/blocks in soil-less crops;
– excess nitrogen increases the severity of root symptoms associated with attacks of certain *Phytophthora* spp. On the contrary, the addition of potash reduces the severity. In addition, depending on the species, high salt concentrations increase or decrease their effects, particularly on young plants;
– the presence of water which is almost always inevitable. High soil moisture and reduced gas exchange are an ecological advantage for the Chromista at the expense of other fungi and micro-organisms which are sometimes competing for soil organic matter. Heavy soils and/or compacted ones are very conducive to their attacks because they deleteriously affect the vigour of the host and create an environment conducive to the spread of exudates required for oomycete germination and growth. In addition, soil moisture contributes to the production and dissemination of zoospores;
– temperature influences behaviour differently. There are species that like cold soils, with temperatures around 15°C, such as *Pythium ultimum* (optimal temperature 15–20°C, minimum 2°C, maximum 42°C), others have higher thermal optima. This is the case of *Pythium aphanidermatum* (optimal temperature 26–30°C, minimum 5°C, maximum 41°C). Note, for example, the optimum temperatures in some species of *Phytophthora*: 27–32°C for *P. parasitica*, 22–25°C for *P. cryptogea*, 28–31°C for *P. drechsleri*, 24–28°C for *P. citrophthora*, and 28°C for *P. capsici*;
– the responsiveness of the host is not constant throughout its life. Thus, succulent or etiolated seedlings are very sensitive while mature plants are less so, but may become so, especially when they undergo various climatic and agronomic stresses (low oxygen availability, water stress, too much concentration of elements in the nutrient solution);
– the intervention of other pest leading to interactions much more destructive to tomato. For example, this is the case between *Pythium aphanidermatum* and root-knot nematode *Meloidogyne incognita*, pre-and postemergence and on seedlings.

In addition, occasionally significant root death occurs naturally during the tomato production cycle. These are important when harvesting the first fruits, a critical period, especially for soil-less crops. Indeed, there is a decline in the root system (physiological death) concomitant with fruit development. This decline is attributed to a redistribution of assimilates from the roots to the fruits. Note that the root exudates, which act on the germination and growth of the Chromista, are also influenced by the plant fruit load, temperature, and light intensity.

Control methods[1]

• During cultivation

In the nursery, it is urgent to drench the entire substrate with an anti-oomycetes fungicide solution, for example based on propamocarb HCl or its association with fosetyl-aluminum. It is also necessary to limit irrigation to avoid soil saturation. Encourage good drying of the substrate after sprinkler irrigation. If the seedlings are produced under protection, ventilate the glasshouses as much as possible and then control the climate inside to avoid excessive moisture. Diseased seedlings and those close by must be destroyed.

During cultivation and after root attacks, plant treatment is different depending on the situation:
– in soil crops, usually very few plants are affected and no treatment is needed;

– in soil-less crops, fungicides can be used in the nutrient solution or locally by watering the base of plants. The doses selected should be low in order to avoid phytotoxicity.

Foliage attacks, occurring mainly in the field, are normally well controlled by antimildew treatments(see p. 448).

Several fungicides are commonly used worldwide to combat the Chromista; the active ingredient chosen will vary depending according to the plant part affected: propamocarb HCl, fosetyl-aluminum, mefenoxam, furalaxyl, dimethomorph, etridiazole.

These fungicides prevent the formation of zoospores at low concentration and kill the Chromista at higher doses.

Alternate applications of products belonging to different chemical families and with dissimilar

[1] See note to the reader p. 417.

476

modes of action helps to prevent the development of fungicide resistance. Sexual reproduction is common in these fungi and is the source of some variability. This allows the Chromista to adapt to their environment, and notably to some factors that exert selective pressure (e.g. fungicides). Thus a risk of selecting strains resistant to some of these fungicides exists. For example, strains of *Pythium* spp. tolerant to metalaxyl and furalaxyl have been reported (especially in *Pythium aphanidermatum*).

Balanced use of fertilizers is imperative, and avoid stressing the plants.

Diseased plants and plant debris should be carefully disposed of during and after cultivation, especially the root systems.

• Subsequent crop

In the nursery, the substrate used should be healthy or disinfected. In countries where the producers make their own, for example from sand, from recovered soil, or from different composts, risks of contamination exist. It is the same for producers who mix their commercial substrates with the ingredients mentioned above. The propagation medium must not be put directly on the ground, especially if it has not been disinfected. It is best to place it on a plastic film or on benches. In contaminated soils which are not disinfected, the seed compost may be impregnated with an anti-Pythiaceous fungicide solution such as propamocarb HCl or the compost may be drenched before use. The latter option may involve oxyquinoline potassium sulphate. Hygienic measures recommended in the nursery (p. 541) may be implemented. In some countries the seeds are coated with an anti-oomycetes fungicide. Dimethomorph limits *Phytophthora nicotianae* attacks on young seedlings.

In soil cultures, crop rotations with cereals and forage grasses are suitable. For example, in the case of *Phytophthora capsici*, rotations of 3 years at least without a sensitive host, are recommended. Heavy and wet soils must be drained. Organic matter can be added to alleviate them. Tomatoes should be planted on ridges to avoid water retention around plants. The soil can be mulched. Fertilization should be balanced. Note that animal composts may limit the action of *Phytophthora nicotianae*.

In some particularly affected soils, treatment with a disinfecting fumigant should be considered. The conventionally used active ingredients (metam-sodium, dazomet, propamocarb HCl) and steam are effective against *Pythium* spp. and *Phytophthora* spp. In production areas where it is possible, a solar disinfection of the soil (solarization) will be suitable. Quite spectacular results were recorded, particularly in some Mediterranean countries. The soil to be disinfected is carefully prepared and moistened and then covered with a polyethylene film 35–50 µm thick, held in place at least 1 month at a very sunny time of the year. This economical method, with a broad-spectrum efficacy, controls the Pythiaceae.

Avoid planting in soil which is too wet or cold. Irrigation performed at this stage of culture should not be excessive. The plants should be grown on stake supports, with a plastic mulch to prevent fruit contact with the ground at the end of cultivation.

Limit sprinkler irrigation to avoid keeping the soil moist for too long. Note that soil applications of mefenoxam 4–8 weeks before harvest are recommended in some countries.

In hydroponics, before any cultivation, the irrigation system must be drained and disinfected. The substrate should be replaced every year.

Be wary of the sanitary quality of the water used for preparing the nutrient solution and for irrigating the plants, especially if it comes from an irrigation canal, a river, a pond, etc. which may have been contaminated. If the nutrient solution is recycled, it can be disinfected. Several methods can be used to cleanse it: chlorination, iodination, ozonation, biofiltration, UV radiation, pasteurization, TiO_2 photocatalysis. The spectrum efficiency of these methods seems quite broad and includes a number of fungi in soil-less crops, notably the Chromista, but not all of them are completely effective. Some have partial efficacy and sometimes result in problems for the production systems such as an interference with mineral nutrition, particularly iron. The most successful methods are ultrafiltration and disinfection by heat. Bleach at 5 ppm has sometimes been used to disinfect irrigation water.

It should be emphasized that a single gene resistance to *Phytophthora nicotianae* (*P. parasitica*) has been demonstrated in India in commercial lines. Also sources of resistance to *P. capsici* have been found by a Russian team in wild species of *Lycopersicon* and in tomato lines, notably CRA66.

Some antagonistic micro-organisms have been tried to control some species of the Chromista: *Trichoderma harzianum*, *T. virens*, *Gliocladium virens*, *Pythium oligandrum*, *Pseudomonas fluo-* *rescens*, *Bacillus subtilis*, *Enterobacter cloacae*, *Burkholderia cepiaciae*. They are already used effectively in some countries.

Table 44 Some morphological characteristics of the main species of *Pythium* reported on tomato

Pythium species[a]	Main spore types	
	Sporangia	Oospores
Pythium acanthicum	(Sub-) globular.	Ornamented, plerotic 18–23 µm (average: 21 µm).
Pythium periplocum	Filamentous, lobed.	Ornamented, aplerotic 20–24 µm (average: 22 µm).
Pythium spinosum	Absent.	Ornamented, often plerotic 15–19 µm (average: 17.2 µm).
Pythium aphanidermatum (syn. P. butleri)	Filamentous, swollen.	Smooth, aplerotic 20–22 µm (average : 20,2 µm).
Pythium arrhenomanes	Filamentous, swollen, lobed.	Smooth, plerotic 22–23 µm (average: 21 µm).
Pythium deliens (syn. P. indicum)	Filamentous, swollen.	Smooth, aplerotic 16–18 µm (average: 17 µm).
Pythium dissotocum	Filamentous, swollen.	Smooth, aplerotic 18–21 µm (average: 19.8 µm).

Pythium species[a] (continued)	Main spore types	
	Sporangia	Oospores
Pythium inflatum	Filamentous, swollen.	Smooth, aplerotic 19–24 μm (average: 21.4 μm).
Pythium intermedium	Absent.	Smooth 16–20 μm (average: 17.5 μm).
Pythium irregulare (syn. _P. debaryanum_)	Globular.	Faintly ornamented, often aplerotic 15–18 μm (average: 15.9 μm).
_Pythium myriotylu_m	Filamentoud, swollen.	Smooth, aplerotic 20–27 μm (average: 24.5 μm).
Pythium paroecandrum	Globular or ellipsoid.	Smooth, aplerotic 15–21 μm (average: 17 μm).
Pythium salpingophorum	Globular.	Smooth, plerotic 15–18 μm (average: 15.5 μm).
Pythium ultimum Trow var. _'sporangiiferum'_	(Sub-) globular.	Smooth, aplerotic 17–20 μm (average: 18 μm).
Pythium ultimum var. _'ultimum'_	Often absent.	Smooth, aplerotic 17–20 μm (average: 18 μm).
Pythium vexans	Ovoid to pyriform, sometimes proliferating.	Smooth, aplerotic 16–19 μm (average: 17.3 μm).
Pythium group F	Filamentous, nonswollen.	Absent.

[a] Other more uncommon species have been associated with tomatoes for one reason or another (isolated from differently aged plants and/or pathogens on them during artificial inoculations): _P. adhaerens_ Sparrow (able to infect fruit), _P. anandrum_ Drechsler, _P. oedochilum_ Drechsler, _P. oligandrum_ Drechsler, _P. perniciosum_ Serbinow, _P. rostratum_ Butler, _P. tracheiphilum_ Matta, _P. uncinulatum_ Van der Plaats-Niterink & Blok, _P. catenulatum_ Matthews, _P. hydnosporum_ (Mont.) Schröter, _P. mamillatum_ Meurs, and _P. middletonii_ Sparrow.

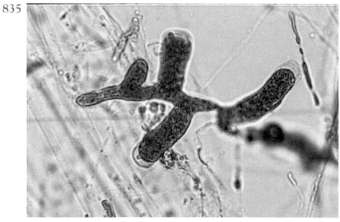

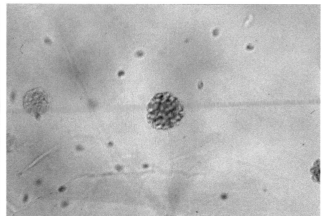

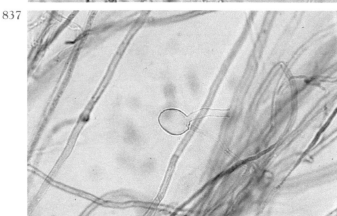

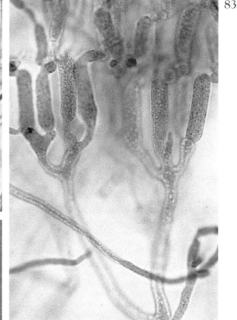

835 The mycelial hyphae, branched and terminal, is actually a filamentous swollen sporangium of *Pythium aphanidermatum*.

836 Filamentous sporangia of *Pythium* sp., demonstrating the many zoospores present inside.

837 A globular sporangium, as formed by *Pythium acanthicum* and *P. irregulare* among others, is clearly visible. It is empty and has a discharge tube.

Structures observed in *Pythium* spp. and characteristics of some mycelial colonies on nutrient media in a Petri dish

838 The shape of appressoria can be a significant criterion for identifying *Pythium* spp. In *P. myriotylum*, they are large and arranged in clusters.

839 These three Petri dishes show the appearance of the colonies of some species of *Pythium* on a nutrient medium, in this case malt-agar.

841 Several antheridia surround this smooth oogonium. *Pythium myriotylum*

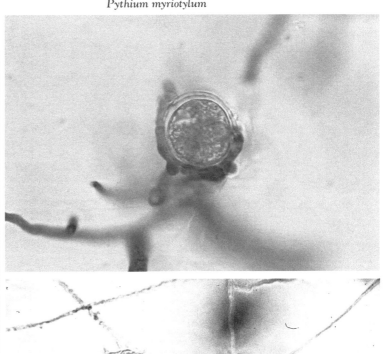

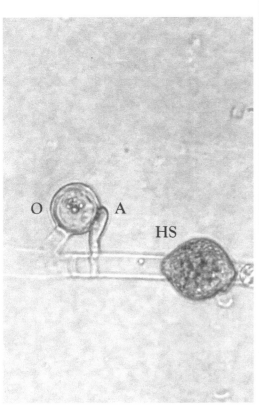

840 For sexual reproduction to occur in *Pythium* spp., an oogonium (O) – smooth in this case – must be fertilized by an antheridium (A). On the right of these two structures, a hyphal swelling (HS) is clearly visible.

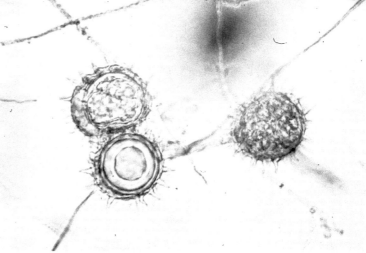

842 *Pythium spinosum, P. periplocum* form ornamented oogonia that are the source of oospores. One of these three structures can be described as an 'aplerotic oospore'.

Table 45 Some morphological characteristics of the main *Phytophthora* species reported on tomato

Phytophthora spp.[a]	Taxonomic group	Main spore types	
		Sporangia Presence of papillae Shape (Dimensions)	**Oospores** Diameter
Phytophthora capsici	II	Papillate Ellipsoid to pyriform (variable)	15–40 μm (average: 26 μm)
Phytophthora citrophthora	II	Papillate Variable shape 30–90 × 27–60 μm (average: 41 × 30 μm)	None produced
Phytophthora mexicana	II	Papillate Ovoid to fusiform 16–77 × 16–33 μm (average: 46 × 24 μm)	24–37 μm (average: 25 μm)
Phytophthora nicotianae (syn. *P. parasitica*)	II	Papillate Spherical, ovoid to ellipsoid, citriform 14–74 × 12–60 μm (average: 43 × 36 μm)	12–34 μm (average: 24 μm)
Phytophthora palmivora (syn. *P. arecae*)	II	Papillate Ellipsoid to spherical 20–72 × 20–48 μm (average: 28 × 30 μm)	23–36 μm (average: 28 μm)
Phytophthora palmivora var. *palmivora*	II	Papillate Ovoid-ellipsoid to ob-pyriform 35–60 × 20–40 μm (average: 55 × 33μm)	16–30 μm (average: 23 μm)
Phytophthora citricola	III	Semi-papillate Ob-pyriform to irregular 30–75 × 21–44 μm (average: 47 × 34 μm)	16–30 μm (average: 22 μm)
Phytophthora infestans	IV	Semi-papillate Ovoid, lemon-shaped 21–38 × 12–23 μm (average: 29 × 19 μm)	24–46 μm (average: 30 μm)
Phytophthoran cinnamomi var. *'cinnamomi'*	VI	Nonpapillate Ovoid to ellipsoid 27–114 × 20–63 μm (average: 75 × 40 μm)	19–54 μm (average: 33 μm)
Phytophthora cryptogea	VI	Nonpapillate Ovoid to irregular 35–63 × 24–35 μm (average: 52 × 30 μm)	24–32 μm (average: 27 μm)
Phytophthora drechsleri	VI	Nonpapillate Ovoid to ob-pyriform 24–38 × 15–24 μm (average: 31 × 21 μm)	17–45 μm (average: 26 μm)
Phytophthora erythroseptica	VI	Non papillate Ellipsoid to ob-pyriform 43–69 × 26–47 μm (average: 44 × 27 μm)	28–35 μm (average: 30 μm)

[a] *Phytophthora cactorum* (Lebert & Cohn) J. Schröt. and *P. richardiae* Buisman have also been reported on *Lycopersicon esculentum*.

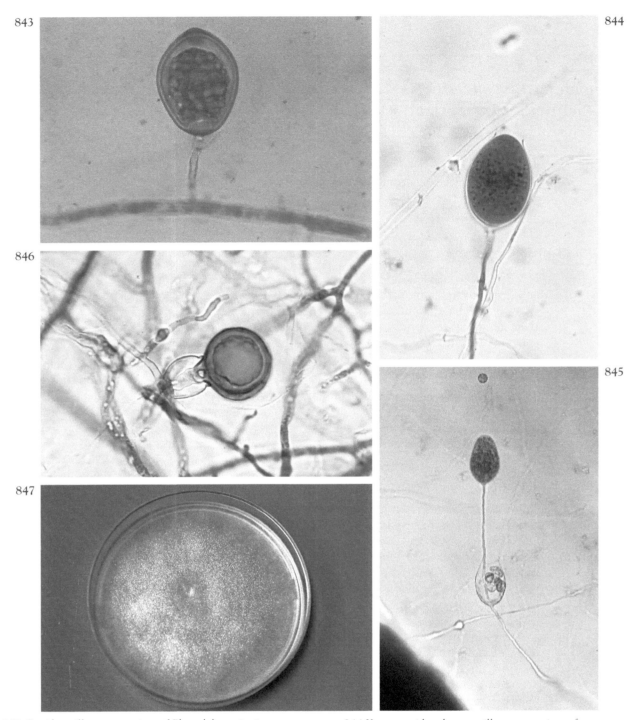

843 Ovoid papillate sporangium of *Phytophthora nicotianae* (syn. *P. parasitica*).

844 Young ovoid and nonpapillate sporangium of *Phytophthora cryptogea*.

846 Globular oogonium with an amphyginous antheridia, as can be seen in the majority of the *Phytophthora* species attacking tomato.

845 Nonpapillate proliferating sporangia; this criterion is common in *Phytophthora cinnamomi* and *Phytophthora erythroseptica*.

847 Appearance of a colony of *Phytophthora nicotianae* (syn. *P. parasitica*) on malt-agar.

Structures observed in *Phytophthora* spp. and characteristics of mycelial colonies on agar in a Petri dish

Pyrenochaeta lycopersici R.W. Schneider & Gerlach (1966)

Corky root

(Anamorph of *Herpotrichia*, Incertae sedis, Pleosporales, Pleosporomycetidae, Dothideomycetes, Ascomycota, Fungi)

Principal characteristics

Few data are available regarding the biological and genetic variability of *P. lycopersici*. Three strain groups have been defined according to their temperature optima: cold, temperate, and warm strains. These groups do not seem to have a high molecular variability from the limited work on the subject.

The fungus *Rhizopycnis vagum* DF Farr has been reported in Italy as the cause of corky roots on tomatoes (see p. 263). In this country, it is sometimes co-isolated from tomato roots along with *P. lycopersici*.

• Frequency and extent of damage

This soil fungus, described initially in Europe, is now present in the production areas of several continents. It is a marker of so called 'sick' soil that has supported tomato crops several times and/or other sensitive crops, mostly vegetables. It is surprising that it has been reported in soil-less systems both organic and inorganic, as it does not have structures enabling it to spread easily. In addition to yield losses that may result (40–70% or even more in crops seriously affected), it often forces producers to implement a range of expensive measures to reduce its impact: soil disinfection, grafting, soil-less cultivation etc.

Its importance is not negligible, especially in 'sick' soils in which tomato crops have been grown too frequently. This is the case in protected crops in plastic tunnels in particular, but also for the production of processingl tomatoes. In heavily infested soils, plant growth is highly affected.

• Main symptoms

P. lycopersici mainly attacks the tomato root system. Fine roots which are particularly sensitive go brown, deteriorate rapidly, and disappear. Locally, affected roots are brown and smooth at first. On the largest of them, there are superficial corky lesions on the cortex, gradually encircling the roots that eventually develop swollen 'sleeves' cracked with a dry appearance. Such corky sleeves are observed both on the main roots and the secondary roots. In general, the affected plants have a limited root system. Note that this fungus is frequently accompanied by a succession of non-specific secondary invaders such as *Rhizoctonia solani* (see Description 14), *Fusarium* spp., or by ones specialized on Solanaceae, such as *Colletotrichum coccodes* (syn. *C. atramentarium*, see Description 10).

The root damage affects the plant's development, which becomes stunted to varying degrees. The lower leaves are chlorotic and wilt during warm periods of the day, but can regain turgidity overnight. In severe cases, leaves can dry out and fall off. (See Photos **476**, **481–486**.)

• Biology, epidemiology

Survival, inoculum sources: *P. lycopersici* will survive for several years in the soil on plant debris as mycelium or chlamydospores. Various alternative hosts cultivated in rotation with tomatoes are able to host and multiply it: lettuce, melon, cucumber, eggplant, peppers, beans, watermelon, spinach, strawberries,

Carthamus tinctorius. The same is true for many weeds. Note that this fungus, although having very slow growth in the soil, can reach a considerable depth.

Penetration and invasion: it penetrates the root tissues and colonizes the cortex with its mycelium, leading to gradual decay and/or corkiness.

Sporulation and dissemination: This fungus exceptionally forms irregular-sized pycnidia on the roots of affected tomatoes which produce ellipsoidal conidia 4–6 × 1–1.5 µm (Photos **848** and **849**) with setae. Given the rarity of these pycnidia, it is conceivable that these structures do not contribute much to the dissemination of *P. lycopersici*. The potential for dispersal of this fungus is very limited and depends mainly on the activities related to tomato cultivation. Thus, dissemination can be through plants and/or contaminated substrates, tools, farming implements, and machinery. Unlike other soil fungi, it is relatively slow to recolonize disinfected soil. This is why disinfection is effective over a relatively long period as long as the method or the fumigant have been carefully chosen.

Conditions encouraging development: the fungus is best known to develop in cold weather. It seems to exist as at least two types of strains: 'cold' strains with a thermal optimum between 15 and 20°C (northern European strains) and warm strains, still pathogenic at 26–30°C, encountered in several countries of the Mediterranean (Tunisia, Lebanon). The monoculture of tomatoes and/or alternating with sensitive crops in the same field increase the concentration of soil inoculum and promotes its development in it.

Control methods[1]

• During cultivation

By the time symptoms of corky roots are detected during cropping, it is too late for control measures to be effective. Indeed, there is no way of eliminating the pathogen from the roots of plants without damaging them permanently. In an attempt to keep the plants alive as long as possible:
– earth up the plants to promote the development of adventitious roots that can supplement the older roots that are affected. In soil-less culture (on peat or peat + pozzolan) in cases of severe attacks, peat can be added locally in the stem base area to allow additional rooting. Sawdust is sometimes used;
– drench the tomato plants during the hottest periods of the day to avoid excessive evaporation, leading to wilting, desiccation, and death of plants;
– closely monitor irrigation. Indeed, if the plants wilt, it is not necessarily related to a lack of water, but rather to root damage caused by *Pyrenochaeta lycopersici*. In some cases, producers tend to increase irrigation in response to wilting, aggravating root lesions by asphyxiation.

Carefully remove and destroy diseased plants and their root systems, during, but especially at the end of cultivation. This will avoid them being buried in the soil and thus enriching the inoculum of *P. lycopersici* already present.

• Subsequent crop

In the event that propagation is performed each year in the same glasshouse, it will be necessary to implement the hygiene measures recommended on p. 541. Seedlings should be developed in healthy substrates (disinfected), and compost not placed on the ground because it may become contaminated by contact with it, especially if the soil has not been disinfected.

[1] See note to the reader p. 417.

848 The picnidia of Pyrenochaeta are brown and globular with a diameter of 150–300 μm. They have brown setae around the ostiole.

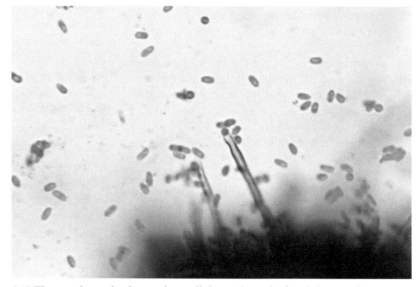

849 The conidia are hyaline and unicellular, with a cylindrical shape and measure 4–8 × 1.5–2 μm.
Pyrenochaeta lycopersici

Crop rotations are useful when done preventively, before the soil is heavily contaminated. In the latter case, their effectiveness is relative because the fungus survives a long time in the soil. In addition, it can multiply through various alternative hosts. Of course, care should be taken not to grow sensitive plants in the rotation. The soil should be well worked including a good sub-soiling which allows roots access to lower and 'clean' layers.

Highly contaminated soil must be disinfected. Several fumigants can be used (dazomet, metam-sodium, dimethyl disulfide). These fumigants are effective to varying degree; some of them, however, pose problems and result in a number of disadvantages:
– destruction of natural micro-organisms antagonistic to pathogens;
– increased sensitivity to pathogens in the disinfected potting soil;
– development of toxicities (excess of exchangeable manganese, excess ammonia resulting from a partial or complete blockage of nitrification).

Note that in some countries before planting, the soil has been treated with fungicide solutions based on thiophanate-methyl or carbendazim.

In sunny regions, solar disinfection of the soil (solarization or pasteurization) is increasingly used to control *P. lycopersici*. It involves preparing and moistening the soil well and covering it with a polyethylene film 35–50 µm thick. It is held in place for at least 1 month at a very sunny period of the year.

This method is economical, efficient, and can control the fungi near to the surface of the soil. Disinfection by steam is also effective.

Without disinfection, it is possible to delay attacks by *P. lycopersici* by establishing the seedlings in warmed up soil and/or by increasing the volume of the 'clean' compost at planting.

In affected soil-less crops, the substrate should be changed and the source of inoculum contaminating the substrate identified. Potential sources to be monitored and examined include: the plants, dust, soil, irrigation water and so on.

Genetic resistance has been identified in accessions of wild *Lycopersicon* such as *L. hirsutum* and *L. peruvianum*. Resistance from *L. hirsutum*, is polygenic with a dominant trend, but could not be transmitted to the cultivated tomato. Meanwhile, the recessive gene '*pyl*' (located on chromosome 3), has been introgressed into tomato from *L. peruvianum*. Many hybrid tomatoes currently on the market possess it.

Also, if the soil is not to be disinfected, grafting can be a highly effective solution, especially for protected crops and in gardens of amateurs. Several hybrid rootstocks, intra- or interspecific, are currently available (see p. 658).

Several fungi and antagonistic bacteria have been evaluated *in vitro* or in the field for the control of *Pyrenochaeta lycopersici*, including *Streptomyces griseoviridis*, *Gliocladium* spp., *Teratospermia sclerotivorum* (Uecker, Ayers WA & Adams PB) S. Hughes (syn. *Sporidesmium sclerotivorum* Uecker, Ayers WA & Adams PB), *Talaromyces flavus*, and *Trichoderma harzianum*.

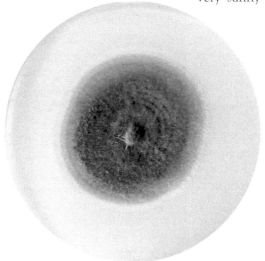

850 *Pyrenochaeta lycopersici* slowly forms grey and usually sterile mycelium on a malt-agar in a Petri dish; this characteristic does not enable its identification.

Rhizoctonia solani J.G. Kühn (1858)
Thanatephorus cucumeris (A.B. Frank) Donk (1956)

Damping-off, fruit rot

(Ceratobasidiaceae, Cantharellales, Incertae sedis, Agaricomycetes, Basidiomycota, Fungi)

Principal characteristics

Many studies have been conducted on this ubiquitous and polyphagous fungus to characterize the diversity of strains. They have been shown to differ in their cultural characteristics, their pathogenicity, and their host specificity. The criterion used to differentiate them is their ability to anastomose, and in this way more than a dozen groups (AG), and even sub-groups have been identified. Knowledge available of the strains on tomato stem is still limited.

Strains belonging to different anastomosis groups have been characterized from plants in the Solanaceae. For example strains belonging to AG-2-1, AG-3, AG-4, AG-4 sub-group HG-1, AG-5. AG-2 is divided into several sub-groups and includes relatively specialized strains on Brassicaceae (AG-2-1) or corn and sugar beet (AG-2-2). The strains belonging to AG-4, also subdivided, with *Thanatephorus praticola* as teleomorph, are more polyphagous when the temperatures are high. The strains of the AG-3 group are often specialized on the Solanaceae.

• Frequency and economic importance

R. solani is globally widespread and widely reported in all areas of tomato production. As suggested earlier, it is capable of infecting hundreds of different hosts. On tomato, this soil fungus is best known as a of cause damping-off in the nursery, root lesions and cankers on the stem base, and symptoms on the stem and fruits. The disease is found in all types of production, from the most extensive crops to the most intensive, in open fields and under protection and mainly in soil crops rather than soil-less ones.

It is frequently observed on the roots of tomato, alone or in combination with other root pathogens such as *Pyrenochaeta lycopersici*, *Colletotrichum coccodes*, and root-knot nematodes of the genus *Meloidogyne*. Lesions on the stem base and on the fruits of tomatoes for processing are also attributed to it. It is considered a biomarker of 'sick' soils used many times to grow tomatoes and/or other sensitive vegetable crops. In recent years, it has been part of the complex of soil pathogens attacking the roots of rootstocks grown under protection (see p. 658), particularly with *C. coccodes*, *Phytophthora nicotianae*, several species of *Meloidogyne*, and *Globodera tabacum*.

• Description of symptoms

R. solani is a soil fungus that grows preferentially on parts of plants in the soil or nearby.

As with *Pythium* spp., it is a well-known cause of damping-off, occurring at pre- as well as postemergence of the seedlings. It can cause reddish-brown lesions on all parts of the germinated seed. After emergence, a brown lesion can surround the stem portion located at ground level, causing collapse and death of the seedling.

It also produces cankers at the stem base of young plants or more developed plants, grown in some wet and heavily contaminated soils. These cankers have a reddish-brown colour and a consistency varying with soil moisture. Eventually, they may completely surround the stems. Roots adjacent to these cankers show browning. In addition, this fungus is sometimes capable of producing reddish root lesions, of various sizes and rather moist; the roots may also become superficially corky.

Aerial symptoms may also occur on plants. Brown, shrivelled, canker lesions have been reported on tomato stems. They develop from injured leaflets in contact with the ground or from infection from contaminated aerial soil particles or basidiospores produced by the teleomorph of this fungus, *Thanatephorus cucumeris*.

The fruits (green, but mainly mature) display some circular lesions when in contact with the ground. Firm to start with, these lesions consisting of light rings alternating with darker ones, gradually soften. As with many conditions that occur on fruits, the affectd areas can be colonized by secondary invaders which increases the damage.

Several fungal structures are used to confirm the presence of this fungus on or near the damaged tissues, regardless of the location:
– discreet whitish to brown filaments of *R. solani* on the germinated seeds, roots, along the stem, and on the fruits;
– ill-defined masses, brown, sometimes visible on the damaged tissues (the sclerotia of the fungus);
– highly agglomerated mycelium, whitish to cream coloured, forming the hymenium of the sexual form *Thanatephorus cucumeris*. It is on the latter that are formed basidia and basidiospores, structures rarely visible on the tomato plant.

(See Photos **440, 441, 457, 522, 550–553, 589–597, 804–806.**)

• Biology, epidemiology

Survival, inoculum sources: *R. solani* is frequently found in many soils that have had numerous vegetable crops. It has saprophytic potential allowing it to survive in the soil in the absence of sensitive hosts. It exists as mycelium and pseudo-sclerotia (Photos **851** and **852**), often in organic matter that it colonizes easily and in the most varied types of plant debris. It grows well in the soil, especially if it has been disinfected and freed of potential antagonistic micro-organisms. This very polyphagous fungus can attack and survive on the the most diverse hosts and their debris. It can be found in some composts and substrates, sometimes in peat, or on a few purchased plants. It is not uncommon in nondisinfected material used in nurseries.

Penetration and invasion: contamination takes place through the mycelium in the soil or from sclerotia. It can superficially colonize all parts of the tomato in the soil or in contact with it. Rare infections can also occur through basidiospores (Photo **853**). In this case, they germinate and give rise to hyphae. Subsequently, the mycelium penetrates the tissues directly through the cuticle or through various injuries. Inter- and intracellular development is often very fast and destructive because of the enzymes produced, particularly in favourable weather conditions. This parasitic process is responsible for damping-off and for rot and lesions visible on the different parts of the tomato.

Sporulation and dissemination: the fungal mycelium colonizes from damaged tissues, and grows on the soil and spreads to other plant parts. Sclerotia, mixed with soil particles contaminating different materials also contribute to its spread. Its sexual form, the basidiospores formed on basidia on the surface of the hymenium, enable the fungus to spread aerially. These spores can be spread over long distances by wind and air currents.

Conditions encouraging development: *R. solani* can grow in moist and heavy soils as well as in lighter, dryer soils, with both acidic or basic pH and at temperatures between 5 and 36°C. Soils too dry or too wet seem to inhibit it. It can affect the tomato throughout its development cycle. It is particularly damaging in the presence of moisture and when temperatures are in the range of 23–27°C or when they are unfavourable to the tomato, especially below 20°C.

Given the diversity of strains present in the field, it is difficult to determine the optimal development conditions of this fungus. Note that attacks can be serious when it forms a complex with other root pests, especially root-knot nematodes. Injuries caused by *Meloidogyne* spp. enable penetration and disease development. For example, development on tomato roots of *Meloidogyne incognita* increases the diseases caused by *R. solani* and *Thielaviopsis basicola*.

489

Control methods[1]

• During cultivation

If *R. solani* occurs during propagation the diseased seedlings and those nearby should be removed. When in doubt a fungicide application to the affected area should be made with one of the active ingredients deemed effective on the fungus (iprodione, pencicuron, mepronil). The environmental conditions should also be controlled to prevent excess humidity and temperature. Irrigation of seedlings should be optimal and not excessive. The surface of the substrate must be able to dry.

Attacks on adult plants are never harmful and do not require special measures. In all cases, plant debris and diseased plants are eliminated during and at the end of cultivation.

• Subsequent crop

To avoid introducing this fungus into the crop, a healthy substrate and good quality seedlings must be used. Seedlings produced for transplanting must not be placed on the soil before planting, especially if it has not been disinfected, or a plastic film must be used to isolate them. The measures recommended on p. 541 should be implemented for seedling production.

In highly contaminated soils, soil disinfection with a fumigant may be considered. The active ingredients conventionally used (metam-sodium, dazomet) and steam are effective with regard to *R. solani*. Be particularly vigilant to prevent soil reinfection after treatment.

Note that the combination of metam-sodium and formaldehyde has been very successful in controlling soil-borne attacks of *R. solani*, but also of *Fusarium oxysporum* f. sp. *radicis-lycopersici* and *Verticillium dahliae*.

In production areas where it is possible, a solar disinfection of the soil (solarization) can be used effectively. Quite spectacular results have been recorded particularly in some Mediterranean countries: the soil to disinfect is carefully prepared and moistened and then covered with a polyethylene film 35–50 μm thick, held in place for at least 1 month at a very sunny period of the year. This economical, efficient method eliminates this soil surface colonizing fungus.

The agricultural measures recommended earlier during cultivation should be implemented. In addition, heavy, wet soils must be drained and a pre-plantation cultivation must be performed to bury some of the sclerotia which are more quickly destroyed when deep buried. Crop rotations can be made with cereals, sweet corn, forage grasses, and onions. Tomatoes should be planted on raised beds and staked. The ground should be mulched so that the tomato parts (especially fruits) are not in contact with the soil and the water which may be present on the surface. In addition, in some cases the ventilation of the stem base of the plants and of the lower vegetation should be improved. Fertilizer application should be balanced, and should never be too low in nitrogen. There are

[1] See note to the reader p. 417.

currently no resistant varieties. Selection work is underway to obtain fruit resistant varieties.

Several antagonistic micro-organisms have been tried to control *R. solani*: *Bacillus lentimorbus*, *B. subtilis*, *Burkholderia cepiacea*, *Pseudomonas aeruginosa*, *P. fluorescens*, *Streptomyces* sp., *Paenibacillus lentimorbus*, *Pochonia chlamydosporia*, *Chaetomium globosum*, *Glomus mosseae*, *Tolypocladium niveum*, *Trichoderma viride*, *T. harzianum*, *T. koningii*. Although promising in some cases, the use of these micro-organisms is not yet reliable enough to be able to recommend them.

Note also other unconventional solutions which have some potential against the fungus: amendments based on *Azadirachta indica* in particular, essential oils of *Callistemon lanceolatus* and *Ocimum canum*, and antioxidants.

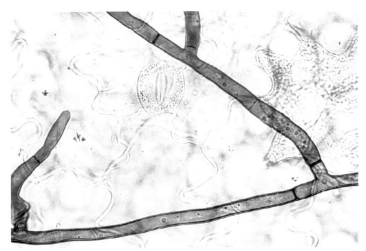

851 The mycelium of *Rhizoctonia solani* is characterized by its robust appearance, a width of between 5 and 15 µm, and a dark brown colour. There is also a slight constriction at the level of the lateral branches, and the presence of septae.

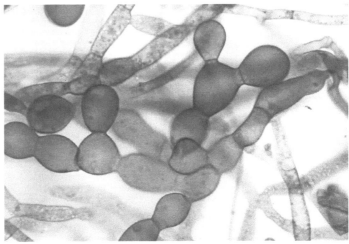

852 Some barrel-shaped structures can be observed on the mycelium and are considered the early stages of sclerotia.

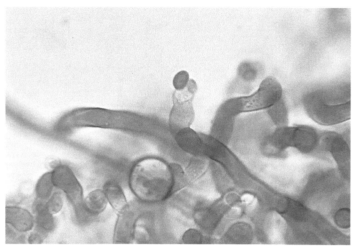

853 Basidia (12–18 × 8–11 µm) on
3–7 sterigmata; these rarely form on
the mycelium. The basidiospores are
asymmetrical and measure 7–11 × 4–7 µm.

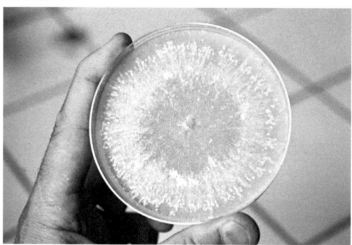

854 This fungus develops rapidly in colonies
with a variable appearance depending on the
strains. The mycelium at first is relatively
compact and light-coloured.

Rhizoctonia solani

Sclerotinia sclerotiorum (Lib.) de Bary (1884)
Sclerotinia minor Jagger (1920)

Sclerotinia drop, watery soft rot

(Sclerotiniaceae, Helotiales, Leotiomycetidae, Leotiomycetes, Ascomycota, Fungi)

Principal characteristics

• Frequency and extent of damage

Both these fairly polyphagous ascomycetes are widespread throughout the world. Although *S. minor* is occasionally reported on tomato, most of the damage on this plant is caused by *S.sclerotiorum*, regardless of the production area. These fungi attack both seedlings and adult plants. The economic losses in some countries are sometimes significant.

• Main symptoms

In nurseries, the seedlings display moist cankerous lesions on the stem, starting either from senescent cotyledons or from the first colonized leaf, or at the stem base. A brown rot develops later, leading inexorably to seedlings' death. It is often covered with a dense white mycelium and some black structures: sclerotia. Some undetected lesions may develop after planting if the seedling selection process was not thorough. On older plants, these fungi colonize wounds and senescent tissue (floral parts, leaflets) that are nutritional bases. They also readily attack plant parts that come into contact with the ground. Thus, moist cankerous lesions can be observed on the stems, on the stem base, or higher up. In the latter case, they are often located near sideshoot removal and de-leafing wounds on staked tomatoes, especially when grown under protection. These symptoms evolve gradually and eventually surround the stem, which becomes beige to whitish and necrotic. The stem may have been colonized after the establishment of the fungus on a floral part which had fallen directly on it or on a leaflet nearby.

White and cotton-like mycelium grows on the affected tissues. Sclerotia can be seen on the stem but also in the pith cavity as the pith has been completely destroyed and the stem is hollow. The sclerotia are the distinguishing features of these two *Sclerotinia* spp.:
– *S. sclerotiorum* (Photo **855**): some large black sclerotia, rather elongated, measuring 2–20 mm × 3–7 mm;
– *S. minor* (Photo **856**) small black sclerotia, irregular and mostly circular, 0.5–2 mm in diameter.

Leaflets and fruits are also affected by a moist and soft rot causing the disintegration of the former and the liquefaction of the latter. Thus, young green fruits show extensive moist and dark lesions with a looped edge.

The teleomorph of these fungi is sometimes visible on the soil surface, mainly in the case of *S. sclerotiorum*. Thus, some small 'cups', which are the apothecia (Photo **857**), form from the largest sclerotia. These produce ascospores, the source of aerial contamination.

(See Photos **549**, **565**, **577–580**, **782**.)

• Biology, epidemiology

Survival inoculum sources: *S. sclerotiorum* and *S. minor* have significant saprophytic potential. They can remain in the soil 8–10 years thanks to the sclerotia (Photos **855** and **856**) they produce on the affected organs and/or the mycelium present in plant debris left on the soil. In addition, they are polyphagous fungi that can be found on many host plants.

S. sclerotiorum is reported on over 400 different plant species, cultivated or weeds. It infects many vegetable crops used in rotation with tomatoes, including lettuce, beans, cabbage, peppers, eggplant, many cucurbits, celery, peas, carrots, rutabaga, potatoes, and sunflowers in the case of tomatoes for processing. A number of weeds are symptomless hosts.

Although less polyphagous, *S. minor* is still reported on over 90 plant species. The severity of attack is highly correlated with the number of sclerotia in the soil.

The many hosts are able to multiply the sclerotia and they serve as sources of inoculum when incorporated in the soil after harvest. Infection by *S. minor* is primarily through the mycelium from sclerotia in the vicinity of the plant part in contact with the ground. These sclerotia must have dried for some time before they can germinate.

Infection of tomato plants by *S. sclerotiorum* can occur in the same way. However, this fungus forms apothecia on its sclerotia. These organs allow its sexual reproduction and generate numerous asci containing ascospores. Thus, millions of ascospores are released from apothecia in the air for 2–3 weeks sometimes over several hundred metres, and these are the source of aerial inoculum. Their germination on the leaves can only occur in the presence of water after rain, sprinkler irrigation, or a dew.

Penetration and invasion: whatever the nature of the inoculum (mycelium, ascospores), these two fungi easily penetrate the living, injured, senescent, or dead organs in contact with the ground, or not, and quickly invade them. Their mycelium grows into the healthy tissues, which they rot through numerous lytic enzymes. For example, *S. sclerotiorum* produces endo-and exopectinases, hemicellulases, and proteases. It also synthesizes oxalic acid, which influences both the expression of its pathogenicity and the sensitivity of its host.

When humidity conditions allow it, the two *Sclerotinia* produce dense white mycelium and sclerotia on the damaged tissues. When crop residues are incorporated into the soil, 70% of sclerotia are found in the top 8 cm.

Sporulation and dissemination: the sclerotia sometimes ensure the transmission of these fungi to other areas, such as when they are transported through soil particles on tillage tools or plants. As noted above, unlike *S. minor* (heterothallic species), *S. sclerotiorum* (homothallic species) produces apothecia, asci (Photo **857**), and ascospores, especially when temperatures are low (between 8 and 16°C).

Conditions encouraging development: the optimum temperature is slightly below 20°C, but both *Sclerotinia* are able to grow at temperatures between 4 and 30°C. They are encouraged by wet and rainy periods and are frequently found on tissues at an advanced stage of development.

Light soils rich in humus are more conducive to the development of *S. sclerotiorum*. It is sensitive to carbon dioxide, which is why it is found in the first few centimetres of soil. The temperature and soil moisture conditions also affect the survival of the sclerotia of these fungi. The apothecia form following rain, a storm, or irrigation which increases the soil moisture.

Control methods[1]

• During cultivation

When symptoms occur due to an attack of one of these *Sclerotinia* spp. on stems (stem base or pruning wounds), on vegetation, or on fruits in field crops, a treatment with a fungicide can be applied. The following active ingredients have been and are still used in some countries: benomyl, thiram, iprodione, procymidone, vinclozolin and its association with maneb, pyrimethanil, fludioxonil + cyprodinyl. Note that strains resistant to benzimidazoles (benomyl, carbendazim) and to quintozene are reported in the literature. Also, fungicides with different modes of action, must be alternated.

Cultural measures must accompany chemical control in order to reduce ambient humidity and to avoid the presence of free water on plants: maximum ventilation and possibly heating of the glasshouses, and irrigation used in the morning or early afternoon, never at night.

Remember that drip irrigation is preferable to other methods of irrigation.

Plant debris must be removed during cultivation, especially the affected plants on which these fungi produce numerous sclerotia, but also at the end of cultivation to avoid their survival in the soil after burial. Nitrogen fertilizers needs to be regulated and it should be neither too high (causing very receptive succulent tissues), nor too low (source of chlorotic leaves constituting ideal nutrient bases).

• Subsequent crop

In the event that propagation is performed each year in the same glasshouse, it is essential to implement the hygiene measures recommended on p. 541.

Note that in the absence of a crop, flooding the infested fields would reduce the number of viable sclerotia in the soil.

The effectiveness of crop rotation is rather disappointing; this is certainly due to the polyphagous nature of the two *Sclerotinia* spp. involved. Heavily contaminated soils, however, justify such rotations, for at least 5 years. It is advisable to switch from tomato crops, for example, to cereals. Note in the US the cultivation of broccoli and burying its debris reduces the number of sclerotia in the soil and the damage caused by *S. minor*. In the US, it is considered that maize, cereals, onions, and spinach are not conducive to the development of either species of *Sclerotinia*. Green fertilizers sensitive to these fungi must not be included in the rotation, although some organic amendments reduce the damage caused by *S. sclerotiorum*.

Repeated cultivation of plants sensitive to one or both species of *Sclerotinia* on the same area inexorably lead to an increase in the soil inoculum. In this case, several means of preventive control can be considered including soil disinfection. Several fumigants may be used: dazomet, metam-sodium. Although effective, some of them nevertheless do pose problems and their use can result in some disadvantages:
– destruction of natural micro-organisms antagonistic to some pathogens;
– increased sensitivity to pests in disinfected potting soils;
– development of toxicities (excess of exchangeable manganese, excess ammonia resulting from a blockage of nitrification).

In the US, metam-sodium is applied by spray or through the localized irrigation system to combat *S. minor*.

In sunny areas of production, solar disinfection of the soil (solarization or pasteurization) has been implemented with some success. It involves covering the carefully prepared and moistened soil with a polyethylene film 35–50 μm thick, held in place for at least 1 month at a very sunny period of the year. This cost-effective and efficient method allows to control the fungi in the top layers of soil.

---

[1] See note to the reader p. 417.

855

856

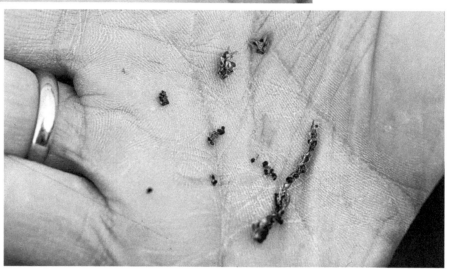

857

855 The sclerotia of *Sclerotinia sclerotiorum* are usually black, irregularly shaped, and rather large (2.5–20 mm).

856 *Sclerotinia minor* produce sclerotia much smaller than those of *S. sclerotiorum*: black, often grouped and 0.5–2 mm in diameter.

857 The production of the teleomorph of *Sclerotinia. Sclerotiorum* apothecia develop from sclerotia. The colour is variable (yellowish-white, light brown to brown).

Steam is also effective against both fungi.

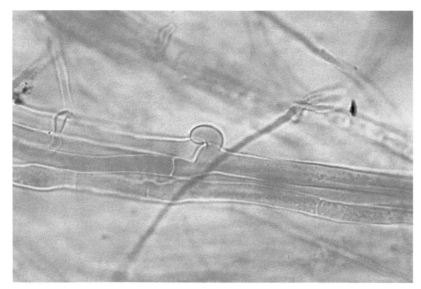

860 The hyaline and septate mycelium of *Sclerotium rolfsii* shows some anastomosis loops under microscopy which are characteristic of Basidiomycetes.

861 On malt-agar medium in a Petri dish, *Sclerotium rolfsii* grows rapidly, creating a white mycelial colony covered with sclerotia 1–3 mm in diameter, which are initially white then slowly turn brown.

Sclerotium rolfsii Sacc. (1911)
Athelia rolfsii (Curzi) C.C. Tu & Kimbrough (1978)

Sclerotium stem rot, Southern wilt

(Atheliaceae, Atheliales, Agaricomycetidae, Agaricomycetes, Basidiomycota, Fungi)

Principal characteristics

• Frequency and extent of damage

This soil-borne basidiomycete is very polyphagous and is particularly common in warm tropical and sub-tropical areas (e.g. in the West Indies, French Riviera, Basque Country). It mainly affects soil-grown tomato crops and a large number of other plants, cultivated or not (several hundred belonging to botanical families representing both mono- and dicotyledons).

The fungus is especially common and feared in field crops in the French West Indies.

• Main symptoms

S. rolfsii attacks all the tomato parts in the ground or located nearby, irrespectively of the age of the plants (very young at nursery stage, or fully grown).

The stem base is particularly vulnerable. A lesion develops over several centimetres and rapidly changes to rot which gradually surrounds the stem base. The affected tissues are moist and turn brown. If the weather and/or the soil is sufficiently moist, a dense white mycelium quickly covers the diseased parts. Smooth spherical (1–3 mm in diameter) structures at first white, then tan to reddish brown, are gradually visible within the mycelium; they are the sclerotia of *Sclerotium rolfsii*. They have an appearance similar to mustard seeds. Subsequently, the fungus also invades the roots and rots them quickly.

The destruction of the plants' stem base and root systems causes irreversible secondary wilting, although initially some tomato plants regain their usual turgidity overnight. In many cases, this recovery is short-lived and the plants wilt again as soon as temperatures rise during the day. During hot weather, wilting is total with all of the vegetation eventually drying out completely.

S. rolfsii is also capable of affecting the fruits and leaves that come in contact with the ground or with already affected plant parts. The fruits show some slightly yellowish symptoms at first, which quickly become wet and soft and deepen while the skin splits. Eventually, they may liquefy under the combined action of secondary invaders. Large wet spots are sometimes found on the leaflets in contact with the soil. They spread quickly and eventually completely rot the leaflets. As with the stem base, mycelium and numerous sclerotia gradually cover the rotten tissues.

Note that this fungus develops especially on young plants with soft tissues. It is therefore not surprising that some attacks occur in the few weeks after planting, especially if weather conditions are hot and humid.

Diseased plants are usually distributed in clusters. In some areas of high humidity, the mycelium spreads on the soil surface to move from one plant to another.

(See Photos **520**, **539–547**, **783**.)

• Biology, epidemiology

Survival, inoculum sources: *S. rolfsii* will survive for several years in the soil on plant debris in the form of aggregated mycelium (Photo **860**), but the primary inoculum is sclerotia (Photo

861). These structures are found in the soil or associated with plant debris. Their persistence is greater in the first few centimetres of soil rather than deeper. The fungus also remains on different organic substrates in its mycelial form in a saprophytic state.

S. rolfsii is a highly polyphagous pathogen that can attack and persist on some 500 plants, cultivated or not, belonging to a hundred or so botanical families. Among the vegetables it affects are pepper, eggplant, various lettuces, melon, cucumber, watermelon, beans, artichokes, beets, carrots, cauliflower, celery, garlic, onion, radish, turnip, and sweet potato. Many ornamental plants are also likely to host it: narcissus, lily, zinnia, and chrysanthemum. Peanuts and soybeans are important potential hosts in the areas where they are grown. It is also found on apple.

Penetration and invasion: the mycelium already present in the soil or sclerotia cause primary infection. Lesions are often located on plant parts in contact with the soil. When near a potential host, S. rolfsii produces lytic enzymes such as polygalacturonase and oxalic acid, which cause cell death. In this way, it can penetrate directly into tissues and invade them after having destroyed the cells through the combined action of the above compounds. Its inter- and intracellular development is very fast, especially if weather conditions are favourable. This parasitic process, which lasts from 2 to 10 days depending on weather conditions, is the cause of the wet symptoms and rot observed on different plant parts. It rapidly forms some mycelium and sclerotia on the damaged tissues. The sclerotia can germinate quickly because they do not need a dormant stage. However, they are able to survive several years in the soil without germinating.

Sporulation and dissemination: the transmission of this fungus is in soil contaminated by sclerotia, during tillage, by contaminated tools and agricultural implements, by water and seedlings produced in already affected nurseries.

The rarely occurring perfect form of this fungus produces basidiospores on the mycelium at the periphery of the lesions. These spores can result in the aerial dissemination of the fungus over long distances. Their epidemiological role is little known, but is probably quite limited given the low incidence of basidiospore production and also the absence of infection in the upper parts of plants.

Conditions encouraging development: S. rolfsii is most common in hot climates and proliferates in response to wet periods and/or irrigation. It is able to survive in a wide range of environmental conditions.

It prefers acid soil and mycelial growth is optimal between pH 3 and 5; the germination of sclerotia occurs between pH 2 and 5. Germination is inhibited beyond pH 7.

Periods of hot, humid weather and asphyxiating soils encourage development. It develops well between 25 and 35°C, its growth is stopped below 10°C and above 40°C. Below 0°C the mycelium is inactive although the sclerotia can withstand temperatures of about minus 10°C. It grows well and produces a large number of sclerotia between 27 and 30°C. Humidity also affects the germination of sclerotia. Contrary to what one might think, sclerotia germination is inhibited in water saturated soils and some authors report that it is best at relative humidities between 25 and 35%. The incidence of the disease can be higher in well-drained and sandy soils. Alternating wet and dry periods stimulate the germination of sclerotia. The presence of organic substrates, such as senescent leaves could increase the severity of the disease.

Control methods[1]

• During cultivation

Controlling this soil fungus during cropping is almost impossible, there are no effective means of stopping its development .

Management of irrigation (optimal amount, localized supply) is still recommended.

Plant debris, healthy or diseased, likely to host or to encourage the development and survival of this fungus in the soil should preferably be eliminated as well as potential weed hosts, during and at the end of cropping.

• Subsequent crop

Its wide range of potential hosts and the limited number of nonhost plants make crop rotations rarely effective against this fungus. In new or unaffected areas, crop rotations with nonhosts such as corn or other cereals may be worthwhile although the sensitivity of corn and wheat is controversial in the literature. While onion is sensitive, winter crops of this vegetable are not conducive to the development of S. rolfsii and may reduce the viability of sclerotia. The sclerotia may be more sensitive to the soil antagonists that develop as a result of the exudates excreted by this species of Allium.

We must accept that the wide host range and its saprophytic ability often substantially mitigates the effectiveness of crop rotations in soils where the fungus is already present.

Heavy and wet soils must be drained. Use deep ploughing to bury and destroy plant debris with sclerotia.

In heavily infested soils, consider eliminating or minimizing the inoculum present by disinfecting before planting For this, several methods may be chosen:
– the use of a fumigant (metam-sodium, dazomet), steam or other products (formaldehyde, PCNB);

– solarization in sunny regions. It involves covering the carefully prepared and moistened soil to be disinfected with a polyethylene film 35–50 μm thick, held in place at least 1 month at a very sunny period of the year. This cost-effective and efficient method allows the control the fungi colonizing the top layers of the soil. In fact, it seems that it is only eliminated in the first few centimetres of soil. This means planting immediately after solarization before this layer is recolonized. This is likely to limit the use of this method. Note that solarization has successfully been combined with the addition of antagonistic fungi such as Trichoderma harzianum; Sclerotium rolfsii does not like an alkaline pH so liming the soil is often recommended. For the same reason, some calcium-rich fertilizers reduce the incidence of this fungus by raising the pH, especially if the amount of soil inoculum is low. The same is true for the surface application of some soluble nitrogen fertilizers (urea, ammonium salts). To maintain the partial effectiveness of the latter, application should be divided and avoid leaching the soil.

The quality of the seedlings is often a guarantee of the quality of the crop. For this reason, the seedlings must be healthy and produced in ways that avoid excessive vegetative growth and etiolation as this may result in planting too deeply. Equipment used for seedling production must be well cleaned and disinfected, for example with a solution of 10% bleach for several minutes. Rinse thoroughly as this product is corrosive. At planting, the blocks should not be buried and the stem base must not be covered with soil. Avoid the presence of dead leaves and weeds near the base of the plants, they can serve as a primary food source for the fungus. Weeds must be destroyed.

Some authors report that a plastic mulch reduces the incidence of the disease on some plants

[1] See note to the reader p. 417.

by creating a barrier between the colonized plant debris in the soil and the lower parts of plants. This mulch can also increase the soil temperature, maintain the soil moisture, and reduce weeds.

Note that fungicides have been tried to control *S. rolfsii*, such as PCNB, thiophanate-methyl, difenoconazole, hymexazol, flutolanil, quintozene, benodanil, and vinclozolin. These products have been used alone or in combination with microbial antagonists, according to the control strategy: in the soil before or at plant-ing time, by soaking the stem base, and so on. These same methods have been used for antagonistic fungi, the latter sometimes being combined with organic matter. Indeed, many of them have shown some antagonism against *S. rolfsii* under controlled conditions, including *Trichoderma viride*, *Trichoderma harzianum*, and *Gliocladium virens*. Their effectiveness was more variable in field experiments.

To our knowledge, there is no tomato variety resistant to *S. rolfsii*.

Thielaviopsis basicola (Berk. & Broome) Ferraris (1912) (chlamydospore type)
Chalara elegans Nag Raj & Kendrick (conidia type)

Black root rot

(Anamorph of *Ceratocystis*, Incertae sedis, Microascales, Hypocreomycetidae, Sordariomycetes, Ascomycota, Fungi)

Principal characteristics

• Frequency and extent of damage

This fungus, which is present in many countries around the world, has rarely been reported on tomato, with the exception of in the US (California). It is not considered an important fungal pathogen of tomato, with only the seedlings being really sensitive.

T. basicola occurs in soil-less crops in Europe although rarely as a serious problem. It has also been recorded in some crops for processing, especially in areas where tobacco crops are grown as the latter is particularly sensitive.

• Main symptoms

This soil fungus may be responsible for significant damage to the underground parts of young tomato plants. It gradually colonizes them, resulting initially in small and discrete light brown lesions. Subsequently, they expand and throughout their length become dark brown to black in colour. Ultimately, a significant proportion of the plant's root system decomposes slowly and is affected by a fairly typical black rot of the roots. Seedling growth is greatly reduced. Note that the diseased seedlings when they are cultivated, grow almost normally, which confirms the lower aggressiveness of the fungus on adult plants. On the roots of the latter, there are more localized lesions, brown to black. In contrast, feeder rootlets are completely destroyed.

Arthroconidia (chlamydospores) are visible and are very characteristic of *T. basicola*, so allow easy identification.

(See Photos **447–451, 460**.)

• Biology, epidemiology

Survival, inoculum sources: *T. basicola* survives for long periods in the soil by chlamydospores (Photo **862**). It is able to colonize the organic matter and infect many host plants, cultivated or not (over 120 species have been recorded, belonging to at least 15 botanical families) that will contibute to its survival and development. Among the vegetable crops affected are beans, peas, cucumber, melon, watermelon, carrot, eggplant, lettuce, and endive. These hosts do not all have the same sensitivity to this fungus and the growth of the pathogen varies accordingly. Also, note that the strains of *T. basicola* differ in their pathogenicity. Contaminated soil dust is an important source of inoculum during propagation. The fungus also survives in the compost used for seedling production. Finally, it has sometimes been found before sowing in some peat-based substrates.

Penetration and invasion: the chlamydospores, and to a lesser extent the endoconidia, germinate close to the roots and penetrate directly through the surface or through wounds. The fungus rapidly colonizes the cortex tissues and vessels and rots them. It produces numerous chlamydospores in the damaged tissues. It also forms some at the root surface, as well as many endoconidia (Photo **863**).

Sporulation and dissemination: the chlamydospores and endoconidia are easily spread by water and soil dust. It is likely that the soil present on agricultural tools contributes to its spread. Infected tomato plants may also be a means of spread.

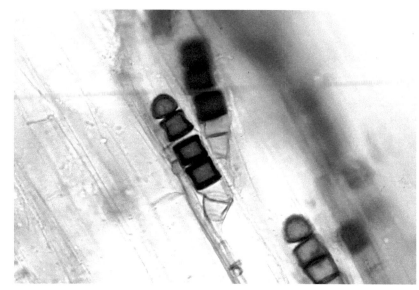

862 In the cells of the cortex of the root, there are some chlamydospores (aleuriospores) shaped like stacked barrels(25–60 × 10–12 μm) that gradually melanize.
Thielaviopsis basicola

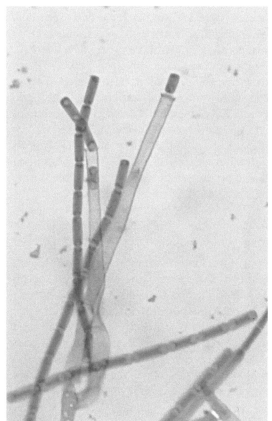

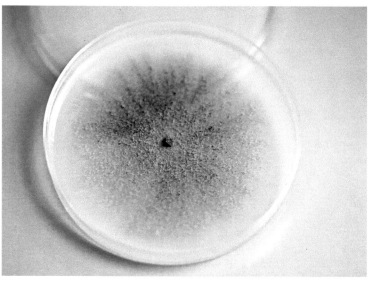

864 *Thielaviopsisbasicola* is difficult to isolate and has a relatively slow growth *in vitro*. On malt-agar, the colony is quite compact and gradually takes on a greyish brown to light purple colour.

863 Very elongated phialides producing cylindrical and hyaline endoconidia (10–23 × 3–5 μm).
Chalara elegans

Conditions encouraging development: *T. basicola* is best known for being dangerous on various plants when growing conditions are sub-optimal. This is the case in cold, wet springs. The fungus has a greater effect in these conditions because the root development of plants is reduced or stopped. Cool moist soil favours the fungus (although its optimum temperature is normally around 17–23°C). Australian strains have a higher optimum, between 23 and 26°C, reflecting an adaptation to the sub-tropical conditions. Soil pH also influences the behaviour of *T. basicola*. At acidic pH (around 5.6), it is usually less active. The addition of calcium in the soil can encourage fungal development.

The presence of *Meloidogyne incognita* on tomato roots helps the development of both *T. basicola* and *Rhizoctonia solani*.

Control methods[1]

• During cultivation

Unfortunately, at nursery stage, there is no way to save plants that have been attacked by *T. basicola*. Because of the risk of spread of the disease from infected plants, they should be removed as soon as possible. This is particularly important for seedling production.

The use of fungicides of the benzimidazoles family (benomyl, thiophanate-methyl) and some inhibitors of sterol biosynthesis (propiconazole) are partially efficaceous against this fungus. They are used in some countries for soaking the planting substrate/soil. However, strains of *T. basicola* resistant to benomyl are reported in the literature. Generally, the damage observed on adult plants is very limited and does not require fungicide treatment.

It is imperative that the tomato root systems are removed from the field and destroyed to avoid increasing the inoculum in the soil in the plant debris. In soil-less crops the substrate should be discarded.

• Subsequent crop

In the nursery, it is important to use 'clean' substrates (see p. 541). Bunches of plants produced for field crops should not be placed on the soil, especially if it has not been disinfected; use a plastic film to avoid contamination. Hygiene measures mentioned in the section on *Sclerotinia* spp. should also be applied (Description 15).

Crop rotations, although not very effective against the fungus, may be implemented to prevent an increase in the rate of soil inoculum. They must be for at least 4–5 years, and must not involve sensitive crops. *T. basicola* has been reported on many plants, including previously mentioned vegetable crops, on field crops and industrial crops (cotton, groundnuts, soybeans, alfalfa, lupin), and on various ornamental plants (chrysanthemum, geranium). The strains of *T. basicola* isolated on tomato are clearly polyphagous, but their host range is not yet fully understood. The species of plant reported are only an indication of the possible host range. Cereals such as rice and sorghum are good components in a rotation as they are not hosts.

Disinfection of the soil with a fumigant (dazomet, metam-sodium), could be considered mainly in the nursery.

The tools used to work on contaminated plots should be cleaned well before use in other healthy areas. Similarly tractor wheels must be cleaned. Thorough rinsing with water and disinfection of such equipment will often be sufficient to eliminate the soil and *T. basicola*.

Field drainage, fertilizers, and irrigation should be optimal. It will be necessary to maintain the soil pH around 6 avoiding lime. Be wary of certain organic materials added to the soil for their effects on pH.

[1] See note to the reader p. 417.

Other main fungi attacking the roots and/or the stem base[1]

Macrophomina phaseolina (Tassi) Goidanish (1947)
Macrophomina root and crown rot, charcoal rot

Syn.: *Rhizoctonia bataticola* (Taubenh.) Butler

(Incertae sedis, Incertae sedis, Incertae sedis, Incertae sedis, Ascomycota, Fungi)

Spongospora subterranea (Wallr.) Lagerh. 1892
Spongospora root gall

Syn.: *Spongospora subterranea* f. sp. *subterranea* J.A. Toml. (1892)

(Plasmodiophoraceae, Plasmodiophorales, Incertae sedis, Pytomyxea, Cercozoa, Protozoa)

Distribution and damage

This soil fungus, highly polyphagous (more than 500 known hosts), with rather limited saprophytic potential, is an occasional parasite encouraged by high temperatures. It is present in many countries where tomatoes are grown. It is especially active in warm, tropical, and temperate production areas. *M. phaseolina* is for example reported on tomato in the US and India. In the latter, it is part of a complex with *Rhizoctonia solani*, several *Fusarium* spp., root-knot nematodes, and is responsible for the tomato wilt syndrome.

This aquatic Protozoa is an obligate parasite and is best known on potatoes as the cause of powdery scab and as a vector for viruses. It is damaging and reported on tomatoes in many European countries, but also on other continents (US, Peru, several African countries, Australia, New Zealand, India, Pakistan). On tomato, its effect is limited.

It is sometimes reported on roots of plants grown in both the soil and soil-less. The eggplant is a reported host.

The symptoms of this disease must not be confused with those caused by *Agrobacterium tumefaciens* (see Description 23).

Symptoms

It is the cause of various symptoms on tomato:

– black and depressed cankers, appearing under the cotyledons at the time of emergence, root and crown rot of seedlings at the nursery stage or after planting;

– rotting of the root cortex and degradation of the lower part of the stem, the latter taking on a greyish to black colour inside.

Often, numerous small black sclerotia (50–200 μm) are present on and in the affected tissues. In addition, the plants, whether they are very young or more mature, are stunted, chlorotic, and shrivel or may even desiccate completely.

(See Photos **492** and **493**.)

White pustules developing into tumours form on the root surface. They gradually turn brown and become superficially corky.

(See Photos **492** and **493**.)

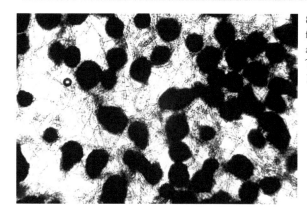

865 The presence of microsclerotia helps with the identification of this fungus. Their number and size vary with the strain, the environment, and the original host, and their diameter is between 60 and 200 μm. This fungus can produce black globose pycnidia with a diameter between 100 and 250 μm.

[1] See also pp. 250 and 256 for information on *Humicola fuscoatra* and *Plectosporium tabacinum* (*Monographella cucumerina*), two fungi reported on tomato roots. Note that another fungus producing sclerotia, *Ozonium texanum* Tirum D.C. Neal & R.E. Webster var. '*parasiticum*' Thirumalachar, can be responsible in India for root and stem base rot on tomato, but also on other vegetables such as peas, eggplant, and potato.

Elements of biology

This fungus is able to survive in the soil for several years in the absence of sensitive hosts, because of its sclerotia (Photo **865**). These arere able to withstand a variety of conditions, including temperatures above 55°C. The plant debris with sclerotia or invaded by the mycelium also contribute to survival.

Inoculum of *M. phaseolina* is maintained by alternative hosts such as (tobacco, corn, soybeans, alfalfa, clover, and weeds.

The mycelium in the soil or from germinating sclerotia makes contact with the roots and enters the cortex. It gradually colonizes the stem base and then the stem. Once in place, the fungus produces numerous sclerotia. It can also form globose pycnidia, with a truncated ostiole, which produce hyaline conidia, ellipsoid to oval, measuring 16–29 × 6–9 µm.

The dissemination of *M. phaseolina* occurs through farming tools during field work. The role of the pycnidia and their conidia in the transmission of the pathogen is not known.

The survival and activity of this fungus in the soil is influenced by mineral fertilizers and organic amendments. It is favoured by warm temperatures of around 28–35°C and water stress.

Because of its resting spores, this Protozoa is able to survive several years in the soil in the absence of sensitive hosts. It can multiply and thus be perpetuated by colonizing the roots of other plants, whether cultivated or weeds (corn, sorghum, tobacco, peas, cauliflower, radishes, turnips, many *Solanum* such as *S. nigrum*). It is sometimes introduced in manure, mud, and so on. Its resting spores then produce either myxamibes, or biflagellate zoospores that infect the root. Once in the tissues many cells hypertrophy, leading to tumour formation. At the same time, polynuclear plasmodia form in the tissues, and these eventually develop into either zoosporangia producing new zoospores responsible for secondary infections, or to cystosores that become the resting spores.

The disease occurs primarily in poorly drained wet soils, at temperatures between 16 and 18°C. It can also occur in drier soils and at high temperatures.

Control methods[1]

Unfortunately, during cropping, no action can effectively control this fungus. A solution of benomyl or products from the same activity group applied to the stem base can be effective. Nevertheless, plant debris should be removed or buried deeply so that the fungus can be subjected to the action of potential antagonists in the soil.

Normally, given the limited impact of the disease, the development of control methods is not required for the next crop. If problems still remain in some plots, the following are recommended:

– select fields free of the fungus. Rotations of 3–4 years with some cereals such as wheat or barley should be carried out;

– avoid high densities of plants and ensure optimal fertilizer use, especially in nitrogen and phosphorus;

– avoid water stress;

– disinfect the soil with a fumigant or steam if it is contaminated.

The disease caused by *S. subterranea* f. sp. *subterranea* is very unusual and rarely serious on tomato and does not normally justify the use of specific control methods.

No action is effective during cultivation.

In hydroponics, the substrate(s) must be replaced and the glasshouse and equipment disinfected.

In soil systems cultivation of tomatoes after potatoes must be avoided. Remove weeds that may harbour the fungus, such as *Datura stramonium*, and *Solanum nigrum*. Irrigation must well managed, avoiding excess watering.

866 The mycelial colony of *Macrophomina phaseolina* on malt-agar in a Petri dish, at first hyaline, gradually becoming grey. Numerous black sclerotia, round to oblong, have covered it.

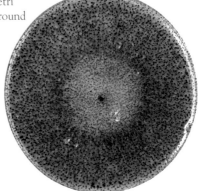

[1] See note to the reader p. 417.

Rhizoctonia crocorum (Pers.) D.C. (1815)

Helicobasidium brebissonii (Desm.) Donk (1958)

Violet root rot

(Former names: *Rhizoctonia violacea* Tul. & C. Tul., *Helicobasidium purpureum* [Tul.] Pat.)

(Helicobasidiaceae, Helicobasidiales, Incertae sedis, Urediniomycetes, Basidiomycota, Fungi)

Olpidium brassicae (Woronin) Dang. P.A. 1886

(Olpidiaceae, Incertae sedis, Incertae sedis, Chytridiomycetes, Chytridiomycota, Fungi)

Distribution and damage

This soil fungus is only rarely observed on tomato although it has been reported in many countries on different plants. Tomatoes are not a preferred host. Highly polyphagous, it is known in Europe as violet root rot, affecting especially asparagus, beets, and carrots. Elsewhere it is commonly called 'violet root rot'.

We have only observed it once in France, in an processing tomato crop.

O. brassicae is globally widespread but only causes a disease on tobacco. It is often seen in the roots of tomatoes grown in both the soil and in soil-less systems. The conditions in soil-less culture are very favourable for its development in the cortical cells. It is the vector of several serious viruses on lettuce: Mirafiori lettuce virus (MiLV), Lettuce big vein virus (LBVV), and Lettuce ring necrosis agent (LRNA).

Symptoms

Many brown purplish lesions are visible on the cortex of colonized roots. These are also superficially covered by a purple/wine coloured mycelium, sometimes sparce, but more often concentrated, in twisted aggregations or in the form of sleeves. Small mycelial clusters called 'corps miliaires' are observed here and there. Ultimately, the cortex rots locally and eventually breaks down in the form of sleeves or in a more generalized manner. Affected plants, in clusters, are slow growing and chlorotic, and they may eventually wither and dry out.

(See Photos **8, 466–473**.)

The presence of *O. brassicae* in roots can be easily confirmed by microscopic observation. This fungus, an, aquatic obligate parasite, forms spores (chlamydospores) in the cortical cells and some resting sporangia capable of liberating many zoospores.

The decomposed roots in soil-less cultures are often invaded by them. It is difficult to establish the true role of this organism, and it may be a weak parasite or simply a secondary opportunist.

(See Photo **456**.)

Elements of biology

Rhizoctonia crocorum survives very well in the soil for many years. Its mycelium and corp miliaires allow it to survive and colonize other hosts which includes crops and weeds.

Among vegetables, celery, fennel, parsley, parsnips, beans, beets, cucumber, cabbage, potatoes, turnips, shallots, onion, spinach, and lettuce can be attacked. Also alfalfa, saffron and many weeds (*Agropyron repens, Rumex acetosella, Taraxacum officinale, Sonchus arvensis, Cichorium intybus*). The mycelium penetrates the roots and develops slowly. It primarily affects the cortex, where its mycelium is clearly visible.

This fungus has a teleomorph with very elongated basidia, seen only exceptionally and whose epidemic role is not known. The dissemination of this fungus in a crop is quite slow. The disease spreads from affected groups of plants very gradually as a result of cultivation. *R. crocorum* is dispersed on tools and equipment contaminated with soil particles. Some vegetable seedlings or vegetatively propagated plants may introduce it into new areas.

It grows at temperatures between 9 and 30°C with an optimum around 26°C. The effects of *R. crocorum* are more severe in soil types that are wet, with a relatively low pH. The monoculture of sensitive species often leads to its development.

Olpidium brassicae is characterized by the presence of zoosporangia filling the cortical cells of the root and by star-shaped resting spores (see p. 252).

The chlamydospores produced by *O. brassicae* ensure its preservation for many years in the soil and on plant debris. It can also survive on several hosts, particularly several species of vegetables such as lettuce and cucumber. It can survive in the irrigation system, in the tanks of nutrient solution, and in the substrates.

Contamination takes place through uniflagellated zoospores from resting spores or sporangia. They are attracted by the roots, encyst on the surface, then penetrate directly into the cells. The fungus produces one or a number of aggregate sporangia that will result in zoospores which give rise to secondary infection.

The zoospores are released outside the root cells through the release tubes. Once in the soil aqueous phase or in the nutrient solution, a flagellum allows them to move and spread to other plants. *O. brassicae* is also spread by soil dust, contaminated seedlings, and circulating water.

This fungus is well adapted to aquatic life and spreads rapidly in hydroponics. It grows very well at temperatures between 10 and 16°C.

Control methods[1]

Controlling this fungus is not normally justified on tomato and is in any case extremely difficult to achieve being often ineffective. Among the measures suggested to protect sensitive plants are:
– careful removal of plant debris, especially the root systems;
– destruction of weeds, especially sensitive species;
– very long preventive crop rotation, with cereals (somewhat effective);
– drainage of heavy soils;
– cleaning tools and equipment coming into contact with the soil of the affected plot;
– in highly contaminated soil, a fumigant may be used.

Controlling this soil fungus is not justified in tomato production.

Nevertheless it is still appropriate to be particularly vigilant about the quality of water used, especially in soil-less culture as it can be the source of infection by various other fungi including those in the Pythiaceae, by *Colletotrichum coccodes*, and *Fusarium oxysporum* f. sp. *radicis-lycopersici*.

[1] See note to the reader p. 417.

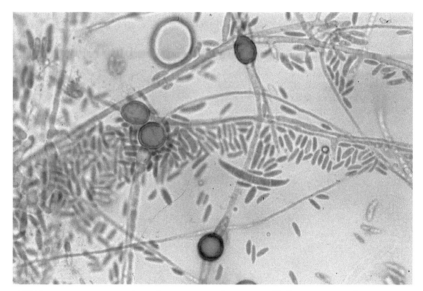

867 *Fusarium oxysporum* f. sp. *lycopersici* produces unicellular microconidia and crescent-shaped macroconidia with a maximum of 3–4 walls. It also forms intercalary or terminal thick-walled chlamydospores in chains or isolated.

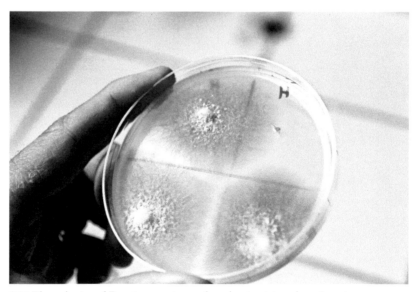

868 The colonies of *Fusarium oxysporum* f. sp. lycopersici, such as *Fusarium oxysporum* f. sp. *radicis-lycopersici*, grow relatively quickly. First white, they become a dark pink to purplish colour with age.

Vascular fungi

Fusarium oxysporum f. sp. *lycopersici* W.C. Snyder & H.N. Hasen (1940)

Fusarium wilt

(Anamorph of *Gibberella*, Nectriaceae, Hypocreales, Hypocreomycetidae, Sordariomycetes, Ascomycota, Fungi)

Principal characteristics

• Frequency and extent of damage

This soil fungus is globally widespread. Described for the first time in Europe in the late nineteenth century, it is now present in dozens of countries on every continent, where its damage varies depending on the race and the crop variety cultivated. In the past, it has caused variable damage sometimes severe especially before resistant varieties were available. The use of such varieties limits its impact in many countries. Unfortunately, many races have become progressively more aggressive challenging the resitance to wilt in some production areas. Thus, three physiological races[1] are now known on tomato:

– race 1, the oldest (described in 1886), is widely present in many production areas;

– race 2, reported in 1945 in Ohio, US, has become serious since the 1960s in several countries: US, Mexico, Brazil, Venezuela, Australia, UK, Netherlands, Israel, Morocco, Iraq, Taiwan, and China;

– the most recent is race 3, which has a more limited range. Described in 1978 in Australia and capable of overcoming the gene 'I' and 'I-2', it now occurs in Brazil, in some states of the US, and in Mexico and Japan.

Several vegetative compatibility groups (VCG) have been observed in *F. oxysporum* f. sp. *lycopersici*. Races 1 and 2 belong to groups VCG 0030 and VCG 0032, race 3 to groups VCG 0030 and VCG 0033. A study conduc-

ted in California of 39 isolates 2 years after the onset of race 3, showed that 22 of them could be associated with the group VCG 0030 (13 to race 3 and 9 to race 2), 11 with group VCG 0031 (7 to race 2 and 1 to race 1) and 6 with a new group VCG 0035 (all race 2). This work suggests that race 3, in California has originated from a local race 2 isolate.

• Main symptoms

Symptoms of Fusarium wilt can occur both on seedlings and on mature plants.

In the nursery, affected seedlings have a reduced growth, and the older leaves may turn yellow, wilt, and curve. Young plants may completely wilt and die. A cut in the stem reveals a marked brown colour of the vessels.

Adult plants suffer symptoms very similar to those described above on seedlings. Thus, some older leaves show unilateral yellowing, i.e. affecting the leaflets located on one side of the leaves and one part of the leaf, a feature of vascular disease. This yellowing of leaf tissues is often associated with wilting and tends to worsen in the warmest parts of the day. Leaves eventually dry out completely, but do not fall. Subsequently, these secondary symptoms spread to other leaflets and the plant becomes more generally affected, often leading to dessication and death.

Longitudinal lesions, first chlorotic and then gradually necrotic are visible on one side of

---

[1] Two systems of nomenclatures are used in the literature to identify races of *Fusarium oxysporum* f. sp. *lycopersici*. The most common is the one used in this description of races 1, 2, 3. The other nomenclature lists the first as race 0, the second one as race 1, and the third one as race 2.

the stem. They can extend over several tens of centimetres. Adventitious roots can form on the stem.

The vascular system is particularly affected. It is a dark brown colour, which differentiates Fusarium wilt from Verticillium. Pith is not affected, unlike 'black pith' or, to a lesser extent, 'bacterial canker' caused respectively by *Pseudomonas corrugata* (Description 25) and *Clavibacter michiganensis* subsp. *michiganensis* (Description 24).

Note that F. *oxysporum* f. sp. *lycopersici* is also responsible for rot on fruit.

(See Photos **618, 651–660**.)

• Biology, epidemiology

Survival, inoculum sources: This fungus is able to survive in plant debris in the soil up to 80 cm deep for more than 10 years, due to its chlamydospores which have thick and resistant walls (Photo **867**). It can be saprophytic enabling it to colonize and survive on various organic compounds. Other plants are possible hosts, especially several species of *Lycopersicon*.

Penetration and invasion: after germination of its chlamydospores, the *Fusarium* enters the plant through natural wounds, such as those present where secondary roots emerge, or through various injuries. Once established, the mycelium develops between cortical cells and reaches the xylem vessels and invades them. As in the case of Verticillium wilt, plants can respond to this vascular invasion by forming gum or tyloses that impede its progress, but also contribute to the wilt symptom.

Sporulation and dissemination: this fungus produces chlamydospores, microconidia, or macroconidia (Photo **867**) in the vessels and in the colonized and affected tissues. It can be disseminated over great distances through contaminated seeds and seedlings, in substrates, in water, and by agricultural machinery moving from one area to another particularly when tyres are covered with soil particles and plant debris. Soil dust containing chlamydospores is easily dispersed by air currents and by splashing water.

Conditions encouraging its development: F. *oxysporum* f. sp. *lycopersici* is a high temperature fungus occuring during warm periods of the year; its optimum temperature is around 28°C. It develops well in sandy and acidic soils. The disease is encouraged by low soil humidity, short days, low light levels, and plant tissues deficient in nitrogen, phosphorus, and calcium. The excessive use of nitrogen fertilizers, especially ammonia-based has the same effect.

Attacks by root-knot nematodes and asphyxiating soil predisposes the tomato to Fusarium wilt.

They also allow disease development on resistant varieties by strains that cannot normally overcome gene 'I' resistance. For example, an interaction between *Meloidogyne incognita* and F. *oxysporum* f. sp. *lycopersici* was observed in India and Australia, where it is particularly damaging.

Control methods[1]

• During cultivation

No method of control or product can effectively control the disease during cropping.

The addition of thiophanate-methyl or benomyl applied to the base of the plant sometimes recommended in some countries but is costly and often ineffective.

Many producers burying the crop residue into the soil after harvest. Buried plant tissues are abundantly colonized by *F. oxysporum* f. sp. *lycopersici* producing numerous chlamydospores. The removal of the plants with their root systems limits this phenomenon and reduces the amount of inoculum in the soil.

• Subsequent crop

It is necessary to use quality seeds and substrate and seedlings that are free of contamination; the producers must monitor quality from seedling delivery onwards. The plants produced should not come into contact with the soil, especially if it has not been disinfected. A plastic film must isolate them from contact with the soil and the hygiene measures recommended in the section on *Sclerotinia* spp. (Description 15) should be implemented.

Avoid production of tomato plants in an area already affected. A routine and lengthy crop rotation, of at least 3–4 years, will help prevent the onset of the disease. However, this is ineffective if *F. oxysporum* f. sp. *lycopersici* is already present in the soil as it will survive for many years.

Soil disinfection by steam or with a fumigant[2] (*chloropicrin*, to a lesser extent other fumigants such as dazomet, metam-sodium) is partially effective. Their effectiveness will depend on the fumigant used, the terms and conditions of application, but especially on the precautions taken to avoid early recontamination by *Fusarium*. At its best, it will last no more than one crop. Indeed, like all *Fusarium*, *F. oxysporum* f. sp. *lycopersici* quickly recolonizes disinfected soil. Some authors advocate fully

covering the ground to prevent recontamination of protected crops (this is imperative in the case of soil-less crops). Solarization has some effects, particularly associated with treatment with dazomet (see p. 487). The combination 1·3·dichloropropene + chloropicrin seem to work against the complex *Meloidogyne incognita* + *Fusarium oxysporum* f. sp. *lycopersici*.

Note that Fusarium wilt suppressive soils have been reported in several countries.

In soil-less crops, it will be necessary to disinfect all equipment used during cultivation, the tanks of nutrient solution (to be covered), and the nutrient solution channels, using solutions of 3% formalin or bleach, which can also be used to disinfect the substrates in soil-less crops.

Fertilizers containing nitrates may be used as they are less favourable to Fusarium wilt than the ammonium types. Be wary of excessive nitrogen fertilizer use as this seems to influence the disease positively. Liming mitigates the effects of Fusarium wilt.

Plant handling and cultivations must be done with care in order to avoid injury to root systems. The tools used for cultivation in contaminated areas must be thoroughly cleaned before use in healthy ones. The same applies to tractors' wheels. Thorough rinsing with water is often sufficient to get rid of the contaminated soil.

By far the most effective way to control Fusarium wilt is to use resistant varieties. Several genes are used:
– the gene 'I', derived from *Lycopersicon pimpinellifolium* ('PI 79 532') and located on chromosome 7, induces resistance to race 1;
– the gene 'I-2' from a hybrid between *L. esculentum* and *L. pimpinellifolium* ('PI 12 6915') and located on chromosome 11, confers resistance to race 2;
– resistance to race 3 is derived from *L. pennellii* (LA 716). It is controlled by a dominant gene 'I-3' located on chromosome 7.

---

[1] See note to the reader p. 417.
[2] Decreased efficacy of metam-sodium and dazomet was found in some soils in Israel, where these fumigants were often used. This is due to an accelerated degradation of methyl isothiocyanate.

Note that in Brazil, several accessions of *L. hirsutum*, *L. chilense*, *L. pennellii*, and *L. peruvianum* have a high level of resistance to three races.

Many varieties marketed today are resistant to Fusarium wilt, especially to race 1 and increasingly to race 2 or even to race 3.

Indeed, some varieties now combine resistance genes to all three races. As noted above, three races exist that can overcome resistance to *Fusarium*. Also, the choice of varieties used must take into account the situation of the races present in the area of production. The use of KNVF type rootstocks may also be considered: its behaviour *vis-à-vis* Fusarium wilt is quite comparable to the resistant varieties (see box 8 on rootstocks p. 658).

Among the alternative methods, note that a water extract of *Oxalis articulata* used to water the base of the plants controls Fusarium wilt. Other plant extracts have the same effects (e.g. *Citrus paradisi*).

The differing effects of several micro-organisms against *F. oxysporum* f. sp. *lycopersici* have been reported. For example, *Penicillium oxalicum* (applied to the seedlings before planting) induces resistance in tomato, thus providing some protection. Other organisms appear to have shown some activities that alter the development of *Fusarium*, including *Trichoderma harzianum*, *Streptomyces griseoviridis*, *Glomus intraradices*, nonpathogenic isolates of *Fusarium oxysporum*, *Pseudomonas fluorescens* in combination with benomyl.

Verticillium dahliae Klebahn (1913)
Verticillium albo-atrum Reinke & Berthold (1879)

Verticillium wilt

(Anamorph of *Hypomyces*, Incertae sedis, Hypocreales, Hypocreomycetidae, Sordariomycetes, Ascomycota, Fungi)

Principal characteristics

Of the two species of *Verticillium* that can attack tomatoes, *V. dahliae* is the most reported. A limited number of vegetative compatibility groups (VCG) has been demonstrated in this species: VCG 1, VCG 2 (including VCG 2A and VCG 2B), VCG 3, and VCG 4 (including VCG 4A, 4B and 4AB). Another group was proposed, which could be included in VCG 4. In addition, one isolate could be the only one belonging to a new group VCG 5. Several isolates from peppers have been recently associated with another new group VCG 6.

Note that in Japan, the nomenclature of VCG is different, and three sub-groups have been differentiated: J1, J2, and J3, but their relationship with the groups defined above is unknown.

For example, in Israel, isolates from VCG 2A group are the most aggressive on tomato, compared with isolates of VCG 4 and VCG 2B groups. Those in group VCG 1, aggressive on cotton, are not pathogenic on tomato. In addition the pathogenicity of *Verticillium dahliae* seems to vary depending on the strain studied, its original host, and its compatibility group. Some strains exhibit some host specificity.

Another species, *V. albo-atrum*, is also reported. Unlike the previous species, it does not form microsclerotia and its temperature optima are lower. Isolates of this *Verticillium* have, however, been found in Tunisia with higher thermal optima.

Isolates of *V. tricorpus*, pathogenic on potato and particularly aggressive on tomato and eggplant have been identified in Tunisia. This species produces chlamydospores and dark resting mycelium, as well as microsclerotia.

• Frequency and extent of damage

Verticillium dahliae, a highly polyphagous fungus, has been reported on tomato in many countries in temperate and sub-tropical zones. It affects hundreds of herbaceous and woody host plants, including several others in the Solanaceae: tobacco, potatoes, peppers, and especially the eggplant, which is particularly sensitive. On tomato, its impact was significant in the past, but is now no longer so in countries where resistant varieties are used.

A race 2^1, able to overcome resistance to *Verticillium* conferred by the 'Ve' gene, has been reported in several countries. A similar situation regarding the 'Ve' gene is also the case for *V. albo-atrum*.

• Main symptoms

The plants affected by Verticillium wilt frequently show a discrete wilting of the lowest leaflets at the hottest times of the day. At first, this wilting is reversible during the night. Subsequently, and as the disease develops, some sectors of the leaf of the leaflets (which are often intervenous and in a 'V' shape) gradually lose turgidity and turn yellow. Significant portions of the leaflets end up a beige to brown

---

[1] The nomenclature of *Verticillium dahliae* races used is not always conventional. Race 1 is race 0, race 2 should be called 'race 1'.

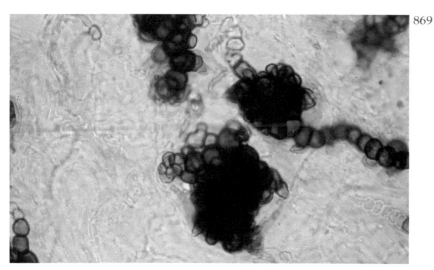

869

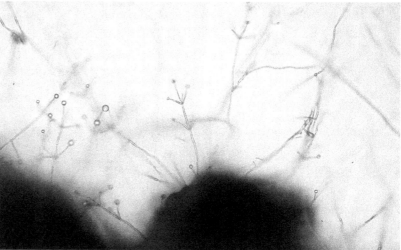

870

871

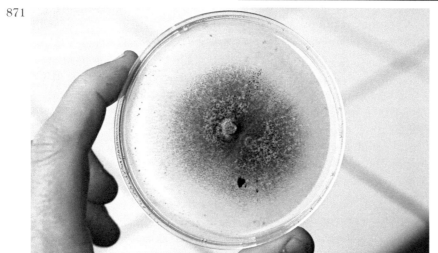

869 This fungus also forms black globular or elongated microsclerotia, measuring 50–200 µm, which differentiates it from *Verticillium albo-atrum*.

871 *Verticillium dahliae*, a slow growing fungus produces *in vitro* on malt-agar some small colonies which are white at first and becoming progressively melanized as a number of sclerotia appear in the agar medium.

870 The conidiophores of *Verticillium dahliae* are branched and in whorls; they produce ellipsoid, hyaline conidia in droplet clusters at their terminal or lateral ends.

colour and become necrotic. The affected leaflets and leaves die prematurely, which can lead to exposure of fruits to sunlight and therefore to the risk of sunscald (see p. 385).

A longitudinal or transverse cut in the vessels of the lower part of the stem or branch shows a slight browning, much less marked than in the case of Fusarium wilt.

Symptoms of Verticillium wilt tend to occurr in the spring and autumn. Unlike *Fusarium*, they do not lead to plant death but contribute to decline and to reduced yields.

(See Photos **661–666**.)

• Biology, epidemiology

Survival, inoculum sources: *V. dahliae* is a poor competitor in the soil, with no saprophytic capability. Despite this, its survival is ensured by mycelial fragments in plant debris and also by its microsclerotia (Photo **869**) that can persist for over 15 years. This fungus has many strains with differing parasitic potential which allows them to be pathogenic on a wide range of hosts.

This host range includes many cultivated plants (tomatoes, peppers, potatoes, cucurbits, artichoke, lettuce, cabbage, strawberry, rose, aster, chrysanthemum, tobacco, cotton) as well as weeds (black nightshade, pigweed) all ensuring its multiplication and survival.

Penetration and invasion: infection occurs either by direct penetration of the root or through various roots injuries by the mycelium that results from the germination of microsclerotia.

Wounds caused by root-knot nematodes and *Pratylenchus* spp. are also points of entry. Once in the plant, the fungus spreads to the vascular system of plants which it gradually colonizes. Xylem vessels react to invasion by forming some gum or tyloses that impedes the pathogen's progress. These defence mechanisms associated with the colonizing and clogging of vessels by the mycelium, contribute to the wilting of plants. Note that *V. dahliae* produces some microsclerotia in the tissues (at temperatures between 10 and 20°C) and also fragile verticillate conidiophores with ovoid conidia.

Sporulation and dissemination: this is possible from compost and on tools soiled with contaminated soil and plant debris. Soil with microsclerotia and/or conidia is easily dispersed by air currents, as well as water splashes and soil insects.

Conditions encouraging development: *V. dahliae* is most pathogenic at cool temperatures, but there are a number of strains with varying temperature requirements. The optimum temperature is between 20 and 32°C. Short day length and low light will make plants more sensitive to the disease. The effect of latter is more severe in neutral to alkaline soils. Monoculture of sensitive plants or rotations which are too short or poorly chosen help to increase its occurrence.

Control methods[1]

• During cultivation

There are no method of treating this disease during cropping. However, it is worth attempting too retard its development by applying a fungicide solution of thiophanate-methyl, a benzimidazole, to the base of the plants or via the drip irrigation system. In addition, light drenching at the hottest times of the day helps to reduce wilting.

The tools used in contaminated areas should be thoroughly cleaned before being used in others that are not contaminated. The same applies to tractor wheels. Thorough rinsing with water is often sufficient to remove soil and with it, *Verticillium dahliae*.

Crop residues, normally buried in the soil develop numerous microsclerotia when colonized by *V. dahliae*. Removal of residues limits

[1] See note to the reader p. 417.

this and helps reduce the inoculum left in the ground.

• Subsequent crop

Crop rotations prevent or delay the onset of the disease. To be effective, they must be long enough (at least 4 years) and must not include sensitive crops such as eggplant, potato, pepper, cucurbits, or strawberries. The monocots, notably cereals, do not seem to be affected by this vascular fungus. Peas, beans, and cabbage could be used in rotations as they do not result in the survival of inoculum in the soil.

The destruction of some weeds such as black nightshade and pigweed (*Amaranthus* spp.) is worthwhile as they are hosts of *Verticillium dahliae*.

Use balanced fertilizer applications to prevent succulent tissues especially in young plants. Irrigation should be optimal during warm periods in order to limit wilting.

Soil disinfection with fumigants (dazomet, metam-sodium) is only partially effective, is expensive, and not sustainable. Disinfection by solarization, recommended against other soil fungi such as *Rhizoctonia solani*, reduces the incidence of *V. dahliae*. The disinfected soil should not be cultivated too deeply before planting, as this might bring microsclerotia from depth into the treated soil layer.

The flooding of contaminated areas for several days may help limit the number of microsclerotia in the soil, by reducing the amount of available oxygen and increasing the amount of CO_2. However, this not generally recommended as it does not significantly reduce the occurrence of the disease. Propagules that remain are still sufficient to allow *V. dahliae* to parasitize its host.

Thiophanate-methyl is registered for use as a tomato 'soil treatment'. The use of carbendazim is no longer permissible.

The most effective control is the use of resistant varieties. Indeed, many varieties currently grown have a monogenic dominant resistance conferred by the 'Ve' gene (derived from *Lycopersicon pimpinellifolium* and located on chromosome 9). They can control the common populations (race 0, incorrectly called 'race 1'). This resistance is effective against *Verticillium dahliae*, but also against *V. albo-atrum*. A race 2 able to overcome this resistance gene was described in the late 1950s, and now occurs in many production areas of the world: North America (US notably California), South America, North and South Africa (Tunisia), Australia and southern Europe (France, Greece).

Grafting onto a tomato rootstocks an intra- or interspecific hybrid (see p. 658) may be used to control Verticillium wilt. Note that these rootstocks have only the gene 'Ve' and can therefore be attacked by race 2.

In addition, several antagonistic microorganisms are reported in the literature to be effective to varying degrees against *V. dahliae*: *Trichoderma viridae*, *Talaromyces flavus*, *Bacillus subtilis*, *Streptomyces griseoviridis*, *Pseudomonas fluorescens*, *Gliocladium roseum*, *Glomus mosseae*, and *Penicillium oxalicum*.

Main fungi responsible for fruit rot

- *Alternaria alternata* (see p. 393)
- *Alternata* f. sp. *lycopersici* (see p. 309): Description 9
- *Alternaria tomato* (see p. 393)
- *Alternaria tomatophila*: Description 1
- *Botrytis cinerea*: Description 2
- *Colletotrichum* spp. (see p. 391): Description 10
- *Didymella lycopersici*: Description 3
- *Geotrichum candidum* (see p. 405)
- *Penicillium* spp. (see p. 400)

- *Phoma destructiva* (see p. 396)
- *Phytophthora infestans* (see p. 374): Description 7
- *Phytophthora nicotianae* (see p. 374): Description 12
- *Rhizoctonia solani* (see p. 411)
- *Rhizopus stolonifer* (see p. 403)
- *Sclerotinia* spp.: Description 15
- *Stemphylium botryosum* (see p. 393)
- *Trichothecium roseum* (see p. 400)
- *Ulocladium syndicated* (see p. 393)

In vitro aspects of
fungi responsible for
fruit rot

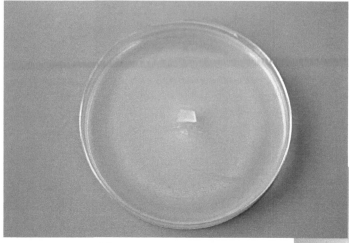

872 *Geotrichum candidum* is quick growing fungus. The colonies are very flat on malt-agar and have a beige tinge.

873 This colony of *Rhizopus stolonifer* produces aerial mycelium on which there are many black pin dots fairly typical of this zygomycete.

874 *Penicillium expansum* grows rather slowly, forming some concentric zones on which the formation of a fairly typical blue-green mould is clearly visible.

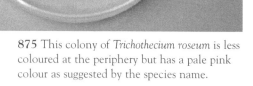

875 This colony of *Trichothecium roseum* is less coloured at the periphery but has a pale pink colour as suggested by the species name.

Main control methods influencing the susceptibility of fruit and the development of pathogens

Table 46 Main control methods for primary or secondary infections by micro-organisms responsible for fruit rot on tomato pre-and postharvest[1]

Possible methods	Expected outcomes
Before and during cultivation	
Use tolerant varieties.	Tomato varieties show differing sensitivities towards fruit rots: – low sensitivity, often nonspecific, e.g. associated with a low fruit susceptibility (for instance ripeness, strength of cuticle, susceptibility of the stem scar, adaptation to stress); – variable resistances, mainly specific to *Phytophthora nicotianae* (partial polygenic resistance, monogenic resistance conferred by the '*Br*' gene, resistance conferred by *Lycopersicon esculentum* var. 'Cerasiform'), *Rhizoctonia solani* (multiple resistance involving at least four major genes in one case or two genes in the other), *Alternaria alternata* (derived from an accession of *Lycopersicon esculentum* var. 'Cerasiform').
Grow plants on raised beds.	Avoiding fruit contact with the soil thus preventing certain diseases (*Phytophthora nicotianae*, *Rhizoctonia solani*, *Sclerotium rolfsii*, *Pythium* spp.).
Choose favourable planting densities to allow good ventilation of the crop.	
Stake plants or grow them on a support system.	Enabling good ventilation of the plants.
Arrange the rows to allow maximal ventilation by the prevailing winds.	
Remove the lower axillary branches.	
Control fertilizer application, especially nitrogen.	The larger fruits, from crops with high nitrogen application, appear to be more sensitive to rot.
Manage irrigation to minimize plant humidity (drip irrigation is preferable), and avoid water stress.	The presence of free water on the foliage and fruits promotes the dissemination of inoculum and infection.
Do not work upon or harvest wet crops.	

[1] See note to the reader p. 417.

Possible methods	Expected outcomes
Before and during cultivation (contin)	
Apply fungicides.	Several fungicides are recommended in the literature to control some rots (especially caused by *Alternaria alternata*): chlorothanonil, difenoconazole, diflolatan, mancozeb, anilazine, azoxystrobin, copper, and carbendazim. One to three applications are made in the weeks before harvest, taking into account the predictions made by the forecasting model Tom-Cast.
Apply natural substances.	Various preparations have been tried for the control of certain rot agents: vapours of essential oils of lemon, extracts of *Withania somnifera*, garlic, onion, *Dennettia tripetala*, and of *Azadirachta indica*.
Apply various micro-organisms.	Some yeast-like fungi (*Pichia onychis*, *P. anomala*, *Debaryomyces hansenii*) and other fungi (*Trichoderma hamatum*, *Gliocladium virens*) have proved to be antagonists and competitors of several rot agents.
Remove the diseased and rotten fruits and destroy them or move them far away from the crop.	These fruits are an important potential source of inoculum, as insects become contaminated when they touch them. Infection can also occur as spores are spread by means of air currents and water splashes.
Clean and disinfect equipment and containers used for harvesting and transporting fruits.	
After harvest	
Clean and disinfect equipment and facilities used for fruit storage.	These measures contribute to the elimination of the nutrient bases that allow the micro-organisms to survive, and thus reduces the pressure of inoculum for the harvested fruits.
Add active chlorine to the wash water, which must be above pH 6.4.	This product is a well known disinfectant whose effectiveness is well established. Other treatments have been tried on fruits, with varying degrees of success: warm water, gamma rays or UV.
Do not submerge the fruits for more than 3 min.	Beyond this period, infiltration takes place resulting in more numerous infections.
Sort and remove rotting fruits.	This avoids storing tomatoes that are potential sources of rots and will minimize spread from rotten to healthy fruit harvest.

Table 47 Main methods reducing the sensitivity of tomato fruits to micro-organisms responsible for rot pre- and postharvest

Possible methods	Expected outcomes
Before and during cultivation	
Do not choose a production schedule that risks chill damage to the fruits or harvesting them at an advanced stage of maturity.	The effects of cold and advanced maturity allow many secondary organisms to invade the fruits.
Avoid injuries related to various techniques of production, handling, packaging, storage (bites, splits, bruises, scratches).	These injuries aid the penetration of many rot inducing micro-organisms, in the field and postharvest. See the specific information and/or implement the recommendations given for growth splits (p. 358), microcracks (p. 360), insect damage (pp. 362 and 380), damage from leaf blight (p. 384), from blossom-end rot (p. 386), and cold (p. 384).
Protect and maintain the maximum level of leaves.	Maintaining a good leaf canopy will protect the fruits from external damage (sun, water).
Harvest with care and use fruit-friendly equipment.	This will avoid causing injuries and, bruising, which would predispose the fruits to subsequent damage by rotting agents.
After harvest	
Keep fruits in cool warehouses but not at a temperature below 12°C.	Storage of fruit at high temperatures promotes the development of rots.
Avoid the presence of free water on the fruits after washing or as a result of condensation; let them dry thoroughly.	Free water on the fruits and high humidity encourages both bacterial and fungal postharvest rots.

BACTERIA

Bacteria are relatively simple small (0.5–1 μm × 1–3 μm) unicellular organisms, which can be detected by a light microscopy using high magnification. They do not have a differentiated nucleus, which classifies them as prokaryotes, but do have nucleic acid but not within a nuclear membrane. The nucleic acid is double-stranded DNA (deoxyribonucleic acid) and this contains the bacterial gene pool. Plasmids, much shorter cytoplasmic strands of DNA, also contain genetic information that can be exchanged with other bacteria during reproduction that involves conjugation. Reproduction most commonly occurs by simple cell division, resulting in the formation of two daughter bacteria (vegetative reproduction by fission). This method of rapid multiplication (generation time varies from 20 minutes to a few hours) allows bacterial populations to grow exponentially when conditions are favourable.

Bacteria have a variable shape and can be spherical, ellipsoidal, or filamentous. The great majority of those attacking the tomato are rod-shaped. They are mobile due to the presence of flagella arranged around the bacterial cell (*Pectobacterium* spp. [Former *Erwinia* spp.]) or at an end (*Pseudomonas syringae* pv. *tomato*, *P. corrugata*, *Xanthomonas* spp., *Pectobacterium carotovorum* subsp. *carotovorum*, *Ralstonia solanacearum*). The presence of these flagella and their location on the bacterium contributes to their identification. Other criteria are also used to distinguish them:
– the composition and structure of the bacterial wall allows the separation of bacteria into two groups: gram-positive (*Clavibacter michiganensis* subsp. *michiganensis*) and gram-negative (*Pseudomonas syringae* pv. *tomato*, *P. corrugata*, *Xanthomonas* spp., *Pectobacterium carotovorum* subsp. carotovorum, *Ralstonia solanacearum*). Other characteristics used in the classification of bacteria include:
– growth rate and appearance of colonies on artificial media including the possible production of diffusible pigments;
– many biochemical characteristics;
– serological and molecular tests.

Many bacteria are commonly associated with the tomato, some as pathogens and others as saprophytes. The saprophytes are found both in the phyllosphere (epiphytes) and rhizosphere. Some of these bacteria occur on other plants, as well as in the soil, in organic matter, and on plant debris on which they can easily survive. Their structure allows them to withstand conditions of stress.

About 250 species, biovars and pathovars, representing 17 bacterial genera have been identified in association with tomatoes. Bacterial pathogens in six genera can affect tomatoes: *Pseudomonas*, *Ralstonia*, *Xanthomonas*, *Clavibacter*, *Pectobacterium* (formerly *Erwinia*), and *Agrobacterium*. *Pseudomonas* are divided into two groups: the fluorescent (*P. syringae* pv. *tomato*, *P. syringae* pv. *syringae*) and nonfluorescent (*P. corrugata*) based on whether a diffusible and fluorescent pigment forms under UV light on Kings medium. Among Pectobacteria, *P. carotovorum* subsp. *carotovorum* is especially important as it damages tomato stem and fruits, during cultivation and also during transport and storage.

All plant parts may be affected by bacterial pathogens:
– leaves and fruits (*Pseudomonas syringae* pv. *tomato*, *Xanthomonas* spp., *Clavibacter michiganensis* subsp. *Michiganensis*, *Pectobacterium carotovorum* subsp. *carotovorum*);
– the roots (*Agrobacterium tumefaciens*, *A. radiobacter*);
– the pith of the stem (*Pectobacterium carotovorum* subsp. *carotovorum*, *Pseudomonas corrugata*);
– the vascular system (*Clavibacter michiganensis* subsp. *michiganensis*, *Ralstonia solanacearum*, *Pectobacterium carotovorum* subsp. *carotovorum* in some cases).

There is a variety of symptoms: spots, wilts, rots, tumours.

Penetration of bacteria into the various parts of the plant occurs in several ways:

– through natural openings such as stomata and hydathodes (*Pseudomonas syringae* pv. *tomato*, *Xanthomonas* spp.);

– through injuries (hail or insects damage) or caused by farming activities such as pruning, leaf removal, harvesting (*Clavibacter michiganensis* subsp. *michiganensis*, *Pectobacterium carotovorum* subsp. *carotovorum*).

Once in the tissues, bacteria develop between plant cells which are killed relatively quickly by enzyme action (proteases, pectinases, amylases, lipases). This provides nutrients for the bacteria and the enzymes overcome the plants defence mechanism. This is especially true of *Pectobacterium* spp. whose pectinolytic enzymes are responsible for soft rot. Bacteria can also disrupt the metabolism of the host plants by producing or interfering with the production of growth regulators. They sometimes produce toxins inducing the formation of a yellow halo around the damaged tissue. Some of their polysaccharides may play a role in the induction of wilting, as is the case for *R. solanacearum*.

Colonized tissues contain large populations of bacteria and these may be spread in several ways:

– by rain drop splashes or dripping water;

– via contaminated seeds (*Clavibacter michiganensis* subsp. *michiganensis*, *Pseudomonas syringae* pv. *tomato*, *Xanthomonas* spp.);

– by man and tools (especially *Clavibacter michiganensis* subsp. *michiganensis*);

– by some animals and insects (*Pectobacterium carotovorum* subsp. *carotovorum*).

Water on plants and in the soil is essential for the proper growth of bacteria and their dissemination, and is a major factor in the development of epidemics of bacterial diseases.

Temperature has less influence on their development, but there are differences depending on the species. For example, *Ralstonia solanacearum* and to a lesser extent the *Xanthomonas* spp. prefer high temperatures, while *Pseudomonas syringae* pv. *tomato* grows best in cool, wet conditions. Temperature definitely plays a role in disease development both geographically and seasonally.

■ Air-borne bacteria

Pseudomonas syringae pv. *tomato* (Okabe) Young, Dye & Wilkie

Bacterial speck

(*Pseudomonas*, Pseudomonadaceae, Pseudomonadales, Gammaproteobacteria, Proteobacteria, Bacteria)

Principal characteristics

•Frequency and importance of damage

This gram-negative bacterium is rod-shaped, and is now present in all areas of tomato production. It causes severe damage when the weather is cold and wet. It occurs mainly on field crops, especially on varieties for canning, making the fruits unfit for marketing and processing. It is also found in unheated glasshouses where condensation of water occurs and temperatures are often low.

It is particularly troublesome on crops of processing tomatoes. Fortunately, the use of resistant varieties has limited its impact.

Two races have been described: race 0 (on universally susceptible varieties) and race 1 on varieties with resistance conferred by the incompletely dominant gene '*Pto*'.

• Main symptoms

Symptoms on leaflets of this air-borne bacterial pathogen are somewhat similar to those described for *Xanthomonas* spp.: small greasy dark stains, quickly become brown to black in colour and appear at random on the leaflets, often the youngest or on the edge. Damaged areas are circular or slightly angular and have a yellow halo of various size. Initially the lesions are about 2–3 mm, but can develop and coalesce in the presence of moisture, affecting large areas of the leaf that eventually becomes necrotic and desiccates.

Lesions of similar appearance are produced on other plant parts, particularly the petioles and stems. On these they are often more extensive and elongated than on the leaflets. When present on the flower stalks and sepals, they can lead to flower drop.

Note that leaflet and stem symptoms can be observed on seedlings in the nursery similar to those described in adult plants. During severe attacks seedlings can die.

Fruit symptoms allow the differentiation of bacterial speck and bacterial spot (*Xanthomonas*): *Pseudomonas syringae* pv. *tomato* is the cause of tiny black spots, mainly on green fruits. These are very superficial, slightly raised, and have a diameter of 1 mm (less than 3 mm at most). A wet, dark green halo sometimes surrounds them. These dark spots persist on the mature red fruits.

(See Photos **14, 148, 236, 241, 245, 250–254, 674, 745**.)

• Biology, epidemiology

Survival, inoculum sources: *P. syringae* pv. *tomato* is able to survive in the soil or on plant debris for at least 7 months. In areas of production, it survives on plants growing from self-sown seeds or from regrowth of old plants. This bacterium seems to persist on the roots and foliage of some weeds (*Amaranthus retroflexus*, *Chenopodium album*, *Galinsoga parviflora*, *Hibiscus trionum*, *Lamium amplexicaule*, *Portulaca oleracea*, *Sinapis arvensis*, *Stellaria media*, *Polygonum lapathifolium*). Its cultivated host range is very narrow and is limited to tomatoes. This bacterium may be present on seeds and survive on them.

Penetration and invasion: bacterial cells on the surface of the tomato (epiphytic development) enter the leaflets through natural openings such as stomata and hydathodes or through various injuries (broken trichomes, growth cracks, damage resulting from the effects of sand or wind). Subsequently, the bacteria invade the tissues (especially the intercellular spaces) and multiply in large quantities. If the conditions are very favourable, the first symptoms appear in less than 1 week. In 24 hours, millions of bacterial cells are produced, which is an asset for their dissemination. Note that this bacterium has been associated with hairs on ovaries (glandular or not) during pollination. These hairs gradually disappear after this stage, leaving entry points in the skin of young fruits which are infection sites for *P. syringae* pv. *tomato*.

Transmission, dissemination: this occurs mainly with water splashes or aerosols, during rainy periods and with the use of sprinkler irrigation. These water droplets can be transported over long distances by the wind. In this way bacterial cells are transported to other plant parts and surrounding plants. Workers moving in crops with wet foliage contribute to dispersion.

Contaminated seeds and seedlings also ensure its dissemination to other farms or to other areas or countries of production.

Conditions encouraging development: moderate temperatures and high humidities are very conducive to outbreaks of *Pseudomonas syringae* pv. *tomato*. This bacterium grows at temperatures between 13 and 28°C; the range 18–24°C is particularly favourable, and its activity is greatly reduced above 30°C. Periods of dew or fog, rain, or sprinkler irrigation on plants leave a film of water which is ideal for infection. Six hours are sufficient for an epidemic to begin. Note that the conditions of the crop play a greater role in the development of the disease than contamination at the nursery stage.

Like many plant bacterial diseases, speck is very difficult to control. This will require the careful implementation of all the recommended measures.

Control methods[1]

• During cultivation

When the first spots are found in a tomato field, it is too late and there are no satisfactory measures to prevent the disease developing.

Copper in the form of salts (copper, copper sulfate, copper hydroxide, cuprous oxide, copper oxychloride) can be used to limit the spread of this disease. Copper has a preventive effect but only on the surface of aerial organs. Its effectiveness is enhanced by mixing with dithiocarbamates fungicides such as maneb or mancozeb. Other products have been, or are still being used to control *P. syringae* pv. *tomato*; for example, sodium hypochlorite (at 125 ml of bleach at 48° active chlorine[2] per hectare for use in the evening as this product is photosensitive), and several antibiotics, particularly streptomycin (used mainly before

planting, notably in the US but banned in many countries).

The intensive use of copper during the 1980s (it is not degraded in the soil and accumulates) led to the emergence of strains resistant to this metal's salts, particularly in the Americas. Similarly, strains resistant to streptomycin emerged during the 1960s. Finally, it should be noted that in many situations, these treatments are not sufficient to prevent outbreaks of speck when climatic conditions are favourable. Note that acibenzolar-S-methyl is now used in some countries.

In addition to copper treatment, sprinkler irrigation should be reduced to a strict minimum and used during the day, at a time when plants will dry out quickly. In addition, work and movement in the crops should be restricted to

[1] See note to the reader p. 417.
[2] The chlormetric system of measuring active chlorine uses degrees: 1% active chlorine converts to 3.16 chlormetric degrees of chlorine.

dry conditions. Good ventilation will reduce leaf wetness and thus prevent infection and reduce the impact of the disease.

During harvest and at the end of cropping, avoid leaving crop residue on the ground or burying them. If it is not possible to do otherwise, bury them deeply to enable decomposition.

• Subsequent crop

At propagation stage

It is essential to use good quality healthy seeds. If in doubt, seed treatment will be necessary. Several methods or means of disinfection can remove the bacterial cells present in the seeds:
– hot, dry air (70°C for 96 hours, which slows the rate of seed germination);
– water at 50°C for 25 minutes;
– fermentation of the pulp and seeds for 4–5 days;
– soaking the seeds in: 0.8% acetic acid for 24 hours; in 5% hydrochloric acid at for 5–10 hours; in sodium hypochlorite at 1.05% for 20–40 minutes; mercuric chloride at 0.05% for 5 minutes.

These methods are particularly effective if seed contamination is external.

Nursery soils must be disinfected by steaming or with a fumigant. Similar treatment is required for the structures and walls of the glasshouses. Note that solarization (see p. 487) reduces bacterial populations of P. syringae pv. tomato up to 30 cm deep in nursery soil, thus significantly reducing the incidence and to a lesser extent, the severity of the disease.

The production of plug plants is an effective means of producing 'clean' plants, especially if good quality substrates are used and the plugs are placed on shelves and never in contact with the soil. Sprinkler irrigation must be limited or managed in such a way as to achieve rapid drying of the seedlings. These must not be handled while still wet. Tools must be disinfected, and workers must wash their hands before and after handling seedlings. Avoid the use of seedlings of doubtful origin. Beware that seedlings can appear to be healthy on arrival, but may be carrying the bacteria which will develop later.

Copper and dithiocarbamate sprays must be applied several times in the nursery to ensure the production of healthy seedlings. Infected seedlings are a serious source of the pathogen in the crop. The frequency of treatments will vary depending on the production area and the risks involved. For example, some growers recommend copper treatments to begin 5 days after emergence and continue every 4–5 days. It is best to buy disease-free plants which have been raised in semi-arid areas which are not conducive to this bacterial disease. In some countries, plants are certified free of this bacterial disease.

In the field

Rotations of 2–3 years with nonhost plants, especially cereals, are necessary to break the disease cycle. Weeds should be removed to avoid possible propagation and survival of the bacterium on them. Be aware that self-sown tomato plants can enable survival and propagation.

Fertilizer application should be balanced, as excessive or under application is conducive to disease development. Sprinkler irrigation should be avoided and individual plant application preferred.

Avoid planting too early in the year, so as not to expose plants to conditions that may be cold and wet. Do not plant tomatoes next to pepper or other tomato crops, especially if they are already affected.

The application of a mixture of copper and a dithiocarbamate, notably mancozeb, will delay the development and spread of this disease. Spraying is done during the season and preferably when plants are dry and just before rains are expected. In California, applications are made before every rain and at a frequency of 10–12 days when the weather is cold and wet. They cease when temperatures reach 32°C. In cooler climates the treatments against this disease are mostly made in early summer.

A source of resistance in the wild species Lycopersicon pimpinellifolium and in some cultivars is an incompletely dominant gene, 'Pto'. Homozygous genotypes often show a higher level of resistance to this bacterial disease. Note that the expression of this resistance is closely associated with another locus, 'Fen', which confers sensitivity to an insecticide of the organophosphate family fenthion. Phytotoxicity is expressed by the appearance of spots on leaflets similar to those caused by the bacteria. This helps the work of breeders who can use this insecticide to screen progeny in breeding programmes. A number of cultivars used for field crops have this resistance. Other wild species are resistant to this bacterium (*Lycopsersicon glandulosum*, *L. hirsutum*, and *L. peruvianum*) and other sources of resistance exist.

Note that a strain able to develop on several cultivars resistant to speck has been reported in California in 1998 and identified as race 1 of *P. syringae* pv. *tomato*. This race has also been identified in Italy in 1995 on a '*Pto*' heterozygous resistant cultivar. Its virulence has been confirmed on the variety 'Ontario 7710' which is homozygous for this gene and has a high level of resistance.

Several antagonistic micro-organisms, some natural stimulators of defence (SDNs), and various products have been tried with varying degrees of success against speck, but also bacterial gall. They are mentioned here in case their use is permitted in a grower's country.

The micro-organisms that can be used are almost exclusively other bacteria: *Pseudomonas putida*, *Pseudomonas syringae* strain Cit7, *Rahnella aquatilis*, *Cellulomonas turbata*, a nonpathogenic strain of *Xanthomonas vesicatoria*.

In addition, foliar applications based on bacteriophages also showed some effectiveness, especially if used during the evening. Persistence problems of bacteriophages on foliage have been found, possibly due to the presence of certain pesticides.

Several rhizobacteria that stimulate plant growth (PGPR) such as *Pseudomonas fluorescens* (associated with antagonism to *P. syringae*) or *Azospirullum brasilense* (by inoculation of seeds and in combination with foliar treatments based on copper in particular) have reduced attacks by this disease. It is the same for SDNs such as acibenzolar-S-methyl. This compound reduces the severity of symptoms on the leaves but also on the fruit. Note that it is also associated in the US with treatments with bacteriophages and also with copper + mancozeb sprays.

Finally, a product combining hydrogen peroxide, acetic acid, and iron limits the level of bacterial populations on plants. Some compost extracts used in aqueous solutions and sprayed on plants limit the severity of the disease, as do *Azadirachta indica* oil, emulsions based on fish, the ammonium lignosulfonate, and potassium phosphate.

876 *Pseudomonas syringae* pv. *tomato*, like all Pseudomonas is a gram-negative bacterium, rod-shaped, with polar flagellae.

Pseudomonas syringae pv. *syringae* van Hall

Syringae leaf spot

(*Pseudomonas*, Pseudomonadaceae, Pseudomonadales, Gammaproteobacteria, Proteobacteria, Bacteria)

Distribution and damage

This gram-negative bacterium is widespread on all continents, affecting many plant species. However, it has only been reported on tomato in a limited number of countries including the US, Iran, and several states in Brazil. It is an opportunistic bacterium that does not constitute a significant threat to tomato crops. It is only very occasionally serious.

Symptoms

It causes symptoms on seedlings in nurseries and adult plants in the field. In Brazil leaves develop wet spots which turn brown and necrotic. These are generally small (1–2 mm in diameter, never more than 10 mm), are often circular, and without a chlorotic halo. The spots may coalesce and form larger irregular necrotic patches. In the US this pathogen causes various small brown leaf lesions with no halo similar to those caused by *Pseudomonas syringae* pv. *tomato*, crowned with a yellow margin. (See Photo **247**.)

Aspects of biology

This bacterium is probably able to survive in the soil, on plant debris, and on various cultivated hosts and on various weeds. It has been reported, for example, to be responsible for leaf spots on pepper seedlings in nurseries in Italy. It is also reported on cereals especially rice, lemon, hazelnut, walnut, several stone fruit trees (apricot, peach, plum, mango) and some pome fruit trees (pear).

Bacterial present on the surface of the leaflets enter these hosts via various injuries often caused by other tomato pests. Because of this it is generally considered to be a weak pathogen. Bacteria invade the tissues and multiply in large quantities. Spread is by water splash resulting from rain or irrigation. Workers in a wet crop also aid dispersion. Contaminated seedlings are a source of the pathogen and contribute to spread.

This bacterial disease develops well in wet weather but not at high temperatures.

Control methods[1]

Many methods of protection recommended for the control of the other two air-borne bacterial pathogens, *Pseudomonas syringae* pv. *tomato* and *Xanthomonas* spp., may limit the development of *P. syringae* pv. *syringae* (see pp. 528 and 534).

[1] See note to the reader p. 417.

Xanthomonas spp.

X. euvesicatoria (Jones *et al.*, 2006)

X. perforans (Jones *et al.*, 2006)

X. gardneri (ex-Sutic, 1957) Jones *et al.*, 2006

X. vesicatoria (ex-Doidge) Vauterin *et al.*, 1995

(syn.: *Xanthomonas campestris* pv. *vesicatoria* [Doidge] Dye.)

(*Xanthomonas*, Xanthomonadaceae, Xanthomonadales, Gammaproteobacteria, Proteobacteria, Bacteria)

Bacterial spot

Principal characteristics

These gram-negative bacteria affect both pepper and tomato. For approximately 70 years, it was considered a single organism, with various races. During the past two decades, it has been shown that several species are responsible for this blight on tomatoes and they may occur in the same area of production. They are four species, classified according to phenotypic and genetic criteria.

Note that strains belonging to the four groups (*Table 48*) have been isolated on tomatoes, and not on peppers. Some strains appear to be specific to tomato, others specific to pepper but some affect both. The biological situation is quite complex and constantly changing with increased information. The description here should be viewed with some caution as it does not distinguish between the four species. It is not yet known whether these have fundamentally different biological properties. For simplicity, *X. campestris* pv. *vesicatoria* is used to describe all of these bacteria.

• Frequency and extent of damage

This bacterium is harmful on tomatoes and but even more so on peppers. The disease is found on almost every continent where tomatoes are grown. In the early twentieth century it was

Table 48 *Xanthomonas* species and races associated with tomato bacterial spot

Species	Other names used in the literature	Group described	Related races
Xanthomonas euvesicatoria	*Xanthomonas axonopodis* pv. *vesicatoria*	*Xanthomonas campestris* pv. *vesicatoria* Group A	T1
Xanthomonas vesicatoria		*Xanthomonas campestris* pv. *vesicatoria* Group B	T2
Xanthomonas perforans	*Xanthomonas axonopodis* pv. *vesicatoria*	*Xanthomonas campestris* pv. *vesicatoria* Group C	T2, T3 and T4
Xanthomonas gardneri		*Xanthomonas campestris* pv. *vesicatoria* Group D	T2

described for the first time in South Africa on tomatoes. It is particularly serious in the field, in tropical, sub-tropical, and temperate regions, where climatic conditions are hot and humid. In temperate regions, particularly in Europe, it does not seem to affect protected crops whether heated or cold. In France, it occurs mainly on the processing varieties, during the summer.

The strains of groups A and B are globally distributed. Those in group C have been found in Mexico, Thailand, and the US, those from Group D in the US, Brazil, Costa Rica, and Yugoslavia. Only strains of group B occur in France.

• Main symptoms

Symptoms on leaves are comparable to those described for *Pseudomonas syringae* pv. *tomato*: small, oily, and translucent lesions develop randomly on the leaflets or on the edge of the leaves and these turn brown or black. They are circular or slightly angular, with a diameter of 2–3 mm, and sometimes have a yellow halo. The centre dries and the decayed tissues fall out. In moist conditions spots can merge and affect large areas of the leaves that eventually become necrotic and desiccate. In some cases, entire leaves turn yellow, wither, and may remain attached to the plant or fall. Young leaves are often more sensitive.

Similar symptoms are visible in other plant parts, especially on petioles, stem, and fruit pedicels (stalks). They are often more extensive and elongated than on the leaflets. When they are present on the floral pedicels, a few flowers may fall.

Oily dark green to black lesions form initially on the green fruits. They give rise later to cracked corky raised pustules, up to 10 mm in diameter. An oily halo sometimes surrounds them. When used for canning, the affected fruits are more difficult to peel.

These symptoms can be confused with fruit damage caused by hailstones. Also it is not uncommon for this bacterium and *Pseudomonas syringae* pv. *tomato* to be present at the same time in some crops. (See Photos **246**, **255–259**, **747**.)

• Biology, epidemiology

Survival, inoculum sources: *X. axonopodis* pv. *vesicatoria*, a rod-shaped gram-negative bacteria, survives for only a short time in the soil but better in plant debris, (present in or on the soil) and on seeds (several months to a year, possibly 10 years). The primary inoculum often comes from the plant's environment or from seeds. This pathogen persists on wheat roots. It can also live in an epiphytic state or infect several cultivated solanaceous plants, weeds, and wild plants. Among the plants listed are: peppers (*Capsicum annuum*, *C. frutescens*), eggplant, *Amaranthus retroflexus*, *Datura stramonium*, *Chenopodium album*, *Portulaca oleracea*, *Galinsoga parviflora*, *Hibiscus trionum*, *Hyoscyamus niger*, *H. aureus*, *Lycium chinense*, *L. halimifolium*, *Physalis minima*, *Solanum dulcamara*, *S. nigrum*, *Nicandra physaloides*, *Lycopersicon pimpinellifolium*, *Nicotiana rustica*. Some of these hosts have been established by artificial inoculation only, and it is possible that hypersensitivity reactions may have been confused with true infection. Evidence regarding the four species and host range is lacking.

Penetration and invasion: after an epiphytic phase the bacteria enter the leaflets through natural openings such as stomata and hydathodes, or through various injuries (broken hairs [trichomes], growth cracks, damage following the effects of sand or wind or insect bites [*Nezara viridula*]). Subsequently, the bacteria invade the tissues and multiply in large quantities. In 24 hours, millions of bacterial cells are produced, which is an asset for dissemination, and the first symptoms appear in 5–6 days if conditions are very favourable.

Transmission, dissemination: this is by water splashes that occur during rain and sprinkler irrigation. The part of the aerosol spray (microdroplets) can be transported over longer distances by the wind, which is why in some plots the progression of the disease is in the direction of prevailing winds. Thus, bacterial cells are splashed to other plant parts and surrounding plants. Workers' movements in wet foliage of the crop also contribute to dispersion. Contaminated seeds and plants also result in spread. Note that this bacterium has been found on several insects: *Aulacophora concta*, *Eucoptaera praemorsa*, *Phanoroptera gracilis*, *Henosepilachna vigintioctopunctata*.

Conditions encouraging development: these include high temperatures and high humidities, hence it is most damaging in the summer. Its optimum temperature is around 26°C, with a range of development between 20 and

35°C. Warm nights, between 23 and 27°C are very conducive to disease development. Cold nights with temperatures below 16°C inhibit its development. As with all bacterial pathogens it requires high humidity often associated with frequent rainstorms as well as dew, and also with sprinkler irrigation. Excessive nitrogen also encourages this disease.

Control methods[1]

Bacterial spot, like many bacterial diseases, is very difficult to control especially once the disease has established. As many control methods are common to bacterial speck, Description 21 should be consulted.

When symptoms are recognized in a tomato crop it is too late to intervene. There are no effective measures to prevent the continued development of this bacterial disease.

Note however that the intensive use of copper and certain antibiotics, notably in nurseries, has led to the emergence of tolerant *Xanthomonas* strains in several states of the US and in Mexico and Brazil. For example, in the latter, resistance to this compound and to an antibiotic, streptomycin sulfate, have been detected (with varying frequency depending on the species) in *Xanthomonas gardneri*, *X. axonopodis* pv. *vesicatoria*, and *X. vesicatoria*. In addition, many strains of *X. euvesicatoria* are tolerant of copper.

As in the case of speck, special attention should be paid to seed quality. These can be disinfected as recommended in Descrption 21.

Two sources of resistance to *X. axonopodis* pv. *vesicatoria* have been reported in the genus *Lycopersicon*. These take the form of hypersensitivity against certain strains:

– resistance to race T1 from the cultivar 'Hawaii 7998' which is conferred by three genes, '*Rx1*', '*Rx2*', and '*Rx3*' (note especially the recent description of the lines 'Ohio 9834' and 'Ohio 9816' partially resistant to race T1);

– a second source of resistance from *Lycopersicon pimpinellifolium* ('PI 128 216' and 'PI 126 932') and *L. esculentum* cultivar 'Hawaii 7981', effective against a second group of T3 pathogenic strains on tomato and conferred by one incompletely dominant gene, '*RxvT3*'.

The accession 'PI 114 490' is resistant to all three races T1, T2, T3; at least two loci are responsible for this resistance. Resistance to race T4 is available from the accession 'LA 716' of *Lycopersicon pennellii*; this resistance has apparently not been confirmed in the field.

Three races of *Xanthomonas* on tomato (T1, T2, T3) have been described, initially on tomatoes from three cultivars. Races T1 and T2, described in 1990, have been reported around the world in the Americas, Europe, and Australia. Race T3 has been found more recently in Florida, Mexico, and Thailand. The emergence of races in certain areas of production could be related to the use of resistant varieties or to their introduction through seeds.

[1] See note to the reader p. 417.

Recently, two new races have been identified in Florida from strains of group C: races T2 and T4. For example, a recent survey conducted in Brazil revealed the existence of several species of *Xanthomonas* in the various tomato production areas: X. *gardneri* (group D/race T2), X. *vesicatoria* (group B/race T2), X. *euvesicatoria* (group A/race T1), and X. *perforans* (group C/race T3). Finally, for information, some tests have been conducted with acid electrolysis water, known for its germicidal activity. This may be effective against X. *vesicatoria* as a disinfectant for the seed surface or as a contact bactericide for aerial vegetation.

Endophytic, vascular and/or soil-borne bacteria

Agrobacterium tumefaciens Smith & Townsend

Crown gall

(*Agrobacterium*, Rhizobiaceae, Rhizobiales, Alpha proteobacteria, Proteobacteria, Bacteria)

Principal characteristics

Several biovars of A. *tumefaciens* have been described. The biovars 1 of *Agrobacterium tumefaciens*, *Agrobacterium rhizogenes*, and *Agrobacterium radiobacter* have been combined as A. *tumefaciens*. Crown gall from the vine is caused by biovar 3 and has been spread globally by contaminated cuttings.

Note that the biovar 1 of A. *radiobacter*, and probably of other bacterial species, is involved in a disease of the root systems of cucumber and tomato called 'root mat'. This disease, now common in several European countries, mostly affects soil-less crops (for more information on root mat see p. 276).

• Frequency and extent of damage

This gram-negative bacterium, present in the rhizosphere of many plants, is widely distributed throughout the world. It attacks plants belonging to more than 60 botanical families, whether woody or herbaceous. Despite this, it is infrequently reported on tomato where it causes tumours especially in soil-less crops. The impact of its attacks on plant growth and yields are negligible.

The biology of these bacteria and infection processes are quite original. The mechanism used to infect a plant requires the integration of part of its genome, a fragment of tumour-inducing Ti plasmid (T-DNA), in the plant genome. That is why it is used in transgenesis as a vector for integrating desired genes into the genome of plants.

• Main symptoms

The appearance of small growths on some roots near the soil surface is the first manifestations of this bacterial disease on many plants, including tomatoes. Initially these young tumours are smooth, whitish, and rather spherical. As they get older, they become corky and gradually turn brown. They have an irregular shape and varying size, some reaching several centimetres in diameter. They can become spongy and then rot.

The presence of tumours on the roots does not affect plant growth and fruit production.

(See Photos **479, 508–511**.)

• Biology, epidemiology

Survival, inoculum sources: A. *tumefaciens* is able to survive in a saprophytic state in the soil for at least 2 years. It can also survive in and on plant debris. Water and contaminated soils can be sources of the pathogen. This bacterium is able to survive in contaminated substrates. It affects many crops, fruit and ornamental, including apple, apricot, cherry, peach, grape, rose, aster, and chrysanthemum. Many vegetables in addition to tomatoes show galls, including beet and turnip. These plants, when grown near tomato fields or produced in the same glasshouse, can be sources of inoculum.

Penetration and invasion: penetration in the plant is often through wounds on the roots or the stem near the surface of the soil or substrate resulting from cultural operations, or damage caused by pests, frost, and so on. Mobile bacterial cells are attracted by chemotaxis to the root following the release of sugars and other compounds into the rhizosphere.

Once in place, the pathogen stimulates the rapid and uncontrolled division of surrounding vegetative cells. The insertion of a portion of its genome into that of the plant leads to the overproduction of growth regulators (auxins and cytokinins, which cause cell proliferation) and opines that serve as nutrients for the bacterium. Most of the genes involved are not on the chromosome of *A. tumefaciens*, but from a tumour-inducing Ti plasmid that is a fragment of extra-chromosomal DNA. Tissues continue to grow locally and tumours grow slowly, resulting in reduced transport of water and nutrients in the plant. The bacterium seems to multiply

in the intercellular spaces at the tumours' periphery.

Transmission, dissemination: *A. tumefaciens* can be transmitted through water and contaminated soil particles to other plants or crops. Workers through agricultural operations and agricultural machinery carrying soil particles can ensure the dissemination of this bacterium. It is found in the substrates and the nutrient solution of soil-less cultures. It is not impossible that it can be disseminated in the latter.

Conditions encouraging development: high temperatures encourage its development.

Control of this disease on tomato is not a major issue because it is infrequent and does not cause considerable damage. In many situations, hygiene should still be used in an attempt to eliminate it from the crop.

Control methods[1]

• During culture

When tumours are found on plants, it is too late to intervene. No products or methods can effectively control this bacterial disease.

In hydroponics, if a few plants are affected in a limited number of blocks/bags, they should be removed from the crop and destroyed. Care must be taken not to leave behind plant debris and particles of substrate.

In soil-less culture, all plants, especially their root systems, should be removed and destroyed as well as the contaminated substrate. Thorough

cleaning of the glasshouse and equipment is necessary in the same way as recommended for *Clavibacter michiganensis* subsp. *michiganensis* (see p. 24).

• Subsequent crop

In the field, a rotation with monocots, cereals such as wheat or corn, for several years may be effective.

In protected crops, care should be taken not to grow tomato crops in close proximity with ornamental species that may be hosts of this pathogen.

[1] See note to the reader p. 417.

Clavibacter michiganensis subsp. michiganensis (Smith) Davis et al.

Syn.: *Corynebacterium michiganense* [E.F.Sm] Jensen.

Bacterial canker

(*Clavibacter*, Microbacteriaceae, Micrococcinae, Actinomycetales, Actinobacteria, Bacteria)

Principal characteristics

• Frequency and extent of damage

This bacterium, which mainly affects some Solanaceae, was described for the first time in North America in 1909. It is the source of a vascular bacterial blight particularly feared by producers in many areas of production. Easily transmitted by seeds and seedlings, it occurs in the field and under protection. In the latter, the methods of culture of the crop contribute to its spread and increase its impact. It is found in soil and soil-less crops, where it is present in the substrate and the nutrient solution. Its damage can be considerable: in uncontrolled situations, almost all plants can be affected. Eventually, the number of dead plants result in yields being greatly reduced.

In many countries, this bacterial disease is probably one of the most feared tomato diseases, particularly under protection. In addition to the damage it causes, it often forces the reorganization of work in the crop. Once on a farm, it is difficult to get rid of without the implementation of very strict control measures.

Aggressiveness and host specificity of this gram-positive bacterium appears to fluctuate depending on the strain studied.

• Main symptoms

Various symptoms caused by C. *michiganensis* subsp. *michiganensis* are seen on the aerial parts of the tomato. They are influenced by many factors: the plant (variety, stage of development), bacterium (aggressiveness of the strain involved), the type of crop (under protection, open field), the prevailing climatic conditions, and cultural practices. This bacterial disease can go undetected at certain times of the year but develops very quickly at others. Early detection of symptoms can help limit its spread and sometimes allows its control.

The bacterium is seed-borne so affected seedlings can occur in the nursery. Seedlings may wither quickly, stay stunted, or be apparently healthy. In the latter case, which is the most dangerous, latent infections may develop after planting and during cropping.

The earliest symptoms are sometimes difficult to see especially if they involve only the vascular tissues. Careful examination of affected plants shows the presence of patches of pale green or water soaked tissue, at the edge of the leaves or between the veins of some leaflets. Subsequently, these tissues become necrotic and dry out. Necrotic tissues enlarge and coalesce, destroying significant portions of leaflets and even whole leaves. Also leaf rolling of the lower leaves and overnight reversible wilting are other indications of the presence of this vascular bacterium. At this stage the root system is healthy.

Longitudinal or transverse cuts made at intervals along the stem, show symptoms depending on the stage of disease progression.

The following can be seen in and around the vascular tissue:
– discreet yellowing and browning of vessels, but especially of surrounding tissues. The latter sometimes only display a diffuse yellowish tint and a few tiny holes which are rather difficult to see;
– a marked browning of the xylem and surrounding tissues, phloem, and pith. The pith can sometimes be brown or dry and mealy.

Severely affected plants produce poorly coloured smaller fruits, which may fall prematurely. A cross section often reveals discolored vascular vessels. Although the disease can occur at any time it is especially when the plant begins to bear fruit that symptoms are strongly expressed. The delay between infection and the onset of symptoms varies depending mainly on weather conditions. It can be quite short, 2–3 weeks, and sometimes at least 3 months. In many cases, several adjacent plants of the same row are affected. This linear distribution results from the transmission of the bacteria during the cultural operations (pruning, leafing out, harvesting).

A few tiny cankerous suberized spots, first white and then beige, may appear on the wet leaflets, petioles, stems, and fruits. They result from infection by aerial bacteria. On fruit, spots have a very characteristic shape reminiscent of a bird's eye. They are white and raised, a few millimetres in diameter (3-4 mm), and have a dark brown centre. When they are close to each other, they can coalesce.

(See Photos **9, 15, 248, 249, 561, 617, 620, 630–642, 644–650, 746**.)

• **Biology, epidemiology**

Survival, inoculum sources: once introduced into a tomato crop in the field or in a glasshouse, C. michiganensis subsp. michiganensis is able to survive several months to over a year depending on temperature and soil moisture. Plant debris, especially if dry, contributes to its survival for up to 7 months even after burial (some authors refer to 5 years), and at least 2 years when on the surface. This bacterium can survive for 2 years in compost and 10 months in dead stems. In the presence of moisture, decomposition of residues is much faster and the bacterium persists for a shorter time. A study of the survival in stem fragments left on the ground or buried in various climatic conditions has shown survival for several hundred days but always a shorter time when stem fragments are buried.

The bacterium is also found on equipment (pots, drip irrigation, containers, troughs, string), tools, and glasshouse structures. In some production areas, the canes used to stake the plants are also a primary source of inoculum, and it remains viable on these for at least 100 days. The bacterium may also be present in the nutrient solution of soil-less crops.

A number of alternative hosts has been reported, such as peppers, eggplant, potato, tobacco, and several weeds such as Solanum nigrum, S. douglasii, S. triflorum, and Chenopodium spp. Clavibacter michiganensis subsp. michiganensis seems to also be able to survive in an epiphytic state on several other plants not belonging to the Solanaceae.

Tomato seeds are likely to disseminate this disease, and the bacterium may be both external and internal.

Penetration and invasion: first infection may occur during propagation either from seed (via small wounds on the cotyledons) or there may be external sources within the environment of young seedlings. Secondary infection can occur at propagation during cultural operations such as repotting, grafting, and handling plants, all of which can lead to minor injuries.

During cultivation, the bacteria enter the plant either through natural openings like stomata or hydathodes or through various injuries located on the roots or leaves (broken trichomes, side shooting, leaf removal). They reach and multiply in the vessels (xylem and phloem and the surrounding tissues [cortex and medulla]), hence the name 'bacterial canker'. They can also move downwards to reach the roots, where they can be released into the nutrient solution of soil-less crops.

Many bacteria are also present on the surface of plants, and these can produce cankerous specks that are seen on the leaflets, stems, and fruit.

Transmission, dissemination: C. michiganensis subsp. michiganensis is widely disseminated as a result of rainfall, sprinkler irrigation, and pesticide sprays. By the removal of side shoots and de-leafing, workers aid spread from plant to plant, resulting in the characteristic linear distribution in crops.

In hydroponics, the nutrient solution also contributes to the spread of C. michiganensis subsp. michiganensis, especially in NFT. Many bacteria are released from the roots and transported by the nutrient solution, where they then penetrate healthy plants through their root systems. However, C. michiganensis subsp. michiganensis does not appear to multiply in the solution. Fine debris and dust from contaminated soil are scattered by the wind, and could be a source of inoculum.

Workers can carry the bacterium from crop to crop (via their hands and clothes), as well as on contaminated tools or equipment (carts, boxes). Symptomless plants are a possible source and a means of distribution when offered for sale.

Conditions encouraging development: canker is encouraged by wet weather periods and high humidities (80% and over). The temperature range for bacterial growth is between 12.8 and 33.7°C, optimum between 24 and 27°C. It is particularly resistant to dry conditions.

Low light and poor nutrition appear to make plants more susceptible. Excessive nitrogen in nitrate form, which results in very vigorous plants, also contributes to disease development. After infection the incubation period can last between 12 and 34 days depending on weather conditions. The intensity of symptoms is less in cold weather and on older plants. Symptoms appear earlier on young plants than on older plants. Symptoms are also more severe on plants grown in sandy soils than in organic soils.

Control methods[1]

There are no totally effective control methods. All possible measures should be taken to prevent its introduction into the crop. Once present the methods used will vary depending on the type of crop production (open field, in soil under protection, soil-less under protection).

• During cultivation

As soon as the first diseased plants are found, the quick implementation of a number of measures is essential. If few plants are affected, they should be carefully removed, and put in plastic sack before taking them from the crop. Once away from the crop they must be destroyed. A quarantine zone should be established around the affected area, and a cordon put in place. No staff should enter this area without permission. It should be worked either by a specialized team, or after the rest of the crop has been done, taking care to wear gloves and overshoes, to disinfect tools during and after use, and to change clothes when leaving the area.

Workers should wash hands frequently with soap and water. Even better is wearing waterproof gloves that they can disinfected by treating the gloved hands in a bag containing a sponge soaked in a disinfectant (Virkon-S for example). Tools can be disinfected in the same way or by dipping them in alcohol or in a bleach solution. Attention must also be given to health and safety regulations.

Footbaths must be installed at each entrance and filled with a disinfectant (Virkon or products shown in *Table 49*, p. 542). These must be functional throughout the season. Workers must use them coming in and out of

[1] See note to the reader p. 417.

the crop, and enter the crop only when it is absolutely necessary.

Staff should be made aware of the symptoms of the disease so that these can be recognized at an early stage. The person in charge of the crop must be told immediately symptoms are found.

Overhead irrigation must not be used and pesticides sprays minimized. It is also important to carefully control irrigation so that areas of the crop do not get too wet. Where the crop is affected, excessive run-off water can spread the bacterium so must be avoided. If the water source is from the surface it should be carefully monitored to check for contamination. It is preferable to use well water or water from the water supply. In hydroponics, where the nutrient solution is recycled, it must be disinfected because of possible contamination from affected plant roots. Experiments have shown that if the solution is kept for several hours at pH 4–4.5, C. *michiganensis* subsp. *michiganensis* is eliminated.

Applications of copper have frequently been recommended but are not very effective, especially as spraying can result in the spread of the pathogen. Various forms of copper are approved for the control of tomato bacterial diseases, including copper sulfate, copper hydroxide, cuprous oxide, and copper oxychloride. Note that acibenzolar-S-methyl is registered in some countries for the control of bacterial canker.

If there are many centres of disease in crops, the implementation of control measures is more difficult and time consuming, and effectiveness much more limited.

At the end of cropping it is essential to implement a thorough clean-up in order to try to eliminate the pathogen from the farm. All plants must be destroyed. Avoid accumulating organic waste (leaflets, fruits, seeds, roots) that will easily contaminate the farm.

In protected crops, substrates from soil-less crops must be discarded. In soil crops, minimize burying debris. The structures of the glasshouses, heating pipes, pillars, and all the materials and tools that have been in contact with diseased plants should be high-pressure washed and thoroughly cleaned. This is followed by disinfection using one of the following: acetic acid, peracetic acid, hydrogen peroxide, benzoic acid, sulfamic acid, potassium monopersulfate, malic acid, didecyldimethylammonium chloride, formaldehyde, or glutaraldehyde.

Following disinfection a rinse with clean water is needed (see *Table 49*, p. 542).

In field crops, plant debris, especially the stems, should be removed, burned, limed, or transported to a dump, and never composted because they allow the bacteria to survive, sometimes for more than 2 years.

• **Subsequent crop**

For the next crop clean seeds must be used. A number of methods of seed disinfection are reported in the literature (fermentation at the time of extraction, treatment with acetic acid, sodium hypochlorite, or hydrochloric acid). These treatments may reduce the bacteria on the seeds; they are unlikely to eliminate them completely. In Israel, tomato seeds are routinely treated in a mixture of copper acetate, acetic acid, pentachloronitrobenzene, 5-ethoxy-3 (trichloromethyl) -1,2,4-thiadiazol, and Triton X-100 for 1 hour at 45°C. This method seems effective against most of the common bacteria affecting tomato.

Seed lots are normally checked for the presence of C. *michiganensis* subsp. *michiganensis*. It is difficult to guarantee a total absence of the bacteria in seeds. Be wary of seeds or seedlings of new lines or varieties for variety trials as the seeds are often produced in small quantities, sometimes in areas where the disease is endemic, and methods of protection inadequate or ineffective.

Propagation is a critical period. Indeed, the bacteria are often introduced through the seeds, and develop insidiously. Latent infection may occur and go undetected. Subsequently, symptomless plants may be used in new crops. Be particularly vigilant during the production of grafted plants as handling can result in the spread of the pathogen. It will be imperative in such situations to use optimal hygiene measures and to start with clean seeds.

In propagation areas or nurseries, the land or location used for this work must be first disinfected using either steam or a fumigant. During propagation avoid very high humidities and any practices that lead to the plants wilting. Plant feeding must not be excessive or seeding density too high. As far as it is possible, the environmental conditions should be kept unfavourable for the pathogen. Nearby weeds must be destroyed. Similarly weed control must be effective at the site of the future crop. Treatment with copper from the first true leaf

Table 49 Examples of commercial bactericides for the treatment of facilities and equipment (glass-houses and tunnels) and other general uses[1]

Product	Ingredients	Application rates
Agrigerm 2000	didecyldimethyl ammonium chloride 100 g/l + formaldehyde 31.5 g/l + glutaraldehyde 40 g/l + glyoxal 32 g/l	3%
Agroxyde II	Acetic acid 194.4 g/l + hydrogen peroxide 157.5 g/l + peracetic acid 58.6 g/l	2%
Bactipal ELV	Acetic acid 19.44 % + peracetic acid 5.86 % + hydrogen peroxide 15.75 %	2%
Four-Sann	Acetic acid 19.44 % + peracetic acid 5.86 % + hydrogen peroxide 15.75 %	2%
Menno Florades	Benzoic acid 9%	1%
Phenoseptyl POV	Orthophenylphenol 250 g/l	2%
Virkon	Sulfamic acid 5% + potassium monopersulfate 22.5% + malic acid 10%	0,50 l/hl

growth stage (200–300 g copper metal/hl in the form of Bordeaux mixture) is recommended by some authors.

All suspect equipment must be disinfected (pots, stakes, string). For this use bleach (containing 12° chlorometric[2]), soaking for 24 hours and rinsing with water) or formalin (2–5% commercial formalin, soaked for 1 hour then stored under a plastic film for 24 hours), or other trade products (see *Table 49*).

In soil cultures, a long rotation of at least 3 years is required. The soil can be disinfected with a fumigant such as chloropicrin. Solarization, as described on p. 487, would greatly reduce the damage caused by bacterial canker, as has been shown in Greece.

In cases where the glasshouses are surrounded by affected crops, it is advisable to have a barrier crop between them and the glasshouse. In addition, if the affected crop is kept wet by spraying, dust from it will not be blown into the glasshouses.

Foot baths should be used at each entrance and filled with a disinfectant. Ensure that they remain functional throughout the season.

Several sources of resistance are reported in the literature. One of them, from *Lycopersicon peruvianum* var. 'humifusum' has been introduced into tomato through *L. chilense*. It is controlled by a dominant gene ('Cm'), which is not allelic to the gene present in *L. hirsutum* f. *glabratum*, and is located on chromosome 4.

Another source of resistance has been demonstrated in a cross between an accession of *L. hirsutum* ('LA 407') and *L. esculentum*. Two loci are responsible for this resistance: 'Rcm 2.0' on chromosome 2 and 'Rcm 5.1' on 5.

In addition, crosses between *L. esculentum* ('LA 6 203') and *L. parviflorum* ('LA 2 133') have revealed a polygenic resistance. *Lycopersicon pimpinellifolium* and *L. racemigenum* have provided a high level of resistance to bacterial canker.

[1] Many other commercial products are available, often using the same active ingredients. Before making a choice, advice should be sought on the product(s) currently approved for this use.
[2] The chlormetric system of measuring active chlorine uses degrees: 1% active chlorine converts to 3.16 chlormetric of chlorine.

Resistant varieties of tomato are described in the literature and further breeding programmes are under way but commercial varieties are not yet available

Few products appear to be effective against these bacteria. The activator of SDNs, aci-benzolar-S-methyl, which is approved in some countries, reduces the severity of the disease and the multiplication of bacteria in plants. Note that several biopesticides may have an anti-*Clavibacter* activity in the field but have only been demonstrated *in vitro*: *Pseudomonas fluorescens*, *P. putida*, *Bacillus subtilis*. Finally some effect has been found using suspensions of garlic and peat humates.

877 Colonies of *Clavibacter michiganensis* subsp. *michiganensis* on LPGA medium.

Pectobacterium carotovorum subsp. carotovorum (Jones) Hauben et al.

Syn.: *Erwinia carotovora* subsp. *carotovora* [Jones] Bergey *et al.*

Bacterial stem rot and fruit rot

(*Pectobacterium*, Enterobacteriaceae, Enterobacteriales, Proteobacteria, Bacteria)

Principal characteristics

• Frequency and extent of damage

Pectobacterium carotovorum subsp. *carotovorum*, a gram-negative bacterium with rod-shaped cells, is globally widespread and attacks a wide range of plants, notably various vegetables. It is reported in tomato crops on all continents. In wet weather in the field, it can cause damage to the stem and/or fruits. It is also damaging in protected crops whether, they are in the soil or soil-less. Fruit damage can occur, during tomato transportation and in storage.

Other bacteria in the genus *Erwinia* or closely related to it, are reported on tomato, affecting mainly stems or fruits: *Erwinia aroideae* (Town.) Holland, *Dickeya chrysanthemi* (Burkholder *et al.*) Samson *et al.* (*Pectobacterium chrysanthemi* [Burkholder *et al.*] Brenner *et al.*, *Erwinia chrysanthemi* Burkholder *et al.*) and *Pectobacterium carotovorum* subsp. *atrosepticum* (van Hall) Hauben *et al.* Comb. Nov. (*Erwinia carotovora* subsp. *atroseptica* [van Hall] Dye).

• Main symptoms

Pectobacterium carotovorum subsp. *carotovorum* causes stem lesions. These are moist and brown and penetrate to the pith which decomposes resulting in a hollow centre to the affected stem. The bacterium can readily spread to unaffected parts of the stem and lesions may surround the stem and be several centimetres long. Decaying tissues are wet and soft.

The internal damage to the stem affects its functioning, especially the transport of water and nutrients. This results in wilting and yellowing of leaves, especially during warm periods and/or from the time the first fruits are harvested. In favourable conditions, the disease

progresses and the plants may die. If growing conditions become warmer, slightly affected plants produce fruit normally.

These stem symptoms, accompanied by wilting, should not be confused with those caused by *Pseudomonas corrugata* which causes black pith (see p. 26).

When it invades the fruits, this bacterium produces a wet rot which is soft and viscous, and affected fruits eventually decompose into a liquid mass. Ultimately, only the epidermis and some remnants of shrivelled tissues remain.

(See Photos **560, 601, 602, 667, 797**.)

• Biology, epidemiology

Survival, inoculum sources: *P. carotovorum* subsp. *carotovorum* is ubiquitous and is present in many soils in which it survives for several years, especially in plant debris and in water. It is also found in the tomato phyllosphere. This bacterium is polyphagous and is likely to remain on a fairly large number of cultivated and noncultivated hosts, mostly herbaceous dicotyledons. It occurs on many vegetables, including other members of the Solanaceae such as potato, eggplant, and peppers that are even more sensitive. It is also found on lettuce, celeriac, cabbage, basil, and fennel. This bacterium can be isolated from the washing water in tanks which has been used for cleaning fruits before processing.

Penetration and invasion: it enters the various parts of the tomato, especially fruits, mainly through injuries (stem scar, mechanical wounds, damage due to insects, effects of sand) and/or as a result of various cultural operations during or after harvest (harvest during wet period, fruit washing). It is a weak parasite and can also be secondary following other pathogens. Once in place, its cellulolytic and pectinolytic enzymes actively contribute to its rapid development in tissues that rot quickly and these may also have a foul odour.

Transmission, dissemination: it is easily spread by water splashes and runoff. Also insects and tools used during cultivation are potential sources and contribute to its dispersion.

Conditions encouraging development: they are mainly hot and humid. Cloudy and rainy periods increase the risk of spread of this bacterium. It seems able to grow at temperatures between 5 and 37°C, its optimum being between 25 and 30°C. In dry soils, whose moisture is below 40%, the development of *P. carotovorum* subsp. *carotovorum* is inhibited and in some situations this can be total.

Poor control of storage temperature, the presence of multiple injuries (moth damage, growth cracks) and use of contaminated water during fruit washing encourage bacterial damage during storage and transport.

Very vigorous plants are more susceptible. A low ratio of potassium to nitrogen in the nutrient solution of soil-less crops appears to make them more susceptible to stem damage.

Control methods[1]

• During cultivation

It is impossible to control the development of this bacterium once it is present in the tomato stem. However, a number of measures, often identical to those used to control black pith, may be used (Description 26).

First all diseased plants should be removed as soon as possible. It also advisable to lower the humidity and reduce the water on the foliage, as well as avoiding wet soil. Avoid overhead irrigation; if this is not possible, apply water in the morning rather than evening, so that the plants dry quickly during the day. Be aware that wet plants will aid dissemination of the bacterium as well as contributing to affected fruits so avoid working in crops when they are wet.

The glasshouse should be well ventilated to dry the foliage which may have become wet as a result of night condensation or overhead irrigation.

Remove crop residue at harvest, and avoid burying them in soil as *P. carotovorum* subsp. *carotovorum* survives very well in the soil.

• Subsequent crop

This bacterial disease is a significant risk for future crops, so preventive measures must be practised to limit its introduction and development.

Crop rotations after serious attacks can reduce the incidence of soil-borne pathogens and this reduction will help to limit the survival of *P. carotovorum* subsp. *carotovorum* in the soil. Because of its wide host range there is limited choice of crops for use in rotation. Those least susceptible include cereals, grasses and corn, soy, or beans. The proximity of susceptible and/or already affected crops as well as weed hosts will add to the risk of attack so should be avoided if at all possible.

[1] See note to the reader p. 417.

Fields in well ventilated locations are preferred as in these the leaves will dry more readily. If possible propagate the plants at a dry time of year. Soil preparation should focus on good drainage. Plant nutrition should be balanced, avoiding excess nitrogen. In hydroponics, it is advantageous to use a nutrient solution with a rather high ratio of potassium to nitrogen. Avoid conditions conducive to the occurrence of injuries.

Fruit should be harvested in dry weather and before they are over ripe, taking care to minimize injury. The fruits are best refrigerated quickly or stored in a cool dry place. It is advisable to consult the recommendations for the control of postharvest diseases described on p. 522.

There are currently no commercially available tomato varieties resistant to *P. carotovorum* subsp. *carotovorum*. Some lines have very occasionally been reported in the literature as having fruit less sensitive to this bacterium. This tolerance is polygenic in nature.

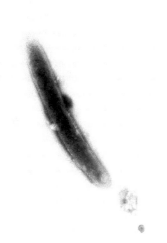

878 *Pectobacterium carotovorum* subsp. *carotovorum* is a rod-shaped, gram-negative bacterium with peritrichous flagellae.

546

Pseudomonas corrugata Roberts & Scarlett

Tomato pith necrosis

(*Pseudomonas*, Pseudomonadaceae, Pseudomonadales, Gammaproteobacteria, Proteobacteria, Bacteria)

Main characteristics

• Frequency and extent of damage

This gram-negative bacterium is very common and is an opportunistic pathogen on tomatoes. Described for the first time in the UK in 1978, it was reported during the 1980s in several European countries (Germany, Italy, Sweden, Portugal, France), New Zealand, and the US. This bacterial disease is widespread in many production areas on every continent, in field and protected crops. Its damage on tomato may be anecdotal or very substantial depending on production situations.

Fairly recent work has shown that *Pseudomonas corrugata* from affected plants usually occurs together with another species, *Pseudomonas mediterranea* Catala et al., and the geographical distribution of either of these two species is not now well known since the two often occur together. The distribution of *P. corrugata* is thought to be worldwide but P. *mediterranea* has been isolated in Italy, Spain, Portugal, Turkey, and France among assumed strains of *P. corrugata*.

It is of interest that strains of *P. corrugata* have been successfully tested as biological agents on different plants, against bacteria (*Agrobacterium tumefaciens*, *Clavibacter michiganensis* subsp. *Sepedonicus*) but also fungi (*Pythium* spp., *Botrytis cinerea*, *Verticillium dahliae*, *Rhizoctonia solani*).

Several other bacteria have been associated with symptoms of tomato pith necrosis in addition to *Pseudomonas corrugata*: *Pseudomonas cichorii* (Swingle) Stapp, *P. viridiflava* (Burkholder) Dowson, *P. fluorescens* (Flügge) Migula biotype I, *Pectobacterium carotovorum* subsp. *carotovorum*, *P. carotovorum* subsp. *atrosepticum*, and *Erwinia chrysanthemi*. Several of them occur in the same country. For example, at least five *Pseudomonas* (*P. cichorii*, *P. viridiflava*, *P. fluorescens*, *P. corrugata*, and *P. mediterranea*) have been implicated in the early 2000s as responsible for tomato pith necrosis in Turkey.

• Main symptoms

Affected plants are often very vigorous and have thick stems. Apical leaflets are not curled in the usual way shown by healthy plants but are chlorotic and wilt at the hottest times of the day.

Moist dark brown to black patches are visible along the stem. They affect the epidermis and may spread over several tens of centimetres or the complete length of the stem. Aerial adventitious roots tend to form on portions of stems affected by pith necrosis. A mucous discharge from leaf scars is often seen.

Transverse or longitudinal cuts in the stem can show that the pith is affected in various ways depending on climatic conditions and severity of symptoms. It can be brown and necrotic, dry and slightly decomposed, or it may stay white and firm in its centre; however, the tissues near the xylem vessels are a yellow to light brown colour and become necrotic. The vessels turn brown and become necrotic. In some extreme cases, the pith is disintegrated, with limited transverse cavities or complete collapse of the brown tissues. Symptoms can also be seen in the petioles, in the main root, and in some side roots.

The disease starts at the base of the stem as harvesting begins and progresses up the plant. Sometimes severely affected plants die. If climatic conditions change and the environment becomes warmer, slightly affected plants recover. The distribution of diseased plants seems to vary depending on the types of crop: affected plants are generally randomly distributed in the field, but in clusters in some glasshouses.

Pseudomonas corrugata can cause rot on fruit. In addition, symptoms of pith necrosis have occasionally been observed on seedlings in nurseries in Italy, causing significant damage.

(See Photos **562**, **598**, **599**, **603–609**, **619**, **621**, **668**.)

• Biology, epidemiology

The epidemiology of tomato pith necrosis is still poorly understood.

Survival, inoculum sources: *P. corrugata*, a ubiquitous bacterium, can survive in the soil and the rhizosphere of a number of crops irrespective of whether the plants are sensitive to this weak pathogen. It can cause symptoms of pith necrosis on pepper and chrysanthemum, and also *Pelargonium*. Artificial inoculations of the stem of some plant species have shown signs of pith necrosis on eggplant, tobacco, melon, cucumber, zucchini, beans, peas, and celery. It is present in the rhizosphere of many plants as well as being present on leaves, shoots, and seeds without producing symptoms: broccoli, cucumber, potatoes, strawberries, alfalfa, rice, tea tree, vine. *P. corrugata* is often present in the rhizosphere of wheat.

Penetration and invasion: the bacterium multiplies well in the tomato rhizosphere. It enters the plant through wounds on the roots, crown, and stem, where it begins the endophytic colonization of the plant. During the parasitic phase it invades the pith, reaches the vessels and subsequently the epidermis. Bacterial exudates may form on the latter.

Transmission, dissemination: bacteria on plant surfaces are easily scattered by rain and overhead irrigation, especially by splashes. Dew and workers during crop culture also contribute to its distribution. *P. corrugata* exuded from the roots of plants in nutrient solutions can be recycled and be redistributed to healthy plants. In some situations, irrigation water plays an important role in the dissemination of this bacterium.

Note that this bacterium is transmitted through the seed and that it has been detected in seed lots in Israel and Egypt. The impact of this mode of transmission is unknown at present.

Conditions encouraging development: depending on countries and production areas, tomato pith necrosis occurs both in protected crops and in open fields, particularly in cloudy and wet climatic conditions. It seems to be encouraged by excessive irrigation and/or nitrogen; diseased plants are often very vigourous, with a significant growth of the foliage, thick stems, and succulent stems. The disease occurs frequently after periods of cloudy weather and/or cold nights, notably accompanied by high humidity in the glasshouses. The presence of water on the leaves, stems, and leaf removal scars, especially those at the base of the stem, all encourage it.

Control methods[1]

• During cultivation

By the time symptoms of this bacterial disease occur it is too late to instigate control measures. There are no effective measures to prevent its progression.

Despite this and knowing the biology of *P. corrugata*, a number of measures should be taken. It is also advisable to consult the description of *Clavibacter michiganensis* subsp. *michiganensis* as many control methods that are recommended for this pathogen can be used to control tomato pith necrosis.

Above all plant nutrition should be regulated in order to minimize plant vigour. This is usually done by reducing the nitrogen input and increasing the potassium.

The glasshouses should be ventilated and/or heated to prevent high humidity and in particular the presence of water on plants. In addition, irrigation must be well controlled and never excessive.

It is desirable to work in the crops when the plants are dry, especially during the removal of leaves at the stem bases.

During or at the end of cultivation, severely affected or dead plants are removed. Avoid burying plant debris in the soil.

Finally, it must be remember that changes in climatic conditions can completely reverse the adverse effects of this disease and affected plants can restart.

• Subsequent crop

This bacterial disease does not consistently occur each year and depends on climatic conditions and the state of its host. There are no specific measures to use during the preparation for the next crop, but production conditions should not induce vulnerable plants or contribute to a conducive environment. In addition, some authors report that in the field, black plastic mulching and soil fumigation reduces the incidence of the disease.

No currently marketed tomato variety is resistant to pith necrosis. Note that this plant seems much more sensitive than some wild species of *Lycopersicon*. Thus unlike tomato, *L. hirsutum*, *L. hirsutum* f. sp. *glabratum*, *L. cheesmanii*, and *Lycopersicon cerasiforme* have been proved to be resistant to two strains of *P. corrugata*.

[1] See note to the reader p. 417

Ralstonia solanacearum (Smith) Yabuuchi et al.

Bacterial wilt

Syn.: *Burkholderia solanacearum* Yabuuchi *et al.*
Pseudomonas solanacearum E.F. Smith.

(*Ralstonia*, Ralstoniaceae, Burkholderiales, Beta proteobacteria, Proteobacteria, Bacteria)

Principal characteristics

This polyphagous gram-negative soil bacterium with a global distribution, is a serious pathogen of tomato crops. It exists as several strains, which differ notably in their ability to metabolize various sugars and denitrify nitrates, as well as by their distinct host range. *R. solanacearum* is traditionally classified into five physiological races and six biovars:
– race 1, strains isolated from various hosts in a number of countries. It is known as the race of the Solanaceae, infecting potato, eggplant, peppers, tobacco, and tomato;
– race 2, adapted to triploid banana trees in the tropics;
– race 3, which attacks the potato, tomato and weeds belonging to the Solanaceae family, in temperate conditions;
– race 4, associated with ginger;
– race 5, affecting the mulberry tree.

Recently, a phylogenetic analysis based on different molecular technologies showed that this bacterium is a complex of species belonging to four phylotypes related to strains from different geographic origins: *Asiaticum* (phylotype I, biovars 3, 4, and 5), *Americanum* (phylotype II, biovars 1, 2, and 2T and races 3, 2), *Africanum* (phylotype III, biovars 1 and 2T), and *Indonesium* (phylotype IV, biovars 1, 2, 2T).

The tomato is attacked by various strains from race 1 referred to as 'tropical' (*lowland*) belonging to four phylotypes and by strains called 'temperate' from race 3 (highland) belonging to phylotype II. Race 3 of *R. solanacearum* can be pathogenic on tomatoes growing in temperate zones, unlike the tropical strains.

• Frequency and extent of damage

This bacterial disease causes severe problems in tropical, semi-tropical, and warm temperate countries. It is very damaging in many Asian countries (China, Japan, Malaysia, Pakistan, Thailand, Vietnam, India), America (US, Mexico, Brazil, Argentina), and Africa (north and south). *R. solanacearum* is now well established in Europe: race 3 has been reported in Belgium, Germany, UK, the Netherlands, France, Italy, Spain, and Greece.

• Main symptoms

As its name implies, the main symptom is the rapid wilting of young leaves at the hottest times of the day, often reversible at night at first. Subsequently, the wilting becomes permanent. Fairly quickly, the affected tissues become necrotic and dry, and many plants eventually die. Less typical symptoms of stunted growth, leaf epinasty, new adventitious roots developing on the stem, and lower leaves yellowing may occur especially in poor conditions for disease development. All these symptoms can appear on both seedlings in nurseries and on adult plants. The roots can be affected by *R. solanacearum* which causes some of them to rot.

In all cases, a longitudinal section of roots and stems show a yellowish to dark brown coloured vascular. Late in the course of the disease, the pith and cortex may develop wet and brown changes.

(See Photos **622–629**.)

• Biology, epidemiology

Survival, inoculum sources: *R. solanacearum* readily survives for several years up to 30 cm deep in many soils. Plant debris is a particularly potent source in soils with a high capacity for water retention. The rate of soil inoculum can be maintained or even increased during the growth of susceptible plants in the same area. This bacterium is able to infect many crops (sunflower, peanut, tobacco, pepper, cassava, potato, banana) and weeds. Over 250 plant species, belonging to at least 50 botanical families, are likely to be attacked. These are mostly dicotyledons. Severity of attack often depends on the quantity of soil inoculum and the aggressiveness of the strain. This bacterium has been isolated from many weeds with and without symptoms: *Amaranthus spinosus, Chenopodium album, Cyperus rotundus, Erechtites valerianaefolia, Euphorbia hirta, Hydrocotyle ranunculoides, Malva* sp., *Physalis minima, Polygonum pensylvanicum, Rumex dentatus, Solanum nigrum, S. dulcamara, Vicia* sp.

Note that *Ralstonia solanacearum* is present in some irrigation water from ponds and various water sources.

Penetration and invasion: during growth in contaminated soil, the roots are invariably injured:

– either naturally, especially at the point of emergence of lateral roots;

– or as a result of the effects, intentional or not, of tools, bites, and damage caused by root-knot nematodes of the *Meloidogyne* genus, and insect bites.

These injuries are potential entry points for *R. solanacearum*. They facilitate its penetration into the root cortex giving access to the vascular tissue in which the bacterial cells multiply more quickly. It is found in the xylem and phloem vessels but also in the parenchyma cells bordering them.

Transmission, dissemination: many bacterial cells are released into the soil from the roots and healthy roots of neighbouring plants become infected. The spread of *R. solanacearum* often takes place through water runoff, contaminated plants, and tools. The workers during side-shooting and leafing operations contribute to its spread. This bacterium can be transmitted by seeds, although this is means of dissemination is refuted by some authorities.

Conditions encouraging development: *R. solanacearum* likes high temperatures between 25 and 35°C. Heavy and wet soils with a neutral pH, are more favourable to infection and disease development. Dry soil conditions and temperatures below 10°C do not favour its development. High levels of nitrogenous fertilizers make plants more susceptible.

The presence of gall nematodes (*Meloidogyne incognita*) in the same soils aggravates bacterial wilt. In Brazil, it has been demonstrated that drip irrigation encourages the disease more than overhead irrigation. All the soils do not have the same disease potential. For example, in Guadeloupe, ferralitic or recent (alluvial, volcanic) soils at pH between 5 and 7 are more easily contaminated by *R. solanacearum*. The bacterium persists indefinitely. Calcarous clays, however, are inhibitory. Smectite clay prevents the survival of *R. solanacerum*, provided there are no Solanaceous plants present and the soil is kept moist at all times.

Control methods[1]

• During cultivation

There is no way to control this disease effectively once it is present in a tomato crop.

Hygiene measures must be used from the outset: affected crops should be worked last and tools, machinery, and workers' footwear disinfected. Avoid causing root injuries. Description 24 on bacterial canker describes these measures more fully.

It is also strongly recommended that the root systems and stems of plants are removed and destroyed at the end of the crop. This measure will avoid gross contamination of the soil by reducing the amount of debris left on the site.

• Subsequent crop

In countries and areas where this bacterial blight is a major threat to tomato crops, control of this disease by prevention requires the combination of several complementary methods of protection.

Propagation will be conducted in plots that have not carried susceptible crops. If there is any doubt, disinfect the soil (steam, chloropicrin) or the substrate, especially if its origin is questionable. Ideally seedlings should be produced in miniplugs or in soil-less culture using a clean substrate. If seedlings are bought, growers should check that they were produced under conditions avoiding any risk of contamination.

Preferably use disease-free areas and use crop rotation. The management of the crop is of great concern with an emphasis on keeping the soil 'clean' and delaying as long as possible the development of soil-borne diseases. But with *R. solanacearum*, this is not easy because of the very large number of potential hosts. The longer the rotation the lower the rate of soil inoculum. The rotation will be of maximum efficacy if it is possible to use bacterial wilt resistant plants. Fescue, cotton, soybeans, grasses, corn, and rice are the best choice. For example, 4 years of sugar cane, 2 years of *Digitaria* grassland, sorghum as green manure, cut several times on site for 5–6 months, will clean the soil. Similarly, high doses of urea, sewage sludge, and probably any organic fertilizer will help to reduce soil inoculum levels.

Soil disinfection does not seem very effective. In Brazil, soil solarization (see p. 487) associated with the input of organic residues (biofumigation) has been tested: the combination reduces the level of soil bacteria and thus the severity of the damage. However, conflicting results have been obtained in India, where solarization resulted in an increase in the soil bacterial population. The incidence of bacterial wilt is proportional to the duration of implementation of this method of disinfection.

Note that some plant essential oils (thymol thyme, *Cymbopogon martini*) used as biofumigation have been effective in reducing population levels of *R. solanacearum* in soil. A similar effect has been obtained with a fertilizer containing silicon.

The various agricultural operations should help to minimize disease incidence. Good field drainage is essential and excessive soil moisture avoided at all times. Water quality should be monitored: streams can become contaminated, but this is not the case with well water. A balanced fertilizer application, without the use of excessive nitrogen is required. Crops should be kept weed free.

The use of resistant varieties is probably the most efficient, economical, and environmentally responsible way to control bacterial wilt. Unfortunately, the challenge of developing resistant varieties is particularly difficult because the resistances identified are often quantitative and strongly influenced by soil type, temperature, pH, and moisture.

Two types of resistance are found in tomato: a resistance from the line 'Hawaii 7996' and another from the University of North Carolina, the AVRDC and INRA Antilles-Guyane. The first, which is the most effective, is dominant and easier to manipulate in a breeding programme. The second is also polygenic (five estimated genes) and partially effective. The latter must therefore be supported by other agricultural measures and practices. Different breeding work has resulted in several resistant varieties or lines of tomato, adapted to varying degrees to tropical conditions: 'Venus', 'Saturn', 'Caraïbo', 'L 3", 'King Kong'.

[1] See note to the reader p. 417

Note that the resistance to bacterial wilt in tomato appears to be associated with the production of fruits of reduced size.

In addition, several resistant rootstocks are used to control the this disease: cultivated lines and cultivars ('Hawaii 7996', 'Kewalo', 'Venus', 'CRA 66', 'Cranita 2.5.7') or selected for that use ('LS-89', 'BF Okitsu 101', 'PFN 1'), various *Solanum* (*S. aethiopicum* 'Iizuka', *S. torvum*, *S. straminifolium*); eggplant lines are similar in their behaviour against *R. solanacearum*. However, bacterial wilt damage has been reported in Japan on some of these rootstocks.

Various bacteria stimulating plant growth or interacting directly with *R. solanacearum* reduce levels of bacterial wilt attack: several rhizobacteria (*Pseudomonas fluorescens*, *Bacillus pumilus*, *B. subtilis*, *Chryseobacterium* sp., *Streptomyces* sp., *Paenibacillus polymyxa*, *Brevibacillus brevis*).

Another bacterium, *Burkholderia* sp. W3, is responsible for the suppression of bacterial wilt observed for over a decade in Japan on tomato crops produced in soil-less culture on pumice.

A hypovirulent strain of *Ralstonia solanacearum*, obtained by mutagenesis in China, has shown a significant reduction in bacterial wilt on tomato.

Few products appear effective against this bacterium. The SND acibenzolar-S-methyl can reduce the severity of the disease. The level of protection obtained, however, fluctuates depending on the degree of natural resistance of the tomato variety and the inoculum pressure. Chitosan has also been trialed with 'some success'.

PHYTOPLASMAS

Phytoplasmas (formerly mycoplasma, or MLO for 'mycoplasma-like organisms') are bacterial micro-organisms, biotrophic obligate parasites, without a cell wall (gram-negative) and with poorly specified mobility. They live in the sap of cells of the conducting tissues (phloem vessels). When examined with an electron microscope they are seen to be spherical, sometimes polymorphic, with a diameter of about 0.2–0.5 μm. They are therefore smaller than the most common bacteria, and cannot be cultivated *in vitro*. They are prokaryotes belonging to the family of Mollicutes, micro-organisms that are parasitic on both animals and plants, and this group also includes the spiroplasmas, the acholeplasmas, the entomoplasmas, and mycoplasmas. Phytoplasmas are all descended from a gram-positive bacterial (*Bacillus subtilis*) common ancestor.

Phytoplasmas have been found in 700 different plant species, cultivated and wild, woody and herbaceous, and especially dicotyledons. They have a very small genome. The symptoms they cause are quite typical, but they may differ between species: reduced growth of plants and shorter internodes, yellowing, reddening (anthocyanization), leaf deformation, morphological aberrations of flowers.

They are spread by insect vectors pricking/sucking as they feed: leafhoppers, Fulgoridae, and psyllids belonging to the Hemiptera order, and more specifically to the families of Jassidae, Cixiidae, and Psyllidae.

Their detection in plants can be done in several ways: by electron microscopy, using monoclonal or polyclonal antibodies in ELISA tests, or PCR.

Candidatus Phytoplasma asteris, Candidatus Phytoplasma solani

Phytoplasmas causing stolbur, tomato big bud, aster yellows

(*Candidatus* Phytoplasma, Acholeplasmatacae, Acholeplasmatales, Mollicutes, Firmicutes, Bacteria)

Principal characteristics

Several phytoplasma diseases, associated with similar symptoms on tomato, have been described in many production areas of the world and under different names: 'stolbur', 'big bud', 'yellows', 'proliferation'. For several decades, there were difficulties in studying the phytoplasmas because of the lack of effective methods to characterize them. It was impossible to know if the same micro-organism was involved in diseases that appeared to be similar on the same or different hosts in various locations. The advent of molecular tools has enabled the classification of phytoplasmas into groups and sub-groups, depending in particular on the sequence of ribosomal 16S RNA analysis. Thus, several phytoplasmas have been reported on tomato in several countries and are detailed in *Table 49a*.

Table 49a Some characteristics of the phytoplasmas affecting tomatoes

Group N° 16Sr	Disease group name	*Candidatus* Phytoplasma species	Common name of disease	Countries where the disease on tomato has been reported
I (2 sub-groups, A and B)	Aster yellows	*Candidatus* Phytoplasma asteris.	Stolbur, big bud, Aster yellows.	Phytoplasmas from subgroup B have a global distribution and are very common in Japan, Europe (Italy and Portugal). Phytoplasmas in Group A detected in the US.
II	Peanut witches' broom	*Candidatus* Phytoplasma aurantifolia.	Tomato big bud.	Australia.
III	X-Disease	*Candidatus* Phytoplasma pruni.	Yellows.	Brazil, Italy.
V	Elm yellows	*Candidatus* Phytoplasma ulmi.	Yellows.	Italy.
VI	Clover proliferation	*Candidatus* Phytoplasma trifolii.	Stolbur.	Eastern Mediterranean, Spain, Jordan, Lebanon, North America.
XII (2 sub-groups, A and B)	Stolbur	*Candidatus* Phytoplasma solani (Photo **879**). *Candidatus* Phytoplasma australiense.	Stolbur.	Common in parts of Europe, especially in Italy and Portugal.

Many vegetable crops are affected by phytoplasmas belonging mostly to the group of Aster yellows and these are worldwide in their distribution. In tomato, several phytoplasmas in different groups cause disease in a country, often on the same plant, as is the case for example in Italy.

To add to the confusion of the nomenclature, several strains of phytoplasmas, responsible for a stolbur symptom (bushy appearance), have been described under various names: 'parastolbur' (stolbur), 'metastolbur' (stolbur), 'northern stolbur' (probably the Potato witches' broom), 'pseudoclassic stolbur' (undefined) and 'pseudostolbur'(a physiological disease). In addition, a phytoplasma disease called 'small leaf' which is different from stolbur, has been reported in France in the 1980s.

Stolbur is widely distributed throughout Europe.

• Frequency and extent of damage

From one year to another, the effect of phytoplasmas on tomato crops can be very mixed. In many situations, a few dispersed diseased plants occur in the crop. Because of their low frequency they do not cause concern and are often regarded as mere curiosities. But considerable damage can occur in tomato crops: the proportion of affected plants may reach 30–40% or, in particularly serious situations, almost all plants. In addition, if infection occurs early, yields are very low or zero, because of the sterility of many trusses, and the small size of the few fruits produced.

In France, the disease is not uncommon in the field and in exceptional cases under protection on plants near the vents. Serious epidemics have been observed occasionally, especially in 2006.

• Main symptoms

Phytoplasmas attacking tomato cause various symptoms on young stems. These often appear during summer (July or early August in Europe) and affect plant growth. Internodes near to the plants apex are shorter with smaller leaves, sometimes referred to as curled. Leaf tissues are often thicker or even brittle. The leaves are discoloured and are yellow (yellows) and/ or purple (anthocyanin). Adventitious roots sometimes appear on the stems. Plants infected early are rather bushy, because of the development of numerous axillary branches.

The flowers are affected markedly and are abnormally straight. They are often sterile with various morphological changes which vary according to strain and the developmental stage of the flowers at the time of infection:
– sepals, whose veins are stained purple, remain completely sealed and the calyx is enlarged (big bud);
– the flowers are sterile and the petals are green, with stamens of the same colour (loss of floral pigment, virescence);
– the sepals may be leaf-like (phyllody);
– sometimes the malformation or absence of petals, stamens, and carpels, the overdevelopment of petioles;
– intense proliferations are observed in plants infected with group 16SrVI phytoplasma.

The few fruits formed are reduced in growth and are dense, develop colour slowly and irregularly, and have a rather thick stem which contrasts with the reduced size of the fruits.

(See Photos 22, 41, 42, 53–59, 105, 176–180, 219, 220.)

• Biology, epidemiology

Survival, inoculum sources: phytoplasmas responsible for aster yellows, stolbur, etc., have a wide host range including various cultivated hosts and weeds, the latter being important reservoirs. The host range varies with the phytoplasma.

The Aster yellows group affects more than 350 different plant species, both cultivated and wild, in some 50 botanical genera. Many weeds and wild plants are hosts of this phytoplasma: clover, *Salsola tragus*, several species of *Plantago* and *Sonchus* (among which S. *asper*), *Taraxacum officinale*, wild lettuce, *Senecio cruentus*, *Argyranthenium frutescens*, *Spartium junceum*. Cultivated host plants include potato, eggplant, peppers, corn, *Ipomoea obscura*, lettuce, carrot, spinach, celery, courgettes, Brussels sprouts, onion, bean, dahlia, *Hydrangea macrophylla*, *Lavandula officinalis*, olive tree, *Primula* sp., *Viola odorata*.

Phytoplasma from the potato stolbur group infect over 45 species in the Solanaceae, and at least 16 species belonging to six other botanical families. They multiply in these plants which are sometimes a source of inoculum. Among the crops, affected are peppers, eggplant, potato, celery, carrots, strawberries, grapes, tobacco, lavender, and avocado. Various weeds are also hosts such as bindweed (*Convolvulus*

879 A diseased plant with spherical structures in the phloem vessels as seen with an electron microscope.

880 The vector of *Candidatus* Phytoplasma solani is a leafhopper: *Hyalesthes obsoletus*.

881 Buckwheat is a weed host of *Candidatus* Phytoplasma solani. It may have various symptoms including limited growth, small and chlorotic leaves, and short internodes.

882 Black nightshade is a host of *Candidatus* Phytoplasma solani. Affected plants are sterile and chlorotic or purple (the plant on the left is not infected).

arvensis and *Calystegia sepium*, see Photo **881**), black nightshade (Photo **882**), nettle (*Urtica dioica*), and clover. Bindweed and nettle play a crucial role in the epidemiological cycle of stolbur as they are hosts of choice for one of the potential vectors insects and in this way strongly influence epidemics of stolbur.

In several countries, many other hosts have been suspected to be phytoplasma reservoirs or 'cul-de-sac' hosts (without knowing precisely their group), and have been responsible for disease outbreaks on tomato: carrot, peppers, endive, strawberries, avocado, pear (Spain), *Artemisia absinthium*, *Cirsium arvense*, *Cichorium intybus*, *Convolvulus arvensis*, *Taraxacum officinale* (Russia), peppers, tomatoes, tobacco, carrot, parsley, celery, grapes, turnips, *Datura stramonium*, *Taraxacum offinale*, *Silene vulgaris* (Hungary), eggplant, peppers, *Cryptotaenia japonica*, *Chrysanthemum coronarium*, *Gentiana* sp. (Japan). Bindweed and nettles seem to be particularly important as sources.

These phytoplasmas also survive in their vectors which are several species of leafhoppers. The cycle of these insects involves eggs which play no role in the survival of phytoplasmas but ensure the sustainability of the insect from one season to another.

Transmission, dissemination: phytoplasmas are transmitted by several species of leafhoppers in the persistent manner when they feed. There are a large number of species of leafhoppers and the number able to transmit phytoplasmas varies with the phytoplasma. For instance:

Candidatus Phytoplasma asteris: 30 species of leafhoppers including *Macrosteles* spp., *Euscelis* spp., *Scaphytopius* spp., *Aphrodes* spp., *Orius argentatus*, *Euscelidius variegatus*.

Candidatus Phytoplasma solani: *Hyalesthes obsoletus* Signoret (Photo **880**) is the most important vector in Europe. This leafhopper, in the Cixiidae family, is a polyphagous species and a vector of the stolbur phytoplasma group 16SrXII-A. It is found on *Convolvulus arvensis*, *Urtica* spp., *Ranunculus* spp., *Senecio* spp., and *Artemisia* spp., rarely on the fruit of woody plants and on vine. The adults are active from May to mid-August in Europe, with one generation per year and they overwinter as a larval form.

Other insects of the same family, *Hyalesthes mlokosiewiczi*, *Pentastiridius leporinus*, have been reported as vectors.

Leafhoppers spread phytoplasmas over long distances during their migration and infect tomato plants in late spring and summer. Once in contact with the leaf, they penetrate the phloem vessels to feed, injecting or withdrawing phytoplasmas in the process. The phytoplasma(s), once in the insect, multiply in the intestinal wall cells and then cross it. They reach the haemolymph, and from there, various organs, including salivary glands which makes leafhoppers infectious. *Macrosteles quadrilineatus* can remain infectious for at least 100 days.

Leaf hoppers are usually casual visitors to infected plants. They are able to fly long distances. The date of symptom appearance, which is usually 30–45 days after infection, depends on the migration period of the vector(s). Migration is a complex phenomenon involving a transfer of populations of insects from place to place in the form of a mass flight. The causes of this are not fully understood but appear to be related to local unfavourable conditions for the leafhoppers. Among the factors that influence migration and the nature of the flights include hunger, overcrowding, host deterioration, day length, an endocrine deficiency in the insect or genetic effects, temperature, and wind. The insects prefer young plants with succulent tissues and in times of drought they move more readily from wild plants to irrigated crops. Hot, dry summers stimulate the migration of certain vectors. Their feeding behaviour has been studied but is still poorly understood. Cold winters help reduce winter populations.

Phytoplasmas are transmitted by grafting. Different species of dodder (*Cuscuta campestris*, *C. epilinum*, *C. trifolii*), plant parasites affecting various plants including tomatoes, are capable of transmitting phytoplasmas, particularly those responsible for stolbur. Note that phytoplasmas were found in broomrape (*Orobanche aegyptiaca*) parasitizing the roots of tomato, suggesting that they could contribute to transmission. Phytoplasmas do not appear to be transmitted by seeds in the Solanaceae.

Control methods

• During cultivation

As in the case of viruses, there is no control method to control phytoplasmas during cropping: an infected plant will remain so throughout its life.

Currently there are very few effective control measures for phytoplasma diseases. If the damage is severe it may be necessary to terminate the crop. Otherwise the crop is left and a reduced yield has to be accepted.

The removal of diseased plants is not a very effective means of control because often by the time the symptoms appear, most infections have occurred and the insect vectors have often left to visit other plants. However, at the end of the crop, care should be taken to remove diseased plants, but especially weeds in the field or on the crop periphery, as they can be reservoirs.

• Subsequent crop

In countries where attacks can occur during propagation, it is necessary to protect seedlings. Given the length of the latent period of the disease, it may be difficult to see symptoms. Plants produced under protection are generally not affected. The best way to protect plants is to propagate them under an agrotextile (nonwoven or mesh fabrics) which will be a mechanical barrier more effective than insecticides. The efficiency of the latter is rather controversial: a number of insecticides known to be very effective against leafhoppers do not prevent infection in the field. Insecticide treatments directed against other insects sometimes reduce leafhopper populations. It has been found that the use of an aluminized mulch can reduce the number of vectors and the incidence of disease.

Careful weeding of nursery plots and the surroundings area (hedges and paths borders) helps to minimize sources. Only known healthy seedlings should be used, and crops must not be planted near to other susceptible ones such as eggplant, peppers, potatoes, or tobacco.

There is currently no tomato variety resistant to Aster yellows and stolbur. There are however, two lines in fifth generation crosses between tomato and *Lycopersicon peruvianum*, 'PR18-4' and 'PR8 -5' which express resistance to pathogens localized in the phloem, such as Tomato yellow top virus (TYTV), virescence agents, and phytoplasma responsible for big bud on tomato.

VIRUSES

Viruses are small infectious entities, invisible by light microscopy. Their structure is very simple: it is limited to a protein shell, the capsid, within which is nucleic acid that is generally RNA (ribonucleic acid) in plant viruses, and rarely DNA (deoxyribonucleic acid). This nucleic acid is the form of one or more chain molecules made up of hundreds or thousands of units called 'nucleotides', each of which is composed of bases (adenine, guanine, cytosine, thiamine). It is therefore protected by the capsid which is itself made up of repeated subunits (capsomeres). This protein which is antigenic to a variable extent, is frequently used to produce specific antisera to detect plant viruses. Some of them, quite rare, also have a lipid envelope such as Tomato Spotted Wilt virus (TSWV).

The use of electron microscopy shows that the virus particles, or 'virions', affecting tomato come in the form of symmetrical structures whose size is measured in nanometers, and they have a variable shape: elongated (TMV, ToMV, RMV) or flexuous (PepMV, ToBTV) rods, bacilliform structures (AMV), spherical (TSWV), isometric (CMV), two incomplete spheres (TYLCV), and so on.

Numerous viruses can be artificially inoculated to tomato, and over a hundred occur naturally on this plant in various production areas. Each of these viruses is generally associated with a number of hosts of various importance. The tomato is sometimes attacked by several viruses at once, for instance with CMV and PVY, and also with ToMV and PVX (double streak). The viruses only develop in living cells so are 'obligate' parasites. Viruses divert the biochemical cellular 'machinery' of the plant cell to their advantage, to ensure their multiplication. They cause various symptoms which can be confusing and make their identification difficult. Thus, different laboratory techniques are used to detect and identify them: indexing of various host plants (use of tests plants), electron microscopy, serology and immuno-enzymatic methods, or molecular techniques (PCR, molecular hybridization).

In general, when tomato cells die, virus particles do also, except for some very stable viruses that retain their infectivity in the soil or in plant debris (PepMV, TMV, ToMV, TBSV). To survive, they must penetrate the cells of healthy plants and for this reason, viruses have several ways of dissemination.

Some viruses are easily transmitted by contact, especially during cultural operations (PepMV, ToMV). Indeed, the farmer working in the crop causes microwounds on plants (epidermal hairs break for example) and the sap from these contaminates his hands, tools, and even clothes. Similar wounds created on healthy plants allow the penetration of viruses into plant cells. This mode of transmission is very effective for several viruses in tomato.

The vast majority of viruses affecting tomato are transmitted by biting insects. Aphids, because of their biological characteristics, are formidable carriers of viruses, notably in the nonpersistent way. In this case, viruses are acquired or transmitted by aphids in a few seconds, during brief test feeds allowing these insects to explore the suitability of the plant on which they land. The aphids will remain viruliferous, that is to say capable of transmitting the viruses, for a short period of time (a few minutes to hours, e.g. CMV, PVY, AMV, TEV). These viruses are called 'nonpersistent' and they are also 'noncirculative'.

There are also other modes of transmission:
– semi-persistent manner, in which viruses are transmitted to plants or acquired from them during longer feeding periods; the acquisition time can be of the order of a day. In this case, as in the previous one, viruses do not remain within the insect;
– persistent manner (BWYV, PLRV), for which the acquisition or transmission of the viruses occurs during prolonged feeding periods into the phloem vessels (from a few hours to 1–2 days for acquisition). Once absorbed, the viruses undergo their lifecycle in the body of the insect before they are transmitted again; these viruses are called 'circulative' viruses. They go through the digestive tract into the body cavity, to concentrate in the salivary glands. Once viruliferous, aphids will remain so for several days or even a lifetime.

Other insects ensure the efficient transmission of several viruses to the tomato: whiteflies (TYLCV, ToCV, TICV, TLCV), thrips (TSWV, INSV, TSV), leafhoppers (BCTV, PYDV, TYDV). The relatively recent introduction of *Bemisia tabacci* in northern Europe in the early 2000s, led to the occurrence of several viruses transmitted by this effective vector.

A number of other organisms are also virus vectors, notably fungi and nematodes. In tomato, TNV is transmitted by a aquatic fungus *Olpidium brassicae*. The latter is frequently seen in the roots of tomatoes grown in soil-less systems and also survives in the soil by producing resting spores. It has mobile flagellated zoospores, which by infecting the roots of healthy plants, enable the transmission and dissemination of this virus. Several kinds of nematodes (*Longidorus, Paratrichodorus, Trichodorus, Xiphinema*) transmit several viruses as a result of their parasitic activities on roots (TRV, TRSV, TBRV).

Among the viruses seriously affecting tomato, tomato mosaic (ToMV) and pepino mosaic (PepMV) can be a serious problem because they are transmitted through tomato seeds. Seed transmission also ocurs in other host plants of these viruses, notably weeds (CMV). Sometimes it is not strictly seed tramsmission in that the virus may be present on the seed coat but not in the embryo.

■ Viruses transmitted by contact

Pepino mosaic virus (PepMV)

(Potexvirus, Flexiviridae)

Principal characteristics

This virus was detected for the first time in Peru in 1974, in cultures of pepino crops (*Solanum muricatum*), grown for their edible fruit called 'pear-melon'. This crop is traditionally grown in the Andes where the fruit is eaten fresh or in salads.

Several strains of PepMV have been described on tomato in the world, including in the US (US1 and US2), Canada, and Europe (EU). These strains differ from the original strain from Peru (PE) isolated on pepino in various ways, including symptoms, aggressiveness, and nucleotide sequences. More recently, three new strains have been isolated: two in Chile (CH1 and CH2) and one in Poland (PK). Note that many of these strains (mostly EU, PE, and US2) have been identified in Spain, sometimes in mixed infections. In addition, the US1 strain has been found in the Canary Islands. In North America, several major genotypes: EU, US1, US2, and CH2 have been isolated. These observations tend to prove that the introduction of PepMV, at least in several countries, could have happened in several stages.

Finally very recent molecular biology studies reduced the number of PeMV strains to five: PE, EU, US1 (same as CH1), US2, and CH2 (common origin to PK).

Virus particles are present in the cytoplasm of leaf cells. They are filamentous, nonenveloped, often flexuous, with a length of 508 × 11 nm.

• Frequency, extent of damage

PepMV was introduced relatively recently in Europe and the first records were from the Netherlands, in 1998. About 40 outbreaks have since been identified in tomato crops in Dutch glasshouses. The disease subsequently spread to UK, Belgium, Germany, and Austria. It was also found in Spain (Murcia, Almeria, and the Canary Islands), Denmark, Finland, Sweden, Estonia, and Italy. Three outbreaks occurred in France in 2000 and 2001. This virus is now present in Morocco and North America (several states in the US and Canada), where it has been observed since 2001. Within 2 years this virus has been found in several continents.

The virus spreads rapidly by contact within a crop and reduces yields. Depending on the country, yield losses range from 5 to 15%. In Spain, losses of around 40% have been recorded. A particularly aggressive strain has had the same effect in Canada. It is known that because of its aggressivenesss, PepMV can infect at least 70% of tomato plants in a glasshouse in about 6 weeks.

• Main symptoms

On tomato PepMV causes various symptoms such as mosaics of varying severity, sometimes yellow, spots or even quite marked chlorotic patches. The yellow spots are the most characteristic symptom of this virus. Deformation of the leaflets and leaves are also seen during low light conditions; they can curl upward or downward and display blistering and dark green enations. The plants apical growth stops and the leaflets and the leaves become rather dark, narrower, and more serrated. These symptoms give them the characteristic appearance of nettle leaves or of damage due to hormone type herbicides. In addition, brown and corky ridges may appear on the stem.

The sepals are sometimes superficially necrotic, and the flowers can brown and abort. The fruits show a mosaic of varying intensity, especially when they are ripe. This mosaic can occur without any other symptoms being present on plants.

Diseased plants tend to become senescent prematurely. They are also often distributed along rows, at least initially.

The expression of symptoms seems strongly influenced by climatic conditions, particularly temperature and light in the glasshouses. The symptoms are easily seen from autumn to spring, but tend to disappear in hot and bright weather. The disease is expressed differently depending on the stage of plant development and on the cultivar. Note that it may go completely unnoticed despite infection of many plants, or only occur on particular leaves and fruit which are forming at the time the plant is infected.

(See Photos **13**, **107**, **131–136**, **156–160**, **279–282**, **710**.)

• Ecology, epidemiology

Survival, inoculum sources: PepMVis a stable and highly contagious virus. It is able to survive in the soil, especially in leaf and root debris, and on glasshouses structures, as also do ToMV and TMV. It can survive more than 90 days in the dried plant material. It can also survive in wet contaminated debris. Contaminated clothing can retain it for at least 14 days.

In addition to tomato, a rather limited number of plants in the Solanaceae are hosts, e.g. *Lycopersicon chilense*, *L. chmielewskii*, *L. parviflorum*, and *L. peruvianum*. In addition, many weeds can be symptomless carriers: *Amaranthus* sp., *Malva parviflora*, *Solanum nigrum*, *Sonchus oleraceus*. Grafted eggplants may have the virus in their rootstocks.

Transmission, dissemination: this virus is very easily transmitted by contact. The virus is transmitted from diseased plants to healthy plants just by contact and the chances of this type of tranmission is increased if there is a slight wind or the plants are moved by workers. Workers become contaminated if they are in an affected crop and then they are efficient vectors. Therefore, all field operations conducted in infected crops greatly influence epidemics of PepMV. In addition, it seems easily transmitted through the nutrient solution in hydroponic systems in soil-less crops, especially where the solution is recycled.

Workers are a very important means of spread as their hands become contaminated when they work in an affected crop but also their tools, clothes, and shoes, as well as crates and boxes used for harvesting.

Seed transmission of PepMV occurs at a low rate and the virus is present on the seed coat but not within the embryo or endosperm. Note that the Peru strain (PE) is not transmitted by seeds. Finally, bees used in glasshouses to pollinate tomatoes can transmit the virus, especially if they are numerous. *Macrolophus caliginosus*, a biocontrol agent, can also spread it. The risk of transmission by hand pollination is much greater than with bees. Grafting is very likely to result in transmission because of the physical handling of the plants.

Control methods[1]

In contrast to fungal diseases, there is no curative method of control for viruses. Generally, an infected plant will remain so all its life.

• During cultivation

Because PepMV is easily transmitted by infected seedling, by the seeds, and maybe fruits, it is important to avoid introducing the virus into a crop in these ways. This virus is now included in the alert list of EPPO, and is the subject of a formal monitoring programme at European community level. As a result, interceptions of infected plants have been made in Europe, particularly plants from the Netherlands and Spain including the Canary Islands. Similarly, the Canadian inspection services have detected PepMV in tomato fruits from Colorado, Arizona, and Texas.

In some countries, any detection of this virus in a glasshouse must result in crop destruction and the implementation of compulsory control measures. These require:
– all suspected incidence to be reported to the relevant authorities;
– contaminated seed lots disinfected or destroyed;
– drastic prophylactic measures applied in the affected glasshouses.

Currently, countries approach the management of this virus in different ways. Two main situations are recognized:
– production areas where eradication measures have not worked, and where the virus is now endemic; producers and technicians have learned to 'live' with this virus, and thus manage the production of infected plants;
– countries still unaffected by the PepMV, where at first everything will be done to eradicate this virus from the first outbreak to prevent further development in the area or the country. Note that it can be extremely difficult to eradicate the virus locally. To do so, it is best to destroy the crop completely. Otherwise, in many cases, attempts are made to contain it in a given area of the glasshouse or farm but this strategy has a high risks of failure given the ease of transmission of the virus.

Whatever the situation, producers are often forced to apply many of the measures described below to avoid the introduction and spread of the virus, or to attempt to eliminate it from the crop.

If attacks occur during propagation and are detected early, there is still a chance that a significant proportion of plants may have become infected. In this situation, the risk of spreading the virus should not be taken, as localized eradication is unlikely to be successful. It is better to remove and destroy all of the plants. All the debris found on and in the soil or the substrate should be removed and burned, and all surfaces of the glasshouse cleaned and disinfected. Note that in production areas where the virus is now endemic, such as the Netherlands and Spain, once infections have taken place at a relatively early stage, the producers encourage the dissemination of PepMV to all plants to achieve a 'cross-protection'. For this, the producers use as inoculum the lower leaves from diseased plants. Thereafter, they only need to ensure optimal management of these virus-infected plants.

Once the first diseased plants have been detected in a crop and the virus has been identified the following is suggested:
– use disposable coveralls, gloves, and boots when in the crop being careful not to touch the plants;
– remove the plants with symptoms by putting them in a plastic bag and avoid contact with other plants during removal from the crop. Many apparently healthy plants should also be removed from either side of the diseased plant(s);
– burn or bury deeply the debris as soon as possible;
– mark and quarantine the marked affected area(s), and work them last, taking care to use only the equipment that is dedicated to these areas.

Footbaths are installed at the entrances of each glasshouse and other areas that are used either for nursery seed beds or for other means of propagation. These should be filled with a disinfectant solution and must be functional for the duration of crop. Many viricidal products can be used for footbaths, driveways, and glasshouses structures (*Table 50*, overleaf).

Thereafter, nobody should enter or leave the glasshouses or treated areas without washing

[1] See note to the reader p. 417.

Table 50 Main viricides used for disinfection of premises, equipment, and tools (glasshouses and tunnels)

Product name*	Active ingredients	Dosage
Agrlgerm 2000	didecyl dimethyl ammonium chloride 100 g/l + formaldehyde 31.5 g/l + glutaraldehyde 40 g/l + glyoxal 32 g/l	3%
Agroxyde II	Acetic acid 194.4 g/l + hydrogen peroxide 157.5 g/l + peracetic acid + 58.6 g / l	2%
Bactipal ELV	Acetic acid 19.44% + 5.86% peracetic acid + 15.75% hydrogen peroxide	2%
Four-Sann	Acetic acid 19.44% + 5.86% peracetic acid + 15.75% hydrogen peroxide	2%
Menno Florades	Benzoic acid 9%	See terms of use
Phenoseptyl POV	Orthophenyl phenol 250 g/l	2%
Virkon	Sulfamic acid 5% + potassium monopersulfate 22.5% + malic acid 10%	0.50 l/hl
Virofree	Magnesium peroxyphtalate 30%	1%

In some countries the same active ingredients may be available but the product names may differ.

their hands with warm, soapy water or even a disinfectant. Wearing disposable gloves is often preferable; in addition, they can be more easily disinfected. A viricide can also be used. Note that workers can decontaminate their hands by dipping them in disinfectant and skimmed milk (milk proteins reduce the infectivity of viral particles), a phosphate buffer, or a detergent solution are suggested for this purpose. Hand disinfection must be repeated frequently during and after working in the nursery or glasshouse.

Affected glasshouses should be locked to prevent access to people from outside of the farm. Any visitors must wear special clothing, overshoes, and latex gloves. Pets should not be allowed in the glasshouse. Similarly, one should avoid moving hives of bumblebees from one glasshouse into another.

It is important to deter workers from bringing in and eating tomatoes of unknown origin as they can introduce the virus. Renewal and cleaning of their clothes should be carefully managed.

Equipment or tools should never be lent to producers on other farms. The tools should be disinfected with a viricide. Sometimes it is recommended to soak them for 20 minutes either in a solution of trisodium phosphate at 10%, or in bleach at 3° chlorine[1].

All debris from pruning and leafing must be removed and destroyed immediately and in no circumstances left in pathways between rows of plants. These pathways as well as cemented corridors, should be disinfected with an approved viricide. The burning of crop residues or their burial with lime (mixed in equal quantities) should be systematically organized, and under no circumstances be left in a heap outside the glasshouse.

Water and nutrient solution should be filtered and disinfected. Pasteurization with ozone and UV is an effective preventive measure.

---

[1] The chlormetric system of measuring active chlorine uses degrees: 1% active chlorine converts to 3.16 chlormetric degrees of chlorine.

At the end of the crop all plants (especially the root systems, in which the virus lasts longer), substrates, equipment, and tools should be removed from the glasshouses, taking care not to break up the green material as this might result in small fragments of plant with the virus. Once the glasshouses are empty, the surfaces should be cleaned using water under pressure. The irrigation system must be disinfected with nitric acid at pH 1–2, for 24 hours, then rinsed with clean water. The drippers must be free of any organic matter and disinfected. Subsequently, the whole glasshouse should be treated with a disinfectant (see *Table* 50) at the recommended concentration and allowed to stand for several minutes to eliminate the virus effectively.

• **Subsequent crop**

Many of the measures detailed above should be implemented during the next crop. After removing the PepMV from the farm, it will be necessary to minimize the risk of its introduction and spread. This will require the use of healthy plants obtained from a source known to be free from the disease. The field site used to raise plants last year must be disinfected (with a fumigant, steam) or preferably, a new site found. As the virus can survive in the soil for several years, long crop rotations are necessary in the field, in the same way as for soil-borne diseases. Be particulary aware of root residues in areas where affected crops have been grown.

The seeds used for planting must be virus free. If in doubt, they should be disinfected in the same way as for ToMV (see Description 31).

Several products are recommended to eliminate the PepMV from seed without adversely affecting germination, alone or in combination: sodium hypochlorite, trisodium phosphate, benzoic acid (Menno Floradis). After these treatments, the seeds should be rinsed in running water, but dried quickly. Appropriate methodology for a particular country should be discussed with an adviser. Heat treatments are also reported as PepMV is very sensitive to high temperatures.

The health status of seedlings and young plants before planting must be regularly monitored. It is desirable that staff working in propagation areas do not work in crops to avoid the risks of virus introduction into the propagation area.

Be wary of plants and seeds from countries already affected by PepMV: guarantees must be given on the health status. Growers should be wary of variety trials conducted on their farms as seeds used for these may be a source of the virus and may not have been thoroughly checked before sowing. There are records of the introduction of the virus through new varieties under test for seed companies.

In countries where the virus has not been eradicated, especially in the Netherlands, grafting on vigorous rootstocks ('Maxifort', 'Beaufort', 'Eldorado') allows some compensation for the effects of the virus as it gives the plants more vigour which, in turn, decreases the virulence of the virus.

If, despite the measures taken, some attacks occur after planting, the measures to be taken during cultivation should be consulted.

Tobacco mosaic virus (TMV)

(Tobamovirus, not classified in a family)

Principal characteristics

The tobacco mosaic, the first virus disease to be described, was first on tobacco in the Netherlands in 1886 and Russia in 1892. TMV has been selected as the type virus of the group Tobamovirus. This virus has been the subject of many fundamental studies particularly at molecular level.

Several strains have been reported: U1 in the US, SP1 and aucuba (Tmj) in France, KMS (Kassanis). The virulence of these strains have been characterized; strains U1, SP1, and KMS are pathotype 0. The strain Tmj, which overcomes the gene 'Tm-1' is pathotype 1 (see also the Control methods section in Description 31 on ToMV).

The virus particles have a rigid rod shape, measuring 300 × 15 nm.

• Frequency, extent of damage

TMV is present in all production areas where sensitive tobacco varieties are grown and also occurs on tomato in many countries. Note that for a long time TMV was associated with the disease on tomato but it is now known that a specialized form of the virus called tomato mosaic virus (ToMV) is responsible for the tomato disease and is much more competitive on this host. This has led to confusion in the literature as some of the early work on TMV may have been on ToMV.

This virus is still serious in countries where resistant varieties are not widely used but is far less common on tomatoes than ToMV. In industrialized countries, its importance is only sporadic but it remains a threat to tomato, chilli, and tobacco breeding programmes.

• Main symptoms

Symptoms caused by TMV on tomato vary according to strains. Light mottling and a green to yellow mosaic can be observed on the leaves. In winter, leaf deformation occurs and the width of the leaflets is greatly reduced, referred to as 'fern-leaf' or'filiform'.

Sometimes the sepals and petals have a wavy edge and the number of pollen sacs is reduced. Infected tomato fruits may be normal or show symptoms of internal browning (see p. 368).

The nature and severity of symptoms vary with plant age and variety. They develop more severely on young, rapidly growing tissues.

Note several other diseases involve different strains of TMV, alone or in combination:
– single virus streak, caused by a strain of TMV, causes brown necrotic symptoms on stems and petioles in certain conditions. The fruits also have brown rings;
– double virus streak, combining the TMV and PVX, can cause brown symptoms, largely on stems and petioles and sometimes on fruits;
– leaf rolling, resulting from the interaction between TMV and the gene 'wt' present in some varieties of tomato. In this case, the infected leaflets have a tendency to roll down.

• Ecology, epidemiology

Survival, inoculum sources: TMV is a particularly infectious and persistent virus. Unlike almost all viruses, high stability allows it to survive in the soil and other substrates for several years, notably in leaf and root debris, and in the environment of infected nurseries. TMV survives longer in moist soils than in dry ones. It can survive for several months on clothes, tools, and the structure of glasshouses and tunnels. The first infections commonly occur during propagation as a result of the use of contaminated soil or tools. In the open field, root contact between nearby plants can result in spread, and roots remaining from a previous crop are frequently the source of the initial disease.

TMV has a wide host range and can infect many species of dicotyledons, and these contribute to its development and survival. It affects nine genera in the Solanaceae and at least 39 species, some cultivated: *Capsicum annuum* (pepper), *C. frutescens*, *Datura meteloides*, *D. stramonium*, *Hyoscyamus niger*, *Nicotiana tabacum*, *N. glauca*, *Petunia hybrida*, *Physalis alkekengi*, *P. angulata*, *P. heterophylla*, *Solanum melongena* (eggplant), *S. tuberosum* (potato), *S. torvum*, *S. nigrum*. It also attacks many other species belonging to more than 40 dicotyledonous botanical families. It can also infect some monocotyledons.

Transmission, dissemination: this virus is easily transmitted by contact. This can be simply by hand contact or by contact with the contaminated clothing of workers or the tools they are using, or possibly from plant to plant by the plants rubbing each other which may be aided by wind: all activities and field operations contribute towards the spread of this virus. TMV survives in the nutrient solution of hydroponic systems as well as in other soil-less crops. Spread of the disease is aided by the recirculation of nutrients throughout these crops.

Some biting insects are likely to pick up the virus during feeding on diseased leaves, and may then transmit the virus to healthy plants, although the efficiency of transmission is surprisingly low. Other insects, typically reported as virus vectors (aphids, whiteflies), do not transmit it. Transmission is ensured by the use of contaminated plants and then subsequently by workers and their tools and by technicians moving from an affected crop to a healthy one.

Transmission of TMV has not been verified on the seeds of Solanaceous plants, and no case has been shown in tobacco. In tomato it has not ben found in the embryo or endosperm of seeds but is found in tissues of maternal origin, the residual layer of the nucellus, and the testa or seed coat which can be externally contaminated. The young seedling, in contact of the seed coats, becomes infected by microwounds that occur during germination or during transplanting.

Control methods[1]

There are no methods of control of TMV or other viruses during cultivation. Generally, an infected plant will remain so all its life.

Measures to control viruses transmitted by contact are the same, regardless of the virus, and the Control method section for PepMV (Description 29) should be consulted.

Information is only presented here if particularly relevant for the control of TMV. The use of resistant varieties is the only efficient and

[1] See note to the reader p. 417.

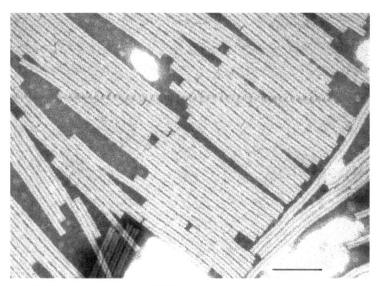

883 The virus particles of TMV are rod-shaped and rigid, measuring 300 × 15 nm.

sustainable way to control this virus. Sources of resistance described for the ToMV (see Description 31) are also effective against TMV. Many hybrids are currently being marketed with resistance and the use of these will help to solve this virus problem.

The use of inoculation with avirulent strains as used for the control of ToMV are not effective for the control of TMV.

Tomato mosaic virus (ToMV)

(Tobamovirus, not classified in a family)

Principal characteristics

ToMV was reported for the first time on tomato in 1909, in the US (Connecticut). Long considered a strain of TMV, it has different properties (serology, viral genome, and host range), which allows it to be considerd a separate virus. Like TMV, ToMV has been the subject of much research.

Several strains of ToMV have been identified on tomato, primarily based on the symptoms they cause: tomato aucuba mosaic, tomato enation mosaic, yellow ringspot, winter necrosis. Strains have also been classified according to their virulence. At least six pathotypes have been defined (see Control methods section).

ToMV virus particles are morphologically identical to those of TMV. They are rigid rods, measuring about 300 × 15 nm.

• Frequency, extent of damage

ToMV is present on all continents. It occurs more frequently than TMV on tomato and pepper. It is serious both in field and protected crops. Although its incidence has decreased significantly with the use of resistant tomato varieties, the recent marketing of new susceptible types has shown that the virus is still a threat.

• Main symptoms

Symptoms caused by ToMV are very varied. In addition to slowing plant growth, various discoloration may appear on leaves: vein clearing, mottling, mosaic, with patches of various shades of green, yellow, or even white (aucuba). The leaves may be distorted to a greater or lesser extent. As with TMV, the leaflets and the leaves may be filiform or fern-like in low light conditions, especially when grown in glasshouses in the winter.

Flower drop may occur. The fruits when ripe, are small and sometimes bumpy. They also undergo yellow discoloration, sometimes in rings, as well as internal localized necrotic symptoms in the vascular tissues (internal browning, see p. 368). Internal symptoms may be present on green or mature fruits when the plant is otherwise healthy looking. The earlier the attack the greater the effect on yield; infection late in the crop have far less effect on yield and quality.

Given the mode of transmission of this virus (by contact), the distribution of diseased plants in the crop is often in lines, often related to cultural operations. Note that generally the symptoms can vary in intensity depending on the strain, cultivar, plant growth stage at the time of infection, temperature, light intensity, the nitrogen content of the soil, and the boron level. Thus, high temperatures contribute to a reduction in foliar symptoms. In contrast, tomato varieties with the genes 'Tm2' or 'Tm2²' can, under conditions of high temperature, produce necrotic reactions in the presence of common strains of TMV and ToMV.

Mixed infections between ToMV and other viruses are very common, especially with CMV and PVY, and the symptoms are then often more severe. Also double virus streak, once relatively common, is a combination of ToMV and potato virus X. This diseases causes severe streak symptoms on the stems of affected plants. (See Photos **28, 32, 101, 106, 111–116, 713, 724–726.**)

• Ecology, epidemiology

Survival, inoculum sources: ToMV is a very stable virus that survives in the soil and other substrates for several years, particularly in leaf and root debris. It can also survive on structures. Infected crops are a major source.

ToMV can infect many different hosts although the literature suggests that it infects fewer species than TMV. Its main hosts are in the Solanaceae, for example, *Capsicum annuum* (pepper) and *C. frutescens*. It is less common on other species of this botanical family such as *Nicotiana tabacum* (tobacco), *Petunia hybrida*, *Physalis alkekengi*, *P. heterophylla*, *P. longifolia*, *P. peruviana*, *P. subglabrata*, *P. virginiana*, *Solanum tuberosum* (potato), *S. americanum*, *S. scabrum*, *S. villosum* and, more recently, *S. muricatum*. It also affects some species of the following botanical families: Asteraceae, Chenopodiaceae, Cornaceae, Gentianaceae, Oleaceae, Pinaceae, Plantaginaceae, and Rosaceae.

Note that ToMV may be experimentally inoculated to at least 145 plant species in 46 genera belonging to 27 botanical families.

Transmission, dissemination: this virus is very easily transmitted by contact. The touching of leaves of infected and healthy plants and slight movement as a result of wind is sufficient to transmit this virus. Contamined hands and clothing are major sources of the virus and a major means of transmission. Almost all cultural operations and activities within the crop, whether under protection or in the the field, contribute to its transmission and dissemination. ToMV is easily transmitted in hydroponic systems of soil-less crops, through the nutrient solution.

ToMV is easily transmitted through the seeds of tomato (external contamination); transmission rates can be high. The virus is also present in large quantities in the viscous coating of fresh seeds and will remain on the seed coat if not eliminated by fermentation or acid extraction. It is found in lesser amounts in the testa and the endosperm but not in the embryo. In the endosperm, the virus can reamin viable for up to 9 years. The young seedling is infected with the virus by contact with the testa, especially at transplanting stage.

Control methods[1]

ToMV cannot be controlled once it is established within a crop and infected plants remain a source of the virus for the whole of their lives. The Control methods section of the description dealing with PepMV (Description 29) should be consulted for the details of measures to be taken against this virus as these are the same for all viruses transmitted by contact.

Information is only presented here if particularly relevant for the control of ToMV. Particular attention should be paid to seed quality as they transmit the virus, sometimes at a very high rate. If there are any doubts, they should be disinfected. For example, ToMV may be inactivated in the seeds by treatment with dry heat (thermotherapy: 80°C for 24 hours, 78°C for 48 hours or 70°C for 72 hours) or trisodium phosphate (Na_3PO_4) at 10% for 30 minutes to 1 hour. A French seed treatment method involves a solution of 2% (v/v) hydrochloric acid (HCl), and 3 g/l pectinase. This enzyme provides good separation of the seeds and pulp. Later, dried seeds are placed in an oven at 80°C for 24 hours (dry heat). The combination of these two methods results in the denaturation of the virus. Several other methods exist: local advisory authorities should be consulted to determine which method is commonly practised in a particular country.

Several resistance genes have been used for the control of ToMV:

– the gene 'Tm-1' (sometimes referred to as 'Tm'), derived from *Lycopersicon hirsutum* confers resistance. This gene has appeared relatively ineffective as it is quickly overcome by ToMV strains of pathotype 1;

– the gene 'Tm-2', located on chromosome 9, obtained from a variety of *Lycopersicon peruvianum* is more stable (although strains of pathotype 2 do overcome it but not quite so readily as with pathotype 1 and the *Tm-1* gene). Note 'Tm-2-nv' is associated with a semi-lethal 'nv' gene and induces necrosis (netted virescence) in plants homozygous for this gene;

– the gene 'Tm-2^2' allele of the gene 'Tm-2' of the same origin, is associated with fertility and quality defects in fruits in the homozygous state.

Like many hypersensitivity genes, 'Tm-2' and 'Tm-2^2' are not effective at high temperatures. In addition, in the presence of large inoculum levels as occurs with the proximity of a ToMV infected susceptible crop, large necrotic lesions can occur on plants heterozygous for the genes 'Tm-2' or 'Tm-2^2'; this reaction is a more general hypersensitive response. Such symptoms generally occur in glasshouses when temperatures range from 18–20°C at night to 35°C during the day.

In addition, as mentioned above, several strains or pathotypes were found capable of overcoming the genes 'Tm-1', 'Tm-2' or 'Tm-2^2', used alone or in combination. Six pathotypes at least could be defined, and their respective virulence are detailed in *Table 50a*.

Note that a strain called 'M97' can overcome the resistance of genes 'Tm-2' and 'Tm-2^2' and may be an additional pathotype.

To make resistance to ToMV more durable, the selection strategy developed has been to combine into a single tomato genotype different genes responsible for different mechanisms of resistance. For example, F1 hybrids, now grown in glasshouses, include the following combinations of genes, Tm-1, Tm-2^2/Tm1$^+$, Tm-2$^+$ or Tm-1, Tm-2/Tm-1$^+$, Tm-2^2, but mainly Tm-2^2, Tm-2$^+$.

Cross protection which consists of artificially infecting tomato plants with a 'weak' strain of ToMV was used in the past to control this virus. The plants were protected by the mild strain from the more damaging effects of aggressive strains. The weak strains used were obtained by random mutagenesis with nitrous acid. The hypoaggressive strain MII-16 was used to protect protected crops of sensitive tomato cultivars in many countries of Europe, America, China, Japan, and New Zealand.

Table 50a Virulence of ToMV pathotypes

ToMV pathotypes	Resistance genes		
	Tm-1	Tm-2	Tm-2^2
Pathotype 0	−	−	−
Pathotype 1	+	−	−
Pathotype 2	−	+	−
Pathotype 1,2	+	+	−
Pathotype 2^2	−	−	+
Pathotype 1,2^2	+	−	+

− : strain unable to overcome the resistance gene.
+ : strain overcoming the resistance gene.

Cross protection has some limitations: it is not effective against TMV, and severe symptoms can be observed if plants also become infected by other viruses such as CMV, for example.

Note that the 'N' gene which confers resistance to TMV in tobacco has been isolated from tobacco and transferred into transgenic sensitive tomatoes giving them resistance to TMV and ToMV.

◼ Other viruses transmitted by contact

Viruses	Symptoms and mode of transmission of the virus	Shape of virus	Principal characteristics
Potato virus X (PVX) **Potexvirus**	Mottling, mosaic, sometimes with necrosis and necrotic spots. Primarily transmission in tomato through contact.	Filamentous particles measuring 515 × 13 nm. 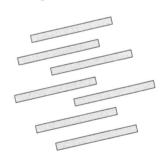	This virus was first described in the UK on potatoes in 1931. It is present in Europe and in all regions of the world where potatoes are grown. It was once quite common in tomato crops and a serious cause for concern, especially when associated with TMV or ToMV. Currently it is not a problem because of the use of ToMV resistant genes. PVX has a wide host range including at least 67 species belonging to 27 botanical families. Among these are pepper, tobacco, potato, *Solanum nigrum*, *Physalis ixocarpa*, *Datura stramonium*, *Nicandra physaloides*, *Hyoscyamus muticus*, artichoke, *Brassica rapa*, *Chenopodium album*, *Convolvulus tricolor*, *Rumex acetosella*, *Plantago major*, the vine. Control methods are as recommended for PepMV, Description 29.
Ribgrass mosaic virus (RMV) **Tobamovirus**	Mottling and green mosaics on leaflets. The fruit can develop internal necrosis (internal browning, see p. 368). Easily transmissible by contact. It has been detected in water from some rivers.	Rigid rod-shaped virus particles, measuring 300 × 18 nm.	It was first reported in the US in 1941 on *Plantago lanceolata*. This virus is probably now worldwide in its distribution. It has been known to occur on tobacco for some time, but on tomato reports are more recent, dating from 1996. See Control methods for the control of PepMV (Description 29).

Viruses	Symptoms and mode of transmission	Shape of virus particles	Principal characteristics
Tomato bushy stunt virus (TBSV) **Tombusvirus**	Spots, chlorotic rings, and line patterns on the leaves. These can also have anthocyanin (red colours) and affected leaves can be deformed. Necrotic streaks may develop on the stems. The apex may become necrotic; many axillary branches then form, contributing to the bushy appearance of diseased plants. Affected plants often have greatly reduced growth and are stunted and bushy, producing small fruits. Transmitted by contact and by water. This virus has been detected in running and stagnant waters in several countries in Europe. Transmitted by pollen in tomato and other species. Man can contribute to its distribution by eating affected fruits. The virus particles are able to pass through the digestive tract unharmed and could eventually become a component in organic manures based on sewage sludge.	Truncated viral particles in an isocahedron shape (isometric), with a diameter of 30 nm. 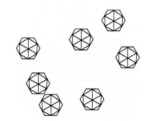	Reported for the first time on tomatoes in UK in 1935, this virus has now been found in several countries in Europe, North Africa, Turkey, Asia, and North and South America. It causes severe attacks, especially in Morocco and in the southeast of Spain, in both glasshouses and field crops. Tomato plant decline described in California is associated with the BS-3 strain of TBSV. In addition, in Colorado and New Mexico, a virus serologically close to TBSV (named 'Lettuce necrotic stunt virus') is responsible for necrotic decline in glasshouse crops. Several variants of TBSV have been reported. This virus is very stable and has a wide host range affecting many plants belonging to different botanical families but especially the Solanaceae: peppers, eggplant, tomato, petunia. The Control methods recommended for PepMV (Description 29) are relevant for the management of this virus disease. Note also that soil solarization will reduce the damage caused by this virus.

■ Viruses transmitted by aphids

Alfalfa mosaic virus (AMV)

(Alfamovirus, Bromoviridae)

Main characteristics

This virus was first reported on alfalfa (*Medicago sativa*) in the US in 1931. It is the type species of the genus Alfamovirus and belongs to the family of Bromoviridae like Cucumovirus, Bromovirus, and Ilarvirus.

Many strains or variants of AMV have been described, with minor differences in their thermal sensitivity, their possible transmission through pollen and seed, and the differential reactions of some inoculated hosts. For example, note AMV-S strains from alfalfa, AMV 425 from clover, and yellow spot mosaic (YSMV).

The virus particles have a variable shape. The smallest are isometric with a diameter of 18 nm, the others are bacilliform; their lengths are variable (29, 38, 49, and 58 nm) (Photo 884).

• Frequency and extent of damage

AMV has probably a worldwide distribution, but it is most common in temperate regions, where alfalfa is grown (this plant is the preferred host of this virus). It is also present in warmer regions, in Africa and America, where various legumes, chillies, and tomatoes are grown. AMV is not considered a major tomato virus in the Americas. On the other hand it is routinely observed in field crops in southern Europe, but with large variations in frequency according to the location and year.

The virus occurs occasionally on tomato in France, mainly in the field. It is found in alfalfa, tobacco crops, but also on crops of other vegetables and aromatic plants. Serious attacks have sometimes been found in tomatoes grown for processing and grown near alfalfa. In this case, nearly 80% of plants showed symptoms.

If the attacks are early, plant growth is strongly affected. The number of fruits is often reduced and quality greatly impaired.

• Main symptoms

AMV causes two main symptoms on leaves:
– a pronounced aucuba mosaic;
– necrotic spots, often starting at the base of leaflets and distributed throughout the leaf, sometimes leading to necrosis of the veins.

Longitudinal and unilateral necrotic lesions may then appear on the stems, up to 20 cm in length. Apical buds and surrounding young leaves are sometimes completely destroyed by necrosis.

In the presence of a particularly aggressive strain, the necrotic lesions can be lethal especially when the plant is infected early. In such cases a reddening and browning of the phloem vessels is seen throughout the stem length.

Fruits can also express a variety of symptoms. Young fruits may fall, be deformed, or have their growth stopped. Older ones sometimes have necrotic spots, external or internal, extended and recessed. These spots may coalesce and cause widespread necrosis. Alternatively the fruits can grow almost normally, but be slightly bumpy with a skin mottled and round brown spots.

(See Photos **420, 421, 610, 723**.)

• Ecology, epidemiology

Survival, inoculum sources: AMV can be artificially inoculated to a large number of hosts, over 400 plant species, herbaceous or woody, belonging to 50 different botanical families. Some of them are symptomless hosts. It can easily survive from one season to the next on some weeds, but also on perennial crops. Alfalfa appears to be the most important source; it expresses symptoms of mosaic, mottling and distortion, but no symptoms are visible during the summer. Many other plants can be a sources in winter or summer. A significant number of the Solanaceae both cultivated and wild may be infected, including tomato, peppers, eggplant, potato, tobacco, *Atropa belladonna*, *Cyphomandra betacea*, *Datura stramonium*, *Hyoscyamus muticus*, *Nicandra physaloides*, *Physalis angulata*, *Sarachia edulis*, *Schizanthus pinnatus*, *Solanum dulcamara*, *S. nigrum*, *Withania frutescens*, and *W. somnifera*. There are also other possible hosts includings lettuce, celery, beans, peas, chickpeas, alfalfa, clover, some woody plants (currant, lavender, *Cercis siliquastrum*, *Hibiscus cannabinus*, *Ruscus hypoglossum*, *Clematis vitalba*, *Glycine latifolia*, *Viburnum* spp., *Buddleia davidii*, *Rhamnus angula*). Asparagus is the only susceptible monocot.

Transmission, dissemination: AMV is transmitted to other plants by aphids, in the nonpersistent manner. The virus is quickly acquired in short feeding periods and transmission is immediate, but only for a short period not exceeding a few minutes to several hours. Twenty-six species of aphids can transmit the AMV including *Myzus persicae*, *Aphis fabae*, *A. gossypii*, *A. craccivora*, *Acyrthosiphon pisum*, and *A. kondoi*. However, transmission rates vary depending on the the plant source, the host being infected, the AMV strain, and the aphid species.

The spread of AMV is mainly by aphids, and it depends upon the biology of these insects. Several biotic and abiotic factors play a key role in this and affect virus dissemination and the development of epidemics:

the wind determines their distribution;
– the temperature affects the growth of the tomato and the multiplication of viruses and aphids respectively in and on plants;
– proximity to other contaminated sensitive crops (notably alfalfa) and many virus-infected weeds promotes spread of infection.

Finally, the geography of the location, cultural practices such as the layout of crops, their orientation in relation to wind direction, wind breaks, and the possible presence of many infected hosts in the area also play a role.

The development of epidemics of AMV is also related to climatic conditions. If the winter is severe, a majority of carry-over virus-infected plants, as well as aphid vectors will be destroyed and the inoculum at the start of the next season reduced.

Transmission by seed occurs in some plants particularly alfalfa, and to a lesser extent peppers. In potatoes, seeds may occasionally be infected. Seed transmission also occurs in weeds such as *Nicandra physaloides*, *Stellaria media*, *Senecio vulgaris*, *Chenopodium album*, and *Holosteum umbellatum*.

Control methods[1]

Plants infected by AMV cannot be cured and remain infected throughout the crop whether it is annual or biennial. Control methods for CMV (Description 35) apply equally to this virus. Indeed, measures to control aphid-borne viruses are the same, regardless of the virus. Only information about resistances to AMV is reported here.

Some varieties of tomato and various wild species of *Lycopersicon* have proved resistant to AMV: *L. hirsutum* (form *typicum*, 'LA 1 777'), *L. hirsutum* (form *glabratum* 'PI 134 417'). The study of AMV resistance heredity of 'PI 134 417' showed that it is determined by a dominant gene, 'Am'. In *L. hirsutum*, it is located on T6 chromosome, and confers total resistance.

[1] See note to the reader p. 417

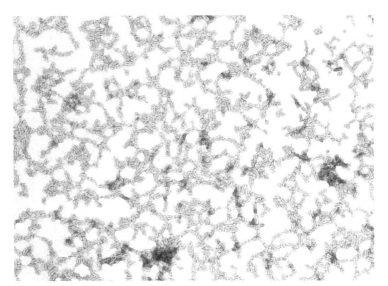

884 The virus particles of AMV have various forms: isometric or bacilliform.

Beet western yellows virus (BWYV)

(Polerovirus, Luteoviridae)

Principal characteristics

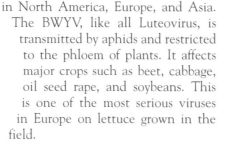
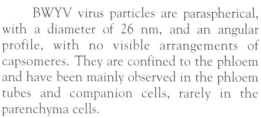

Described for the first time in California in 1960, this virus is probably present in all continents. It has been reported primarily in North America, Europe, and Asia. The BWYV, like all Luteovirus, is transmitted by aphids and restricted to the phloem of plants. It affects major crops such as beet, cabbage, oil seed rape, and soybeans. This is one of the most serious viruses in Europe on lettuce grown in the field.

BWYV virus particles are paraspherical, with a diameter of 26 nm, and an angular profile, with no visible arrangements of capsomeres. They are confined to the phloem and have been mainly observed in the phloem tubes and companion cells, rarely in the parenchyma cells.

• Frequency and extent of damage

This virus is rare on tomato, and is not considered a major virus of this crop.

• Main symptoms

In tomato, as in other Solanaceae crops, BWYV causes chlorotic spots that appear between the veins of the older leaves. Gradually, the spots spread and intensify, leading to a yellowing that can sometimes affect some upper leaves. Affected leaflets are sometimes distorted.

The incubation period is usually between 15 and 25 days after inoculation by aphids according to environmental conditions (light intensity), but also the stage of plant growth at the time of infection and varietal sensitivity. This virus disease occurs mainly in the summer on tomatoes.

(See Photo **145**.)

• Ecology, epidemiology

Survival, inoculum sources: BWYV has a wide host range since it infects about 150 species belonging to 23 botanical families. Potential sources of viruses are many, both in crops and in wild species. Beet, chard, spinach, all species of cabbage, oilseed rape, turnip, radish, lettuce, spinach, soybeans, peas, chickpeas, beans, some Cucurbitaceae, potato, pepper, and *Physalis floridana*, show variable symptoms.

Among the weeds susceptible to this virus, a small number develop inter-veinal yellowing or reddening, but most are symptomless carriers (infected plants with no visible symptoms). The most common of these are wild lettuce, groundsel, and sow thistle (Asteraceae), shepherd's purse and *Lepidium* sp. (Brassicaceae), as well as purslane, plantain, mallows, and pigweed, chickweed.

A succession of susceptible crops (lettuce, cabbage, cauliflower, rapeseed, spinach) during the winter and the presence of many weed species which are potential hosts contributes to the survival of inoculum throughout the year.

Transmission, dissemination: the virus is transmitted by several species of aphids in a persistent circulative manner, bringing in a highly specific mechanism. The minimum acquisition time is 5 minutes and the inoculation time is 10 minutes. The virus enters the insect through the alimentary canal and then passes into the haemocoele in which a latency period of 12–24 hours is necessary before it reaches the salivary glands and can be transmitted. In fact, there must be about 48 hours between the acquisition phase of the virus from the phloem vessels of the plant and a complete cycle of the virus in the aphid, before its effective transmission. The vector aphid then remains infectious for more than 50 days. The virus is retained by the aphids during their successive moults, but is not transmitted to their offspring. Ten species are likely to transmit BWYV but *Myzus persicae* and *Macrosiphum euphorbiae* are the most important. *Aphis craccivora*, *A. gossypii*, and *Acyrthosiphon solani* also transmit BWYV. The virus is not seed transmitted.

The particular nature of the mode of transmission of this virus by aphids and the abundance of susceptible plant species results in very effective spread of BWYV, often over long distances. In addition, the development of epidemics is often proportional to the population size of vector aphids.

Control methods[1]

Controlling BWYV is very difficult given the large number of susceptible plant species, the persistence of the virus in the vector, and the relative abundance of aphid vectors. However, the virus is only occasionally damaging in tomato crops. Therefore, no special measures are normally required except in production areas where it occurs repetitively.

Infected plants cannot be cured and always remain infected. Control measures are the same as for CMV (see Description 35). Unlike many viral diseases that are aphid transmitted in a nonpersistent manner, aphicide treatments are usually relatively effective in controlling the development of epidemics of BWYV, by limiting aphid populations in crops and in their environment. To do this, they must be used preventively (pirimicarb, methomyl, rotenone, cyhalothrin, fluvalinate, pymetrozine.). Once this summer yellow virus is well established in a crop it is often too late to intervene effectively.

No resistance research has been conducted on Solanaceae.

[1] See note to the reader p. 417.

Cucumber mosaic virus (CMV)

(Cucumovirus, Bromoviridae)

Principal characteristics

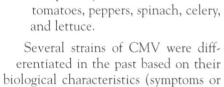

CMV, ubiquitous and affecting very many plant species, was first described on *Cucumis sativus* in the US in 1916. It is considered the reference species of the Cucumovirus genus. It has one of the largest virus host ranges listed and seriously affects vegetable crops in the field – especially Cucurbits – but also tomatoes, peppers, spinach, celery, and lettuce.

Several strains of CMV were differentiated in the past based on their biological characteristics (symptoms or heat sensitivity), serological criteria, and molecular properties. The current official nomenclature refers to two groups, I and II, the first group with two subgroups IA and IB.

Finally, some isolates have a supernumerary small-sized RNA, called 'satellite RNA', which can alter the expression of CMV symptoms on some hosts. In France and Italy, RNA-5, potentially necrogenic on tomato, is sometimes found in weeds and in crops.

The virus particles are isometric and have a diameter of 28 nm (Photo **885**).

• Frequency and extent of damage

CMV is present worldwide, its impact on tomatoes varies from one country to another. It can be severe in temperate regions of production, especially in the Mediterranean climate. It is also found in tropical climates, for example in Africa. It causes severe diseases in Asia, including China, Japan, and Taiwan, alone or in a complex with other viruses. It is the same in Europe, especially in Italy, Spain, and France, where the virus is widespread in crops.

CMV, (together with PVY), is common in tomato crops in various European countries especially in field crops during the summer. It is sometimes found under protection in summer, but also in late winter; in such cases the plants have originated from seedlings produced at a time of year when viruliferous aphids were present in the nursery during propagation.

Mixed infections with two or three other viruses are common in the field, especially with PVY, AMV, and TSWV, and can lead to symptoms of greater severity.

• Main symptoms

CMV is the cause of a wide variety of symptoms on tomato influenced by the particular stage of development of the host, climatic conditions, the nature of the strain, and especially the presence or absence of necrogenic RNA satellites. The leaves are mainly affected:

– mottling or yellow to green mosaic on young leaves;

– a deformation and a reduction in the size of the leaflets. These are sometimes fernleaf-like in appearance (fernleaf) or from time to time become very filiform and reduced to just their veins. These are referred to as shoestring symptoms. It is not uncommon for deformed leaves to occur at several levels on the plant separated by normal leaves. This shows the transient nature of the appearance of CMV symptoms on a plant. When all apical leaves are affected, the plant growth is slowed or stopped. The infection leads to reduced fertility and fruit size in fruits that are alreading developing;

– symptoms include necrotic spots which may coalesce to cover the leaf or a few leaves. They may progress to the petiole, and develop on the stem in the form of brownish longitudinal streaks which can completely encircle the stem. Ultimately, the lesions extends to the whole length of the stem and a lethal necrosis results. Fruits may show olive or brown ring spots and some blistering. These necrotic symptoms are due to some isolates of CMV that have a necrogenic satellite RNA. There are several natural variants of the latter, which have biological properties clearly different depending on the host. Thus, some of them (D, I17N) reduce symptoms in tobacco, but produce a generalized necrosis on tomatoes, causing serious epidemics, particularly in southern Europe and the Balkans. Others (R), however, cause a reduction of symptoms in tomato.

Young tomato plants are more easily infected with CMV than plants that have reached the flowering stage. In addition, the earlier the infection, the greater the effect of CMV on growth and development and therefore on yield. Early affected plants may be particularly stunted and bushy. The incidence of the more severe forms of this virus, the filiform and necrotic strains, are very variable from one year and one season to another. The distributions of these symptoms in the crops are different: plants affected by filiform strains are randomly distributed, while those expressing necrosis form clusters.

(See Photos **20, 21, 30, 34, 49, 60–64, 117–123, 416–419, 727, 800.**)

• Ecology, epidemiology

Survival, inoculum sources: CMV infects over 700 species in 92 different botanical families, belonging to both monocotyledons and dicotyledons. The vegetable and ornamental crops (perennial) are particularly affected. In the Solanaceae, CMV infects: pepper, *Capsicum frutescens*, *Cyphomandra betacea*, six species of *Datura* and *Physalis*, *Lycopersicon pimpinelifolium*, several *Nicotiana* including tobacco, *Petunia hybrida*, eggplant, and *Solanum nigrum*. Other botanical families with vegetables hosts are in the Cucurbitaceae (melon, cucumber, squash, zucchini), the Asteraceae (lettuce, escarole), Apiaceae (celery, carrot, parsley), and Chenopodiaceae (beets, spinach). Ten monocot families may also be affected including the Agavaceae, Amaryllidaceae, Iridaceae, Liliaceae, Musaceae, Orchidaceae, and Poaceae.

Many common wild plants, species such as purslane, black nightshade, groundsel, sow thistle, wild lettuce, dead nettle, speedwell, madder, and dog's mercury are infected. The most important are *Portulaca oleracea*, *Senecio vulgaris*, *Solanum nigrum*, and those which are perennial such as madder, or shepherd's purse. Note that *Stellaria media* transmits the virus in its seed. In Spain, a study of 51 weed species belonging to 19 botanical families found that 25 of them were infected with CMV. Among them, *Convolvulus arvensis*, *Malva sylvestris*, and *Sonchus tenerimus* play an important role in disease epidemics because they are perennial and often associated with tomato crops.

Note that these various plants enable CMV to survive during the winter, and in the spring and throughout the period of cropping they constitute very important sources of the virus which can result in disease epidemics.

Transmission, dissemination: first infections often occur with the arrival of the first aphids as CMV is transmitted by aphids in a nonpersistent manner. Thus, the aphid vector can acquire the virus on an infected plant and transmit it to a healthy plant in seconds, in very short feeding periods. So called 'test feeding' allows the aphid to determine that it has found a suitable host for its development. The aphid can transmit the virus immediately after it has acquired the virus and can continue to do so for 10 minutes or more. It completely loses this ability after 2–4 hours maximum, more rapidly if it is continously test feeding. It can recover the ability to transmit by feeding on another infected plant. One consequence of this mechanism is that aphids not associated with the crop, which are only passing by, are perfectly capable of transmitting the virus, complicating control measures with insecticides.

Over 90 species of aphids may acquire and transmit CMV. Their potential as vectors is different; the best vectors seem to be *Myzus persicae*, *Aphis gossypii*, *A. craccivora*, *A. fabae*, and *Acyrthosiphon pisum*. Some species are only capable of transmitting a certain number of strains.

The aphids infect many plants in the vicinity of the source plant. When carried by the wind, aphids can effectively spread CMV over rather large distances. In addition, the high efficiency of transmission makes dissemination of this virus very quick without major populations of aphids being involved. Several biotic and abiotic factors play a key role in the biology of aphids, affecting the spread of viruses and the development of virus epidemics:

– wind determines their distribution;
– temperature affects plant growth, the multiplication of the virus, and development of aphid colonies;
– proximity to other sensitive crops and contaminated weeds especially promote infection.

To these factors must be added the local parameters such as crop layout, crop shelter against the prevailing winds by hedges, local climatic conditions, and their influence on some source plants.

The development of CMV epidemics is thus linked to climatic conditions. If the winter is severe, a majority of source plants and aphids will be destroyed and the inoculum present at the start of cropping will be reduced.

CMV is not transmitted by seeds on tomato or in any other seeds of Solanaceous plants. However, it is in many plants belonging to different botanical families (Fabaceae, Caryophyllaceae) and, in particular, in many weeds (e.g. *Stellaria media*). This virus could be mechanically transmitted during cultural operations, pruning, and leafing (demonstrated by necrogenic strains associated with a RNA satellite). In this case, the distribution of diseased plants is more in a line, as opposed to transmission by aphids which give rise to random or cluster distribution.

Control methods[1]

• During cultivation

Plants infected by CMV remain so, although symptoms may sometimes tend to diminish.

If attacks occur in nurseries or during cultivation and they are detected early, the few plants showing symptoms of the virus should be rapidly removed. Indeed, the elimination of virus-infected plants in the early stages of an epidemic can usually help to slow or even stop it. But beware, the symptoms may first show 2 weeks after infection. Infected plants may be symptomless and escape detection and be a source of the virus contributing to the development of an epidemic.

Aphicide treatments are often needed to control aphid populations on tomato. Unfortunately, they are not effective in preventing virus outbreaks because transmission is in a nonpersistent manner. In fact, vector aphids often arrive from outside the crop and transmit the virus during brief feeding periods even before the aphicide has time to act. Moreover, the current problems of control of aphids on tomatoes, sometimes related to phenomena of resistance to insecticides, do not help this situation.

A number of auxiliary measures can help in the control the aphid populations (see pp. 649 and 650). Note that the aphids alone can result in damage on tomatoes.

All weeds that can serve as a source of inoculum must be removed from the crop and nearby surrounding areas. This is often difficult to achieve given the large number of potential hosts of CMV.

• Subsequent crop

A package of measures aimed at preventing, or at least to minimizing the introduction of the virus and its spread in the tomato plots should be used.

In countries where infection is very early, it will be necessary to protect nurseries and young plants. For this, barriers (unwoven, knitted fabrics) can be used. This mechanical barrier delays infection. Glasshouses can be made insect-proof. The use of yellow traps can help to monitor population levels and implement a rational programme.

Weeds must be removed from areas where field crops are being propagated as well as the immediate surroundings (border hedges and paths), to remove sources of the viruse and/or vectors. Avoid setting up a tomato crop near crops that are sensitive to CMV like spinach and cucurbits. Crop rotation may be considered where

---

[1] See note to the reader p. 417.

there are known to be populations of weeds that carry the virus.

Some sources of resistance to CMV, often partially effective, have been reported in several wild species of *Lycopersicon*:
– a dominant gene called '*cmr*' which is located on chromosome 12 of *L. chilense*;
– different accessions, including three in *Lycopersicon esculentum*, two in *L. hirsutum*, one in *L. chmielewskii* and *L. pimpinellifolium*. The effectiveness of this resistance seems to vary according to the strain.

Note that in order to protect tomatoes against the lethal necrosis, acquired immunity tests were carried out in France with the strain CMV-R. It contains an RNA-5, which does not cause symptoms after mechanical inoculation. Glasshouse tests showed that this strain provided effective protection against an aggressive strain. Similar results were obtained in different countries with other strains with an RNA-5. In field trials there was a significant reduction in the percentage of plants with mosaic symptoms using this strain.

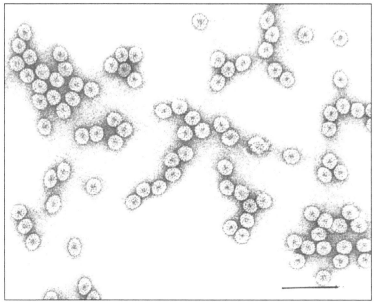

885 CMV particles are isometric and have a diameter of 28 nm.

Potato virus Y (PVY)

(Potyvirus, Potyviridae)

Principal characteristics

PVY was reported for the first time on potato in the UK in 1931. This virus, which is the type species of the Potyvirus, shows significant variability expressed in several of its hosts. The strains are commonly classified into three major groups, defined according to symptoms induced on tobacco, *Physalis floridana*, and potato:

– Y^O group includes all virus Y so-called 'ordinary' or 'common' strains. These strains, which attack tomatoes, are known since the 1930s and are present worldwide;

– Y^N group includes all strains that induce a mosaic and necrosis in tobacco. They also affect tomatoes. These strains were first recognized in the 1940s, and are very unevenly distributed on different continents. Currently a resurgence of this group of strains has occurred in several European countries. Variants inducing ring necrosis on potato tubers have been given the name PVY^{NTN} and are now widely recognized;

– other groups or strains have been described from time to time, including the Y^C group. In general, tomato and tobacco are affected by all strains of PVY, acting as universal hosts.

The virus particles are flexuous rods and measure 73 × 11 nm (Photo **886**).

• Frequency and extent of damage

PVY is probably present in all worldwide potato growing areas where it can also attack tomato, pepper, and tobacco. It is found mainly in the field, but also more frequently in protected crops. It is very damaging in warm regions on tomato and pepper.

Y^O group strains are distributed worldwide, those of the Y^N group (including strains PVY^{NTN}) are present in Europe and the former Soviet Union, in a part of Africa including the Maghreb, South America (Argentina, Chile), and Asia. Y^C strains are probably present in Australia, India, and Europe. Although difficult to quantify, the economic losses caused by this virus may occasionally be significant.

• Main symptoms

Two main types of symptoms are seen on tomato:

– various discolorations on the young leaves, which starts as a discrete mottling and develops into a green mosaic. Bands of dark green tissue localized along the veins (vein banding) and diffuse chlorotic spots are also observed. These symptoms are often caused by strains of the PVY^O group;

– necrotic strains which cause reddish-brown leaf spots which quickly become necrotic. Symptoms of necrotic streaks are sometimes present on the petioles, and even the stem.

The nature and intensity of these symptoms can vary depending on the stage of plant development, cultivar, environmental conditions, but also the nature of the strain involved.

Mixed infections PVY–CMV are relatively common, especially during the summer. These sometimes cause more spectacular symptoms, both in the field and protected crops.

(See Photos **110, 124–127, 270–274, 422.**)

• Biology, epidemiology

Survival, inoculum sources: the host range of PVY is largely limited to Solanaceous crops and weeds: potatoes, peppers, tomatoes, eggplant, and tobacco, and various weeds (*Solanum nigrum*, *Portulaca oleracea*, *Senecio vulgaris*, *Physalis* spp.). The weed hosts in particular enable it to overwinter. Other plants are likely to host it: *Cyphomandra betacea*,

Datura ferrox, D. innoxia, Nicandra physaloides, Nicotiana glauca, N. rustica, Petunia spp., *Physalis angulata, P. heterophylla, P. mendocina, P. virginiana, Schizanthus retusus, Solanum andigenum, S. atropurpureum, S. aviculare, S. cardenensii, S. gracile, S. indigena, S. jasminoides, S. khasianum, S. laciniatum, S. sisymbrifolium, S. xanthocarpum,* and more recently *Hyoscyamus niger.*

Species belonging to other families of dicotyledons are also susceptible including *Dahlia variabilis, Rudbekia amplexicaulis, Senecio vulgaris, Brassica* sp., *Quisqualis indica, Cassia occidentalis, Medicago arabica, Melilotus officinalis, Melilotus* sp., *Plantago lanceolata, Ranunculus asiaticus, Tropaeolum majus,* and *Viola tricolor.*

In temperate regions, potatoes are the main virus reservoir, but other perennial plants such as *Cyphomandra betacea, Nicandra physaloides, Solanum nigrum,* and *S. dulcamara* can also be important. Similarly, weeds such as *Solanum atropurpureum* and other *Solanum* spp. in tropical and sub-tropical areas may be a source.

Transmission, dissemination: at least 40 species of aphids can transmit PVY, in the nonpersistent manner. *Myzus persicae, M. certus, Aphis gossypii, A. fabae, A. craccivora, Acyrtosiphon pisum, Brachycaudus helichrysi, Macrosiphon euphorbiae, Phorodon humili, Rhopalosiphum insertum,* and *R. padi* are the most important. The virus is rapidly acquired by the vector insect in a few seconds and a small number of virus particles are sufficient for efficient transmission. However, transmission by aphids requires an additonal protein of viral origin to be present along with the virus particles, known as the helper component (HC). Only virus particles acquired in the presence of the helper are retained in the maxillary stylets and are likely to be transmitted. However, the transmission efficiency depends on both the vector aphid and the virus strain.

This virus is also mechanically inoculated to several plants in the Chenopodiaceae, Amaranthaceae, Asteraceae, and Fabaceae families, in addition to Solanaceae. It does not seem to be spread by contact or by seed.

Control methods[1]

• During cultivation.

Generally, an infected plant will remain so all its life, although symptoms may tend to diminish. There are no curative treatments.

If attacks occur during propagation and they are detected early, the few plants showing symptoms of PVY should be discarded and the remainder planted after a sufficient incubation period has elapsed to be certain that all infected plants have been detected.

Aphicide treatments are often necessary to control aphid populations on tomatoes (see pp. 210 and 649), but are often ineffective in preventing virus outbreaks. Vector aphids often come from outside the field and transmit the virus during brief feeding periods not long enough for the aphicide to have time to act. In addition, the current problems of aphid control which are often related to insecticide resistance makes control by this method even

more difficult. Note however, that there are a number of other ways of controlling the aphid population (see p. 650).

All weeds that can serve as sources of the virus must be removed from the crop and surrounding areas.

• Subsequent crop

It is useful to remind the reader of all the measures which will aim to prevent or at least to minimize the introduction and development of PVY in tomato plots.

In countries where infection is very early, it will be necessary to protect plants during propagation, and infection may be delayed with the use of mechanical barriers such as nonwoven or mesh fabrics. Where glasshouses are used they can be made insect-proof.

Careful weeding of propagation areas and the immediate surroundings (hedgerows and paths)

[1] See note to the reader p. 417.

will help to eliminate sources of the virus and of vectors.

Tomato crops should not be planted near to other host crops especially the Solanaceae, e.g. potato, pepper, pepper, and tobacco. Remember that tobacco, potato, and to a lesser degree pepper, are likely to harbour different strains.

Spraying mineral oil on plants could reduce the incidence of PVY in crops. They reduce the spread of PVY not because they are toxic to aphids, but because they interfere with the attachment and/or release of the virus particles on the aphid's stylets. To be effective, mineral oil must cover both sides of the leaf and need to be applied at very regular intervals in order to maintain this cover.

The use of reflective mulches can repel vectors. In contrast, yellow sticky traps attract them.

The use of resistant varieties is often the most effective means of control for viruses. Resistance has been demonstrated in *Lycopersicon hirsutum*

'PI 247 087' and was effective against 36 strains or isolates from different geographical or botanical origin. Resistance to PVY is effective and stable when used with artificial inoculation and also when co-inoculated with other viruses. It is effective over a wide temperature range and also when various inoculum concentrations are used. Two mechanisms of resistance have been identified:

– inhibition of the multiplication of common strains of PVY in the inoculated leaf. This mechanism of resistance is expressed in the seedling at the age of 12 days, but can be circumvented by the strains of one pathotype (PVY LYE 84-2);

– a partial inhibition of the multiplication and movement of PVY in the plants 41 days old. This mechanism is expressed in mutated strains.

The resistance of 'PI 247 087' is controlled by at least one recessive gene called '*pot-1*'. Note that this parent also has resistance to TEV.

886 The virus particles of PVY are flexuous rods and measure 73 × 11 nm.

Tobacco etch virus (TEV)

(Potyvirus, Potyviridae)

Principal characteristics

This virus has been described for the first time on *Datura stramonium* in the US in 1921.

Biological variability of TVE strains is known and influence the intensity of symptoms induced, the efficiency of virus transmission by aphids, and the morphology of nuclear inclusions present in the host. However, there are no antigenic differences in SDS-immunodiffusion tests.

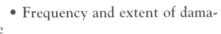

The particles are filamentous and flexuous, 705–730 nm × 12–13 nm.

• Frequency and extent of damage

This virus has a more limited distribution than PVY. It is found throughout the Americas and in far east Asia (China, Korea, Philippines, Taiwan, Thailand). This Potyvirus is also present in the Mediterranean where it causes considerable damage, especially on peppers in Turkey. TEV is also present in tobacco on the African continent, including Madagascar, and it occurrs occasionally in South Africa. It has also been reported in India for many years. Records on tomato are from a limited number of countries (US, Venezuela).

In Europe its distribution is not well known. It has been reported in Germany on tobacco, although there is some doubt about the identification of the virus. It is found in Spain on some Solanaceous plants.

• Main symptoms

On tomato, TVE produces a fairly marked mottling and slight deformation of leaflets and leaves. It may also be responsible for a yellow mosaic and leaf shrivelling. Plant growth is severely limited with early infection and the affected plants are stunted. Yield may be affected; decreases of more than 25% have been described. The fruits are sometimes mottled, small, and unmarketable.

• Biology, epidemiology

Survival, inoculum sources: the host range of TEV is relatively limited. It affects several other Solanaceous crops (potato, pepper, tobacco, petunia) and weeds in various families (*Solanum aculeatissimum*, *S. carolinense*, *S. nigrum*, *S. seaforthianum*, *Cassia obtusifolia*, *Chenopodium album*, *Cirsium vulgare*, *Datura stramonium*, *Linaria canadensis*, *Physalis* spp.). The virus in weed populations enable it to over-winter.

Transmission, dissemination: at least 13 species of aphids may transmit TVE, in the nonpersistent manner including *Aphis fabae*, *A. craccivora*, *A. gossypii*, *A. spiraecola*, *Lipaphis erysimi*, *Macrosiphum euphorbiae*, and *Myzus nicotianae*. The peach green aphid (*Myzus persicae*) seems to be one of the most common vectors in temperate regions. The virus is rapidly acquired by the vector insect; a few seconds and a small number of virus particles are sufficient for efficient transmission. There is no latency period. However, the virus can be retained from 1–4 hours on the aphid's stylets. In addition, transmission by aphids requires the presence in infected cells of a protein of viral origin, the helper component (HC). Only virus particles acquired in the presence of this factor are retained on the maxillary stylets and can be transmitted. This virus is not spread in tomato seeds.

Control methods[1]

Generally, an infected plant will remain so and there are no curative treaments.

Control methods for PVY (Description 36) apply equally for the control of this virus.

Measures to control aphid-borne viruses are the same, regardless of the virus. Only information specific to TEV is presented here.

Avoid planting a tomato crop near TVE susceptible plants, especially those in the Solanaceae (mainly pepper and tobacco, and to a lesser extent potato).

As with PVY, *Lycopersicon hirsutum* 'PI 247 087' is resistant to TEV. This resistance is controlled by a recessive gene and is expressed by a strong inhibition of the accumulation of virus particles in the inoculated leaves and by their lack of long-distance migration in the plant. This resistance is effective against isolates from different geographic origins.

---

[1] See note to the reader p. 417.

■ Other viruses transmitted by aphids

Viruses	Symptoms and mode of transmission	Shape of the virus particles	Principal characteristics
Broad bean wilt virus (BBWV) **Fabavirus**	Mosaic in-line or concentric patterns on leaves. The latter are reduced in growth and are slightly deformed. Necrotic lesions may occur on the leaves and extend to the petioles and stem. Many aphids, nonpersistent manner. More than 25 species able to transmit BBWV including *Acyrthosiphon pisum, Aphis craccivora, A. fabae, A. nasturtii, Macrosiphum euphorbiae, M. solanifolii,* and *Myzus persicae.* The virus is transmitted in the seeds of broad bean.	Isometric particles 30 nm in diameter with an angular hexagonal outline. 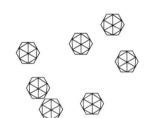	BBWV was first described on broad bean in Australia in 1947. It has occasionally been described in several countries around the world, but it is on the edge of the Mediterranean that it occurrs most frequently on broad bean, lentil, and chickpea. It is not uncommon on pepper and tomato in France, Italy, and Spain. Two groups of strains have been identified by serology; these are known as BBWV-1 and BBWV-2. Over 200 species of plants can be infected, including tobacco, petunia, eggplant, *Capsicum frutescens,* and even vine. Control methods are as recommended for CMV, Description 35.
Colombian datura virus (CDV) **Potyvirus**	Young leaflets with mosaic and poorly coloured fruits. The plants have reduced growth. Aphids (*Myzus persicae*), nonpersistent manner.	Filamentous particles of Potyvirus type, with a length of about 720 nm. 	It is in Colombia, as its name suggests, that CDV was first reported affecting *Datura.* For some time it was rare but in the last 10 years has been found in glasshouses in Germany and the Netherlands. It affects a limited number of hosts in addition to datura and tomato, including petunia, *Juanulloa aurantiaca.* Control methods are as recommended for CMV, Description 35.
Eggplant severe mottle virus (ESMoV) **Potyvirus**	Marked mosaic on leaflets. Aphids, in the nonpersistent manner (*Aphis craccivora* and *Myzus persicae*).	Long filamentous particles 808–842 nm. 	Isolated in Nigeria on eggplant in 1987, this virus does not appear to have been described in any other country. It affects only the eggplant and tomato. Its experimental hosts are almost exclusively Solanaceous. Control methods are as recommended for CMV, Description 35.

Viruses	Symptoms and mode of transmission	Shape of particles	Principal characteristics
Pepper veinal mottle virus (PVMV) **Potyvirus**	Chlorotic to necrotic spots and mosaic on leaflets which are also deformed; longitudinal necrotic lesions may develop on the stem. Transmitted by various aphids in the nonpersistent manner (*Aphis gossypii, A. craccivora, Myzus persicae*).	Filamentous rods, often flexuous, 770–850 nm in length. 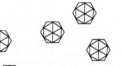	This virus, almost entirely African, was described for the first time in Ghana in 1971 on pepper and petunia. It is now found from eastern to western Africa, and from Tunisia to southern Africa. There is some evidence that it may occurr in Malaysia, India, China, and Yemen. Several variants have been described that differ in the symptoms they produce, their host specificity, and virulence. Many Solanaceous plants are likely to be hosts such as tobacco, eggplant, *Datura stramonium, D. metel, Physalis angulata,* and *Solanum nigrum.* Control methods are as recommended for BWYV, Description 34.
Peru tomato virus (PTV) **Potyvirus**	Whitening of veins, mosaic on leaflets which are also deformed. Some strains cause necrotic lesions on leaves and stems. Mottling of fruits. Transmitted by *Myzus persicae* aphid in the nonpersistent manner.	Filamentous particles 741–778 × 11–13 nm.	As the name suggests, PTV has been described in Peru, in 1971 on tomato. At present it appears to be confined to this country. It is almost always associated with other Solanaceous plants such as pepper, *Capsicum baccatum, C. chinense, Physalis peruviana, Solanum nigrum,* and *Nicandra physaloides.* Control methods are as recommended for CMV, Description 35.
Potato leafroll virus (PLRV) **Polerovirus**	Yellowing of the leaves starting at the periphery followed by leaf curl. Affected leaves become rigid and brittle. Early affected plants are stunted. Transmission is by more than 10 species of aphids in the persistent circulative manner, especially *Aphis nasturtii, Aulacorthum solani,* and *Myzus persicae.*	Isometric particles with a diameter of 24 nm.	This Polerovirus was reported for the first time in the Netherlands in 1916 on potato. It is now present in almost all European countries where potatoes are grown. Several strains have been described, in particular the tomato yellow top strain which is highly aggressive on tomato. In addition to tomatoes, it is found on pepper, *Solanum nigrum, Nicandra physaloides, Datura stramonium, Amaranthus caudatus, Lamium purpureum,* and *Capsella bursa-pastoris.* Note that the two lines PR 8-5 and PR 8-4, from a cross between tomato and *Lycopersicon peruvianum,* have a nonspecific resistance against phloem viruses, notably the tomato yellow top strain. Control methods are as recommended for BWYV, Description 34.

Viruses	Symptoms and mode of transmission	Shape of particles	Principal characteristics
Tobacco vein banding mosaic virus (TVBMV) Potyvirus	Mosaic, sometimes with a dark green edge to the veins. Several species of aphids in a nonpersistent manner, *Myzus persicae* being the most efficient vector.	Filamentous rods of Potyvirus type. 	TVBMV was first observed in 1966 on tobacco in Taiwan and is now present in China and Japan. It was detected in the US in 1990. The virus attacks tobacco, potato, tomato, *Datura stramonium*, *Physalis floridana*, and *Solanum nigrum*. Control methods are as recommended for PVY, Description 36.
Tobacco vein mottling virus (TVMV) Potyvirus	Affects some American varieties of tomato without causing symptoms. Several species of aphids in a nonpersistent manner (*Myzus persicae*, *M. nicotianae*, *Aphis craccivora*, *Macrosiphum euphorbiae*).	Potyvirus-like virus particles, measuring 765 × 13 nm. 	Located mainly in North America and occasionally Africa. This virus has been reported for the first time on tobacco in 1972. Tobacco, *Solanum carolinense*, tomato, and *Rumex* spp. constitute its main natural hosts. Control methods are as recommended for PVY, Description 36.
Tomato aspermy virus (TAV) Cucumovirus	Apical growth is stopped resulting in a proliferation of numerous axillary buds and plants have a bushy appearance. The leaves are mottled, severely deformed, and sometimes filiform. Fruit set is greatly reduced and the few fruits formed are small and deformed. Seed production is low to zero (this contrasts with CMV which also induces filiform leaves). Transmitted by 22 aphid species including *Myzus persicae*, in nonpersistent manner. One isolate is known to be seed transmitted in *Stellaria media*.	Morphology of particles is similar to those of CMV. 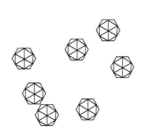	This virus was described for the first time in the UK in 1939 and called 'Chrysanthemum aspermy virus'. It is present in virtually all regions of the world, especially in those where chrysanthemums are grown. Several variants have been described. It has a fairly wide host range involving several botanical families. It infects pepper, various *Apium* spp., spinach, *Chrysanthemum indicum*, *Zinnia elegans*, *Ajuga repens*, and *Tropaeolum majus*. Cross protection tests with an attenuated strain, TAV-M, have shown a reduction of symptoms. Control methods are as recommended for CMV, Description 35.
Tomato mild mottle virus (TMMV) Potyvirus	Mottling, moderate mosaic on young leaves. Reduced yield. Aphids (*Myzus persicae*), nonpersistent manner.	Flexuous particles, 719 nm long. 	Discovered in Yemen in 1991 on three Solanaceous plants including tomatoes, this virus is also present in other countries in the Middle East and has been detected on tomatoes in Ethiopia. Tomato, *Datura stramonium*, *Solanum nigrum*, and *Nicandra physaloides* are natural hosts. Control methods are as recommended for CMV, Description 35.

■ Viruses transmitted by whiteflies

Tomato chlorosis virus (ToCV)

(Crinivirus, Closteroviridae)

Principal characteristics

This virus appears to have originated in the US, where it was first identified in 1998 in Florida, even though symptoms now know to be of ToCV were observed on tomato in the same state in 1989, and at that time called 'yellow leaf disorder'.

The viral particles measure 800–850 × 12 nm.

• Frequency and extent of damage

After Florida, ToCV was described in Louisiana and in Colorado. The first report in Europe was in 1997, in Spain. Later it was found in several Mediterranean countries, chronologically Portugal, Italy, Greece, and France. It was also found in Morocco, the Canary Islands, South Africa, and even Taiwan (1998). It has been present in Israel since 2003. It can be detected in field and protected crops.

In France, its distribution is limited to the south, where crops are affected to a limited extent (from 5 to 30% of the plants). Sometimes the entire crop may be affected and then yields are reduced because of the limited size of fruits.

• Main symptoms

The first symptoms appear on lower and intermediate leaves of plants randomly distributed in the crop. Chlorotic mottling or irregular chlorotic spots appear between the veins of the leaflets and spread gradually. Small reddish to brown necrotic lesions may also be visible. Subsequently, the upper leaves gradually become yellow. After a few weeks, some plants exhibit fairly sustained inter-vein chlorosis affecting many leaves, while the veins of the leaflets remain dark green and contrast with the rest of the foliage. The plants eventually grow old prematurely; old leaves thicken, curl, and become brittle. Sometimes they dessicate.

Affected plants are usually less vigorous than healthy plants. Flowers and fruits do not show symptoms, but their growth and maturation is delayed. The incubation period is 3–4 weeks.

Visual identification of this disease is difficult because symptoms may suggest a nutritional problem. There are several deficiencies that induce similar inter-veinal yellowing of the lower leaves (magnesium, potassium, nitrogen). Pepino mosaic virus (PepMV, Description 29) also causes chlorosis of the leaves, in addition to other symptoms which differentiate it from ToCV. The same is true for Tomato yellow leaf curl virus (TYLCV, Description 41), and Tomato infectious chlorosis virus (TICV, Description 40), which occur as mixed infections in some countries.

(See Photos **151–155**.)

• Ecology, epidemiology

Survival, inoculum sources: as with many viruses, ToCV is likely to survive in a number of plant species. but its natural host range is restricted to around 30 species. It is known to infect potatoes and peppers. Some authors report that some ornamental plants (*Zinnia elegans*) and some other vegetable crops are sensitive. It is also found on weeds (*Datura stramonium, Solanum nigrum, S. nigrescens, Physalis peruviana, P. ixocarpa*). It is likely that it is able to persist in such plants which are then reservoirs of the virus.

Transmission, dissemination: the virus is transmitted by whiteflies, in the semi-persistent manner. Time of acquisition by the insect vector is 48 hours, sometimes less, and it remains viruliferous for approximately 3 days. When 'injected' the virus remains restricted to the phloem vessels, where it multiplies. Several species of whiteflies are capable of transmission: *Trialeurodes vaporariorum* (Westwood), *Bemisia tabaci* (Gennadius) – the most efficient vector of A, B (= *Bemisia argentifolii* Bellows & Perring) and Q biotypes, and *Trialeurodes abutilonea* (Haldeman). Virus is not passed on to new generations of these whiteflies. This virus has not been experimentally transmitted by mechanical inoculation. However, it certainly spreads in production areas where viruliferous whiteflies abound, especially in glasshouses.

Control methods[1]

• During cultivation

This is not an easy disease to control and there are no curative measures. There is always a risk of introduction and reintroduction of the virus in infected tomato seedlings or in different plants from affected areas with viruliferous whiteflies. ToCV is the subject of legislative control in several European countries: The Order of the Decree of July 8, 2002 makes it mandatory to report the presence or suspected presence of ToCV, and introduce strict measures to control whiteflies as well as removing all affected plants.

Given the lack of specificity of symptoms, early detection is not always easy, which makes it difficult to identify this virus, and remove the first affected plants. If there are suspicious symptoms, a specialist laboratory should be consulted.

Control of whitefly vector populations in a glasshouse is not easy.

At the end of cultivation, the plants are removed as quickly as possible, whiteflies destroyed, and a there should be an interval between crops of several days.

• Subsequent crop

It is necessary to implement a package of control measures aimed at the prevention, or at least minimization, of the chance of the introduction of the virus, but especially its vectors, and their entry into tomato crops.

Closely monitor the quality of plants, and in particular their origin. Be particularly wary of plants from countries or regions already affected by ToCV. Nurseries should be insect-proof. Agrotextiles can be used to isolate plants (nonwoven, mesh fabric). The mechanical barrier thus created reduces the risk or delays infection.

Methods to detect and chemically or biologically control whiteflies should be implemented (see p. 649).

---

[1] See note to the reader p. 417.

Careful weeding of the crops and their surroundings is needed in order to eliminate virus reservoirs and/or sources of vectors. New crops must not be planted next to crops already affected.

Currently there are no tomato varieties resistant to this virus.

The Control methods section for TYLCV (Description 41) should also be consulted. Several additional measures to control this virus which is also transmitted by whiteflies, can be used in the case of ToCV.

Tomato infectious chlorosis virus (TICV)

(Crinivirus, Closteroviridae)

Principal characteristics

This virus was first isolated in 1993 in California on tomatoes when it was thought to be ToCV.

In purified preparations very elongated particles about 850–900 × 12 nm were observed. These are characteristic of the Crinivirus group and more broadly of Closteroviridae.

• Frequency and extent of damage

TICV is now present in North Carolina. It was found in Spain, Italy, and Greece, especially in Crete in the early 2000. Since then this virus has been observed very occasionally in France (2003 in the Nice area). It was reported in Taiwan and Japan in 2001 and Indonesia in 2002.

It can cause considerable crop loss (c.f. ToCV). Thus, serious epidemics have been reported in Greece where 100% of the plants grown in glasshouses, and even in the field, expressed symptoms. Under these conditions, the amount of fruit produced is greatly reduced.

• Main symptoms

The symptoms caused by the TICV are identical with those of ToCV so the two can be confused. Affected plants show inter-veinal yellowing on older leaves, sometimes associated with reddening of the leaf and the presence of necrotic lesions. The leaflets are deformed and rolled.

If infection occurs early, the plants are less vigorous and in some particularly serious cases, they can remain stunted. The fruits are small and rather limited in number.

TICV may occurr in complex with other viruses, especially with the ToCV.

(See ToCV symptoms, Photos **151–155**.)

• Ecology, epidemiology

Survival, inoculum sources: knowledge about this virus is still limited. It seems able to infect several other Solanaceous plants such as potato, *Physalis ixocarpa*, and *Petunia hybrida*. Other crops are also likely to be sources: three Asteraceae (lettuce, artichokes, and *Zinnia elegans*), and a buttercup (*Ranunculus* sp.). TICV has been found in several weeds including *Chenopodium album*, *C. murale*, *Picris echloides*, *Senecio vulgaris*, *Sonchus oleraceus*, and *Capsella bursa-pastoris*.

It can be transmitted experimentally by whiteflies to more than 26 experimental hosts from at least eight botanical families.

Transmission, dissemination: as suggested earlier, TICV is a phloem virus transmitted by whiteflies in a semi-persistent manner. Unlike ToCV, only *Trialeurodes vaporariorum* can transmit TICV.

The acquisition of the virus occurs during insect feeding on a diseased plant, and it can persist in the whitefly for 24 hours. Subsequently, the transmission rate of TICV to a plant depends on the number of viruliferous whiteflies present on the latter. For example, in the presence of 1, 10, and 40 whiteflies per plant, transmission rates are respectively about 8, 58, and 83%. The minimum transmission period is approximately 1 hour, but its effectiveness increases with time. For example, after 1, 3, and 48 hours in the presence of viruliferous whiteflies on a healthy plant, transmission rates were 6, 20, and 88%. While most insect vectors lose the virus during the first day, it can sometimes remain in the insect for up to 4 days.

TICV is not transmitted mechanically.

Control methods[1]

Once in a crop, this yellow disease is not easy to control. In contrast to fungal diseases, there is no curative method during cultivation. Generally, an infected plant will remain so all its life.

Note that there is a risk of the introduction of this virus as well as viruliferous whiteflies when plants are brought in from an area where the disease is known to occur. TICV is subject to an order of compulsory control in several European countries. The EU Order of July 8, 2002 makes it mandatory to report the presence or suspected presence of TICV, and to control whiteflies as well as removing infected plants.

The recommendation for the control of TYLCV (Description 41) and ToCV (Description 39) apply equally to this virus disease including the control of the vector whiteflies.

[1] See note to the reader p. 417.

Tomato yellow leaf curl virus (TYLCV) and other associated viral species

(Begomovirus, Geminiviridae)

(Former subgroup III of the Geminiviruses group)

Principal characteristics

TYLCV is not new as the first attacks were reported on tomato in Palestine in 1939. It appears to have been spread from infected plants or as a result of the migration of its vector insect, *Bemisia tabaci*.

The emergence of the yellow leaf curl syndrome in many countries has led to the description of several strains of TYLCV: TYLCV-Ch (China), TYLCV-Is (Israel, also including strains from Egypt and Lebanon), TYLCV-Ng (Nigeria), TYLCV-Sar (Sardinia), TYLCV-SSA (southern Saudi Arabia), TYLCV-Tz (Tanzania), TYLCV-Th (Thailand), and TYLCV-Ye (Yemen).

Subsequently, by studying the genome, these strains and some new ones have been reclassified as internationally recognized species: Tomato yellow leaf curl virus (TYLCV, including the original virus and Israel, Almeria strains), Tomato yellow leaf curl Sardinia virus (TYLCSV), Tomato yellow leaf curl Malaga virus (TYLCMalV, which is a recombinant between TYLCV and TYLCSV), Tomato yellow leaf curl China virus (TYLCCNV), Tomato yellow leaf curl Kanchanaburi virus (TYLCKaV), and Tomato yellow leaf curl Thailand virus (TYLCTHV).

There are other 'potential' species not yet fully validated internationally: Tomato yellow leaf curl Nigeria virus (TYLCNV), Tomato yellow leaf curl Kuwait virus (TYLCKWV), Tomato yellow leaf curl virus Saudi Arabia (TYLCSAV), Tomato yellow leaf curl Tanzania virus (TYLCTZV), Tomato yellow leaf curl Yemen virus (TYLCYV). Note that the situation is similar for other viruses, including ToLCV.

Other viral species associated with yellow leaf curl syndrome are listed in Description 42.

It is highly likely that other species will be described, resulting in a constantly changing situation of the Begomoviruses on tomato and probably in the taxonomy of these viruses.

Thus as knowledge stands, all these viral species show comparable biological properties. To avoid repetition, only TYLCV is described. This information should be sufficient to be able to interpret and manage the disease when it is caused by other Begomoviruses. In addition, all viruses transmitted by *Bemisia tabaci* are considered quarantine viruses.

The TYLCV particles are present in the nuclei of cells of leaves. They are paired and angular, and are about 20–30 nm in diameter (Photo 887).

• **Frequency and extent of damage**

TYLCV and viral species associated with the yellow leaf curl syndrome are now common in many countries on all continents.

TYLCV is present in several southern Mediterranean countries, the near and Middle East, and in some countries in Sudano-Sahelian Africa such as Senegal and Burkina Faso. It was detected on the island of Réunion in 1997. It is also found in Morocco and the Canary Islands, with TYLCSV. The virus is also present in Italy, Sicily, and Sardinia (since the late 1980s) and Calabria (since 1991). It has been found in Spain since 1993 particularly in Andalusia together with TYLCSV. Further distribution resulted in crop loss in 1995 in Murcia and Alicante, and in Portugal in the Algarve region. TYLCV or other viral species are also reported in eastern Europe, Asia (Iran, Turkmenistan, China, Nepal, Thailand, Taiwan, Japan) and Australia. TYLCV has reached the Americas: Dominican Republic (1997), Cuba (1998), Florida (1999), Jamaica (1999), Mexico (1999), Guadeloupe (2001). In France it is now present in the Pyrenees-Orientales, where it is found in wild plants outside the glasshouses.

Several factors have contributed greatly to the spread of TYLCV in the world:
– the characteristics of the vector whitefly of Begomovirus, *Bemisia tabaci* (various biotypes), which is very polyphagous with over 500 plant species as hosts and has very high population levels;
– the sensitivity of tomato varieties grown and continuous production;
– the migration of viruliferous whiteflies, from old tomato crops and other sensitive plant species to new crops;
– the large-scale sale of infected tomato plants.

Economic losses caused by TYLCV are often very important. Several studies point to reductions in yields of 50– 60%, compared with healthy controls. Early infection has the greatest effect.

• Main symptoms

As Begomoviruses are strictly associated with the phloem this affects terminal growth in particular. Affected plants are weak and bushy, because of the development of numerous axillary branches, and a reduction in the length of the internodes. When infection is early the plants are dwarfed and produce no fruit.

The shape and/or colouring of the leaves is abnormal and they are smaller. They gradually curl upwards, giving them appearance of a spoon. They can be yellow and tend to harden, and sometimes to take on a purplish colour, especially on the veins. This is clearly seen on the underside of the affected leaves.

The flowers drop prematurely, so the production of fruit can be greatly reduced, especially if the plants were infected very early. The fruits are often few and small.

Note that whiteflies on tomatoes, even without any virus, can cause various symptoms, see pp. 213 and 370.

(See Photos **16**, **37**, **38**, **48**, **167–171**.)

• Ecology, epidemiology

Survival, inoculum sources: TYLCV is able to survive on a diverse range of hosts contrary to what was originally thought. It is a particular problem on tomatoes but also on some other Solanaceous plants including, tobacco, occasionally pepper, *Datura stramonium*, and *Solanum nigrum*. Other species belonging to different botanical families have been identified as potential natural hosts including beans, *Eustoma grandiflorum* (syn. *Lisianthus russelianus*), *Malva parviflora*, *M. nicaensis*, *Cynanchum acutum*, *Euphorbia pulcherrima*, *E. heterophylla*, *Dittrichia viscosa* (syn. *Innula viscosa*), *Nicandra physaloides*, and *Achyranthes aspera*.

For example, in Lebanon, in addition to beans and eggplant (immune according to some authors), *Amaranthus* sp., *Sonchus oleraceus*, *Plantago* sp., and *Mercurialis annua* can be a reservoir for TYLCV. In southern Spain, *Datura stramonium*, *Solanum nigrum*, *S. luteum*, and *Mercurialis ambigua* show spoon leaf symptoms which are characteristic of infection by TYLCV. Also, TYLCV has gradually replaced TYLCSV in Spain during 1996–1998. It is also more efficiently transmitted by local biotypes of *Bemisia tabaci*. Four new symptomless carriers have recently identified in Spain: *Chenopodium murale*, *Convolvulus* sp., *Cuscuta* sp., and *Conyza sumatrensis*.

TYLCV has been transmitted experimentally by grafting or via *B. tabaci*, to several dozen hosts belonging to very different botanical families.

Transmission, dissemination: TYLCV is transmitted exclusively by *Bemisia tabaci*, a whitefly also known as cotton, tobacco, sweet potato, or silver leaf 'white fly'.

Bemisia tabaci is polyphagous and very common in the tropics and sub-tropics. During the last decade, it has spread to the more temperate regions and is now reported in more northern countries such as Netherlands, Germany, Denmark, and Sweden, but only under protection. In Italy and Spain it is seen in the field. In 1981, *B. tabaci* was identified for the first time in various northern European countries on protected crops. It is now well established on various plant species grown in glasshouses, and may be found in the vicinity of glasshouses in southern Europe in summer. In general, populations of B. tabaci are smaller in spring and early summer and the incidence of TYLCV is therefore negligible on early fields crops. Populations rapidly increase in late summer and, in these situations, the disease becomes severe by the autumn.

Several biotypes of Bemisia tabaci have been described but can only be identified in the laboratory, through the characterization of their enzymatic profiles and by molecular biology. These biotypes also differ in some of their properties, notably in their ability to transmit different strains of TYLCV. Biotype B, also called 'Bemisia argentifolii', was largely instrumental in the spread of the virus. It is characterized by much higher fertility and is more polyphagous than other biotypes. It is also often more efficient in transmitting TYLCV. Other biotypes perform this function less effectively, for instance the Q biotype, a native of the Mediterranean region, and the biotype IC from Ivory Coast. Note that the B biotype is more adapted to transmit TYLCV than the Asian biotype which is better able to transmit ToLCV. Recent work has shown that biotypes B and Q were found many Mediterranean countries (France, Italy, Spain, Canary Islands, Morocco). In addition, the Q biotype is more common. Its predominance is explained by a greater tolerance to insecticides and extreme temperatures (hot or cold). Its biological properties should allow B. tabaci to survive in areas considered to be at the limit of its geographic zone.

TYLCV is transmitted in the persistent circulative manner. The acquisition or transmission of viral particles occurs during prolonged feeding in the phloem vessels. The acquisition period can vary from 10 to 20 minutes minimum, to a few hours or even 1 or 2 days. The nymphs are as effective as adults in acquiring TYLCV.

Once acquired, the particles cycle in the body of the insect before being transmitted again, described as 'circulative'. They go through the digestive tract and the body cavity, to concentrate in the salivary glands. The latent period lasts only a few hours (8–24 hours), and the whitefly is then able to transmit the virus. The virus can be transmitted after a minimum feed time of 15–30 minutes and is optimal if it lasts 6 hours. The symptoms appear on plants at least 2–3 weeks after the initial inoculation.

On tomato, whiteflies cause various symptoms, see pp. 213 and 370, where there is also other information on the biology of these insects.

Once viruliferous, whiteflies will remain so for several days (retention period) or a lifetime (35–40 days) (varying with different authors). Note that virus particles are retained after metamorphosis, and transmitted to offspring. It has recently been demonstrated that TYLCV is transmitted transovarially to its vector offspring over at least two generations. This certainly could have a significant epidemiological impact.

As the larvae are almost motionless, it is mainly the adults who spread of TYLCV. In addition, females appear to be more efficient vectors than males.

Finally, the TYLCV is not transmitted by contact between plants, nor in experiments by mechanical inoculation. Seed transmission has not been reported. Note that TYLCV-infected tomato fruits, imported from countries that have the disease, could potentially allow local Bemisia tabaci to acquire virus particles and transmit them to healthy plants.

Control methods[1]

• During cultivation

There is no curative method to effectively control the TYLCV during cultivation. If the disease is suspected, diagnsis should be checked by a specialized laboratory that performs appropriate tests (ELISA, PCR).

If the number of diseased plants in the culture is limited, they should be removed immediately. It is also necessary to control whitefly populations and thus to use insecticides (acetamiprid, buprofezin, deltamethrin, methomyl, a preparation based on *Paecilomyces fumosoroseus*, pymetrozine, pyriproxyfen, a preparation based on *Verticillium lecanii*). Even if they are not always very effective, they may limit the development of the disease.

Protected crops should be grown in insect-proof structures, by blocking the openings with insect-proof nets, nonwoven material of P17 type for example.

At the end of cultivation, diseased plants must be removed quickly which will prevent the vector whiteflies multiplying and being a threat to future crops.

It is also wise to have an interval of a few weeks between crops in the glasshouse. If this cannot be managed, insecticides should be applied before the crop is removed in order to reduce the population of the vector.

• Subsequent crop

Note that the TYLCV, like all Begomovirus transmitted by *Bemisia tabaci*, is subject to European regulations (Directive 92/103/EEC of 1 December 1992, decree of September 2, 1993). These stipulate that all tomato seedlings sold by producers in areas known to be affected by TYLCV must be accompanied by a plant passport.

If seedlings are purchased, their quality should be checked. Destroy any suspect seedling. The use of tomato seedlings infected with TYLCV is a significant threat, especially in areas of production where the virus is not yet present. Control should include both the appearance of the seedlings and the presence of whiteflies. As *Bemisia tabaci* parasitizes many vegetable and ornamental species, vigilance should be exercised when introducing these. In countries where infection is very early, nurseries and seedlings should use screens, including during transport and storage before planting, using agrotextiles (nonwoven and mesh fabrics). The mechanical barrier thus created at least delays infection. The glasshouse openings and entrances must be insect-proof. Double doors systems with overpressure is another precaution that helps to exclude the vector. Use preventive insecticide treatments as well as biological control (see p. 650). It is also an advantage if propagation areas are well away from production areas.

Yellow sticky cards placed in glasshouses and checked each week will enable vector populations to be monitored and give an early warning of a danger period, particularly if there are known sources of the virus nearby. Careful weeding of nursery plots and their surroundings (border hedges and paths) will eliminate plant sources of viruses and/or vectors. The establishment of a nursery or a tomato crop near TYLCV-affected tomatoes must be avaoided. Wherever possible, the planting of tomato crops is best done when whitefly populations are low. The first diseased plants should be removed immediately.

[1] See note to the reader p. 417.

Resistance to TYLCV, probably polygenic, has been identified in several accessions of wild relatives of tomato. Breeders have been working on them for the last 10 years. Tolerances have been found and introduced into tomato from several *Lycopersicon* species, including *L. pimpinellifolium*, *L. hirsutum*, *L. peruvianum*, and *L. chilense*.

Other resistances to *Bemisia tabaci* have been investigated in several species of *Lycopersicon*. Various accessions are of interest, particularly because of the density of glandular trichomes and various exudates which interfere with whiteflies before they have transmitted the virus, for example by interfering with oviposition.

Moreover, the acquisition and transmission of TYLCV by *Bemisia tabaci* would then be reduced.

F1 hybrids tolerant to TYLCV have recently introduced. The origin of their partial resistance is not specified, and no assurance can be given that it is effective against the different Begomovirus described. The choice is still restricted for protected crops, but should expand rapidly. For many tropical regions, breeders try to combine resistance to TYLCV with those for bacterial wilt. In more temperate zones, more resistant varieties are now available (see p. 656).

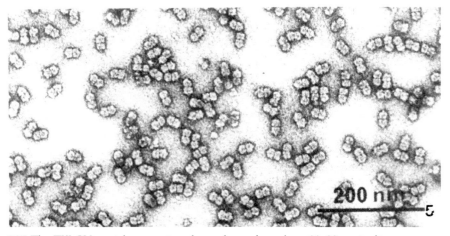

887 The TYLCV particles are twinned, angular, and are about 20–30 nm in diameter.

▣ Other viruses transmitted by whiteflies

Viruses	Symptoms and mode of transmission	Shape of particles	Principal characteristics
Cowpea mild mottle virus (CPMMV)	Relatively mild symptoms: mottling in Nigeria, slightly chlorotic leaflets in Israel. *Bemisia tabaci*, in a semi-persistent manner. Unlike other Carlavirus, it is not transmitted by aphids.	Filamentous rods, 650 nm long characteristics of Carlaviruses.	The first description was in 1973, on *Vigna unguiculata* in Ghana. It was associated with tomato fuzzy vein in Nigeria and tomato pale chlorosis in Israel, in both cases in the early 1980s. CPMMV mainly affects the Fabaceae (*Vigna unguiculata*, *Arachis hypogea*, *Glycine max*). It is also found on eggplant and *Solanum incanum*. Control methods are as recommended for TYLCV, Description 41.

The emergence in many countries of the world of the yellow leaf curl syndrome has resulted in the characterization of several dozen species of Begomovirus. It is highly likely that other species will be added because of the constantly changing situation of the Begomoviruses on tomato. As these viruses have the same biological properties and differ only in their host range (their biological properties being often identical to those of TYLCV [Description 41]) it is not necessary to describe all the viral species. Some viral species are given briefly later in the table. Note that, as in the case of TYLCV, many viral species close to ToLCV, such as TLCV, ToMoV, have been characterized and some of them included in detail.

Viruses	Symptoms and mode of transmission	Shape of particles	Principal characteristics
Eggplant yellow mosaic virus (EYMV) **(Syn. Tobacco leaf curl virus)** **Begomovirus**	Symptoms identical to those caused by TYLCV and TLCV. Irregular chlorotic spots on leaflets, they curve and curl gradually. Plants are stunted. Transmitted by *Bemisia tabaci*.	Paired particless measuring 30 nm long and 18 nm diameter. 	This virus has been reported very occasionally in Asia, in Thailand in 1985, from samples of eggplant, tomato, and tobacco. These species are its natural hosts. Control methods as recommended for TYLCV, Description 41.
Serrano golden mosaic virus (SGMV) **(Syn. Pepper golden mosaic virus)** **Begomovirus**	Inter-veinous chlorosis of young leaflets, the apex of the plants may become necrotic. Deformation of fruits. *Bemisia tabaci*, in the semi-persistent and persistent manner. Not transmitted by seeds.	Paired particles measuring 30 nm long and 18 nm diameter. 	This virus was reported for the first time in Mexico in 1989, on *Capsicum annuum*. It was called 'Texas pepper virus' in 1990 in the US. It is currently confined to several Central American countries. SGMV seems to affect only the pepper and tomato. Control methods are as recommended for TYLCV, Description 41.

Viruses	Symptoms and mode of transmission	Shape of particles	Principal characteristics
Tobacco leaf curl virus (TLCV) **Begomovirus** Note also the existence of the following species: Tobacco leaf curl Japan virus, Tobacco leaf curl Kochi virus, Tobacco leaf curl Yunnan virus, Tobacco leaf curl Zimbabwe virus.	Symptoms quite similar to those produced by the TYLCV, rolling of the leaves with chlorosis. Leaves become wrinkled and deformed. In old plants, they are often brittle. Those infected early are stunted. *Bemisia tabaci*, in persistent or circulative manner. In contrast to TYLCV, three other species of whiteflies can transmit TLCV: *Aleurotrachelus socialis*, *Trialeurodes natalensis*, and *Bemisia tuberculata*. TLCV is not transmitted by its vectors to their offspring. It is not transmitted either by seeds or by contact, nor experimentally by mechanical inoculation.	Paired particles measuring 30 nm long and 18 nm diameter. 	TLCV was described for the first time on tobacco in Tanzania in 1931. It is now present on all continents and in most areas where this plant is grown. It is more common in the tropics and sub-tropics. Several TLCV isolates have been described, mainly on tobacco, most of which are mixtures of strains or variants. This virus is able to survive and multiply on various plants (virus reservoirs), especially on other members of the Solanaceae such as tobacco, pepper, and *Datura stramonium*. It also infects various species of the Asteraceae (*Ageratum conyzoides*, *Zinnia elegans*) and the Caprifoliaceae (*Lonicera japonica*, a naturally occuring species in Japan, especially its '*aureoreticulata*' variety which is used as an ornamental in Europe). Control methods are as recommended for TYLCV, Description 41. Note that *Ricinus communis* and *Helianthus annuus*, when planted around the nurseries, might be attractive barriers to adult whiteflies and thus reduce their number on tobacco seedlings.
Tomato leaf curl virus (ToLCV) **Begomovirus** Note also the existence of at least 15 closely related viral species: Tomato leaf curl Bangalore virus (ToLCBV), Tomato leaf curl Bangladesh virus (ToLCBDV), Tomato leaf curl China virus (ToLCCNV), Tomato leaf curl Sudan virus (ToLCSDV), Tomato leaf curl Indonesia virus (ToLCIDV).	The symptoms are similar to those caused by TYLCV: the leaflets are small, chlorotic, and rolled. *Bemisia tabaci*, in a persistent manner.	Paired particles measuring 30 nm long and 18 nm in diameter. 	Found on tomato in Australia in 1971, ToLCV-Au was called 'Tomato Australian leaf curl virus'. Several very similar viral species have been described in different countries. This virus is not known to affect any plant other than tomato. Control methods are as recommended for TYLCV, Description 41. *Lycopersicon hirsutum* ('PI 390 658' and 'PI 390 659') and *L. peruvianum* ('PI 127 830' and 'PI 127 831'), are resistant to the vector whitefly. Resistance to the virus was found in *L. pimpinellifolium* ('A 1 921'), but mainly in *L. hirsutum* subsp. *glabratum* ('B 6 013'). This last resistance has been introduced into tomato, resulting in the line H-24, used in India.

Viruses	Symptoms and mode of transmission	Shape of particles	Principal characteristics
Indian tomato leaf curl virus (IToLCV) Begomovirus	The symptoms are similar to those caused by TLCV and TYLCV: small, curved, rolled, and chlorotic leaflets, stunted plants when infected early. *Bemisia tabaci* (notably B biotype), in the persistent manner.	Paired particles measuring 30 nm long and 18 nm in diameter.	Native to India where it was first described in 1948, this virus appears to be limited to that country. Its host range is rather limited: *Acanthospermum hispidum*, *Ageratum conyzoides*, *Euphorbia geniculata*, and *Parthemium histerophorus*. Control methods are as recommended for TYLCV, Description 41. Note that a partial resistance was found in *Lycopersicon hirsutum* f. *glabrum* 'B 6 013', and bred into tomato line H-24.
Sinaloa tomato leaf curl virus (STLCV) Begomovirus	Chlorosis and yellowing of young leaflets and leaves, which roll and are somewhat red. Affected plants have very short internodes and reduced growth. *Bemisia tabaci*, in the persistent manner.	Paired particles measuring 30 nm long and 18 nm in diameter.	Discovered for the first time in Brazil on pepper and tomato in 1989, this Begomovirus is now present in Mexico, the US, Costa Rica, and probably in Nicaragua. Its natural host range seems limited to these two plants. Control methods are as recommended for TYLCV, Description 41.
Tomato mottle virus (ToMoV) Begomovirus	Chlorotic mottling is present on young leaves, the older ones are more yellow and rolled. Early infected plants are stunted. *Bemisia tabaci* B biotype, in the persistent manner.	Paired particles measuring 30 nm long and 18 nm in diameter.	Reported on tomato in Florida in 1989, this virus has spread across North and Central America (Mexico, Nicaragua, Caribbean, Puerto Rico, Cuba). It only infects tomato but *Solanum viarum* is a source in Florida. Control methods are as recommended for TYLCV, Description 41. It should be added that some plastic mulch (coloured or aluminized) and imidacloprid-based insecticide treatments reduce the incidence of this disease.
Taino tomato mottle virus (TToMoV) Begomovirus	Symptoms identical to those of TYLCV. Probably *Bemisia tabaci*.	Paired particles measuring 30 nm long and 18 nm in diameter.	The TToMoV was described in Cuba, for the first time in 1995. Control methods are as recommended for TYLCV, Description 41.

Viruses	Symptoms and mode of transmission	Shape of particles	Principal characteristics
Tomato golden mosaic virus (TGMV) **Begomovirus**	Important leaf deformation, bright yellow leaf mosaic. The dwarf plants have numerous axillary branches giving the plants a bushy appearance. Transmitted by *Bemisia tabacci*. The increase in damage due to this virus coincided with the introduction of the B biotype of the whitefly.	Paired particles measuring 30 nm long and 18 nm in diameter. 	TGMV was first reported in 1975 in Brazil. Its distribution at present is limited to Brazil and possibly also Venezuela. It has only been found affecting tomatoes, although experimentally it can be transferred to tobacco and to various other *Nicotiana* spp., as well as to petunias, *Physalis floridana*, and *Datura stramonium*. Control methods are as recommended for TYLCV, Description 41.
Tomato leaf crumple virus (TLCrV) **Begomovirus**	Marked leaf roll and deformation of the leaves which appear crumpled and with a chlorotic mottling. Transmitted by *Bemisia tabaci* but not by seeds or by contact.	Paired particles measuring 30 nm long and 18 nm in diameter. 	The virus was discovered in 1990 on tomato in Mexico. Note that the *Chino del tomato virus* (CDTV), described in the same country, is in fact the same virus. The TLCrV occurs in glasshouse and field crops. It was reported in the US and more recently in Nicaragua. Its natural host range is rather limited: tomato, pepper, *Datura stramonium*, and *Malva parviflora*. Control methods are as recommended for TYLC, Description 41. Note that *Lycopersicon pimpinellifolium* 'TO 121' and 'LA 1 478' are both TWLV and TLCrV tolerant.
Tomato yellow dwarf virus (ToYDV) **Begomovirus**	Yellowing of the leaves, which roll gradually. Early infected plants are stunted. *Bemisia tabaci*, in the persistent manner.	Paired particles measuring 30 nm long and 18 nm in diameter. 	ToYDV was described on tomato in Japan in 1974. It reached Taiwan in the early 1980s. In addition to tomatoes, it has been transmitted to *Datura stramonium* and *Nicotiana glutinosa*. Control methods are as recommended for TYLC V, Description 41.

Viruses	Symptoms and mode of transmission	Shape of particles	Principal characteristics
Tomato yellow mosaic virus (ToYMV) (Syn. Potato yellow mosaic virus [PYMV]) Begomovirus	As with many other Begomovirus, the leaflets become small, chlorotic, and curved with red edges. Early infected plants are stunted, appearing bush-like. Fruiting is greatly reduced and the fruits are much smaller. (See Photos **39, 40, 172–175**.) B biotype of *Bemisia tabaci* (*B. argentifoli*), in a persistent manner.	Paired particles measuring 30 nm long and 18 nm in diameter. 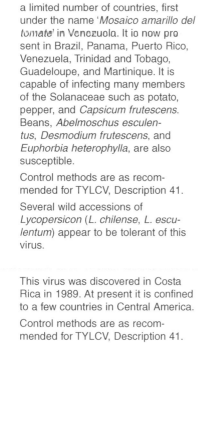	This virus has been described in a limited number of countries, first under the name '*Mosaico amarillo del tomate*' in Venezuela. It is now present in Brazil, Panama, Puerto Rico, Venezuela, Trinidad and Tobago, Guadeloupe, and Martinique. It is capable of infecting many members of the Solanaceae such as potato, pepper, and *Capsicum frutescens*. Beans, *Abelmoschus esculentus*, *Desmodium frutescens*, and *Euphorbia heterophylla*, are also susceptible. Control methods are as recommended for TYLCV, Description 41. Several wild accessions of *Lycopersicon* (*L. chilense*, *L. esculentum*) appear to be tolerant of this virus.
Tomato yellow mottle virus (ToYMoV) Begomovirus	Yellow mottle mosaic on leaflets, rolling and deformation of the leaves. Plants may be stunted. *Bemisia tabaci*, in a persistent manner.	Paired particles measuring 30 nm long and 18 nm in diameter.	This virus was discovered in Costa Rica in 1989. At present it is confined to a few countries in Central America. Control methods are as recommended for TYLCV, Description 41.
Tomato yellow vein streak virus (ToYVSV) Begomovirus	Yellow mosaic of leaves with wavy edges. Chlorotic streaks along the veins. B biotype *Bemisia tabaci* (*B. argentifolii*), in the persistent manner.	Paired particles measuring 30 nm long and 18 nm in diameter.	In Brazil, the ToYVSV was observed in 1996 on tomato. It has only been recorded on this host. Control methods as recommended for TYLCV, description 41.

 In 2000 two new viruses causing necrotic symptoms on tomato leaves were described in Spain and Mexico respectively, and called 'Torrado' and 'Marchitez'. The first was also detected in the Canary Islands, Poland, Hungary, Panama, and New Zealand. The second seems confined to Mexico. These two diseases are caused by two very closely related new viruses that do not belong to a known viral genre; the names proposed for these viruses are Tomato torrado virus (ToTV) and Tomato marchitez virus (ToMarV).

Note that ToTV, with icosahedral virus particles measuring 28 nm in diameter, is transmitted by whiteflies *Bemisia tabaci* and *Trialeurodes vaporariorum* but with transmission of ToMarV by *B.tabaci* only.

It is of concern that if these viruses are transmitted by both species of whitefly they could cause epidemics similar to those caused by TYLCV.

Viruses transmitted by thrips

Tomato spotted wilt virus (TSWV)

(*Tospovirus*, Bunyaviridae)

Principal characteristics

TSWV was observed for the first time in 1915 on *Lycopersicon esculentum* in Australia. It is the type species of the Tospovirus genus. For many years, two different viruses, initially separated into two serologically distinct groups, were included under the name 'TSWV'
– L-TSWV (for 'lettuce') that appeared the most common;
– I-TSWV (for '*Impatiens*'); isolates of this group are now considered to belong to a different virus (described in 1990), called 'Impatiens necrotic spot virus' (INSV, see Description 44).

In addition, several variants have been reported in the literature, differing in the severity of their symptoms, the efficiency of transmission by their vectors (several species of thrips), and their ability to overcome some resistance genes. Note that some variants have been obtained by nitrous acid treatment to obtain attenuated forms which have been used to protect plants against aggressive strains.

The purified virus particles of TSWV are almost spherical and rough, with a diameter of 70–120 nm. The particles are surrounded by a membrane unit which has on its surface protuberances and projections of glycoprotein, 5 nm thick. The morphology of Tospoviruses properties is unique among plant viruses (Photo 888).

• Frequency and extent of damage

TSWV has a very wide host range. Its distribution is almost totally worldwide and it causes serious disease on tomatoes in many countries on several continents. It occurs mainly in temperate and sub-tropical regions, TSWV became more common in almost every continent in the 1980s as did its main vector, the Californian thrips.

It causes severe damage to crops in several US states, in various countries in eastern Europe (Bulgaria, Hungary, Poland), and Greece. It seems to be increasing in incidence in some countries in Asia and Oceania.

TSWV affected various crops in many northern European countries before the Second World War, but has more or less disappeared. During the 1980s, the widespread introduction of a new vector, *Frankliniella occidentalis* thrips, has greatly changed the situation. It is now considered a concern for many horticultural crops. In some areas, the close proximity of vegetable or flower crops with tomatoes gives rise to serious epidemics, sometimes causing significant damage. Crops grown in the field or under protection in the summer may be affected.

The age of plants at the time of infection has a major effect on yield: earlier attacks have the greatest impact. However, late attacks can also be serious as they are able to reduce the production of marketable fruits greatly, due to abnormal ripening.

• Main symptoms

Symptoms of TSWV on tomato are very varied.

The leaves may show a mosaic of variable intensity. There can also be small brown chlorotic to necrotic spots on the leaves, as well as brown rings. Red to brown patches tend to concentrate and coalesce at the base of the leaflets. In some cases, the leaves are reddened by the production of anthocyanin. Elongated necrotic lesions (streaks) are occasionally seen on the petioles and stems.

Subsequently, affected areas become chlorotic and also a tan colour. Necrotic spots coalesce and cause desiccation of entire leaflets, and all of these symptoms give rise to the common name 'tomato spotted wilt'. After several weeks, the diseased plants can be completely necrotic and wilting sometimes occurs.

Plant growth is often severely affected. If it is not totally stopped it can be unilateral, in which case there may be apical curvature (the 'vira cabeça' of the Brazilians). Note that plants can also be chlorotic.

When infection occurs before the development of the first truss, no fruit are formed. If the fruits are already being formed they may become deformed, small, and may have bronzed patches, with dry cracked necrotic areas, or with large arcs and sometimes concentric rings. At a later stage of development the fruits may ripen but be poorly coloured, with internal browning. It is obvious that these various changes make the fruits unmarketable.

(See Photos **35, 209, 212–218, 243, 275– 278, 412–415, 694, 728, 803.**)

• Biology, epidemiology

Survival, inoculum sources: TSWV, like CMV, has a very large natural host range. More than a thousand have been identified, belonging to at least 86 botanical families, mostly dicotyledons. It survives easily near to tomato crops, notably on various weeds such *Amaranthus* spp., *Anagallis arvensis*, *Capsella bursa-pastoris*, *Chenopodium amaranticolor*, *Convolvulus arvensis*, *Fumaria officinalis*, *Oxalis corniculata*, *Picris echioides*, *Poa annua*, *Solanum nigrum*, *Sonchus* spp., *Stellaria media*, *Taraxacum officinale*, and *Veronica* spp.

Families with the most host species affected are Asteraceae (247) and Solanaceae (172). Of particular note among these are *Cyphomandra betacea*, *Lycopersicon hirsutum*, *L. pimpinellifolium*, 15 species of *Solanum* including *S. aethiopicum*, *S. muricatum*, *S. quitoense*, *Nicotiana tabacum*, *N. acuminata*, *N. alata*, *N. glutinosa*, *Physalis heterophylla*, *P. minima*, *P. peruviana*, *P. ixocarpa*, *Browalia* sp., *Datura stramonium*, *Dubosia leichardtii*, *Hyoscyamus niger*, *Lycium procissimum*, *Salpiglossis* sp., *Schizanthus* sp., and *Streptosolen jamesonii*.

Many cultivated plants are hosts: 166 species have been recorded in Europe since the emergence of TSWV in 1987, belonging to 34 families, including 7 monocots:
– vegetable species, both herbs and processing crops (eggplant, peppers, potatoes, tobacco, lettuce, endive, beans, beans, peas, melon, cucumber, squash, spinach, cabbage, artichoke, chard, beet, celery, parsley, lavender, cilantro, tarragon, basil, sage);
– ornamental species (daisy, anemone, arum, begonia, marigold, chrysanthemum, dahlia, zinnia, cyclamen, gladiolus, gerbera, lily, petunia, buttercup, jasmine, impatiens).

It is also occurs in tropical food crops: *Vigna* sp., peanut, chayote, and pineapple.

The multitude of hosts makes control of this virus disease difficult.

Transmission, dissemination: TSWV can be transmitted by several species of thrips in a persistent manner (circulative-propagative). At least 10 species of thrips have been identified as vectors including *Frankliniella fusca* (Hinds), *Frankliniella occidentalis* (Pergande), *Frankliniella schultzei* (Trybom), *Frankliniella intosa* (Trybom), *Frankliniella tenuicornis* (Uzel), *Thrips tabaci* (Lind.), and *Thrips palmi* (Karny). *Scirtothrips dorsalis* (Hood) was identified as a vector in India, and *Thrips setosus* (Moulton) in Japan. In Florida, *Frankliniella bispinosa* has recently been identied as a vector. It occurs in tobacco crops, and is as efficient as *F. occidentalis* in transmitting TSWV. Note that the transmission efficiency of TSWV by different species of vector thrips is not identical.

In northern Europe *F. occidentalis* is the main, and sometimes only, vector. It is very effective and much more so than *Thrips tabaci*. A native of western US and Canada, *F. occidentalis* has spread over the whole of North American and Europe since 1985. This thrips is currently considered the main vector in five continents, especially in protected crops in temperate countries (US, Europe). It is very prolific and particularly difficult to control.

Only the larvae may acquire the virus, in a 15 minutes minimum feeding period. Plants are inoculated mainly by adults. In *F. occidentalis* as in *T. tabaci*, females accumulate more viruses but it is the males that transmit with the greatest efficiency. Larvae penetrate epidermal cells, inject saliva causing lysis of the cell contents, which they then absorb. Five to 15 minutes are enough to complete acquisition, and the latency period lasts at least 4 days. The virus is retained during metamorphosis of the insect, and may multiply in it. Thrips are small and move over short distances. They are sometimes caught in rising air currents which can result in their transport for several hundred metres or more. Adults, whose lifetime varies from 30 to 45 days, are viruliferous until they die. Thrips do not transmit TSWV to their offspring. Symptoms can appear from 7 to 14 days after inoculation.

In addition to being vectors of viruses, thrips are pests and as such cause various symptoms on tomato. These are described on p. 171 where there is also information on their biology.

TSWV is transmitted by mechanical inoculation, but not by casual contact. It is also transmitted by seeds and by vegetative propagation. Seed transmission is somewhat controversial: old results report it can be transmitted at a rate of about 1%, the virus being present in the testa but not in the embryo; recent work contradicts this. Seedlings may be infected during propagation.

The risks of infection of a plant by TSWV depend in particular on the attraction of the plant to the thrips vectors. These show preferential feeding according to the plant, the climate, and the season.

Control methods[1]

• During cultivation

There is no curative method to eliminate TSWV, an infected plant will remain so all its life.

Regulations have been used in order to limit the spread of TSWV. These generally have two objectives: to limit the entry of TSWV from imported plant material (placed on the list of quarantine pests), and to prevent the spread of the virus when present (classified as a pest control requirement). In 1992 a European directive confirmed TSWV as a quarantine disease together with the attendant pest control requirements For implementation an effective method of identification of TSWV was required and the ELISA method was selected.

If attacks occur in the nursery and detected early, the few plants showing symptoms of TSWV are rapidly destroyed. Blue traps are used to monitor populations of thrips. By carefully monitoring populations the most effective use of insecticides is possible.

Insecticide treatments can reduce the thrips populations on tomato, and thus reduce the incidence of TSWV. A number of products, especially the organohalogen insecticides or carbamates such as abamectin, deltamethrin, methomyl, formetanate, and acrinathrin, are used as applications to the soil and the aerial parts.

[1] See note to the reader p. 417

The reported strategy is to treat the plants three times at 4–5 day intervals, alternating insecticides with differing modes of action[1] and making certain that the sprays penetrate to the foliage and give as good a cover as possible. Applications are best made in early morning or late afternoon, when thrips' activity is greatest. Soil treatment can control nymph and pronymph stages which often burrow into the ground.

It is important to remember that insecticide treatments are not effective in controlling outbreaks of the virus, especially in the field. This is often because the vector thrips may come from outside the field and transmit the virus during feeding, before the insecticide has time to act.

At the end of cultivation, remove all diseased plants. It is advisable to leave the crop area unplanted for 3–4 weeks because of the larval stages that are still in the soil, and there may be live adults on plant debris which will disperse and spread the disease. Glasshouses should be disinfected before the next crop is planted.

• Subsequent crop

Be particularly careful to avoid planting infected seedlngs. Low levels of infection at this stage significantly affect the development of epidemics, particularly with field crops. This advice is particularly relevant in countries where early infection is known to occur during propagation.

In order to minimize infection at this stage:
– destroy weeds and nymphs found on the ground. To achieve this disinfection may be worthwhile;
– produce seedlings in an insect-proof glasshouse or protect them by the use of agrotextiles (unwoven, knitted fabrics). The mechanical barrier will delay infection;
– weed the area around the nursery to eliminate sources of viruses and of vectors. This is especially important in the case of this virus. The same precautions should be taken when propagation is done in the field;
– ensure that the seedlings used are healthy.

Rotations of crops in field situations using those that are not hosts of the vector are also well worthwhile. Tomato crop should not be planted next to susceptible crops, or near to one that is a very good host of the vectors (especially ornamental species: anemone, chrysanthemum, buttercup). As with aphids, it has been shown that aluminized or plastic film applied to the soil reduces thrips' populations and so reduces the incidence of TSWV on tomato. In addition, corn barriers around the crop are sometimes effective in reducing the incidence of TSWV epidemics in Brazil.

Insecticide treatments as detailed above apply equally in this situation; in addition, various oils and products forming films on the surface of plants affect the feeding of *F. occidentalis* and its reproduction, and in this away can reduce the severity of TSWV.

Predators and parasites are used in the glasshouse to control thrips biologically, notably *Amblyseius cucumeris* and *A. barkeri* (Phytoseiid) mites, hemipteran bugs of the *Orius* genus (Anthocorid), and other bugs, notably in the Mirid family such as *Dicyphus tamaninii*.

More information is given on p. 650.

Different species of *Lycopersicon* were evaluated for their resistance to TSWV in Hawaii, France, and the US. Among them, *L. peruvianum* has a very high level of resistance controlled by the dominant gene 'Sw-5'. This induces a hypersensitivity reaction to most isolates of TSWV. Hybrids resistant to TSWV are available and their numbers are increasing in various crops. However, some symptoms particularly those that apear to be systemic, have been observed occasionally in Spain on resistant plants. Very competitive strains of the virus isolated from these plants have proven capable of overcoming the 'Sw-5' gene. However, they do not seem able to attack the strains of *L. peruvianum* ('PI 128 660 R', 'PI 128 660 S'), on which they produce local lesions but without systemic infection. Such strains of TSWV have occurred in South Africa and Australia. It is likely that attacks sometimes observed in the field, particularly in Spain, are due to incomplete dominance of 'Sw-5'. Hybrids sold are probably all heterozygous for this gene.

The gene 'Sw-5' also confers resistance to Groundnut ringspot virus (GRSV) and Tomato chlorotic spot virus (TCSV).

---

[1] *Frankliniella occidentalis* strains resistant to several insecticides (dimethoate, acephate, oxamyl, fenpropathrin) are reported; for this reason it is advisable to alternate active ingredients with different modes of action. In addition, some aphicides have some adverse effects on thrips.

Two French lines have been developed with resistance and are now in use. These have been backcrossed to stabilize the desired horticultural characters of the varieties. They have genes for resistance to various diseases, one mainly in the glasshouse and the other in the field.

Finally, resistance to TSWV transmission by thrips is found in *Lycopersicon hirsutum* f. *glabratum* and in cultivars *L.esculentum* 'Manzana', 'Brazil', and 'Anahu'.

Several strategies for transgenesis, based on the genes encoding the nucleocapsid or the NSm protein (supposed movement protein), have proven effective against TSWV, but are not thought to be available in currently marketed tomato genotypes.

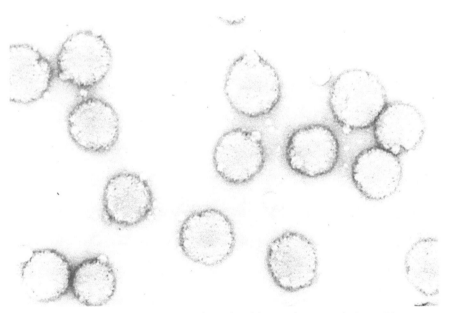

888 The virus particles of TSWV are spherical and have a diameter of 85 nm. The presence of a lipid envelope gives them a characteristic appearance.

Other viruses transmitted by thrips

Viruses	Symptoms and mode of transmission	Shape of particles	Principal characteristics
Groundnut ringspot virus (GRSV) **Tospovirus**	Mottling and mosaic on young leaflets. *Frankliniella occidentalis, F. schultzei.*	Identical in morphology to those of INSV.	GRSV originated in Africa where in 1996 it was found on *Arachis hypogea*. It is present in South Africa, Argentina, and Brazil, where it causes symptoms as severe as those of TSWV. In addition to *Arachis hypogea*, it infects peppers, tomatoes, and coriander. Control methods are as recommended for TSWV, Description 43. Note that the '*Sw-5*' gene in tomato confers resistance to GRSV by hypersensitivity.
Impatiens necrotic spotvirus (INSV) **Tospovirus**	Spots and brown, inter-veinal necrotic lesions. *Frankliniella occidentalis, F. fusca.*	Spherical or oval virus particles, with a diameter 70–120 nm.	INSV was reported for the first time in 1987 on *Impatiens* sp. growing in the US. It was thought to be a strain of TSWV but is now known to be different and distinct. It spread rapidly from California to several countries in the Americas, including Canada, Mexico, and Costa Rica. It reached the Netherlands in 1989 and was found in several European countries. It affects lettuce and chicory. This virus has a wide host range especially on ornamentals in glasshouses (*Ranunculus* sp., cyclamen, *Gloxinia speciosa*). In addition to tomatoes, it can also infect pepper, tobacco, eggplant, and other *Solanum* spp., as well as many plants belonging to other botanical families. Control methods are as recommended for TSWV, Description 43. Note that the TSWV resistance genes used in tomatoes are effective against INSV.
Parietaria mottle virus (PMoV) **Ilarvirus**	Spots, necrosis mostly at the base of the leaflets. Longitudinal necrotic lesions on stems, and death of terminal bud. Chlorotic rings on fruits browning later, becoming corky and causing their deformation. TI12 strain isolated in Italy causes stunting in plants. Probably transmitted by thrips.	Isometric particles, nonenveloped, with a variable diameter (24, 29, or 36 nm). 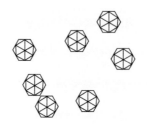	Symptoms were first observed in 1972 on tomato in Italy on *Parietaria officinalis* but the virus was not described until 1987. It spread thoroughout Italy and then to southeastern France and Greece in the late 1990s. Tomato strains are different strains from those that affect *Parietaria*. PMoV affect the two hosts mentioned above. Control methods are as recommended for TSWV, Description 43.

Viruses	Symptoms and mode of transmission	Shape of particles	Principal characteristics
Tobacco streak virus (TSV) Ilarvirus	Spots, chlorotic rings on leaflets becoming progressively necrotic, particularly near the veins. Necrotic streaks are visible on the stem, they extend to the branches. The flowers also become necrotic and may fall. The fruits are sometimes covered with necrotic rings. *Thrips tabaci, Frankliniella* spp. It is also transmitted by seeds in beans and several weeds such as *Datura stramonium, Chenopodium quinoa, Melilotus alba.* Mechanical transmission via contaminated pollen and thrips.	Isometric not enveloped, with a diameter of 26–35 nm. 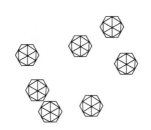	Globally widespread, it naturally or experimentally infects many hosts such as cotton, tomatoes, asparagus, beans, soybeans, grapes, strawberries, alfalfa, tobacco, ornamental plants (*Dahlia* spp., *Rosa setigera*, gladiolus). It can be found on several other plants (*Chenopodium quinoa, Datura stramonium*) that surround tomato crops encouraging epidemics. Control methods are as recommended for TSWV, Description 43.
Tomato chlorotic spot virus (TCSV) Tospovirus	Chlorotic spots on leaflets. Reduced yields. *Frankliniella occidentalis, F. fusca.*	Morphology of the virus particles is similar to that of INSV. 	This virus was discovered on tomatoes in Brazil in 1989. It is now present in Argentina. In addition to tomatoes, it has been found affecting peppers. Control methods are as recommended for TSWV, Description 43. Note that the TSWV resistance genes used on tomatoes are effective against TCSV.

Viruses	Symptoms and mode of transmission	Shape of particles	Principal characteristics
Beet curly top virus (BCTV) **Curtovirus**	The leaves have lighter coloured veins, are wrinkled and rolled up while the stalks curl down. They are occasionally rough and eventually turn yellow. Tissue of the leaves is thicker, and chlorotic. The veins can be chlorotic, purple, and are then prominent. Proliferation of axillary buds gives the plant a stunted and bushy appearance. Several species of leafhoppers in the persistent and circulative manner: *Circulifer tenellus* (US), *C. opacipennis* (Turkey, Iran), *Agallia albidulla* (Brazil), *A. ensigera* (Argentina), *Empoasca decipiens*. It is not transmitted by seeds in tomato.	Paired particles of about 18 × 36 nm, typical of the Geminiviridae family. 	Described on beet in 1909 in the US, BCTV is present throughout the American continent in arid to semi-arid regions, mostly in the field. It is found in Africa, Iran, and several Mediterranean countries. Several strains have been described, particularly distinguishable by their pathogenicity and their geographical origin. It has a wide host range affecting tomato, pepper, *Capsicum frutescens*, tobacco, potato, petunia, *Physalis wrightii*, beans, and Cucurbits. With this disease, weeding, double row planting, and the application of insecticide limits infection rates. Several species of *Lycopersicon* are resistant to BCTV. Control methods are as recommended for managing phytoplasma diseases, also transmitted by leaf hoppers, Description 28.
Eggplant mosaic virus (EMV) **Tymovirus**	Leaf deformation and veinal chlorosis on young leaflets; various green to yellow mosaics, sometimes white. Necrosis and desiccation of the leaf. Plants affected early are stunted. Poorly coloured fruits, mottled and stained. Several strains transmitted by a beetle, *Epithrix* sp. The other modes of transmission are still unclear.	Isometric particles about 30 nm in diameter.	Also known as '*Abelia latent virus*', EMV was reported for the first time on eggplant in Trinidad. It has also been isolated from eggplant in Tobago, Brazil, Bolivia, Colombia, Peru, Argentina, US, and perhaps in Turkey. It is also present in the field on tomato and *Solanum seaforthianum*.

Viruses	Symptoms and mode of transmission	Shape of particles	Principal characteristics
Eggplant mottled dwarf virus (EMDV) **Nucleorhabdovirus**	Vein yellowing and pale green young leaflets. These are also small, deformed and rolled. Plant growth is reduced or stopped. Affected plants become sterile and developing fruits are discolored, crinkled, and swollen (see Photos **33**, **36**, **108**, **137–142**, **161**, **162**, **711**). This virus is transmitted by leafhoppers from the *Agallia* genus.	Bacilliform and enveloped measuring 80–90 × 220–230 nm (Photo **889**). 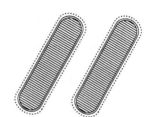	This virus was first described in 1969 on eggplant in Italy. In several Mediterranean countries it has sometimes been called 'Tomato vein yellowing virus' or 'Pittosporum vein yellowing virus'. It first spread in Turkey, Greece, North Africa, Portugal, France, and the Canary Islands. Is now also found in Jordan, Afghanistan, Iran, and Bulgaria. The first record on tomato was in the 1980s. It occurs naturally on both eggplant and peppers as well as tomatoes. EMDV is confined to the Solanaceae, attacking tobacco, eggplant, peppers, potatoes, *Solanum nigrum*, and *S. sodomaeum*. It also affects cucumber, *Hibiscus rosa-sinensis*, *Lonicera* sp., *Pittosporum tobira*, and *Capparis spinosa*. Control methods are as listed for the control of CMV, particularly in terms of elimination of virus-infected and source plants, Description 35.
Potato yellow dwarf virus (PYDV) **Nucleorhabdovirus**	Chlorotic spots gradually extending, veinal chlorosis on young leaflets. Leafhoppers (*Agallia constricta*, *A. quadripunctata*, *Aceratogallia sanguinolenta*) in the multiplicative circulative manner.	Bacilliform or bullet-shaped virus particles, enveloped, measuring 380 × 75 nm. 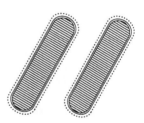	This virus was reported for the first time in 1922. It was found on potatoes in the US but is now known to be present in several Asian countries, Japan, Taiwan, and Saudi Arabia. Its host range is quite limited: potato, tomato, clover, chrysanthemum, *Nicotiana alata*, *Tagetes erecta*, *Zinnia elegans*, *Mirabilis jalapa*. It is a quarantine virus in Europe. See Control methods for phytoplasma diseases, also transmitted by leaf hoppers, Description 28.
Tobacco yellow dwarf virus (TYDV) **Mastrevirus**	Infect certain varieties of tomatoes in Australia without causing symptoms. Leafhopper (*Orosius argentatus*), in the persistent manner.	Paired particles, nonenveloped 35 nm long and 18 nm in diameter. 	TYDV was described in 1937 on tobacco in Australia. Its distribution appears to be limited to that country. In addition to tobacco, it is found on tomato and *Datura stramonium* var. '*Tatula*'. Note that the bean disease called 'summer dieback' is caused by a particular strain of TYDV . See Control methods for phytoplasma diseases, also transmitted by leaf hoppers, Description 28.

Viruses	Symptoms and mode of transmission	Shape of particles	Principal characteristics
Tomato pseudo curly top virus (TPCTV) **Topocuvirus**	Chlorosis of the edge of the leaves, accompanied by vein clearing. The leaflets are curved or even spoon shaped. There is also a proliferation of axillary buds. Transmitted by a Hemiptera insect (*Micrutalis festinus*), in the semi-persistent manner.	They are paired, measuring 18 × 30 nm.	This virus was discovered on tomato in the US in 1950. It is now fairly well established in Florida. Its natural host range is small: tomato, *Solanum nigrum*, *Ambrosia* sp. Control methods are as recommended for TYLCV, Description 41. TYLCV-resistant plant lines are also resistant to TPCTV.

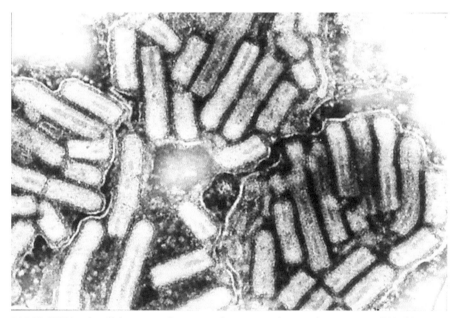

889 EMDV has bacilliform particles wrapped in an envelope and measuring of 80–90 × 220–230 nm.

Viruses transmitted by nematodes and fungi

Viruses	Symptoms and modes of transmission	Shape of particles	Principal characteristics
Tobacco rattle virus (TRV) **Tobravirus**	Dwarf plants with leaflets showing a marked mosaic. Latent infections are possible. Several species of nematodes belonging to the *Trichodorus* spp. and *Paratrichodorus* spp. genera. These nematodes remain infectious for several months or even years. Low rate of transmission by seed, in some weeds.	Rod-shaped viruses, rigid and tubular, of varying size (46–114 or 180–197 × 22 nm, see Photos **890, 891**). 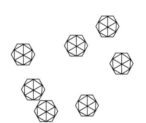	TRV was described for the first time in Germany, in 1931. It is now present in several European countries, the Americas, New Zealand, and Australia. It has a wide host range including several Solanaceae (pepper, tobacco, petunia, potato, *Solanum nigrum*, *Physalis angulata*), as well as many other plants (lettuce, artichoke, sunflower, *Senecio vulgaris*, *Calendula officinalis*). Only the control of the population of vector nematodes in the soil helps to limit outbreaks of this virus – Descriptions 49–52.
Tobacco ring spot virus (TRSV) **Nepovirus**	Chlorotic spots and rings appearing near the veins of leaflets, becoming progressively necrotic. Mottling can also be observed on the leaves. Nematodes, several species of the *Xiphinema* genus. This virus is reported to be transmitted by pollen in some species.	Isometric, nonenveloped, measuring 25–29 nm diameter (Photos **890–892**).	TRSV, reported on tobacco for the first time in 1927, is present in North America and South America, Europe, Asia, and Nigeria. There are unconfirmed reports of it in Australia and New Zealand. Several strains have been described, which differ in the symptoms they induce and their serological properties. Pepper, tomato, tobacco, *Physalis ixocarpa*, *Narcissus* sp., melon, *Glycine max*, *Rosa* sp., are its main natural hosts. *Lycopersicon esculentum* var.'*cerasiforme*' is tolerant, and cultivars 'Rutgers' and 'Cheyenne' are resistant to this virus. In Europe it is subject to quarantine measures. Only the control of the population of vector nematodes in the soil helps to limit outbreaks of this virus – Descriptions 49–52.
Tomato black ring virus (TBRV) **Nepovirus**	Spots of varying size, chlorotic and necrotic rings. The leaflets may be deformed and have chlorotic veins. Plant growth is sometimes reduced with early infection. Nematodes (*Longidorus elongatus* and *L. attenuatus*). It is transmitted by seeds in several species, as well as by pollen.	Isometric particles, nonenveloped with a diameter of about 26 nm, and an angular shape with 5 or 6 sides. 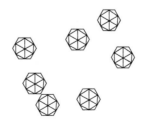	TBRV, which was observed on tomato in 1946 in UK, is present throughout Europe, Russia, Turkey, and in North and South America. It has also been reported in Japan. This virus has a wide host range, infecting pepper, tobacco, tomato, potato, eggplant, beet, lettuce, and beans. It is very stable and has been found in rivers and lakes in Turkey. Only the control of the population of vector nematodes in the soil helps to limit outbreaks of this virus – Descriptions 49–52.

Viruses	Symptoms and modes of transmission	Shape of particles	Principal characteristics
Tomato ringspot virus (ToRSV) **Nepovirus**	Small necrotic spots, marbling on leaflets and systemic necrosis. Nematodes (*Xiphinema americanum* and its subspecies). Seed-borne in other species.	Nonenveloped isometric particles, 25 nm in diameter, with angular contour.	Found on tobacco in the US in 1936, ToRSV has a nearly worldwide distribution. Its natural hosts are tomato, pepper, tobacco, and several species of the *Pelargonium*, *Rubus*, *Prunus*, *Trifolium*, *Taraxacum*, *Stellaria* genera. ToRSV is subject to quarantine in Europe. Only the control of the population of vector nematodes in the soil helps to limit outbreaks of this virus – Descriptions 49–52.
Tobacco necrosis virus (TNV) **Necrovirus**	Necrotic rings on leaflets. Transmitted by the *Olpidium brassicae* fungus.	Nonisometric particles, enveloped, with a diameter of 26 nm (Photos **890–893**).	Reported in UK since 1935, this virus is now widespread throughout the world. Several strains have been described, with differing pathogenicity. It affects many Solanaceae (pepper, tomato, potato, tobacco, petunia), but also plants from other botanical families: carrot, celery, lettuce, chrysanthemum, melon, cucumbers, beans. It is not harmful on tomato and it has only been very occasionally isolated on this plant, especially in glasshouses in Italy. See information on *Olpidium brassicae* – Description 18.

890-1 The virus particles of TRV are rod-shaped, rigid and tubular, with a variable length (46–114 or 180–197 × 22 nm).

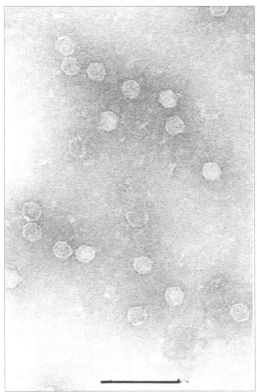

890-2 The TRSV particles are isometric, nonenveloped, and measure 25–29 nm in diameter.

890-3 TNV develops isometric nonenveloped virus particles, with a diameter of 26 nm.

Viruses with unknown mode of transmission

Viruses	Symptoms and modes of transmission	Shape of particles	Principal characteristics
Datura yellow vein virus (DYVV) **Nucleorhabdovirus**	Vein clearing, mottling, chlorosis of young leaflets. Leaf deformation. Unknown mode of transmission.	Enveloped particles, bullet-shaped, measuring 166 × 77 nm. 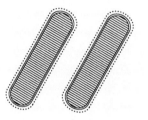	This virus was described for the first time in Australia, in 1982, on *Datura stramonium*. Its range seems limited to this country, where it has sporadically been observed on tomato.
Moroccan pepper virus (MPV) **Tombusvirus**	Mottling, chlorotic spots on leaflets which are also deformed. Dwarf plants. Unknown mode of transmission; transmission from a contaminated soil is possible.	Isometric particles, nonenveloped, 30 nm in diameter. 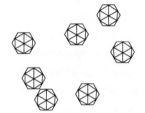	MPV has only been reported in Morocco in 1977, on pepper. It also affects tomato and *Pelargonium zonale*.
Pelargonium zonate spot virus (PZSV) **Uncategorized (close to Ilarvirus)**	Patterns, wavy lines, chlorotic rings on leaflets. Ultimately, these symptoms become necrotic. Chlorotic rings, concentric on the stem and fruit, eventually becoming necrotic. The growth of affected plant can be reduced, leading to stunting. See Photos **143**, **802**. Unclear mode of transmission. Some insects that eat pollen are suspected vectors, as well as water or nutrient solution. The virus might be transmitted by pollen, especially in tomato.	Paraspherical viral particles, nonenveloped, 25–35 nm in diameter (29 nm modal value). 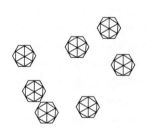	This virus was reported for the first time in Italy, in 1969 on *Pelargonium*. It also affects tomato, artichoke, *Chrysanthema coronarium*, *C. segetum*, and *Diplotaxis erucoides*. PZSV has also been reported in Spain in 1996 and more recently in France.

Viruses	Symptoms and modes of transmission	Shape of particles	Principal characteristics
Pepper ringspot virus (PepRSV) **Tobravirus**	Formation of chlorotic rings on the leaf, and yellow lines along the veins. Unknown mode of transmission.	Rigid rod like particles 22 nm in diameter, with two length categories: 197 and 52 nm.	This virus was first described in 1962 in Brazil. It is now sporadic in this country as well as in Central America. It attacks mainly pepper, but also tomato, artichoke, *Bidens pilosa*, *Gloxinia sylvatica*, and *Eustoma grandiflorum*.

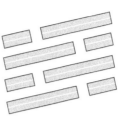

VIROIDS

Viroids are a group of subviral agents infecting plants. They are characterized by the very small size of their genome (270–470 nucleotides in length), which has only about one-tenth of the genetic information of the smallest known viruses. The genome is a single-stranded circular RNA molecule, not encapsidated and noncoding. Because of their noncoding, viroids rely entirely on proteins produced by the host for all processes involved in their replication cycle, the invasion of the host, and the development of symptoms of disease.

The origin of crop-infecting viroids is a mystery. One of the most attractive hypothesis is that viroids are asymptomatically present in wild plants and have been occasionally transferred to crops where they cause disease. Such a theory would, for example, explain the emergence at the beginning of the century of potato spindle tuber virus disease in potato crops. This disease is not present in cultivated or wild *Solanum* sp. in the Andean region where the potato originated. Although this hypothesis, like any hypothesis, remains questionable, it raises interesting questions about the possible role of viroid reservoirs in weeds and their possible role.

Viroids are found in all parts of infected plants, including roots, although they are generally more concentrated in the actively growing parts. Depending on the family, they can replicate in the nucleus of infected cells (specifically in the nucleolus) or in chloroplasts. The invasion of the host plant by viroids seems to follow the same path as infection by plant viruses: entry into the host by an initial injury, movement from cell to cell through plasmodesmata, and long-distance invasion in the phloem. This long-distance phloem transport could involve ribonucleoprotein complexes between the viroid's RNA molecule and phloem proteins from the host plant.

Transmission from plant to plant by a viroid can be in several ways. Viroids are frequently transmitted mechanically, either by contact between plants, or by contact with hands or tools during cultural operations. Furthermore, in a very limited number of cases, a suggested hypothesis involves transmission by aphids. For this to be possible, viroid molecules would need to be encapsidated into virus particles necessitating coinfection in the same plant, and such a situation has been suggested for Potato spindle tuber viroid (PSTVd) and Potato leafroll virus. Viroids are readily transmitted through plant material used in vegetative propagation of infected plants. In some cases, seed transmission has also been demonstrated (such as in the case of PSTVd in some hosts).

The symptoms caused by viroids do not fundamentally differ from those caused by viruses. Stunting of plants is a dominant feature of infected plants. There are a variety of other symptoms: epinasty, leaf discoloration, vein clearing or necrosis, leaf distortion, local or generalized necrotic lesions on the leaves, and in extreme cases infected plants may die. The nature and intensity of symptoms may vary depending on the host or on the variety, the strain of the viroid, and finally the environmental conditions. Unlike many viral diseases, the symptoms caused by viroids are usually accentuated by high temperature and light levels. In addition to these symptoms, ultrastructural changes of infected cells (cytopathic effect) may be seen, especially with changes in the walls and chloroplasts, and possibly the appearance of inclusion bodies in the cell membrane.

As these infectious entities only consist of a naked protein capsid-free nucleic acid, detection may only be performed using immunological or immunoenzymatic techniques conventionally used for the detection of viruses. Historically, the detection of these agents first involved a biological test based on the inoculation of sensitive plants, especially tomatoes. Although this method can be reliable, there is a difficulty of implementation which makes identification of the causative agent uncertain. It is therefore necessary to consider other methods. The identification of viroids can be achieved by electrophoresis of a partially purified fraction of the nucleic acids of a plant. This low sensitivity technique has been improved by the development of 'return' or 'dimensional' electrophoresis techniques based on the circular structure of the viroid molecule to separate it from the mass of cellular RNAs, thus increasing the sensitivity. This method allows unambiguous detection of viroid at around 50 ng/g tissue, but its implementation is cumbersome and does not allow the analysis of many samples at the same time. The study of the genome structure of viroids and the ability to clone and then to transcribe all or part of these genomic sequences has led to the development of more sensitive and reliable detection techniques, such as molecular hybridization and polymerase chain reaction (PCR; amplification of the DNA chain). The first is now routinely used in testing laboratories because it uses highly sensitive and reliable cold markers (nonradioactive) such as biotin or digoxigenin. Most viroids can currently be detected by reverse transctiption (RT) PCR, with the possibility now to dispense with electrophoresis for detection of amplicons (real-time PCR) or to perform multiplex tests for the simultaneous detection of multiple agents.

There are currently at least 28 species of viroids, and new viroids are regularly described. Less than a dozen have been associated with natural outbreaks on tomato: Tomato planta macho viroid (TPMVd), Tomato apical stunt viroid (TASVd), Potato spindle tuber viroid (PSTVd), Tomato chlorotic dwarf viroid (TCDVd), Citrus exocortis viroid (CEVd, syn. Indian Tomato buchy top viroid, I-TBTVd), and Columnea latent viroid (CLVd). They are all classified in the Pospiviroidae family and the Pospiviroid genus. Symptoms caused by these viroids are often comparable: dwarf plants, proliferating apex, chlorotic, deformed and narrower apical leaflets, with necrotic and/or brittle tissues.

They vary in their pathogenicity and appear to be limited to the genera *Solanum*, *Lycopersicon*, *Nicandra*, *Physalis*, *Capsicum*, *Nicotiana*, *Petunia*, and *Datura*. It should also be noted that other viroids (see box below) which have so far never been observed affecting tomatoes naturally, are however able to multiply in this species or in the Solanaceae family.

Viroids	Symptoms	Modes of transmission	Other characteristics
Citrus exocortis viroid (CEVd) **Syn. : Indian bunchy top viroid (I-TBTVd)**	In South Africa, plants show severe stunting and leaf epinasty. In India, in addition to the above symptoms, leaf distortion and vein necrosis have been observed.	The viroid has only rarely been found on tomato and its mode(s) of transmission have not been specifically studied. Like other viroids, it is probably transmitted mechanically from plant to plant during cultural operations.	The tomato bunchy top disease was first described in South Africa in the early 1930s, without the causative agent being characterized at the time. In the early 1980s, the disease was observed in the state of Maharashtra in India. Characterization of the viroid showed that in this case, it was a strain of CEVd. It has since been found several times on tomato in the Netherlands. The CEVd naturally infects citrus and vines and has recently been found in ornamental plants, like verbena, which could serve as a reservoir for transmission to the tomato.
Columnea latent viroid (CLVd)	Affected plants are stunted and have chlorotic and deformed (epinasty) leaves, with possible veinal necrosis.	There is little information on the mechanism of transmission of CLVd in tomato, but this viroid appears to be transmitted in the seeds. As with other Pospiviroids, mechanical transmission from plant to plant during cultivation is, however, very likely.	The CLVd was initially described as a viroid which was usually found in various ornamentals and was symptomless (*Brunfelsia undulata, Columnea erythrophae, Nematanthus wettsteinii*). It has more recently been observed several times in naturally infected tomatoes in the Netherlands. Asymptomatic ornamental hosts could potentially act as a reservoir.
Potato spindle tuber viroid (PSTVd)	The viroid is considered to be closely related or identical to Tomato bunchy top viroid. Plants are stunted and their leaflets have a pronounced epinasty, accompanied by veinal necrosis.	PSTVd is transmitted by vegetative propagation, by leaf contact between plants and on agricultural implements. Pollen and seeds are also likely to transmit it. Experimental data show that it can be transmitted by aphids, particularly *Myzus persicae*, probably with low efficiency, but only after heteroencapsidation in virus particles of potato leaf roll virus.	Martin described potato spindle tuber disease in 1922 in New Jersey, US. It then became widespread in potato producing regions in northern and northeastern US, then in Canada and in eastern Europe. It was probably introduced from time to time into western European countries but seems to have been eradicated. This virus has been detected in tomatoes in South Africa and New Zealand in 2001 (with losses of up to 60%), and in 2002 in the Netherlands. In general, PSTVd seems largely associated with the Solanaceae family, although it is capable of infecting species in other botanical families and has recently been found in very different hosts such as avocado. Several recent studies seem to highlight the frequent infection by PSTVd of an ornamental species, *Solanum jasminoides*, whose role as a potential reservoir remains unclear. PSTVd has occasionally been observed on tomatoes in glasshouse crops located in the southeast of the UK, as well as in the Netherlands and New Zealand. Note that the tomato is now used worldwide as the experimental host of PSTVd.

Viroids	Symptoms	Modes of transmission	Other characteristics
Tomato apical stunt viroid (TASVd)	Affected plants show pronounced stunting and shortening of the internodes. The leaflets may develop various symptoms: severe epinasty, marked leaf deformation, yellowing, veinal necrosis, brittle tissues. The fruits are pale red, small, and often discolored.	The viroid is mechanically transmitted during cultural practices. The distribution of diseased plants is often initially in lines before the symptoms become more widespread in the crop. It can be transmitted by grafting but not by *Myzus persicae* nor by *Bemisia tabaci*. Root transmission does not occur even when affected plants are growing closely together. Recently it has been shown that TASVd is able to invade the embryonic tissues of the seed and its transmission rate by the seeds can be very high. Bumble bees (*Bombus terrestris*), can transmit TASVd during pollination.	TASVd strains have been recorded on tomato, and at least five times over the years: in the Ivory Coast (1981), Indonesia, in a few plastic tunnels, near the coast in Israel (1999–2000), most recently in Senegal in the field and under protection (2005), and in protected crops in the region of Kebili, Tunisia (2005). The biology, epidemiology, and impact of this viroid is poorly understood. Its host range appears to be relatively limited, although there may be differences between strains. Note that a strain of TASVd was also found in asymptomatic plants of an ornamental Solanaceous plant, *Solanum pseudocapsicum*. This strain was experimentally transferred to tomato and induced typical symptoms.
Tomato chlorotic dwarf viroid (TCDVd)	The plants have a stunted appearance, small chlorotic leaves, with vein and petiole necrosis (Photos **24** and **25**).	Currently no information is available on whether this viroid is transmitted in tomato seeds.	The viroid has been observed in Canada on tomato plants grown in glasshouses, and more recently in the Netherlands in plants from the US. It thus seems to be mainly North American. Note that this viroid has been identified at least once in the UK.
			Experimental transmission to other hosts have demonstrated the sensitivity of a number of other members of the Solanaceae (potato, *Nicandra physaloides*) but it has so far only been found occurring naturally on tomato.
Tomato planta macho viroid (TPMVd)	The symptoms are similar to those induced by PSTVd, but are more severe: very marked stunting of plants, pronounced leaf epinasty, sometimes with veinal necrosis. Necrotic lesions can be seen on stem. Plants produce many flowers and fruits but the development ceases early, their size does not exceed that of a large marble, hence the 'male plant' name given to the disease. Crop losses can be very severe.	In experimental transmission, TPMVd has been shown to infect other Solanaceaeous plants, including eggplant and potatoes, but unlike PSTVd, affected plants are symptomless. In nature, this viroid appears to be transmitted mainly by contact between plants or on tools during cultivation. Presumably, at least one vector aphid could transmit this viroid, without the mechanisms involved being known. It is not transmitted by seeds in tomato.	The tomato planta macho disease has been described in Mexico in the states of Morelos and Mexico during the 1970s, and a virus was incorrectly thought to be the cause. However, in 1982, it was shown that the causative agent was in fact a viroid different from PSTVd, which was named 'Tomato planta macho viroid', or TPMVd.
			Epidemics observed in the crops are probably related to an accidental transfer of the viroid from a wild reservoir. The viroid spreads rapidly in the crops by mechanical transmission.

⚠ Several other viroids, which have never been found occurring naturally on the Solanaceae are capable of infecting other plants in this family (including tomatoes) and sometimes to induce severe symptoms when transmitted experimentally. They are mainly two Pospiviroid members, the Mexican papita viroid (MPVd) and chrysanthemum stunt viroid (CSVd), and a Hostuviroid, the hop stunt viroid (HSVd).

Again, the symptoms induced by these agents on tomato are typical viroid infections, with sometimes severe stunting and leaf epinasty, possibly associated with veinal necrosis. The accidental transfer of these viroids to the Solanaceae and their propagation within a crop, in particular through mechanical transmission during cultural practices, could therefore sometimes lead to the occasional potentially serious outbreak.

Control methods

Protection methods used to control viroid outbreaks affecting tomato vary according to the biological properties of the viroid and the situation. They can be divided into three categories:

– **general measures** applicable to all viroid diseases; these consist of removing infected plants and implementing preventive measures to eradicate the disease by preventing its spread in the crop and/or to new crops. Description 31, Tomato mosaic virus (ToMV), gives many of these control measures. It should be noted that for several viroids (PSTVd, CEVd, TASVd, CSVd), recent work has highlighted the existence of mainly symptomless infection in ornamental species, and these plants could possibly act as a reservoir for subsequent transmission to tomato crops. The presence of potential sources in neighbouring crops should be taken into account.

– **specific measures for viroids potentially transmissible by seed**, particularly the case of TASVd, and probably CLVd. Great care must be taken with seed quality especially in areas where these viroids have been reported. This is particularly important in the tropics. Depending on the disease incidence, it may be necessary to develop and implement a detection method for monitoring seed quality.

– **the measures used for vegetatively propagated viroids**, mainly PSTVd. For this viroid, the epidemic cycle can only be prevented by the use of uncontaminated tubers (which usually requires setting up a system of certification of potato crops) and it can only be achieved by severe eradication procedures. Tomato production is not directly involved except that tomatoes should not be grown near to potatoes fields, and if at all possible these two crops should not be grown on the same farm.

In general, the efforts of breeders to identify viroid sources of resistance have so far been unsuccessful, but the intensive screening (at least for the PSTVd) of genetic resources has failed to demonstrate such resistance.

However, thanks to recent advances in plant biotechnology, transgenic plants with lower susceptibility have been obtained, and even possible resistances to PSTVd.

The two approaches that have shown most promise are to develop plants expressing viroid sequences in antisense orientation, or plants expressing a double-stranded RNA specific yeast RNAse, PAC1 nuclease. These plants are still at an experimental stage, but initial data obtained demonstrate the feasibility of these new approaches in providing a means of control of viroids. The acceptability to the consumer of such transgenic plant resistance is still to be established.

NEMATODES

The phylum Nemata or Nematoda includes approximately 26 000 described species, feeding on other nematodes, humans, bacteria, fungi, and plants. It consists of animals that have digestive, nervous, excretory, and reproductive systems. However, nematodes are free of respiratory and circulatory systems. Their size can vary from a few millimetres to more than 8 metres.

Phytophagous nematodes are tiny, transparent cylindrical worms called 'nematodes'. They are usually invisible to the naked eye, but are easily seen with the aid of a microscope. Like many animals, they have a fairly complete digestive system starting with a mouth and ending with an anus. All plant pathogenic nematodes are equipped with a hollow mouth stylet to feed from cells and to absorb the contents as well to inject salivary secretions. In many species, there are male and female nematodes, and they differ primarily in their reproductive system. In others, such as *Meloidogyne incognita*, *M. arenaria*, or *M. javanica*, the males are absent. Nematode reproduction can be sexual, hermaphrodite, or parthenogenetic and leads to the formation of eggs.

These animals are commonly found in soil and water, feeding on the surface of roots and buried plant parts. Survival and multiplication can occur on susceptible alternative hosts. Eggs and some larval stages are able to persist for several years in the soil in a dormant state. The nematodes' development cycle is relatively simple: eggs, which hatch, are sometimes stimulated by root exudates. They give birth to larvae that grow and moult through four stages. The last larval stage generates a male or female adult nematode. A complete cycle takes 2–3 weeks if environmental conditions are favourable. Infective larval stages and adults are parasitic in differing ways depending on the species. More than 20 kinds of phytophagous nematodes have been reported on tomato as well as saprophagous nematodes. Most crop damaging nematodes are mentioned on p. 269. Only a few species cause substantial damage to the Solanaceae and these can be limiting factors in the production of tomatoes in the world.

Several ectoparasite nematodes may also be damaging to a lesser or greater extent. Often migratory, they just feed on the root surface cells, usually without penetrating into the tissues (short stylet genera: *Helicotylenchus*, *Paratrichodorus*, *Trichodorus*; long stylet genera: *Belonolaimus*, *Longidorus*). Their life cycle occurs mainly in the soil.

This is not the same for endoparasitic nematodes. These tend to be sedentary and sometimes penetrate tissues to a significant depth as part of their life cycle. Once in the tissue they may remain localized and cause galls (*Meloidogyne* spp.) or cysts (*Globodera* spp.), or move within the tissues, causing brown discoloration (*Aphelenchoides*, *Ditylenchus*, *Pratylenchus* genera). The injection of saliva into the cells during their nutrition is the cause of much of the damage. In addition to the changes they cause on tomato roots, some species of nematodes interact sometimes with other soil-borne pathogens by promoting their parasitism, and more rarely by reducing it. Damage on the roots depends on the density of nematodes in the soil, the vigour of the crop, cultural operations, and environmental conditions.

The development of nematodes in soil is influenced by moisture, aeration, and temperature. The presence of a film of water is essential for the larvae and adults to move with undulating movements into the soil or onto the plant parts attacked. Nematodes are in contaminated soil particles on tools, agricultural equipment, (*Pratylenchus* spp., *Meloidogyne* spp.), and can spread by drainage and sometimes as a result of splashing by irrigation water.

Many nematodes only very occasionally attack tomatoes, and often with a low incidence. Only the most common and most damaging species are described, or those which are representative of major types of phytophagous nematodes associated with tomatoes. These descriptions provide generic knowledge on the biology of the major nematodes encountered on tomato, and on the protection methods used to control them.

Globodera

Root cyst nematodes

(*Globodera*, Heteroderidae, Tylenchida, Nematoda)

Principal characteristics

Several species of *Globodera* have been reported on tomato, including *G. rostochiensis* Wollenweber, *G. pallida* Stone (potato cyst nematodes), and *G. tabacum* (Lownsbery & Lownsbery) Behrens (tobacco cyst nematodes). The latter species is in fact composed of three sub-species: *Globodera tabacum sensu stricto*, *G. virginiae* (Miller & Gray), and *G. solanacearum* (Miller & Gray) Behrens.

• Frequency and extent of damage

Cyst nematodes are less harmful on tomato than *Meloidogyne* spp. and they damage Solanaceous crops only in a limited number of countries. They occur in both field and protected crops.

Globodera rostochiensis, found on potato, tomato, and eggplant, is present in many countries around the world. It causes significant losses on tomato, particularly in Russia, Poland, Bulgaria, Norway, Mexico, and Sweden. Several pathotypes of this nematode have been reported on potatoes, as well as *G. pallida*.

G. tabacum is described mainly on Solanaceae (mainly tobacco and tomato) in several production areas, but in a smaller number of countries: many European countries, Asia (China, Korea), Africa (Madagascar, Morocco), and America (US, Argentina, Colombia).

• Main symptoms

These nematodes produce many cysts throughout the root system which are at first white and then gradually darken. The root system shows brown discoloration and reduced growth. Affected plants are small and wilt at the warmest times of the day. Parasitic synergies have been observed between *G. rostochiensis* and *Rhizoctonia solani*, *Verticillium dahliae*, *V. albo-atrum*, *Fusarium oxysporum* f. sp. lycopersici, *Colletotrichum coccodes*, and *F. oxysporum* f. sp. *radicis-lycopersici*. Paradoxically, *R. solani*, has a negative influence on the development of this nematode.

(See Photos **480, 498–502**.)

• Biology, epidemiology

Survival, inoculum sources: eggs and young larvae protected inside the cysts are able to survive several years in the soil without plants. Cysts are highly resistant survival bodies, those of *G. rostochiensis* can persist for more than 30 years in the ground and are not adversely affected by drought or the effects of some chemical nematicides.

Globodera affects a small number of hosts, mostly in the Solanaceae. For example, some 90 species of the *Solanum* genus are hosts of *G. rostochiensis*: Solanaceous crops (potato, tomato, eggplant) or weeds (*Solanum dulcamara*, *S. nigrum*, *Datura stramonium*). The host range of *G. pallida* is very similar to that of *G. rostochiensis*, but its rate of reproduction is reduced on tomato and eggplant. *Solanum nigrum*, *S. integrifolium* cv. 'Akanasur' (eggplant rootstock), and *S. ptycanthum*, among others, are hosts of *G. tabacum*.

Penetration and invasion: when weather conditions are favourable, the young larvae, attracted by root exudates, migrate towards the roots to penetrate to the central cylinder. The larvae then change either into males or females. The latter, after mating, produce eggs (up to 500) and grow, while adult males return to the soil. The females' body wall eventually hardens and give rise to cysts.

Transmission, dissemination: cysts eventually separate from the roots, and many young larvae escape from them and spread to other roots. To do this, they can be transported passively by runoff, drainage, or irrigation water, but spread is possible in the dust of contaminated soil, carried by wind to neighbouring fields. Infested plants, tools, farm implements, and vehicles are also means of distribution.

Conditions encouraging development: *Globodera* spp. prefer cool temperate climates.

Control methods[1]

Because these parasitic nematodes only affect the Solanaceae, many nonhost crops can be grown in rotations. Crops such as potatoes or eggplant should not be used; only the cultivation of nonhosts crops for 3–5 years will sufficiently reduce the soil level of the nematode population.

Soil solarization is also effective (see p. 487). There are no resistant varieties of tomato but a high-level resistance to *G. rostochiensis* is known in *Lypersicon chilense* and *L. peruvianum* var. *humifusum*.

 Several methods proposed to control *Meloidogyne* spp. can be used to control *Globodera* spp., see Description 50.

[1] See note to the reader p. 417.

Meloidogyne spp. Goeldi

Root-knot nematodes

(*Meloidogyne*, Heteroderidae, Tylenchida, Nematoda)

Principal characteristics

Root-knot nematodes were discovered during the 1850s. The genus *Meloidogyne* contains about hundred species which are sedentary and extremely polyphagous endoparasites that attack cultivated Solanaceae which are very sensitive, as are many other vegetables. Among the species reported on tomato, the most important are M. *incognita* (Kofoid & White) Chitwood (the most widespread species), M. *arenaria* (Neal) Chitwood, M. *javanica* (Treub) Chitwood, and M. *hapla* Chitwood (northernmost species). Other species have been reported more occasionally: M. *chitwoodi* Golden *et al.*, M. *floridensis* Handoo, M. *ethiopica* Whitehead, M. *acronea* Coetzee, M. *minor* Karsen, and M. *mayaguensis* Rammah & Hirschmann. The latter species is a more recent discovery and has been described in the Americas and Africa.

Natural or 'selected' isolates of M. *incognita*, M. *arenaria*, and M. *javanica*, able to overcome the resistance conferred by the 'Mi' gene, have been identified. Note that several races described in the literature are established on the basis of their behaviour on an international range of differential hosts. These races have no connection with any virulence genes or other resistance genes to root-knot nematodes mentioned later. Four races have been described in M. *incognita*.

• Frequency and extent of damage

Root-knot nematodes occur worldwide, affecting thousands of different plants. As mentioned earlier, several species attack tomatoes. The Solanaceae appears to be a universal host for many of these microscopic worms. *Meloidogyne* are by far the most common and most damaging on tomatoes. They cause serious crop losses in many countries, both in field and protected crops. In the latter, the losses are often more substantial because of the inability to rotate crops.

They are found in the soil on many farms, especially when the soil has been cropped with susceptible species such vegetables. Their damage is most severe where the management of crop rotation, and therefore the soil health, is not good. The loss of methyl bromide and the lack of a suitable replacement have lead to a rise of nematode problems in many vegetable crops.

• Main symptoms

White galls, gradually browning, characterize the presence of these nematodes on the roots. The nature and significance of galls depend on the species and the concentration of soil inoculum. Those produced by M. *hapla* are small and the nematode invades the apical meristems of the main roots and thus affects a smaller proportion of all the roots. Those caused by M. *arenaria* are the size of a pearl and affect almost all of the roots. The other two major species, M. *incognita* and M. *javanica*, cause large galls sometimes covering the entire root system. Longitudinal swellings are also found, and can be convoluted and extensive.

A cross section of galls shows the mature females, thus confirming the parasitism of these nematodes.

The root symptoms adversely affect the absorption of water and nutrients and as a result the plants are smaller. The leaves may be chlorotic, and wilting sometimes occurs during the hottest parts of the day. The underside of some leaves sometimes may be reddened as a result of anthocyanin production. The lower leaves of severely affected plants show early senescence. Fruit size and yields are reduced.

In many situations, root-knot nematodes are not the only organisms to attack the tomato root system and they often predispose them to attack by soil fungi such as *Rhizoctonia solani*, *Colletotrichum coccodes*, *Pyrenochaeta lycopersici*, and *Sclerotium rolfsii*. In such cases, root damge occurs more quickly and is often much more serious. *Meloidogyne* spp. also interact with tomato vascular pathogens such as *Ralstonia solanacearum*, *Fusarium oxysporum* f. sp. *lycopersici*, and *Verticillium dahliae*, increasing the incidence and severity of symptoms.

In contrast, the interaction *Meloidogyne–Fusarium oxysporum* f. sp. *lycopersici* sometimes leads to the opposite effect. Similarly, several *Fusarium* spp. such as *F. oxysporum* and *F. dimerum* reduce the number of root galls, which results in greater plant vigour.

(See Photos **478**, **503–507**.)

• Biology, epidemiology

Survival, inoculum sources: these nematodes survive in the soil for 2 years, in the form of egg masses protected by a white to brown mucilaginous matrix. This matrix is visible on the surface of the galls with a low power dissecting microscope. *Meloidogyne hapla*, unlike other species, survives in frozen soils.

Meloidogyne spp. have a very wide host range (more than 5500 plants) many of which are crop plants, and in this respect their multiplication and survival is very efficient. Crop plants that are hosts include chrysanthemum, carnation, rose, kiwi, peach, banana, peppers, eggplant, lettuce, melon, cucumber, artichoke, carrot, celery, beans, and sweet potato.

Penetration and invasion: the larvae of the second stage are attracted by root exudates and other compounds. They penetrate the roots and migrate through the cortex between cells, to the vascular system. During their feeding, they secrete enzymes that cause the development of giant cells which contribute to their nutrition. Larval development continues along with the root swelling. Eventually, a gall surrounds a large pear-shaped female. She produces many eggs (300–3000, between 400 and 500 on average) that are exuded outside the root, within a mucilaginous matrix. Several generations occur during a season, and the infestation may reach 100 000–200 000 larvae/kg of soil

Transmission, dissemination: many eggs and larvae can be transported passively from diseased plants by runoff, drainage, and irrigation water. The larvae move actively over short distances in wet soils. Dissemination can occur via dust particles from contaminated soil carried by wind to neighbouring fields. The contaminated plants, tools, farm implements, and machinery are also sources for potential spread.

Conditions encouraging development: in general, nematodes are active in moist, warm soils and their development is slower in cold soils. *Meloidogyne arenaria* and to a lesser extent M. *incognita* prefer relatively high temperatures (18–27°C) which occur in light and sandy soils. M. *javanica* tolerates higher temperatures, while M. *hapla* prefers cooler ones. In general, their activity is greatly reduced or even stopped below 5°C and above 38°C. The density of soil inoculum, and various other stress factors (compacted or dry soil, nutritional deficiency, attacks of various pests) also influence nematode attacks and the severity of symptoms.

Control methods[1]

• During cultivation

No control method is really effective during tomato cultivation.

If attacks occur in the nursery, the affected plants must be destroyed, otherwise their use will contribute to the dissemination of nematodes and the possible contamination of healthy soil.

During cultivation earthing up the plants is recommended, in order to compensate for the partial failure of gall-infected root systems. This will encourage the development of adventitious roots that will partially compensate for the root loss.

The plants should be drenched during the hottest periods of the day to prevent or reduce wilting.

In the field, it is imperative that the root systems of affected plants are removed and destroyed to avoid increasing the nematode population in the soil. If removal and destruction are not possible, the roots, when lifted, should be left in the open so that they are fully exposed to the sun. Similarly, several successive cultivations of the soil during the summer will help to expose the nematodes to heat and kill them.

Note that composting, which helps to get rid of any aerial pests present on fruit, leaves, and stems, is not as good for the treatment of roots. Composting does not completely eliminate nematodes from affected plant debris and galls.

• Subsequent crop

To be most effective for the control of root-knot nematodes, all the control methods described must be used. Nematological analysis of the soil should be done before the next crop to assess population levels, allowing the most appropriate action to be taken.

Nematodes are soil-borne pests thus the various measures proposed to control them are aimed at limiting or reducing population levels in the soil.

Crop rotations and some nonsusceptible cover crops are often advised to reduce the incidence and help to manage nematode population levels in the soil. Rotations are not always easy to implement because of the polyphagous nature of some nematodes such as *Meloidogyne* spp. or *Pratylenchus* spp. Indeed, it is not always easy to find nonhost plants which can be used in the rotations. To be effective, rotations should last at least 4 years. Several crops or cover plants are reported to be less conducive to the development of *Meloidogyne* spp., including soybean, onion, garlic, corn, winter cereals, peanuts, arugula, *Paspalum notatum*, *Cynodon dactylon*, *Eragrostis curvula*, *Chloris gayana*, *Digitaria decumbens*, *Panicum maximum*, *Crotalaria* spp., *Mucuna pruriens*, sesame, nematode-resistant genotypes of *Vigna unguiculata*. In Martinique, short rotations with the forage legume *Mucuna pruriens* have reduced population levels of *M. incognita* and *Rotylenchulus reniformis* in the soil. Fallow is sometimes recommended, but it can lead to erosion. It is necessary with fallow areas to work the soil surface several times during the year.

Nematodes are often controlled by flooding affecting fields for 7–9 months. The flooding can be continuous or interrupted by periods of drying. Under these conditions, the soil is depleted of oxygen and accumulates nematode toxic substances such as organic acids and methane. This method is only effective if done in a warm period of the year, and includes the risk of disseminating the nematodes at the same time. Weeds should be destroyed which can be achieved by working the soil surface at regular intervals.

Pre-planting ploughing repeated several times and early planting on mounds are additional measures sometimes recommended to limit damage. The same applies to the use of larger volumes of 'clean' soil for propagation of seedlings which delays infestation after planting. Note that agricultural tools used for work in contaminated areas should be thoroughly cleaned before being used in clean ones. The same applies to tractors' wheels. Power washing with water is often sufficient to remove all of the soil and the nematodes.

The application of compost or green manure to the soil just before planting of the tomato

---

1. See note to the reader p. 417.

Table 51 Main products used to control nematodes attacking tomatoes

Active ingredients	Spectrum of activity	Other information
Fumigants – chloropicrin – 1,3-dichloropropene (1,3-D) – dazomet – dimethyl disulfide (DMDS) – metam-sodium – methyl isothiocyanate	Products often quite versatile (fungicides, insecticides, and herbicides in addition to being nematicides). They are generally more efficient in well-drained and porous soils. They act directly on the nematodes.	More effective than nonfumigants, these products are mainly used in the nursery. 1,3-D has little effect on weeds. A number of them are also used in the field.
Nonfumigants – ethoprophos – aldicarb – carbofuran – fenamiphos – oxamyl	Products more specific against nematodes and soil-borne insects. They are active, either directly on the nematodes in the soil, or through the paratized plant. Often, they temporarily inhibit the development of nematodes without killing them.	Less expensive, they are mostly used in the field. Their effectiveness is lower than that of previous products. They are used either before planting or used as a spray in the following weeks (if oxamyl).

crop may also help reduce nematode damage. Compost made from coffee pulp is known to reduce the number of galls and egg masses of M. incognita. Similarly, cattle-cakes made from *Azadirachta indica*, *Chrysanthemum coronarium*, *Ricinus communis*, *Sorghum sudanense*, rye, and oats have the same effect. The addition of organic matter to the soil (compost, manure) can increase the water capacity, which encourages certain micro-organisms which compete with the nematodes.

Note that the addition of chitin to the soil has a significant effect on the reduction of M. *hapla*.

In addition, weeds must be perfectly controlled in future crop areas as many of them are likely to harbour and allow nematode multiplication. It is necessary to fully control the plants' nutrition and irrigation.

Healthy seedlings are essential. They are best grown on benches using a clean substrate. They can be raised on the ground, provided that the latter is covered with a clean and totally intact plastic sheet. If there are any doubts about the soil quality it should be disinfected as a precaution. In areas of extensive production, propagation should not be done in fields that have previously grown susceptible plants.

Various nematicidal products, listed in *Table 51*, are still used to kill soil nematodes. Their choice will depend on the local pesticide legislation in force and on the economics of the treatment.

The use of these products has several disadvantages: many of them are toxic to humans and the environment; they have little or no specificity and upset the soil biological balance; they are expensive and sometimes require specific equipment.

In some countries they are used as a last resort when the other methods are not effective. In countries where sunshine is important, disinfection of soil by solarization is an option, especially for the treatment of contaminated fields at a low cost. This technique involves covering the soil which has first been well cultivated and moistened, with a 35–50 μm thick polyethylene film. This remains in place for 4–8 weeks at a very sunny time of the year. It increases the soil temperature and encourages the activity of microbial antagonists. This helps to reduce the level of inoculum in the soil of many plant pathogenic micro-organisms, especially certain nematodes. Many of them are killed at temperatures between 44 and 48°C. The use of nematicides and compost is sometimes associated with solarization in order to increase its effectiveness, especially against *Meloidogyne* spp.

Whatever the method used to disinfect the soil, it is often recommended that it should be done soon after harvest as many nematodes are then still present in the soil upper layers. In addition, the soil should not be cultivated too deeply after disinfection as this risks moving untreated soil upwards into the root zone of the new crop.

Resistant varieties are currently available. Resistance to *Meloidogyne* spp. comes from the wild species *Lycopersicon peruvianum*. It is conferred by a single dominant gene 'Mi' (now called 'Mi-1') and results in a hypersensitive reaction at the site of nematode penetration. The larvae can no longer bind to the root and complete their cycle. This gene, located on chromosome 6, controls three of the most common species: M. *incognita*, M. *arenaria*, and M. *javanica*. Unfortunately, this resistance is not effective against M. *hapla*, although its rate of reproduction is reduced on plants with the gene 'Mi'. It is not effective against M. *mayaguensis*.

Virulent biotypes of the three species controlled by the 'Mi' gene have been described in many production regions: California, Japan, Morocco, Spain, Greece, and France. It should be noted that these virulent populations are sometimes separated into two groups:
– selected virulent populations, from fields where resistant varieties have been grown on several occasions. Virulence is acquired gradually and is stable; it involves several genes in these nematodes;
– naturally virulent populations (called 'B races'), not recently exposed to the resistant varieties.

It has been suggested that the genetic virulence mechanism of the first populations is similar in all three *Meloidogyne* species, but is different from the naturally virulent populations.

Races able to overcome the 'Mi' gene have been found in several countries in Europe and the Mediterranean Basin: M. *javanica* in Spain, Greece, Crete, Cyprus, Morocco, Tunisia, and M. *incognita* in France and Greece. Note also that the B races of M. *incognita* have been described in the Ivory Coast, and of M. *arenaria* in Senegal.

Despite this, the 'Mi' gene, which has been used for over 50 years, is still effective in many crops. Note that the reproduction of M. *incognita* on genotypes with 'Mi' in the heterozygous state would be more likely than on genotypes with this gene in the homozygous state and this may affect the resistance durability conferred by this gene. In addition, different levels of efficacy have been reported occasionally between resistant parents and their hybrids, probably due to an incomplete transfer of the 'Mi' gene. Finally, this resistance can be overcome by temperatures of about 28°C and above.

Where resistance appears to have broken down, the possible reasons for this should be investigated and the following actions taken:
– do not grow a resistant variety or rootstock on the area for several years as the risk of adaptation of these nematodes is increased;
– if a loss of efficiency of nematode resistance is detected, it will be necessary to investigate the situation to determine whether there is a strain able to overcome this resistance, or that M. *hapla* is the cause as it is not affected by this gene, or with loss of efficacy related to high temperatures.

The emergence of virulent strains and the limitations of the resistance conferred by 'Mi' at high temperatures, were reasons to search for other sources of resistance to root-knot nematodes. Thus, seven resistance genes to *Meloidogyne* spp., named 'Mi-2' to 'Mi-8', have been identified in accessions of *Lycopersicon peruvianum* and one in *L. chilense*. They confer resistance different from that conferred by the gene 'Mi'. Some are effective against M. *hapla* and/or are still functional at 33°C. Thus, 'Mi-2' and 'Mi-6' confer resistance to M. *incognita* at 32°C, 'Mi-3' to virulent isolates of the latter species, 'Mi-4' and 'Mi-5' confer resistance to M. *incognita* and M. *javanica* at 32°C, 'Mi-7' to virulent isolates of M. *incognita* at 25°C, as does 'Mi-8'. Resistance to M. *hapla* and heat stability, has been observed in an accession of *L. peruvianum*, but also in *L. chilense*.

Crosses between *L. esculentum* and *L. peruvianum* are difficult and the future use of these genes remains uncertain.

In addition, there is a number of nematode traps plants such as the *Tagetes* spp. (*T. erecta*, *T. patula*) that are hosts, of M. *hapla* in particular. These plants are still little used in rotations with tomatoes.

A number of root-knot nematode pathogenic micro-organisms have been tested on various plants: these include fungi such as *Arthrobotrys irregularis*, A. *dactyloides*, A. *conoid*, *Glomus fasciculatum*, *Hirsutella minnesotensis*, *Paecilomyces marquandii*, and *P. lilacinus*, but also bacteria such as *Bacillus penetrans*, *B. thuringiensis*, and *Streptomyces costaricains*. For example, *Verticillium chlamydosporium* infects the second instar and the eggs of M. *hapla*.

Finally, several plant extracts (from leaves and roots) disrupt the development of galls nematodes including *Azadirachta indica*, *Chromolaena odorata*, *Deris elliptica*, *Euphorbia antiquorum*, *Inula viscosa*, *Peganum harmala*, *Ruta graveolens*, *Senecio cineraria*, and *Swietenia mahogani*.

Pratylenchus spp. Filipjev

Root rot nematodes (lesion nematodes)

(*Pratylenchus*, Pratylenchidae, Tylenchida, Nematoda)

Principal characteristics

Several species of *Pratylenchus* may occur on tomato, in particular *P. penetrans* (Cobb) Filipjev & Schuurmans Stekhoven, *P. crenatus* Loof., *P. brachyurus* (Godfrey) Filipjev & Shuurmans Steckhoven, and *P. pratensis Filipjev*.

• Frequency and extent of damage

Nematodes that are responsible for root rot, such as *Globodera* spp., are less damaging on tomato than *Meloidogyne* spp. *Pratylenchus* nematodes, known collectively as root rot nematodes, are worldwide in their distribution, affecting over 500 plant species. Only in a few countries are they reported on the Solanaceae and the damage they cause varies from country to country. The effect of *Pratylenchus* spp. on tomatoes is little known and it is most likely negligible.

• Main symptoms

Many lesions of varying size, mostly longitudinal, appear on the roots. Their colouring varies from yellowish to reddish brown, due to the formation of phenolic compounds in tissues. Eventually, large parts of the cortex decompose. When the attacks are severe, a large proportion of the root system may disappear.

Plant growth is not then vigourous and a variable number of leaves turn yellow or even fade at the hottest times of the day.

Observation of roots under a binocular microscope often shows nematodes in the tissues or nearby, with a clearly visible stylet.

(See Photos **474, 475**.)

• Biology, epidemiology

Survival, inoculum sources: these nematodes are likely to overwinter in the soil, on and in live or dead roots. Many hosts harbour them, ensuring their conservation and multiplication at any time of the year. For example, *P. penetrans* has been listed on more than 400 different plants, making it probably the most common species and that with the most consequential economic impact. Among the many crops that may harbour these *Pratylenchus* spp. are potato, many vegetables, sweet potato, sorghum, oats, sunflower, alfalfa, rice, corn, strawberry, various fruit species, and several weeds.

Penetration and invasion: the nematodes (larvae or adults) get in the roots by forming holes with their mouth stylet. They gradually invade the cortex, destroying cells and digging holes through feeding. In *Pratylenchus* spp., all stages are mobile within the cortex tissues, which is why they are called 'endomigratory' nematodes. Eventually, the damaged tissues contain large quantities of eggs, larvae, and adults. A significant proportion of these return to the soil.

Transmission, dissemination: from the diseased plants, many eggs, larvae, and adults can be passively transported to other plants by runoff, drainage, and irrigation water. The larvae move actively over short distances in wet soils. Dissemination can occur via dust from contaminated soil being carried over by wind to neighbouring plots. The contaminated plants, tools, farm implements, and vehicles also ensure dissemination.

For *P. penetrans*, the duration of a complete cycle (from one adult to another adult) can take 30–86 days, mainly depending on temperature conditions.

Conditions encouraging development: their temperature requirements vary depending on the species: the higher the soil temperature,

641

the more *P. crenatus* and *P. penetrans* are active between 18 and 30°C. *P. crenatus* prefers heavy, muddy soils, while *P. penetrans* appreciates sandy soils. It would survive less well in dry soils than in the same moist soils. It better reproduces at a pH between 5.2 and 6.4.

In general, tomato adverse growing conditions encourages parasitism by these nematodes.

Control methods[1]

Several of the proposed methods to control *Meloidogyne* spp. can also be used to control many other nematodes, particularly *Pratylenchus* spp. (See Description 50).

• During cultivation

No control method makes it possible to monitor attacks by *Pratylenchus* spp. on tomato during cultivation.

If damage occurs in nurseries, the affected plants should be removed. Otherwise, their field planting will contribute to the spread of nematodes and to the contamination of healthy soils.

The tools and tractor wheels used during work in contaminated plots should be thoroughly cleaned before use in healthy plots. Thorough rinsing with water of all the material is often enough to get rid of the soil and of the nematodes.

At the end of cultivation, it is imperative that the plants root systems are removed from the plot and destroyed, to avoid enriching the soil with nematodes.

• Subsequent crop

To be effective, all the control methods listed below need to be implemented against migratory endoparasitic nematodes, as with root-knot nematodes.

Crop rotation should be chosen wisely, given the many potential hosts of these nematodes. Often, however, they only offer very limited action. Spinach, beets, and Cucurbitaceae are relatively tolerant to *P. penetrans*, whereas onions appear extremely sensitive. *Sorghum* spp., *Tagetes* spp., members of the mustard family, and *Agrostis palustris* limit its development. Asparagus does not seem to host it. Winter fallows reduce population levels of nematodes in the soil.

It is essential to use healthy seedlings. They will be preferably produced on shelves and in a disinfected substrate. They can be placed on the ground, provided that the latter is covered with clean and not torn plastic mulch. If there are any doubts about the quality of the nursery soil, it must be disinfected. Many nematicide products can be used (see Description 50).

In countries where sunshine is important, disinfection by solarization may be considered, especially to clean plots at a lower cost. This technique is detailed on p. 638.

Tomato resistances to *Pratylenchus* spp. do not appear to have been explored, probably because of their low impact on this Solanaceae.

[1] See note to the reader p. 417.

Main other nematodes affecting tomato

Rotylenchulus reniformis Linford & Oliveira
Reniform nematode
Syn.: *Rotylenchus reniformis* (Verma and Prasad, 1970)
(*Rotylenchulus*, Hoplolaimidae, Tylenchida, Nematoda)

Belonolaimus longicaudatus Rau
Sting nematode
(*Belonolaimus*, Belonolaimidae, Tylenchida, Nematoda)

Distribution and damage

This nematode affects a large number of cultivated herbaceous and woody plant species, especially in tropical to sub-tropical areas (Americas, Africa, Asia, Pacific Islands). It has been described in over 30 countries. Although its damage on tomato may be occasionally significant, it is not considered a serious bioagressor of the Solanaceae.

This formidable migratory ectoparasite nematode is reported in some countries. In the US, it is most prevalent in warm, sandy soils, especially in Florida, but it has been reported in several other states. It is also reported in the Caribbean islands, Puerto Rico, Bermuda, and Australia. It is one of the largest phytophagous nematodes, with a long stylet. Adults can reach a length greater than 3 mm. There are host specificity differences between populations.

Symptoms

As a result of many bites, the root system shows root damages and has a limited size. Plants can be stunted and chlorotic. Yields are lower.

The females of this nematode, as well as their eggs are visible under the microscope when root tissues are dissected in water.

Like many ectoparasitic nematodes, *B. longicaudatus* causes necrotic lesions in the roots cortical tissues, which tend to deepen when secondary invaders are involved. The architecture of the tomato root system is strongly modified: the main roots lose a high percentage of lateral feeder roots, giving them an appearance of a 'rat tail'.

Plants are often small with chlorotic apex leaflets.

Elements of biology

This nematode is able to survive in the soil on plant debris and several alternative hosts (onion, beet, cabbage, sweet pepper, eggplant, melon, cucumber, carrot, lettuce, beans, radishes). The females feed and grow by penetrating their heads in the root tissues, while the tail remains outside. Subsequently, more than 1 week after infection of the roots, they produce about 50 eggs surrounded by a gelatinous matrix. Four moults, one of them inside the egg, allow the adult stage to be reached in 24–29 days, depending on environmental conditions. Water is essential to the movement of nematodes in the soil, but excess or lack leads to death. They are mostly scattered through the soil particles transported by various vectors: irrigation water, plants, agricultural implements, workers, and animals. Mild temperatures, above 25°C, encourage the development of this rather tropical nematode.

Belonolaimus longicaudatus can survive on many cultivated hosts (cereals, peanuts, corn, cotton[a], fruit trees, vegetables [carrot, melon, cucumber[a], watermelon[a], cabbage[a], cauliflower, turnips[a], celery, eggplant[a], peppers[a], lettuce[a], beans, potatoes], ornamental plants) or noncultivated. It has therefore no difficulty in surviving in an infested plot. Its very long stylet penetrates the cortex and endoderm of root tips. It injects enzymes, leading to changes and later altering root development. The cycle of this nematode lasts 18–24 days. The eggs hatch after approximately 5 days and juvenile nematodes pass through three moults before becoming adults.

Like many nematodes, dripping water, seedlings, agricultural implements, and so on ensure its dissemination.

It does not seem to like soil too wet and hypoxic conditions in the rhizosphere. Reproduction is important between 25 and 30°C, and it seems able to persist at 35°C, unlike other species. It is optimal in soils at 7% moisture, whereas levels of about 30% inhibit it.

[a] The sensitivity of these plants can vary depending on nematode populations.

Species specific control methods[1]

Many protection methods recommended in Description 50 can be used to control *R. reniformis*:
– solarization, in addition to traditional disinfection methods, seems effective in controlling this nematode;
– some altered animal manure, sawdust of *Azadirachta indica* (neem), and *Mangifera indica* give good results;
– biopesticides such as *Paecilomyces lilacinus* associated with certain chemical nematicides and *Bacillus thuringiensis* reduce development.

This nematode is sensitive to many insecticides and fumigants. Some Egyptian tomato varieties mat be resistant to *Rotylenchulus reniformis*. The peanut and sorghum are less sensitive hosts.

Many control methods recommended in Description 50 can be used to control *B. longicaudatus*.

Crop rotations are difficult to implement because of the polyphagy of this nematode. However, many plants do not seem to harbour it or are poor hosts, including alfalfa, asparagus, and tobacco.

---

[1] See note to the reader p. 417.

Table 52 Main control methods used against pathogens and nematodes affecting tomatoes grown in glasshouses and in the field

Main protection measures	Air-borne fungi	Soil-borne fungi	Vascular fungi	Air-borne bacteria	Endophytic, vascular and/or air-borne bacteria	Phytoplasma	Viruses and viroids transmitted by contact	Viruses with air-borne vectors	Viruses with soil-borne vectors	Nematodes
Crop rotation (cereals, green manure, sorghum)	+/– to +	+ to ++	+ to ++	+/– to +	+ to ++	0 to +/–	+/– to +	+/–	+/– to +	+/– to ++
Avoid planting near to susceptible or already affected crops	+ to ++	+/–	+/–	+	+/–	+	+/–	+ to ++	+/–	+/–
Level soil and drainage	+	++	+	+	++	0	0	0	+	+/–
Soil disinfection (fumigant, steam, solarization, biocontrol)	+/–	+/– to ++ depending on the fungus and the method used	+/–	+/–	+/– to +	0	+/–	0	+/– to ++ depending on fumigant	+/– to ++ depending on method and nematode
Disinfection, replacement of the substrate in soil-less crops	+/–	++	+/– to +	+/–	+/– to ++	0	+/– to +	0 to +/–	+	++
Clean equipment for all cultural operations	+	++	++	+	+	0	+	0	+	++
Disinfection of cultural and harvesting implements	+/– to +	+ to ++	+/– to +	+/– to +	+/– to ++	0	+/– to +	0	+/– to +	+ to ++
Plastic mulching	0 to +/–	0 to + (some)	0	0	0	0	0	+/– to +	0	0
Aluminized mulching	0	0 to + (some)	0	0	0	+/–	0	+/– to +	0	0
Use healthy or tested seeds	++ *A. tomatophila*, *D. lycopersici*, *S. lycopersici*	0	0	++ *P. tomato*, *Xanthomonas* spp.	++ *C. michiganensis*	0	++ TMV, ToMV, PepMV	0	0	0
Use resistant varieties or rootstocks (the most common pathogens controlled by resistance)	++ *M. fulva*, *O. neolycopersici*, *Stemphylium* spp., *P. infestans*	++ *P. lycopersici*, *F. oxysporum* f. sp. *radicis-lycopersici*	++ *F. oxysporum* f. sp. *lycopersici*, *V. dahliae*	+ *P. tomato*	++ *R. solanacearum*	0	++ TMV, ToMV	++ TYLCV, TSWV	0	++ *M. incognita*, *M. javanica*, *M. arenaria*
Check seedlings health and quality	+ to ++	++	+	+	+	0	++	++	+	+
Insect-proof structures (nets covering the openings)	0	0	0	0	0	++	0	++	0	0
Insect-proof mesh covering plants	0	0	0	0	0	++	0	++	0	0
Install a foot bath at the entrance of every structure	+/–	++	++	+/–	++	0	+	0	+	+

0: measure of no interest; +/–: measure of limited value; +: recommended measure; ++: essential measure

891 Although the removal of plant debris during cultivation is not always possible, if left it allows various pathogens/pests to survive and even to multiply near to the crop.
Remove plant debris

892 Removal and disposal of debris is particularly effective in humid tropical regions. It is very bad practice to leave debris near to the current crop, as is the case in this photo.
Move plant debris away from crops

894 Collection of plant debris is planned for in this glass-house. Boxes scattered throughout the crop are gradually filled up during the day.
Provide ways of collecting plant debris

893 This pile of debris near the plastic tunnels can be a source of pests and pathogens.
Do not store the plant debris close to the crop

895 Keeping these old bags of peat near the glasshouse may allow some soil fungi to survive and to contaminate the crop as a result of dust being transported by wind and air currents.
Do not store old substrates close to the crop

Some examples of tomato production showing how to handle crop debris and old substrate

Main protection measures	Air-borne fungi	Soil-borne fungi	Vascular fungi	Air-borne bacteria	Endophytic, vascular and/or air-borne bacteria	Phytoplasma	Viruses and viroids transmitted by contact	Viruses with air-borne vectors	Viruses with soil-borne vectors	Nematodes
Adhere to recommended planting densities	+	0	0	+	0	0	0	0	0	0
Manage fertilizer use, especially nitrogen	+	+/–	+	+	+	0	0	0	+/–	+/–
Avoid excess water in the soil (check with an auger, and a tensiometer)	+	++	+/–	+	++	0	0	0	+	+/–
Water preferably in the morning or during the day (so that the plants dry quickly)	++	0	0	++	0	0	0	0	0	0
Avoid sprinkler irrigation	++	+/–	+/–	++	+/– to +	0	0	0	0	0
Use safe or disinfected water after recycling (particularly in soil-less systems)	+/–	+ to ++	+/– to +	+/–	+/– to +	0	+	0	+/– to +	+/– to +
Ventilate structures and use heating if necessary (to reduce humidity and control temperature)	++	0	0	++	0	0	0	0	0	0
Remove weeds (crop and surroundings)	+/– (some)	+/– (some)	+/– (some)	+/– (some)	0	++	+	+	0	+/–
Establish a quarantine area	+/–	0	0	+/–	0 to ++ C. michiganensis	0	++	0	0	0
Remove plant debris (during and at end of cultivation)	++	++	++	++	++	0	++	+/–	++	++
Do not work in wet crops	++	0	0	++	+	0	+/–	0	0	0
Possible direct chemical protection	+/– to ++	+/– to ++	+/–	+/– to +	+ C. michiganensis	0	0	0	0	+/–
Use a disease forecasting model	+	0	0	0	0	0	0	0	0	0
Anti-vector chemical protection possible	0	0	0	0	0	+/–	0	0 to + depending on vector and virus	+/–	0
Use non conventional methods (biopesticides, SDN)	Possible	Possible	Possible	Possible	Possible	0	0	0	0	Possible

0: measure of no interest; +/–: measure of limited value; +: recommended measure; ++: essential measure

896 Many weeds are present in this glasshouse. They can harbour viruses, their potential vectors, and other pests.
Eliminate weeds in and around the glasshouses

898 *Clavibacter michiganensis* subsp. *michiganensis* is responsible for a reduction in yield of many plants in the rows in this crop. To avoid spreading the bacterium during cultural operations, the affected area has been marked so that work can be organized to minimize spread.
Quarantine affected areas of crops

897 Young tomatoes have been planted in this tunnel, right next to plants already infected with Tomato yellow mosaic virus (ToYMV). Whiteflies present on these plants will quickly transmit the virus to the young plants.
Never plant a new crop next to an old one which is already affected

899 In some glasshouses, workers use knives and a container of disinfectant. This reduces possible contamination of the knife which will minimize spread of the bacterium during pruning and leaf removal.
Disinfect pruning tools

Production situations to be avoided or carefully managed

Table 53 Efficiency of the principal control methods used against the main pathogens/nematodes on tomato

Control methods	Mites	Aculops lyco-perisici	Whiteflies	Scales	Miners	Moths	Aphids	Bugs	Thrips
In the presence of high populations of pests, treat plants before removal.	*	*	*	*				*	*
Remove and destroy plant debris and crop residue.		*		*	*			*	*
Wash with water and treat the structure, supports, and concrete walkways with an insecticide or contact miticide.	*	*		*				*	
Disinfect equipment used in the glasshouse (drip system, crates).	*	*		*				*	
Disinfect the reused substrate or soil.				*	*				*
Preheat the glasshouse before planting and use an insecticide or miticide.	*				*				*
Monitor and control the health of plants during propagation and at planting in the glasshouse.	*	*	*	*	*		*		*
Use an insect-proof glasshouse during propagation.			*		*		*	*	*
Use insect-proof barriers on glasshouse doors and ventilators.			*			*	*	*	*
Keep the glasshouse & surrounding areas weed free.	*	*	*		*		*	*	*
Monitor and detect early pest infestation with yellow sticky traps placed over the crop immediately after planting (aphids).			*		*		*		*
Detect and monitor early pest infestation with sticky blue panels placed above the crop (thrips).									*
Use pheromone traps outside the glasshouse.						*			
Use biocontrols (see *Table 53a*)	*		*		*	*	*	*	*
Select and use pesticides with care. Very important where biocontrols are being used (see *Table 53a*).	* (r)	*	* (r)		* (r)	*	*(r)	*	* (r)

(r) resistance to insecticides and miticides are known in these pests.

Table 53a Efficacy of biocontrol agents (insects and mites in clear, micro-organisms in grey) for the control of major pests on tomatoes

Possible auxiliaries	Mites	Aculops lycoperiaici	Whiteflies	Scales	Miners	Moths	Aphids	Bugs	Trips
Aphelinus abdominalis							*		
Aphidius colemani							*		
Aphidius ervi							*		
Aphidoletes aphidimyza							*		
Dacnusa sibirica					*				
Diglyphus isaea					*				
Encarsia formosa[a]			*						
Eretmocerus eremicus[a]			*						
Eretmocerus mundus[b]			*						
Feltiella acarisuga		*							
Macrolophus caliginosus		*c	*		*c		*c		*c
Phytoseiulus persimilis		*							
Bacillus thuringiensis sub-species azawai						*			
Bacillus thuringiensis sub-species kurstaki						*			
Paecilomyces fumosoroseus			*						
Verticillium lecanii[d]			*				*		

[a]: mainly effective on *Trialeurodes vaporariorum*; [b]: primarily effective against *Bemisia tabaci*; [c]: secondary efficacy; [d]: *Verticillium lecanii* effective against whiteflies and aphids can vary from one strain to the other.

Efficacy can change according to geographical areas.

4

Resistance to diseases and pests in tomato

The first control method used against pests and diseases was genetic resistance. Thus, since the beginning of agriculture, plants with improved agronomic characteristics able to crop in the presence of parasites and pathogens have been selected and used. From 1900 onwards this selection has been on a scientific basis. Efforts by breeders since the 1950s have led to the creation of many resistant varieties.

Sources of resistance

In tomato, genetic control is generally based on the exploitation of monogenic dominant resistance derived from wild species related to the tomato. This plant is one of the model species for the use of single-gene resistance in cultivated varieties.

The two wild species that have so far provided the most resistance genes in cultivated varieties are *Lycopersicon pimpinellifolium* and *L. peruvianum*. Taking the tomato as the female parent, crosses with *L. pimpinellifolium* are easily made. In contrast, crosses with *L. peruvianum* are difficult and require the use of specific techniques such as *in vitro* embryo selection. It is the same with the *L. chilense*, which is closely related to *L. peruvianum*, which has been used in recent work on the breeding of resistance to several Begomoviruses. The hybridization of each of these two species with tomato produces embryos that abort in the seeds well before fruit ripening. It is therefore necessary to extract them in an immature stage (30–34 days after the hybridization), without waiting for the 55–60 days required for fruit ripening of the female parent tomato. The embryos are cultured *in vitro* on nutrient medium until seedlings emerge. Another technique is to pollinate the tomatoes with a mixture of tomato pollen and pollen from the wild species. This pollination produces many seeds which include some interspecific hybrids. The F1 hybrids obtained are generally self-sterile. To breed a successful tomato variety it is usually necessary for breeders to backcross with a commercial variety and repeat this process for several generations. The first backcrossing involves using one of the techniques used to obtain the F1 plants. The subsequent backcrossings are straightforward.

Another wild species, *L. hirsutum*, has provided resistance for tomatoes without much hybridization difficulty. Its role has become very important as the male parent of the F1 hybrids used as rootstocks in tomato and in eggplant. These hybrids are produced by crossing a tomato variety carrying resistance dominant genes to several diseases, particularly soil-borne ones ('V', 'I,' 'I-2', 'Fr', 'Tm-2^2' and 'Mi') with an ecotype of *L. hirsutum*, which adds vigour and dominant resistance, including resistance to corky root disease and *Didymella lycopersici*.

The origins of resistance in cultivated varieties are shown in *Table 54*. To these, resistance to *Alternaria alternata* f. sp. *lycopersici* controlled by the 'Asc' gene should be added, present in almost all varieties both old and new.

Table 54 Main *Lycopersicon* species as sources of pathogen/nematode resistance genes used in cultivated tomato varieties

Species and pests controlled	Genes involved
Lycopersicon pimpinellifolium (or ecotypes of *Lycopersicon esculentum* var. '*cerasiforme*')	
Verticillium dahliae	'*Ve*'
Fusarium oxysporum f. sp. *lycopersici*	'*I*', '*I-2*'
Stemphylium spp.	'*Sm*'
Mycovellosiella fulva (*Fulvia fulva*)	'*Cf-2*', '*Cf-5*', '*Cf-6*', and '*Cf-9*'
Phytophthora infestans	'*Ph-2*'
Pseudomonas syringae pv. *tomato*	'*Pto*'
Ralstonia solanacearum	oligogenic resistance
Tomato yellow leaf curl virus (TYLCV)	oligogenic resistance

Lycopersicon hirsutum	
Mycovellosiella fulva (Fulvia fulva)	'Cf-4'
Oidium neolycopersici	oligogenic resistance
Tobacco mosaic virus (TMV)	'Tm-1'
Tomato yellow leaf curl virus (TYLCV)	oligogenic resistance ('Ty2')
Pyrenochaeta lycopersici	oligogenic resistance used in rootstocks
Lycopersicon peruvianum	
Fusarium oxysporum f. sp. radicis lycopersici, FORL)	'Frl'
Pyrenochaeta lycopersici	'pyl'
Tobacco mosaic virus (TMV)	'Tm-2', 'Tm-2^2'
Tomato spotted wilt virus (TSWV)	'Sw-5'
Tomato yellow leaf curl virus (TYLCV)	oligogenic resistance
Meloidogyne spp.	'Mi'
Lycopersicon chilense	
Leveillula taurica	'Lv'
Tomato yellow leaf curl virus (TYLCV)	oligogenic resistance ('Ty-1' and 'Ty-3')
Lycopersicon pennellii	
Fusarium oxysporum f. sp. lycopersici	'I-3'

Breeding programmes using other resistances from the *Lycopersicon* spp. mentioned above are near to their end point. Exploitable resistance research must now be conducted with other species of *Lycopersicon*, as well as with more distantly related species of the *Solanum* genus. Four species (*S. lycopersicoides*, *S. juglandifolium*, *S. ochranthum*, and *S. sitiens*), grouped in the 'juglandifolia' series have strong morphological and chromosomal analogies with species of *Lycopersicon*. Thus, F1 hybrids are easily obtained with *S. lycopersicoides*, an interesting species for its cold tolerance and resistance to Cucumber mosaic virus (CMV), *Clavibacter michiganensis* subsp. *michiganensis*, and *Botrytis cinerea*. Fertility and incompatibility problems, however, make the backcrossing process difficult.

Level of effectiveness and durability of resistance

Resistance is inherited and reduces or eliminates the effects of parasitism. Its effectiveness depends on the combination of two factors:
– the level of expression of resistance;
– sustainability and stability of resistance over time.

These two factors are determined independently.

• Determination of the level of resistance

Resistance and sensitivity are the two extremes of the reactions of the host plant. According to the mechanism involved, the level of resistance may vary, hence the distinction between absolute resistance and partial resistance.

Absolute resistance, also called 'vertical', is due to a phenomenon of immunity or a mechanism of hypersensitivity. In the case of immunity, the plant is completely free from parasitism. This resistance can result from the pathogen being unable to infect the host or, in the case of viruses, the absence in the host of an element or function essential for viral replication. Resistance to *Mycovellosiella fulva*, conferred by the 'Cf-2' gene is a good example of immunity; it is now overcome by race 2 present in many areas. The term 'immunity' is often misused to denote resistance manifested by the absence of visible symptoms, but that does not exclude the penetration of pathogens.

When a mechanism of hypersensitivity occurs, the infection process remains localized and inactivated by the death of infected tissues.

At high temperatures, the mechanism is slower and symptoms may appear. Two examples are provided by the '$Tm-2^2$' resistance gene to Tomato mosaic virus (ToMV) and Tobacco mosaic virus (TMV) and the 'Mi' gene conferring resistance to *Meloidogyne* spp.

Partial resistance, also called 'horizontal', is characterized by limited infection of the host, the slowing of growth and development in the tissues, and a reduced number of infectious units produced. Partial resistance to *Phytophthora infestans*, which is controlled by the '$Ph-2$'gene, illustrates this situation. The overall result of these phenomena is a slower progression of the disease on the plant and of the epidemic in the crop.

Partial resistance supported by good cultural practices and general methods of plant protection may prevent the development of an epidemic.

The notion of partial resistance should not be confused with tolerance which is an agronomic concept. The latter characterizes the behaviour of a plant in which the parasite lives and reproduces, as in a sensitive plant showing typical symptoms of disease, but whose yield is not affected. However, it is common to speak of 'tolerant varieties' in virology to characterize plants that allow active virus multiplication without producing typical disease symptoms and whose performance is not affected.

• **Determination of resistance and stability over time**

The stability of high-level resistance may be very variable depending on the controlling genes. It is only after many years of use that the duration of resistance or the practical importance of a pathogen's adaptation to a given resistance can be reliably assessed.

The speed of emergence of new pathotypes is extremely high in some pathogens such as *Mycovellosiella fulva*. In contrast, after many years of use in a wide variety of environmental conditions, some resistances have never been overcome like those to *Stemphylium* spp. On the other hand, there are many examples of resistances which, although overcome, continue to have a significant practical value in certain agricultural contexts.

Efficiency levels and stability of resistances present in currently available varieties are listed in *Table 55*.

Table 55 Tomato resistance genes, their efficacy, stability, and frequency of use in tomato varieties

Pathogens/pests	Resistance genes	Efficiency	Stability	Frequency in cultivated varieties
Verticillium albo-atrum and *V. dahliae*	'Ve'	+++	++	+++
Fusarium oxysporum f. sp. *lycopersici*				
pathotype 0 (ex-1)	'I'	+++	++	+++
pathotype 1 (ex-2)	'$I-2$'	+++	+++	+++
pathotype 2 (ex-3)	'$I-3$'	?	?	+
Fusarium oxysporum f. sp. *radicis-lycopersici*	'Frl'	+++	+++	++
Pyrenochaeta lycopersici	'pyl'	+	?	+
Alternaria alternata f. sp. *lycopersici*	'Asc'	+++	+++	+++
Mycovellosiella fulva (syn. *Fulvia fulva*)	Accumulation of genes 'Cf' => '$C5$'	+++	++	++
Phytophthora infestans	'$Ph-2$'	+	++	+
Stemphylium spp.	'Sm'	+++	+++	++
Leveillula taurica	'Lv'	+++	?	+
Oidium neolycopersici	Oligogenic resistance	++	++	+

Pathogens	Resistance genes	Efficiency	Stability	Frequency in cultivated varieties
Pseudomonas syringae pv. *tomato*	'Pto'	+++	++	++
Ralstonia solanacearum	Oligogenic resistance	++	++	+
Tobacco mosaic virus (TMV) and Tomato mosaic virus (ToMV)	'Tm-2²'	+++	++	+++
Tomato yellow leaf curl virus (TYLCV)	Oligogenic resistance	++	?	+
Tomato spotted wilt virus (TSWV)	'Sw-5'	+++	++	+
Meloidogyne spp.	'Mi'	+++	++	++

+: low; ++: medium; +++: high; ?: unknown.

Available resistance in cultivated varieties and rootstocks

In tomato about 15 pathogens can now be controlled by genetic resistance (*Table 55*). The effectiveness of these resistances is highly variable, including their level of expression, their stability over time, and their efficacy against the development of new and virulent races of the pathogens.

The varieties selected for their resistance to pathogens are mainly used for protected crops, which are more frequently and severely affected. As their production potential is very high, the high cost of seeds resulting from expensive breeding programmes is easily economical.

Some long identified resistance is only present in a few varieties. In the case of *Pyrenochaeta lycopersici*, the reason is complex: this resistance is partial with recessive monogenic heredity, and the 'pyl' gene that controls it must be present in both parents of F1 hybrids. In addition, the practice of grafting on multi-resistant rootstocks with a high level of resistance to *P. lycopersici* in particular, reduces the value of the selection of F1 hybrids resistant to this soil-borne fungus.

Partial resistance to *Phytophthora infestans*, difficult to identify by early selection tests, has been pursued by only a few breeders. In addition, there are many effective fungicides on the market to at least minimize the effects of this air-borne pathogen.

Other resistances which are of more limited use are for use in specific production areas (e.g. tropical humid for *Ralstonia solanacearum*) or are the subject of relatively recent programmes.

The availability of dominant monogenic resistances allows the development of multiple resistances in F1 hybrids; the hybrids for protected crops generally have four to five resistances and up to seven. For field crops, fixed varieties have in general two to four resistances. Increasing numbers of F1 hybrids in these two different types of production have these resistances and contribute to the sustainable protection of tomato (see *Tables 56* and *57*, overleaf).

Available rootstocks are not very numerous and can be separated into two groups (see Box 8 and *Table 58*, p. 661):
– F1 hybrids in the cultivated tomato with multiple resistance to soil-borne diseases including a partial resistance to *Pyrenochaeta lycopersici*;
– F1 inter-specific hybrids between the cultivated tomato and *Lycopersicon hirsutum*. These hybrids have a strong root system, withstanding lower temperatures than the cultivated tomato. *Lycopersicon hirsutum*, native to the Andes, also has a high level of dominant resistance to *Pyrenochaeta lycopersici*.

Table 56 Examples of resistance patterns to pests, pathogens, and genetic disorders in the main types of glasshouse-grown tomatoes

Fruit type	Pests and physiological diseases controlled														
	TM	V	C3	C5	N	Fr	F/F1	F2	St	Wi	P	Ph	Oi	TSWV	TYLC
Cherry															
Elongated (plum)															
Fleshy															
Cocktail															
Beef steak															
Truss															

◻ resistance

Table 57 Examples of resistance patterns encountered in the elongated (plum shaped) fruiting varieties of glasshouse-grown tomatoes

Resistance patterns	Pests/pathogens and physiological diseases controlled														
	TM	V	C3	C5	N	Fr	F/F1	F2	St	Wi	P	Ph	Oi	TSWV	TYLC
1															
2															
3															
4															
5															
6															
7															
8															
9															
10															
11															
12															
13															
14															
15															
16															
17															
18															
19															
20															
21															
22															

◻ resistance

The shaded areas allow the characterization of almost all known resistance patterns on tomato: for example, the varieties with pattern 1 are resistant to Tobacco mosaic virus and Tomato mosaic virus (TMV and ToMV), and to *Verticillium dahliae*.

TM: Tobacco mosaic virus and Tomato mosaic virus (TMV and ToMV); **V**: *Verticillium dahliae*; **C3** and **C5**: *Mycovellosiella fulva* (syn. *Fulvia fulva*); **N**: Root-knot nematodes; **Fr**: *Fusarium oxysporum* f. sp. *radicis-lycopersici*; **F/F1** and **F2**: *Fusarium oxysporum* f. sp. *lycopersici*; **St**: *Stemphylium* spp.; **Wi**: silvering; **P**: *Pyrenochaeta lycopersici*; **Ph**: *Phytophthora infestans*; **Oi**: *Oidium neolycopersici*; **TSWV**: Tomato spotted wilt virus; **TYLC**: Tomato yellow leaf curl virus.

900 Propagating grafted plants. Rootstocks and scions are ready to be used.

902 On the roots of the rootstock, the effects of a soil-borne disease complex of *Colletotrichum coccodes*, *Rhizoctonia solani*, root-knot nematodes, and cyst can be seen, questioning the value of grafting as an alternative method of protection in this case.

901 The grafting has been completed and was successful. The grafted plants are now left to grow before being delivered to growers.

8 Grafting and resistant rootstocks

Grafting tomato plants is a well established technique which is used in many countries including Asia (Japan, Korea, China), Europe (Netherlands, Italy, Germany, France, Spain, Greece) and other Mediterranean countries (Morocco, Tunisia). The main advantage of this method is to control several soil-borne pests and diseases, especially in glasshouses where soil sickness may occur. It is used for protected crops as well as in those grown in the field. Although limited in the past, grafting has been revived along with other alternative treatments, because of the discontinuation of methyl bromide use. In addition, it has allowed soil-less cultivation of 'long life' tomato varieties, susceptible to *Fusarium oxysporum* f. sp. *radicis-lycopersici*.

Characteristics of rootstocks

Generally, cultivated rootstocks have a strong root system making grafted plants vigorous. They are also more tolerant of water stress and are resistant to soil-borne but also air-borne pests and diseases (see *Table 58*, p. 661). Three types of rootstocks are used: lines and hybrids from intra-specific crosses, inter-specific hybrids of KNVF or PNVF[1] type (between *Lycopersicon esculentum* and the wild species *L. hirsutum*), and species of perennials and wild *Solanum* spp. Among wild species, S. *torvum*, a tropical, very hardy vigorous species is used and tomatoes grafted onto it yield well. However, it requires higher than usual soil and air temperatures. *Solanum aethiopicum* is a species grown in Ivory Coast as '*ndowa*'. Both of these species are good rootstocks for both tomatoes and eggplant. They are used in tropical areas mainly to control *Ralstonia solanacearum*. They are however, both susceptible to *Corynespora cassiicola*, a damaging air-borne fungus in high temperature areas.

The above rootstocks, some of which have been available for several decades, usually controls the main soil-borne pests of tomatoes but also eggplant:
– *Pyrenochaeta lycopersici* (K or P), cause of corky root and responsible for severe root rot;
– *Meloidogyne* genus (N), root-knot nematodes, only *M. arenaria*, *M. incognita*, and *M. javanica* are controlled by rootstocks;
– *Verticillium dahliae* (V), cause of Verticillium wilt; a vascular disease affecting several solanaceous crops and particularly damaging to eggplant;
– *Fusarium oxysporum* f. sp. *lycopersici* (F), causes Fusarium wilt, another vascular but tomato specific disease;
– *Fusarium oxysporum* f. sp. *radicis-lycopersici* (Fr), whose resistance was introduced later into rootstocks, following the emergence of this pathogen which destroys tomato roots and crowns.

In tropical areas, several types of rootstocks can be used to control bacterial wilt of tomato and eggplant as mentioned above:
– tomato lines identified for their high resistance, some of them also offering resistance to nematodes and Fusarium wilt;
– some *Solanum* spp., of which the most compatible with tomato is S. *aethiopicum*, but also S. *torvum* or S. *stamonifolium*.

Note that some of the previous rootstocks are also resistant to some air-borne pathogens such as *Mycovellosiella fulva* (syn. *Fulvia fulva*) and also Tomato mosaic virus (ToMV) where the '$Tm2^2$' gene is used to confer resistance to rootstocks.

Several techniques are used for grafting tomato or eggplant on rootstocks (Photos **900** and **901**), particularly slit and double slit grafting, Japanese grafting (or 'simple English' grafting, in which graft and rootstock are joined edge to edge), and side grafting (bevelled graft inserted into a cut made in the axil of a rootstock leaf). The plants obtained are often more vigorous, resulting in 'vegetative' plants with a poor fruit set of the first truss, and an increased sensitivity to marbling or *Botrytis cinerea*. In addition, they need to be planted with the graft union above the soil so that susceptible adventitious roots from the scion cannot grow into the contaminated soil. The importance of training the plants quickly should be stressed, to prevent breakage at the graft. Finally, nitrogen fertilizers and irrigation should be limited.

Among the pest and disease problems, other than those resulting from soil-borne organisms, there are some that are specific to grafted plants:
– a problem of physiological origin affecting eggplant grafts. It results in the appearance of glassy patches on the leaves, rather localized at the periphery and delimited by the veins. Although not

---

[1] The designations 'KNVF' or 'PNVF' or 'KNVFFr' are actually acronyms, each letter indicates the pathogen /pest controlled by the rootstock.

obvious at first, they turn brown and become necrotic. They probably result from an imbalance between the absorptive capacity of the roots of the rootstock and the transpiration rate of the scion variety. Symptoms are particularly common in warm soils with plenty of water but with a humid air and a low air temperature;

– another physiological condition of grafted eggplants results in slower growth, limited flowering, and fragile or brittle stalks and leaves. In more serious cases, longitudinal blackish lesions are visible on the stem and are quite similar to those associated with black pith or with *Pectobacterium* spp. attacks.

These problems are not thought to occur on grafted tomatoes.

The question of resistance

Grafting has been used extensively and successfully to manage the population of soil-borne pests and pathogens, particularly in soils where disinfection with a fumigant, often methyl bromide, is not possible. The prohibition of use of this fumigant has changed the situation. Grafting has become one of the most commonly used alternatives. On some farms, the 'monoculture' of grafted plants is common practice and as a result there is severe selection pressure on the soil pest and pathogen populations created by the use of these rootstocks. This is a new situation and inoculum levels of certain soil-borne pests and pathogens are currently of particular importance, especially in some glasshouses. As a consequence of this situation, in the past few years severe dieback of grafted eggplants crops has occurred. Nothing similar has been found in tomatoes so far.

In addition, it should be noted that rootstocks used for eggplants and tomatoes are not always effective for the control of soil-borne organisms. In many cases they appear to be tolerant rather than resistant. This is particularly the case with *Meloidogyne* spp., *Pyrenochaeta lycopersici*, and *Verticillium dahliae*.

Recent investigations into the wasting of eggplant rootstocks in France have identified several soil-borne organisms that are involved in the death of roots (Photo **902**):

– three fungi whose pathogenicity has been confirmed (*Colletotrichum coccodes*, *Rhizoctonia solani*, and *Phytophthora nicotianae*);

– at least two nematodes (*Meloidogyne arenaria* and *Globodera tabacum*). In the case of *M. arenaria*, the B race of the root-knot nematode has been occasionally found;

– a vascular fungus, *Verticillium dahliae*, also normally controlled by a resistance expressed by the rootstocks. The study of the virulence of several strains on a host range has revealed the presence of race 2 of *Verticillium dahliae* on several farms in which severe attacks of Verticillium wilt have repeatedly occurred.

It seems clear that the selection pressure exerted by these rootstocks on soil-borne organisms gradually encourages the emergence of new more virulent strains of pests and pathogens that would normally be controlled by rootstocks resistance.

These rootstocks have not been used in several Mediterranean countries. In Italy, abnormally high development of galls was observed on the roots of rootstock 'Beaufort' in the presence of *Meloidogyne arenaria* and *Meloidogyne incognita*. The low level of resistance observed was not due to high temperatures or a particular virulence, but the nematode population. To overcome this situation, *Solanum torvum* was used, although this species did not prevent symptoms of root-knot nematodes and corky root.

Among the soil-borne pests not affected by grafting, symptoms very similar to corky root were observed on rootstocks in Piedmont, Campania and Sicily from the early 2000s. It has been shown that *Colletotrichum coccodes* was responsible as it is able to affect seriously the inter- and intra-specific hybrids of tomato used as rootstocks. The authors have concluded that grafting alone cannot be used to control all soil fungi, especially *Colletotrichum coccodes*. Also *Verticillium dahliae* has been found in France affecting plants grafted onto *Solanum torvum*. Finally, attacks of '*carotovorum*' and '*atrosepticum*' subspecies of *Pectobacterium carotovorum* have also been reported. These bacteria sometimes cause longitudinal symptoms several centimetres long on the stem of several rootstocks. These are moist, soft, and dark green to brown. Vessels go brown and the pith gradually deteriorates and becomes hollow.

In Cyprus, the grafting of eggplant on 'Brigeor' should ensure good control of root-knot nematodes and *Pyrenochaeta lycopersici* but it does not adequately control Verticillium wilt. For this reason, areas to be planted with eggplants grafted onto Brigeor are first solarized.

In addition to solarization, a few other methods of protection have been linked with grafting to achieve a higher level of effectiveness against previously reported problems. For example, steam, but also organic amendments based on *Azadirachta indica* and *Ricinus communis* oilcakes have proven effective for controlling *Meloidogyne javanica* on 'Beaufort'.

These examples show that the rootstocks used to produce eggplants, and to some extent tomatoes, are not always very effective in controlling all their terrestrial pests/pathogens, especially root-knot nematodes, *Pyrenochaeta lycopersici* and, more recently, *Verticillium dahliae*. Other pathogens where the use of rootstocks is not effective include *Colletotrichum coccodes*, *Phytophthora nicotianae*, and *Rhizoctonia solani*. Moreover, some rootstocks are sensitive to attacks by *Pythium* spp. in soil-less culture. In addition, in these production systems, root mat disease occurs.

It seems clear that grafting is no longer effective in some situations in Europe. Although these are limited in number there is clearly a need to find alternative ways of pest and disease control of soil-borne organisms. Some success has been achieved with the combined use of grafting and other means such as solarization, and the use of organic amendments. Such systems must be further developed in order that the production of tomatoes and eggplants continues in a sustainable way.

Table 58 Resistances reported in the main rootstocks (past and present) used for tomato and eggplant grafting

Rootstocks	*Pyrenochaeta lycopersici*	*Meloidogyne* spp. [a]	*Verticillium dahliae*	*Fusarium oxysporum* f. sp. *lycopersici*	*Fusarium oxysporum* f. sp. *radicis-lycopersici*	*Ralstonia solanacearum*	TMV and ToMV	Other information
Intra-specific hybrids [b] : good germination, less vigorous rootstocks and scions, superior resistance to nematodes, low resistance to *Pyrenochaeta lycopersici*.								
'Mogeor'	*	*	*	*	*		*	'Ve' gene in the homozygous state, resistance superior to that of interspecific hybrids.
'Robusta'	*	*	*	*	*		*	Cladosporium resistance.
'Esperanza'		*	*	*	*		*	
'Energy'	*	*	*	*	*		*	Cladosporium and stemphylium blight resistance.
'Kyndia'	*	*	*	*			*	
'PG1' et 'PG2'		*	*	*	*		*	
'PSF 861.61'	*	*	*	*	*		*	
Inter-specific hybrids : inconsistent germination, strong vigour, *Didymella lycopersici* resistance, greater resistance to *Pyrenochaeta lycopersici*.								
'Beaufort'	*	*	*	*	*		*	Resistance to races 0 and 1 of *Fusarium oxysporum* f. sp. *lycopersici*.
'Big Power'	*	*	*	*	*		*	
Body	*	*	*	*	*		*	Resistance to stemphylium blight and Cladosporium (C5).
'Brigeor'	*	*	*	*	*		*	
'Eldorado'		*	*	*	*		*	
'Emperor'	*	*	*	*	*		*	
'He-man'	*	*	*	*	*		*	Only tolerance to *Pyrenochaeta lycopersici*.
'Hires'	*	*	*	*			*	*Didymella lycopersici* resistance.
'Homerum'	*	*	*	*	*		*	Cladosporium resistance (C5).
'KNVF'	*	*	*	*				*Didymella lycopersici* resistance.
'KNVF2'	*	*	*	*				
'Maxifort'	*	*	*	*	*		*	
'Multifort'	*	*	*	*	*		*	FOL 3 races resistance.
'PNVF'	*	*	*	*				*Didymella lycopersici* resistance.
'Resistor'	*	*	*	*	*		*	
'Spirit'	*	*	*	*	*		*	
'TmKNVF2'	*	*	*	*			*	*Didymella lycopersici* resistance.
'TmKNVF2Fr'	*	*	*	*	*		*	*Didymella lycopersici* resistance.
Other rootstocks								
Solanum torvum	*	*	*			*		Poor seedling emergence. Resistance to *Fusarium solani* and *Phytophthora nicotianae*.
Solanum aethiopicum						*		Susceptible to crown rot caused by *Pythium aphanidermatum*.

[a]: resistance applies only to the *Meloidogyne incognita*, *Meloidogyne arenaria*, *Meloidogyne javanica*.

[b]: tolerant or resistant plant material.

According to Italian research, the intra-specific hybrids appear more effective than inter-specific hybrids for the control of root-knot nematodes. This information is not guaranteed, as the resistance used in some rootstocks is not available.

Hopes of further resistances in the short term

Following the success and effectiveness of genetic control together with its safety to the environment, important breeding programmes are underway in various parts of the world. They should eventually allow the control of an increased number of pests/pathogens.

The molecular biological techniques now used to tag genes that are of interest to breeders have helped in the selection of resistance which has previously been difficult to identify by traditional methods. In this way partial resistances, oligogenic or polygenic resistance, and the combination of a greater number of resistance genes has been possible.

In this way there is the possibility that resistance may become available to all of the known plant pathogenic micro-organisms. Many of them relate to the control of some strains which have adapted to the currently available resistance in cultivated varieties. The micro-organisms that can be controlled by resistances from wild species of *Lycopersicon* are presented below.

• A new pathotype of *Fusarium oxysporum* f. sp. *lycopersici* that overcomes the 'I' and 'I-2' genes is now present in different regions. This new race, generally designated 'race 3' is controlled by the dominant gene 'I-3' from *L. pennellii*. It has been introduced into tomato stocks; F1 commercial hybrids resistant to the three races have been produced.

• Many studies have reported partial resistance to *Alternaria tomatophila* from a variety of wild species including *L. hirsutum* which is now present in different commercial varieties. This resistance should not be confused with a resistance effective against *A. alternata* f. sp. *lycopersici* which is controlled by the 'Asc' dominant gene. This resistance, present in almost all grown varieties, ancient and modern, is referred to as 'resistance to *Alternaria*' in some seed catalogues. This statement is misleading as it does not confer resistance to foliage blight caused by *A. tomatophila*.

• The 'Ph-1' gene which confers resistance to *Phytophthora infestans* has been overcome before being even used in selection processes, and it is the same for the 'Ph-2' gene, incompletely dominant and controlling partial resistance. Breeders have then sought a more effective resistance. Hopes are now focused on the 'Ph-3' gene derived from *L. pimpinellifolium*.

• For *Oidium neolycopersici*, breeders aim to combine resistance genes from wild species to ensure a high level of resistance to different recently found races.

• Sources of resistance to *Clavibacter michiganensis* subsp. *michiganensis* have long been known. These are partial resistances, with environmental conditions strongly influencing expression levels. They originate from wild species and are the basis of breeding programmes throughout the world, especially to develop processing varieties.

• Different sources of resistance to *Xanthomonas* spp. have been identified. The selection work is aimed at combining the resistance to different races and to obtain one resistance expressed on foliage as well as on fruit. Resistance from *L. pimpinellifolium* has proved to be particularly useful.

• For the Y (PVY) Potato virus, the work focuses on resistance from *L. hirsutum* which is controlled by two genes, 'Pot-1', recessive, and 'Pot-2', dominant. This resistance is effective against strains that cause mosaic and also those that are necrotic.

• A dominant gene, stable at high temperatures and controlling Alfalfa mosaic virus (AMV) has been detected in a source of *L. hirsutum*. This gene, called 'Am', interests breeders in Mediterranean countries where AMV is sometimes serious.

• For the Tomato yellow leaf curl virus (TYLCV), the selection tends to increase the resistance level of hybrids by combining genes from different wild species. Breeding programmes are also conducted to control other Begomoviruses.

• The 'Mi' gene, has long been used to combat root-knot nematodes of the *Meloidogyne* genus, but is overcome in many regions. Two other genes from *L. peruvianum* have been introduced into tomato lines. The 'Mi-2' gene controls 'Mi' virulent strains but is, like the latter, ineffective at high temperatures. However, 'Mi-3' will control strains adapted to 'Mi' and is stable at high temperatures. Plant breeders are interested in this gene, particularly for tomato crops in hot countries with sandy soil, in which root temperatures are high.

Breeding programs, assisted by molecular markers, are now being used to combine these new resistances with those already available that have a proven record in the commercial market.

In the longer term, resistance to other pests and diseases will be introduced into breeding programmes. It is expected to include the control of the race of *Verticillium albo-atrum* capable of overcoming the 'Ve', gene, *Phytophthora nicotianae*, *Colletotrichum coccodes*, *Botrytis cinerea*, *Pseudomonas syringae* pv. *tomato* (race overcoming the 'Pto'gene), and Pepino mosaic virus (PepMV). Research is also being conducted to control various insects genetically, but it is too early to predict the practical results.

GLOSSARY

Acervulus: asexual fruiting in the form of a wide open conceptacle producing short conidiophores and conidia; it characterizes Melanconiales.

Aecidiospore: binucleate spore produced in an aecidium.

Aecidium (ecie, ecidie): structure formed by rust fungi in which binucleate spores are produced.

Aggressiveness: quantitative component of the pathogenicity of a micro-organism.

Alternative host: a host plant which an organism may parasitize to replace its preferred host, thus allowing it to complete its cycle.

Anamorph: asexual form of a fungus, also called 'imperfect stage', frequently resulting in the formation of conidia.

Anastomosis: merger of mycelial hyphae belonging to the same thallus or to complementary thalluses.

Antheridium: male fungal structure ensuring gamete formation.

Anthocyanin: (with): describes a plant organ that has taken an abnormal purplish hue.

Anthracnose: disease caused by fungi which anamorph form of reproduction is an acervulus (Melanconiales).

Antibody: specific protein produced by an animal in response to an antigen.

Antigen: foreign molecule, frequently of protein nature, that induces antibody formation when injected into an animal.

Apothecium: disc or trumpet -shaped structure on which ascospores are formed in Ascomycetes.

Appressorium: end of a hypha or a germ tube enabling fungus fixation on its host and its penetration.

Ascocarp: Ascomycetes sexual fruiting.

Ascomycetes: group of fungi producing their sexual spores, the ascospores, through asci.

Ascospore: spore resulting from sexual reproduction in Ascomycetes, forming within an ascus.

Ascus: bag-like cell in which typically eight ascospores are formed, and which characterizes the Ascomycetes.

Avirulent: (for example, a strain of a micro-organism) unable to infect a specific cultivar.

Basidium: cell on which basidiospores differentiate or are formed.

Basidiomycetes: group of fungi producing their sexual spores, the basidiospores, on basidia.

Basidiospore: spore resulting from sexual reproduction in basidiomycetes, forming on a basidium.

Basipetal: a production mode for conidial chains in which the youngest of them are located at the base, formed from the conidiophore.

Binucleate: containing two nuclei.

Capsid: protein shell of viruses that contains their nucleic acid.

Canker: necrotic lesion, localized to varying degrees.

Chlamydospore: spore from asexual reproduction, with a thick wall ensuring its protection and conservation in adverse conditions.

Chlorotic: (for example, a plant organ that has) turned an abnormally yellow colour.

Cleistothecium: fully enclosed ascocarp, sometimes covered with fulcrum, bursting open when ripe.

Conidiophores: specialized hypha on which one to several conidia form.

Conidium: spore resulting from asexual reproduction and formed at the end of a conidiophore.

Contamination: the first step in developing a disease in which the pathogen enters the host through its own resources.

Cortex: parenchymal tissue located between the epidermis and phloem of the stem and roots.

Cotyledon: embryonic leaf located in the seed, contributing through its reserves to the early development of the seedling.

Cultivar: cultivated variety.

Cuticle: waxy, impermeable layer covering the epidermal cells of leaves, stems, and fruit.

Damping-off: rapid decaying and disappearance of seedlings frequently associated with stem base and/or root symptoms.

Dissemination: final stage of a disease in which the inoculum is released, dispersed over a variable distance, ensuring healthy plant contamination.

Dominant: describes a gene that is expressed phenotypically when present in a plant.

Enation: leaf growth forming on certain portions of the veins.

Endemic: said of a disease whose frequency remains low and relatively constant over time, sometimes with a localized distribution.

Epidemiology: the study of the emergence, development, and dispersion of a disease in relation to the environment.

Epidermis: outer layer of cells covering the plants, sometimes covered with a cuticle.

Endoconidium: conidium formed within a mycelial hypha.

Enzyme: protein substance which catalyzes a specific biochemical reaction.

Epiphyte: describes an organism living on the surface of a plant.

Fasciation: malformation characterized by an increase in size, hypertrophy, a flattening and/or fusion of multiple organs affecting a shoot, a floral organ.

Flagellum: thin elongated structure which ensures the mobility of bacteria and fungal zoospores.

Fulcrum: filamentous appendix forming on the cleistothecia of powdery mildews.

Fumigant: pesticide acting as a gas, used especially against organisms in the soil.

Fungicide: substance killing fungi or inhibiting their growth or the germination of their spores.

Fruiting(s): production of spores or spores produced by a fungus.

Gene: hereditary unit located on a chromosome, a plasmid, or a cytoplasmic organelle, encoding a protein.

Haploid: refers to a cell or organism that has only one full set of chromosomes.

Haustorium: extension of the mycelium inside the cells, whose role is to collect the fungi necessary nutrients while maintaining their host.

Hermaphrodite: an individual that has both male and female sexual organs.

Heterothallic: said of a fungus whose male and female gametes are formed on different thalli.

Homothallic: said of a fungus whose male and female gametes are formed on the same thallus.

Hyaline: colourless, transparent.

Hybrid: descending from two parents of different genotypes.

Hydatode: specialized structure of the leaves' epidermis where some water is secreted or exuded.

Hyphae: isolated filament of a mycelium.

Immune: said of a plant that, when facing a particular pathogen, is not contaminated.

Incubation: period between contamination and appearance of the first symptoms of a disease.

Infection: process by which a micro-organism enters and multiplies in a plant.

Inoculum: parts, structures of a micro-organism that can infect a plant.

> **Primary inoculum**: inoculum at the origin of an epidemic.

> **Secondary inoculum**: inoculum ensuring disease progression.

Isolate: pure culture of a micro-organism obtained without particular cloning.

Larva: juvenile form of some animals before the adult stage.

Latency: period between contamination and the differentiation of early fruiting.

Latent: (for example, describes a pathogenic micro-organism) present on a plant but remaining invisible and/or inactive.

Monogenic resistance: resistance determined by a single gene.

Mutation: hereditary genetic change occurring in a cell.

Mycelium: filament or cluster of filaments constituting the basic structure of fungi.

Nematicide: substance causing nematode death.

Oogonium: structure containing one or more female gametes in Oomycetes.

Oospore: spore after sexual reproduction between an antheridium and an oogonium in Oomycetes (*Pythium*, Phytophthora).

Ostiole: circular orifice allowing pycnidia and perithecia to release their spores.

Parasite: an organism living at the expense of another living organism.

Parthenogenetic: refers to a mode of reproduction that does not involve a sexual phenomenon.

Pathotype: strain of a micro-organism with one or more pathogenesis genes.

Perithecium: globular or somewhat flattened conceptacle containing asci, frequently opened by an ostiole.

Phialide: terminal cell of a conidiophorea, or conidiophore having one or more end openings through which conidia are issued basipetally.

Phloem: vascular tissue consisting mainly of sieve tubes, companion cells, and parenchyma, providing transport and storage of sap.

Physiological race: see 'Pathotype'.

Prokaryotic: single-celled organism without a nucleus.

Propagule: basic structure of an organism that can be disseminated and reproduce a disease.

Protoplast: plant cell without a wall.

Pycnidium: spherical fruiting often containing conidiophores and conidia, frequently opened by an ostiole.

Recessive: describes a gene that is not expressed phenotypically when accompanied by a second dominant form of the same gene.

Rhizosphere: microenvironment of roots in the soil.

Saprophyte: an organism drawing its nourishment from decaying organic matter.

Sclerotia: compact mass of compacted mycelium, often brown to black, adapted to survival in adverse conditions.

Selection pressure: environmental pressure exerted on a micro-organism population, leading to changes in the genetic makeup of some individuals in this population.

Serum: component of the blood containing the antibodies.

Sorus: compact mass of spores beneath the epidermis of the leaves of plants affected by rust.

Specialized form (sp.f.): form distinct from the other forms of the same organism in its host specificity, e.g. for *Fusarium oxysporum*, the specialized forms *F . oxysporum* f. sp. *lycopersici*, *F. oxysporum* f. sp. *melonis*, and *F. oxysporum* f. sp. *lactucum* are known.

Sporangiophore (sporocystophore): structure bearing sporangia in fungi.

Sporangium: fungal structure producing asexual spores, often zoospores.

Spore: fungal reproducing unit consisting of one or more cells.

Sporodochia: dense clustering of conidiophores which may be confused with acervuli.

Stomata: opening of the epidermis surrounded by guard cells allowing gas exchanges.

Strain: pure culture of a micro-organism 'selected' from an isolate, sometimes with one or more specific biological characteristics.

Stroma: condensed mycelium in which can form different fungal fruiting bodies.

Thallus: all hyphae of a fungal colony.

Teleomorph: sexual form of a fungus, also called 'perfect stage', leading after nuclear fusion to the formation of ascospores, basidiospores.

Tolerance: characteristic of plants experiencing attacks of a pathogen without yield potential loss.

Toxin: poisonous substance of biological origin.

Uredospore: spore produced in sori by rusts.

Virion: mature virus.

Viroid: the smallest known infectious agent, consisting of a single nucleic acid.

Virulence: qualitative component of pathogenicity.

Viruliferous: (an insect or a nematode) carrying a virus, and therefore likely to transmit it.

Xylem: vascular tissue responsible for the transport of raw sap.

Zoospore: fungus spore bearing one or two flagella and capable of moving in water.

MAIN WORKS OF REFERENCE

AGRIOS G.N., 2004. *Plant Pathology*, 5th edition, Elsevier/Academic Press, 922 p.

ALFORD D., 1994. *Atlas en couleur des ravageurs des végétaux d'ornement — Arbres, arbustes, fleurs*, Versailles, INRA Éditions, 464 p.

ANONYMOUS, 1998. *Integrated pest management for tomatoes*, 5th edition, University of California, Agricultural Sciences, Publication 3274, 118 p.

BENNETT W.F., 1993. *Nutrient deficiencies and toxicities in crop plants*. APS Press, 202 p.

BLANCARD D., 1984. *Maladies et accidents culturaux de la tomate*, coédition P.H.M. Revue horticole-CTIFL, 88 p.

BLANCARD D., 1988. *Maladies de la tomate — Observer, identifier, lutter*, Versailles, INRA Éditions, 212 p.

BLANCARD D., 1998. *Maladies du tabac — Observer, identifier, lutter*, Versailles, INRA Éditions, 376 p.

BLANCARD D., LOT H., MAISONNEUVE B., 2003. *Maladies des salades — Identifier, connaître, maîtriser*, Versailles, INRA Éditions, 376 p.

BRIDHE J., STARR J.L., 2007. *Plant nematodes of agricultural importance : a colour handbook*. Manson Publishing Ltd, 152 p.

CRÜGER G., BACKHAUS F., HOMMES M., SMOLKA S., VETTEN H.-J., 2002. *Pflanzenschutz im gemüsebau*, Ulmer, 318 p.

EVANS K., TRUDGILL D.L., WEBSTER J.M. (eds), 1993. *Plant parasitic nematodes in temperate agriculture*, CABI Publishing, 656 p.

FARR D.F., BILLS G.F., CHAMURIS G.P., ROSSMAN A.Y., 1989. *Fungi on plants and plant products in the United States*, APS Press, 1251 p.

FLETCHER J.T., 1984. *Diseases of greenhouse plants*, New York, Longman Inc, 351 p.

HAWKSWORTH D.L., KIRK P.M., SUTTON B.C., PEGLER D.N., 1996. *Dictionary of the fungi*, CABI Publishing, 616 p.

HEUVELINK E., 2005. *Tomatoes*, CABI Publishing, 339 p.

JONES J.B., STALL R.E., ZITTER T.A., 1993. *Compendium of tomato diseases*, APS Press, 73 p.

KOIKE S., GLADDERS P., PAULUS A.O., 2007. *Vegetable diseases: a color handbook*, Manson Publishing Ltd, 448 p.

MALAIS H.M., RAVENSBERG W.J., 2006. *Connaître et reconnaître: la biologie des ravageurs des serres et de leurs ennemis naturels*. Koppert B. V. – Reed Business, 290 p.

MARCHOUX G., GOGNALONS P., GÉBRÉ SÉLASSIÉ K., 2008. *Virus des Solanacées – Du génome viral à la protection des cultures*, Versailles, Éditions Quæ, 848 p.

MESSIAEN C.M., BLANCARD D., ROUXEL F., LAFON R., 1991. *Les maladies des plantes maraîchères*, 3rd edition, Versailles, INRA Éditions, 568 p.

MURPHY F.A., FAUQUET C.M., BISHOP D.H.L., CHABRIAL S.A., JARVIS A.W., MARTELLI G.P., MAYO M.A., SUMMERS M.D., 1995. *Virus taxonomy, classification and nomenclature of Virus*, 6th report of the International Committee on Taxonomy of Virus, New York, Springer-Verlag Wien, 586 p.

RICHARD C., BOIVIN G., 1994. *Maladies et ravageurs des cultures légumières au Canada. Un traité pratique illustré*, Société canadienne de phytopathologie, Société d'entomologie du Canada, 590 p.

SEMAL J., 1989. *Traité de pathologie végétale*, Les presses agronomiques de Gembloux, 621 p.

SHEPHERD J.A., BARKER K.R., 1990. *Plant parasite nematodes in subtropical and tropical agriculture*. CABI Publishing, 648 p.

SHERF A.F., MACNAB A.A., 1986. *Vegetables diseases and their control*, New York, John Wiley & Sons, 728 p.

SHEW H.D., LUCAS G.B., 1991. *Compendium of tobacco diseases*, APS Press, 68 p.

SHURTLEFF M.C., AVERRE III C.W., 1997. *The plant disease clinic and field diagnosis of abiotic diseases*, APS Press, 245 p.

TROTTIN-CAUDAL Y., GRASSELLY D., MILLOT P., VESCHAMBRE D., 1995. *Tomate sous serre et abris : maîtrise de la protection sanitaire*, CTIFL, 174 p.

INDEX

Note: Page numbers in **bold** refer to figures, those in *italic* to tables. The letter 'n' following page numbers refers to material in the footnotes

A

frequency/extent of damage *33, 263, 484*
resistance *653–4, 656*
symptoms and identification **234, 260,** *261,* **26349**
pyrimethanil *426*
Pythium spp. *471–81*
  biology and epidemiology *473, 475–6*
  control *476–8*
  diagnostic features *239–40, 243, 251*
  distribution *40*
  frequency/extent of damage *33, 251, 471–2*
  fruit disease *388*
  group F *472, 474, 479*
  identification in roots *251,* **252**
  morphology *475, 478–9,* **480–1**
  symptoms *472–3, 474*
Pythium acanthicum *474, 478,* **480**
Pythium aphanidermatum *240,* **240,** *474, 475, 478,* **480**
Pythium arrhenomanes *474, 478*
Pythium deliens *474, 478*
Pythium dissotocum *474, 478*
Pythium inflatum *474, 479*
Pythium intermedium *474, 479*
Pythium irregulare (syn. *debaryanum*) *474, 479,* **480**
Pythium myriotylum *240, 474, 475, 479,* **480**
Pythium paroecandrum *240, 474, 479*
Pythium periplocum *474, 478,* **481**
Pythium salpingophorum *474, 479*
Pythium segnitium *474*
Pythium spinosum *474, 475, 478,* **481**
Pythium ultimum **238,** *474, 475, 479*
Pythium ultimum var. '*sporangiferum*' *474, 479*
Pythium vexans *474, 479*

Q

quarantine *647,* **648**

R

'rain check' *360*
Ralstonia solanacearum (bacterial wilt) *550–3*
  biology/epidemiology *551*
  control *552–3*
  diagnostic criteria *324*
  distribution *40*
  resistance *652, 655*
  symptoms *325,* **326–7,** *550–1*
'Recento' variety *54*
resistance *24*
  absolute (vertical) *653–4*
  Alternaria early blight *422*

bacterial canker *542–3*
Botrytis cinerea (grey mould) *427*
breeding programmes *652, 653, 662–3*
fruit rot *521*
Fusarium wilt *513–14*
future developments *662–3*
late blight *450–1*
Leveillula taurica *440*
level of effectiveness *653–4, 654*
M. fulva/Cladosporium *437, 653*
nematodes *639, 653*
Oidium neolycopersici *443*
partial (horizontal) *654*
patterns in glasshouse-grown tomatoes *656*
Phytopohthora spp. *477–8*
Pseudomonas syringae pv. *tomato* *529–30, 655*
Pyrenochaeta lycopersici *487*
Ralstonia solanacearum *552–3*
  and root-knot nematodes *273*
rootstocks *469, 553, 567, 658–60, 661*
stability *654, 654*
Stemphylium blight *455*
Verticillium wilt *518, 652, 654*
viruses *572–3, 573, 585, 590, 603*
Xanthomonas spp. *534*
see also fungicides, resistance; pesticide, resistance
resistance genes *652–3, 652–5*
'Revido' variety *54*
Reynoutria sachalinensis *428, 440*
rhizobacteria *530*
Rhizoctonia crocorum **256,** *257, 508–9*
Rhizoctonia solani *245, 488–92*
  anastomosis groups *488*
  antagonistic *470*
  biology and epidemiology *489,* **491–2**
  control *490–1*
  damping-off *243, 488*
  distribution of disease *40*
  frequency and impact *33, 251, 488*
  fruit rot *388,* **410,** *411*
  internal stem *311,* **313**
  mycelium **313, 491**
  root and stem base lesions *245, 251,* **252, 280,** *281, 291,* **292,** *293, 488–9*
Rhizopus spp. *388*
Rhizopus stolonifer **346, 402,** *403,* **520**
Rhizopycnis vagum *263*
'Ri' plasmid *276*
ribgrass mosaic virus (RMV) *111, 575*
ring spots **96**
RNA, satellite *582, 585*